AF411167

Age of Information: Concept, Metric and Tool for Network Control

Age of Information: Concept, Metric and Tool for Network Control

Editors

Anthony Ephremides
Yin Sun

MDPI • Basel • Beijing • Wuhan • Barcelona • Belgrade • Manchester • Tokyo • Cluj • Tianjin

Editors
Anthony Ephremides
University of Maryland
College Park
USA

Yin Sun
Auburn University
Auburn
USA

Editorial Office
MDPI
St. Alban-Anlage 66
4052 Basel, Switzerland

This is a reprint of articles from the Special Issue published online in the open access journal *Entropy* (ISSN 1099-4300) (available at: https://www.mdpi.com/journal/entropy/special_issues/age_of_information).

For citation purposes, cite each article independently as indicated on the article page online and as indicated below:

LastName, A.A.; LastName, B.B.; LastName, C.C. Article Title. *Journal Name* **Year**, *Volume Number*, Page Range.

ISBN 978-3-0365-7292-5 (Hbk)
ISBN 978-3-0365-7293-2 (PDF)

Contents

About the Editors

Anthony Ephremides

Anthony Ephremides has been at the University of Maryland for over 50 years, conducting research and teaching in the areas of information theory, communication networks, and system theory. His work has pioneered the areas of ad hoc wireless networks, cross-layer integration in networks, energy-efficient communications, and most recently, the field of semantic communications as exemplified in information freshness and communication for computing and actuation. He is a native of Greece where he received his undergraduate education before completing a Ph.D. at Princeton University in 1971. He has mentored over 40 doctoral students and received numerous awards. He has been engaged in numerous professional activities including being a member of the Board of Directors of IEEE and a Life Fellow of the institute. He has published and lectured extensively and maintains active collaborations with colleagues around the world.

Yin Sun

Yin Sun is an Assistant Professor in the Department of Electrical and Computer Engineering at Auburn University, Alabama. He received his B.Eng. and Ph.D. degrees in Electronic Engineering from Tsinghua University, in 2006 and 2011, respectively. He was a Postdoctoral Scholar and Research Associate at the Ohio State University from 2011–2017. His research interests include Wireless Networks, Machine Learning, Semantic Communications, Age of Information, Information Theory, and Robotic Control. His articles received the Best Student Paper Award of the IEEE/IFIP WiOpt 2013, Best Paper Award of the IEEE/IFIP WiOpt 2019, runner-up for the Best Paper Award of ACM MobiHoc 2020, and 2021 Journal of Communications and Networks (JCN) Best Paper Award. He co-authored a monograph Age of Information: A New Metric for Information Freshness, published by Morgan & Claypool Publishers in 2019. He received the Auburn Author Award of 2020. He received the National Science Foundation (NSF) CAREER Award in 2023. He is a Senior Member of the IEEE and a Member of the ACM.

Article

Scheduling to Minimize Age of Incorrect Information with Imperfect Channel State Information

Yutao Chen and Anthony Ephremides *

Department of Electrical and Computer Engineering, University of Maryland, College Park, MD 20742, USA; cheny@umd.edu
* Correspondence: etony@umd.edu

Abstract: In this paper, we study a slotted-time system where a base station needs to update multiple users at the same time. Due to the limited resources, only part of the users can be updated in each time slot. We consider the problem of minimizing the Age of Incorrect Information (AoII) when imperfect Channel State Information (CSI) is available. Leveraging the notion of the Markov Decision Process (MDP), we obtain the structural properties of the optimal policy. By introducing a relaxed version of the original problem, we develop the Whittle's index policy under a simple condition. However, indexability is required to ensure the existence of Whittle's index. To avoid indexability, we develop Indexed priority policy based on the optimal policy for the relaxed problem. Finally, numerical results are laid out to showcase the application of the derived structural properties and highlight the performance of the developed scheduling policies.

Keywords: age of incorrect information; multi-user system; scheduling policy

Citation: Chen, Y.; Ephremides, A. Scheduling to Minimize Age of Incorrect Information with Imperfect Channel State Information. *Entropy* **2021**, *23*, 1572. https://doi.org/10.3390/e23121572

Academic Editor: Mario Martinelli

Received: 3 November 2021
Accepted: 23 November 2021
Published: 25 November 2021

Publisher's Note: MDPI stays neutral with regard to jurisdictional claims in published maps and institutional affiliations.

1. Introduction

The Age of Incorrect Information (AoII) is introduced in [1] as a combination of age-based metrics (e.g., Age of Information (AoI)) and error-based metrics (e.g., Minimum Mean Square Error). In communication systems, AoII captures not only the information mismatch between the source and the destination but also the aging process of inconsistent information. Hence, two functions dominate AoII. The first is the time penalty function, which reflects how the inconsistency of information affects the system over time. In real-life applications, inconsistent information will affect different communication systems in different ways. For example, machine temperature monitoring is time-sensitive because the damage caused by overheating will accumulate quickly. However, reservoir water level monitoring is less sensitive to time. Therefore, by adopting different time penalty functions, AoII can capture different aging processes of the mismatch in different systems. The second is the information penalty function, which captures the information mismatch between the source and the destination. It allows us to measure mismatches in different ways, depending on how sensitive different systems are to information inconsistencies. For example, the navigation system requires precise information to give correct instructions, but the real-time delivery tracking system does not need very accurate location information. Since we can choose different penalty functions for different systems, AoII is adaptable to various communication goals, which is why it is regarded as a semantic metric [2].

Since the introduction of AoII, several studies have been performed to reveal its fundamental nature. The authors of [3] consider a system with random packet delivery times and compare AoII with AoI and real-time error via extensive numerical results. The authors of [4] study the problem of minimizing the AoII that takes the general time penalty function. Three real-life applications are considered to showcase the performance advantages of AoII over AoI and real-time error. In [5], the authors investigate the AoII that considers the quantified mismatch between the source and the destination. The optimization problem is studied when the system is resource-constrained. The authors of [6] studied the AoII minimization problem in the context of scheduling. It considers a system where the central scheduler needs to update multiple users at the same time. However, the central scheduler cannot know the states of the sources before receiving the updates. By introducing the belief value, Whittle's index policy is developed and evaluated. In this paper, we also consider the problem of minimizing AoII in scheduling. Different from [6], we consider the generic time penalty function and study the minimization problem in the presence of imperfect Channel State Information (CSI). Due to the existence of CSI, Whittle's index policy becomes infeasible in general. Hence, we introduce another scheduling policy that is more versatile and has comparable performance to Whittle's index policy.

The problem of scheduling to minimize AoI is studied under various system settings in [7–11]. The problem studied in this paper is different and more complicated because AoII considers the aging process of inconsistent information rather than the aging process of updates. Meanwhile, none of them consider the case where CSI is available. The problem of optimizing information freshness in the presence of CSI is studied in [12,13]. However, they focus on the system with a single user and mainly discuss the case where CSI is perfect. The scheduling problems with the goal of minimizing an error-based performance measure are considered in [14–16]. Our problem is fundamentally different because AoII also considers the time effect. Moreover, we consider the system where a base station observes multiple sources simultaneously and needs to send updates to multiple destinations.

The main contributions of this work can be summarized as follows. (1) We study the problem of minimizing AoII in a multi-user system where imperfect CSI is available. Meanwhile, the time penalty function is generic. (2) We derive the structural properties of the optimal policy for the considered problem. (3) We establish the indexability of the considered problem under a simple condition and develop Whittle's index policy. (4) We obtain the optimal policy for a relaxed version of the original problem. By exploring the characteristics of the relaxed problem, we provide an efficient algorithm to obtain the optimal policy. (5) Based on the optimal policy for the relaxed problem, we develop the Indexed priority policy that is free from indexability and has comparable performance to Whittle's index policy.

The remainder of this paper is organized in the following way. In Section 2, we introduce the system model and formulate the primal problem. Section 3 explores the structural properties of the optimal policy for the primal problem. Under a simple condition, we develop Whittle's index policy in Section 4. Section 5 presents the optimal policy for a relaxed version of the primal problem. On this basis, we develop the Indexed priority policy in Section 6. Finally, in Section 7, the numerical results are laid out.

2. System Overview

2.1. Communication Model

We consider a slotted-time system with N users and one base station. Each user is composed of a source process, a channel, and a receiver. We assume all the users share the same structure, but the parameters are different. The structure of the communication model is provided in Figure 1.

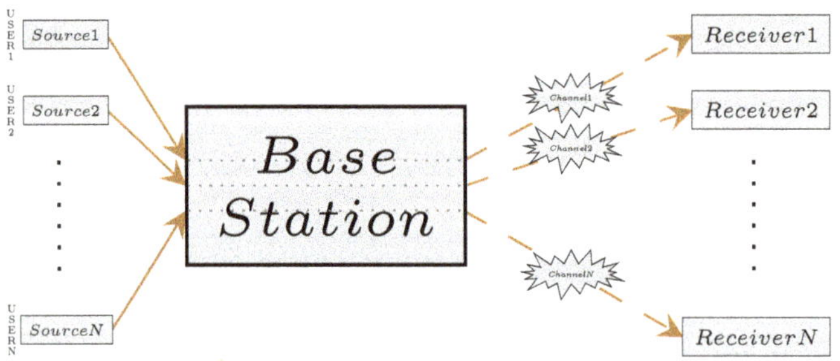

Figure 1. The structure of the communication model.

For user i, the source process is modeled by a two-state Markov chain where transitions happen between the two states with probability $p_i > 0$ and self-transitions happen with probability $1 - p_i$. At any time slot t, the state of the source process $X_{i,t} \in \{0,1\}$ will be reported to the base station as an update, and the base station will decide whether to transmit this update through the corresponding channel. The channel is unreliable, but the estimate of the Channel State Information (CSI) is available at the beginning of each time slot. Let $r_{i,t} \in \{0,1\}$ be the CSI at time t. We assume that $r_{i,t}$ is independent across time and user indices. $r_{i,t} = 1$ if and only if the transmission attempt at time t will succeed and $r_{i,t} = 0$ otherwise. Then, we denote by $\hat{r}_{i,t} \in \{0,1\}$ the estimate of $r_{i,t}$. We assume that $\hat{r}_{i,t}$ is an independent Bernoulli random variable with parameter γ_i, i.e., $\hat{r}_{i,t} = 1$ with probability $\gamma_i \in [0,1]$ and $\hat{r}_{i,t} = 0$ with probability $1 - \gamma_i$. However, the estimate is imperfect. We assume that the error depends only on the user and its estimate. More precisely, we define the probability of error as $p_{e,i}^{\hat{r}_i} \triangleq Pr[r_i \neq \hat{r}_i \mid \hat{r}_i]$. We assume $p_{e,i}^{\hat{r}_i} < 0.5$ because we can flip the estimate if $p_{e,i}^{\hat{r}_i} > 0.5$. We are not interested in the case of $p_{e,i}^{\hat{r}_i} = 0.5$ since $\hat{r}_{i,t}$ is useless in this case. Although the channel is unreliable, each transmission attempt takes exactly one time slot regardless of the result, and the successfully transmitted update will not be corrupted. Every time an update is received, the receiver will use it as the new estimate $\hat{X}_{i,t}$. The receiver will send an $ACK/NACK$ packet to inform the base station of its reception of the new update. Since an $ACK/NACK$ packet is generally very small and simple, we assume that it is transmitted reliably and received instantaneously. Then, if ACK is received, the base station knows that the receiver's estimate changed to the transmitted update. If $NACK$ is received, the base station knows that the receiver's estimate did not change. Therefore, the base station always knows the estimate at the receiver side.

At the beginning of each time slot, the base station receives updates from each source and the estimates of CSI from each channel. The old updates and estimates are discarded upon the arrival of new ones. Then, the base station decides which updates to transmit, and the decision is independent of the transmission history. Due to the limited resources, at most $M < N$ updates are allowed per transmission attempt. We consider a base station that always transmits M updates.

2.2. Age of Incorrect Information

All the users adopt AoII as a performance metric, but the choices of penalty functions vary. Let X_t and $\hat{X}_t$ be the true state and the estimate of the source process, respectively. Then, in a slotted-time system, AoII can be expressed as follows

$$\Delta_{AoII}(X_t, \hat{X}_t, t) = \sum_{k=U_t+1}^{t} \left(g(X_k, \hat{X}_k) \times F(k - U_t) \right), \tag{1}$$

where U_t is the last time instance before time t (including t) that the receiver's estimate is correct. $g(X_t, \hat{X}_t)$ can be any information penalty function that captures the difference between X_t and $\hat{X}_t$. $F(t) \triangleq f(t) - f(t-1)$ where $f(t)$ can be any time penalty function that is non-decreasing in t. We consider the case where the users adopt the same information penalty function $g(X_t, \hat{X}_t) = |X_t - \hat{X}_t|$ but possibly different time penalty functions. To ease the analysis, we require $f(t)$ to be unbounded. Combined together, we require $f(t_1) \leq f(t_2)$ if $t_1 < t_2$ and $\lim_{t \to +\infty} f(t) = +\infty$. Without a loss of generality, we assume $f(0) = 0$, as the source is modeled by a two-state Markov chain, $g(X_t, \hat{X}_t) \in \{0,1\}$. Hence, Equation (1) can be simplified to

$$\Delta_{AoII}(X_t, \hat{X}_t, t) = \sum_{k=U_t+1}^{t} F(k - U_t) = f(s_t),$$

where $s_t \triangleq t - U_t$. Therefore, the evolution of s_t is sufficient to characterize the evolution of AoII. To this end, we distinguish between the following cases.

- When the receiver's estimate is correct at time $t + 1$, we have $U_{t+1} = t + 1$. Then, by definition, $s_{t+1} = 0$.
- When the receiver's estimate is incorrect at time $t + 1$, we have $U_{t+1} = U_t$. Then, by definition, $s_{t+1} = t + 1 - U_t = s_t + 1$.

To sum up, we get

$$s_{t+1} = \mathbb{1}_{\{U_{t+1} \neq t+1\}} \times (s_t + 1). \tag{2}$$

A sample path of s_t is shown in Figure 2. In the remainder of this paper, we use $f_i(\cdot)$ to denote the time penalty function user i adopts.

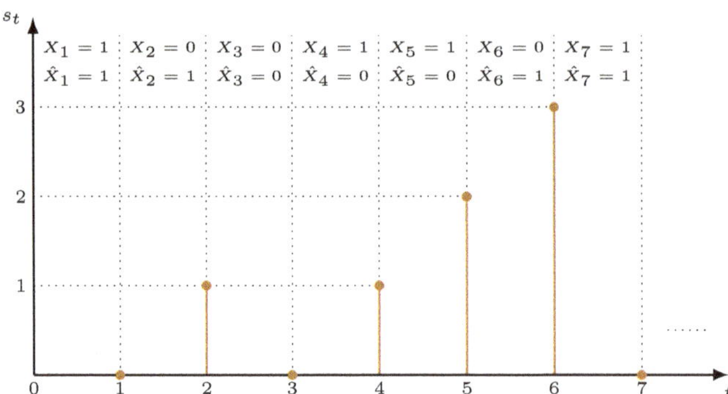

Figure 2. A sample path of s_t.

Remark 1. *Under this particular choice of the penalty function, s_t can be interpreted as the time elapsed since the last time the receiver's estimate is correct. Please note that s_t is different from the Age of Information (AoI) [17], which is defined as the time elapsed since the generation time of the last received update. We can see that AoI considers the aging process of the update, while AoII considers the aging process of the estimation error. At the same time, s_t is also fundamentally different from the holding time, which, according to [18,19], is defined as the time elapsed since the last successful transmission. We notice that the receiver's estimate can become correct even when no new update is successfully transmitted. Moreover, the information carried by the update may have become incorrect by the time it is received. We also notice that [18,19] consider the problem of minimizing the estimation error. However, by adopting AoII as the performance metric, we study the impact of estimation error on the system.*

2.3. System Dynamic

In this section, we tackle the system dynamic. We notice that the status of user i can be captured by the pair $x_{i,t} \triangleq (s_{i,t}, \hat{r}_{i,t})$. In the following, we will use $x_{i,t}$ and $(s_{i,t}, \hat{r}_{i,t})$ interchangeably. Then, the system dynamic can be fully characterized by the dynamic of $x_t \triangleq (x_{1,t}, \ldots, x_{N,t})$. Hence, it suffices to characterize the value of x_{t+1} given x_t and the base station's action. To this end, we denote, by $a_t = (a_{1,t}, \ldots, a_{N,t})$, the base station's action at time t. $a_{i,t} = 1$ if the base station transmits the update from user i at time t and $a_{i,t} = 0$ otherwise. We notice that given action a_t, users are independent and the action taken on user i will only affect itself. Consequently

$$Pr(x_{t+1} \mid x_t, a_t) = \prod_{i=1}^{N} Pr(x_{i,t+1} \mid x_{i,t}, a_t) = \prod_{i=1}^{N} Pr(x_{i,t+1} \mid x_{i,t}, a_{i,t}).$$

Combined with the fact that all the users share the same structure, it is sufficient to study the dynamic of a single user. In the following discussions, we drop the user-dependent subscript i. We recall that $\hat{r}_{t+1}$ is an independent Bernoulli random variable. Then, we have

$$Pr(x_{t+1} \mid x_t, a_t) = P(\hat{r}_{t+1}) \times Pr(s_{t+1} \mid x_t, a_t). \tag{3}$$

By definition, $P(\hat{r}_{t+1} = 1) = \gamma$ and $P(\hat{r}_{t+1} = 0) = 1 - \gamma$. Then, we only need to tackle the value of $Pr(s_{t+1} \mid x_t, a_t)$. To this end, we distinguish between the following cases

- When $x_t = (0, \hat{r}_t)$, the estimate at time t is correct (i.e., $\hat{X}_t = X_t$). Hence, for the receiver, X_t carries no new information about the source process. In other words, $\hat{X}_{t+1} = \hat{X}_t$ regardless of whether an update is transmitted at time t. We recall that $U_{t+1} = U_t$ if $\hat{X}_{t+1} \neq X_{t+1}$ and $U_{t+1} = t + 1$ otherwise. Since the source is binary, we obtain $U_{t+1} = U_t$ if $X_{t+1} \neq X_t$, which happens with probability p and $U_{t+1} = t + 1$ otherwise. According to (2), we obtain

$$Pr(1 \mid (0, \hat{r}_t), a_t) = p,$$
$$Pr(0 \mid (0, \hat{r}_t), a_t) = 1 - p.$$

- When $a_t = 0$ and $x_t = (s_t, \hat{r}_t)$, where $s_t > 0$, the channel will not be used and no new update will be received by the receiver, and so, $\hat{X}_{t+1} = \hat{X}_t$. We recall that $U_{t+1} = U_t$ if $\hat{X}_{t+1} \neq X_{t+1}$ and $U_{t+1} = t + 1$ otherwise. Since $X_t \neq \hat{X}_t$ and the source is binary, we have $U_{t+1} = U_t$ if $X_{t+1} = X_t$, which happens with probability $1 - p$ and $U_{t+1} = t + 1$ otherwise. According to (2), we obtain

$$Pr(s_t + 1 \mid (s_t, \hat{r}_t), a_t = 0) = 1 - p,$$
$$Pr(0 \mid (s_t, \hat{r}_t), a_t = 0) = p.$$

- When $a_t = 1$ and $x_t = (s_t, 1)$ where $s_t > 0$, the transmission attempt will succeed with probability $1 - p_e^1$ and fail with probability p_e^1. We recall that $U_{t+1} = U_t$ if $\hat{X}_{t+1} \neq X_{t+1}$ and $U_{t+1} = t + 1$ otherwise. Then, when the transmission attempt succeeds (i.e., $\hat{X}_{t+1} = X_t$), $U_{t+1} = U_t$ if $X_{t+1} \neq X_t$ and $U_{t+1} = t + 1$ otherwise. When the transmission attempt fails (i.e., $\hat{X}_{t+1} = \hat{X}_t \neq X_t$), we have $U_{t+1} = U_t$ if $X_{t+1} = X_t$ and $U_{t+1} = t + 1$ otherwise. Combining (2) with the dynamic of the source process we obtain

$$Pr(s_t + 1 \mid (s_t, 1), a_t = 1) = p_e^1(1 - p) + (1 - p_e^1)p \triangleq \alpha,$$
$$Pr(0 \mid (s_t, 1), a_t = 1) = p_e^1 p + (1 - p_e^1)(1 - p) = 1 - \alpha.$$

- When $a_t = 1$ and $x_t = (s_t, 0)$, where $s_t > 0$, following the same line, we obtain

$$Pr(s_t + 1 \mid (s_t, 0), a_t = 1) = p_e^0 p + (1 - p_e^0)(1 - p) \triangleq \beta,$$
$$Pr(0 \mid (s_t, 0), a_t = 1) = p_e^0(1 - p) + (1 - p_e^0)p = 1 - \beta.$$

Combines together, we obtain the value of $Pr(s_{t+1} \mid x_t, a_t)$ in all cases. As only M out of N updates are allowed per transmission attempt, we realize a necessity to require transmission attempts always help minimize AoII. It is equivalent to impose $Pr(s_{t+1} > s_t \mid (s_t, \hat{r}_t), a_t = 0) > Pr(s_{t+1} > s_t \mid (s_t, \hat{r}_t), a_t = 1)$ for any $(s_t, \hat{r}_t)$. Leveraging the results above, it is sufficient to require $p < 0.5$. As all the users share the same structure, we assume, for the rest of this paper, that $0 < p_i < 0.5$ for $1 \leq i \leq N$.

2.4. Problem Formulation

The communication goal is to minimize the expected AoII. Therefore, the problem can be formulated as the following

$$\underset{\phi \in \Phi}{\arg\min} \quad \lim_{T \to \infty} \frac{1}{T} \mathbb{E}_\phi \left(\sum_{t=0}^{T-1} \sum_{i=1}^{N} f_i(s_{i,t}) \right) \tag{4a}$$

$$\text{subject to} \quad \sum_{i=1}^{N} a_{i,t} = M \quad \forall t, \tag{4b}$$

where Φ is the set of all causal policies. We refer to the constrained minimization problem reported in problem (4) as the Primal Problem (PP). We notice that the PP is a Restless Multi-Armed Bandit (RMAB) Problem. The optimal policy for this type of problem is far from reachable since it is PSPACE-hard in general [20]. However, we can still derive the structural properties of the optimal policy. These structural properties can be used as a guide for the development of scheduling policies and can indicate the good performance of the developed scheduling policies.

3. Structural Properties of the Optimal Policy

In this section, we investigate the structural properties of the optimal policy for PP. We first define an infinite horizon with an average cost Markov Decision Process (MDP) $\mathcal{M}_N(w, M) = (\mathcal{X}_N, \mathcal{A}_N(M), \mathcal{P}_N, \mathcal{C}_N(w))$, where

- $\mathcal{X}_N$ denotes the state space. The state is $x = (x_1, \ldots, x_N)$ where $x_i = (s_i, \hat{r}_i)$.
- $\mathcal{A}_N(M)$ denotes the action space. The feasible action is $a = (a_1, \ldots, a_N)$ where $a_i \in \{0, 1\}$ and $\sum_{i=1}^N a_i = M$. Note that the feasible actions are independent of the state and the time.
- $\mathcal{P}_N$ denotes the state transition probabilities. We define $P_{x,x'}(a)$ as the probability that action a at state x will lead to state x'. It is calculated by

$$P_{x,x'}(a) = \prod_{i=1}^N P(\hat{r}_i') P_{s_i, s_i'}(a_i, \hat{r}_i),$$

  where $P_{s_i, s_i'}(a_i, \hat{r}_i)$ is the transition probability from s_i to s_i' when the estimate of CSI is $\hat{r}_i$ and action a_i is taken. The values of $P_{s_i, s_i'}(a_i, \hat{r}_i)$ can be obtained easily from the results in Section 2.3.
- $\mathcal{C}_N(w)$ denotes the instant cost. When the system is at state x and action a is taken, the instant cost is $C(x, a) \triangleq \sum_{i=1}^N C(x_i, a_i) \triangleq \sum_{i=1}^N \left(f_i(s_i) + w a_i \right)$.

We notice that PP can be cast into $\mathcal{M}_N(0, M)$. Since $w = 0$, the instant cost is independent of action a. Therefore, we abbreviate $C(x, a)$ as $C(x)$. To simplify the analysis, we consider the case of $M = 1$. Equivalently, we investigate the structural properties of the optimal policy for $\mathcal{M}_N(0, 1)$.

Remark 2. *For the case of $M > 1$, we can apply the same methodology. However, as M increases, the action space will grow quickly, resulting in the need to consider more feasible actions in each step of the proof. Hence, to better demonstrate the methodology, we only consider the case of $M = 1$ in this paper.*

It is well known that the optimal policy for $\mathcal{M}_N(0, 1)$ can be characterized by the value function. We denote the value function of state x as $V(x)$. A canonical procedure to calculate $V(x)$ is applying the Value Iteration Algorithm (VIA). To this end, we define $V_\nu(\cdot)$ as the estimated value function at iteration ν of VIA and initialize $V_0(\cdot) = 0$. Then, VIA updates the estimated value functions in the following way

$$V_{\nu+1}(x) = C(x) - \theta + \min_{a \in \mathcal{A}_N(1)} \left\{ \sum_{x' \in \mathcal{X}_N} P_{x,x'}(a) V_\nu(x') \right\}, \tag{5}$$

where θ is the optimal value of $\mathcal{M}_N(0, 1)$. VIA is guaranteed to converge to the value function [21]. More precisely, $V_\nu(\cdot) = V(\cdot)$ when $\nu \to +\infty$. However, the exact value function is impossible to get since we need infinite iterations and the state space is infinite. Instead, we provide two structural properties of the value function.

Lemma 1 (Monotonicity)**.** *For $\mathcal{M}_N(0, 1)$, $V(x)$ is non-decreasing in s_i for $1 \leq i \leq N$.*

Proof. Leveraging the iterative nature of VIA, we use mathematical induction to prove the desired results. The complete proof can be found in Appendix A. □

Before introducing the next structural property, we make the following definition.

Definition 1 (Statistically identical). *Two users are said to be statistically identical if the user-dependent parameters and the adopted time penalty functions are the same.*

For the users that are statistically identical, we can prove the following

Lemma 2 (Equivalence). *For $\mathcal{M}_N(0,1)$, if users j and k are statistically identical, $V(x) = V(\mathcal{P}(x))$ where $\mathcal{P}(x)$ is state x with x_j and x_k exchanged.*

Proof. Leveraging the iterative nature of VIA, we use mathematical induction to prove the desired results. At each iteration, we show that for each feasible action at state x, we can find an equivalent action at state $\mathcal{P}(x)$. Two actions are equivalent if they lead to the same value function. The complete proof can be found in Appendix B. □

Equipped with the above lemmas, we proceed with characterizing the structural properties of the optimal policy. We recall that the optimal action at each state can be characterized by the value function. Hence, we denote, by $V^j(x)$, the value function resulting from choosing user j to update at state x. Then, $V^j(x)$ can be calculated by

$$V^j(x) = C(x) - \theta + \sum_{x'-x'_j} \left\{ \left(\prod_{i \neq j} P_{x_i, x'_i}(0) \right) \sum_{\hat{r}'_j} \left[P(\hat{r}'_j) \left(\sum_{s'_j} P_{s_j, s'_j}(1, \hat{r}_j) V(x') \right) \right] \right\}.$$

If $V^j(x) < V^k(x)$ for all $k \neq j$, it is optimal to transmit the update from user j. When $V^j(x) = V^k(x)$, the two choices are equally desirable. In the following, we will characterize the properties of $\delta^{j,k}(x) \triangleq V^j(x) - V^k(x)$ for any j and k.

Theorem 1 (Structural properties). *For $\mathcal{M}_N(0,1)$, $\delta^{j,k}(x)$ has the following properties*

1. $\delta^{j,k}(x) \leq 0$ *if $\hat{r}_k = p^0_{e,k} = 0$. The equality holds when $s_j = 0$ or $\hat{r}_j = p^0_{e,j} = 0$.*
2. $\delta^{j,k}(x)$ *is non-increasing in $\hat{r}_j$ and is non-decreasing in $\hat{r}_k$ when $s_j, s_k > 0$. At the same time, $\delta^{j,k}(x)$ is independent of $\hat{r}_i$ for any $i \neq j, k$.*
3. $\delta^{j,k}(x) \leq 0$ *if $s_k = 0$. The equality holds when $s_j = 0$ or $\hat{r}_j = p^0_{e,j} = 0$.*
4. $\delta^{j,k}(x)$ *is non-increasing in s_j if $\Gamma_j^{\hat{r}_j} \leq \Gamma_k^{\hat{r}_k}$ and is non-decreasing in s_k if $\Gamma_j^{\hat{r}_j} \geq \Gamma_k^{\hat{r}_k}$ when $s_j, s_k > 0$. We define $\Gamma_i^1 \triangleq \frac{\alpha_i}{1-p_i}$ and $\Gamma_i^0 \triangleq \frac{\beta_i}{1-p_i}$ for $1 \leq i \leq N$.*
5. $\delta^{j,k}(x) \leq 0$ *if $s_j \geq s_k$, $\hat{r}_j \geq \hat{r}_k$, and users j and k are statistically identical.*

Proof. The proof can be found in Appendix C. □

We notice that $\Gamma_i^{\hat{r}_i}$ can be written as

$$\Gamma_i^{\hat{r}_i} = \frac{Pr(s_i + 1 \mid (s_i, \hat{r}_i), a_i = 1)}{Pr(s_i + 1 \mid (s_i, \hat{r}_i), a_i = 0)} < 1,$$

where s_i can be any positive integer. Consequently, $\Gamma_i^{\hat{r}_i}$ is independent of any $s_i > 0$ and indicates the decrease in the probability of increasing s_i caused by action $a_i = 1$. When $\Gamma_i^{\hat{r}_i}$ is large, action $a_i = 1$ will achieve a small decrease in the probability of increasing s_i. In the following, we provide an intuitive interpretation of why the monotonicity in Property 4 of Theorem 1 depends on $\Gamma_i^{\hat{r}_i}$. We take the case of $\Gamma_j^{\hat{r}_j} \leq \Gamma_k^{\hat{r}_k}$ as an example and assume that there are only users j and k in the system. Then, according to Section 2.3, the dynamic of s_j and s_k can be divided into the following three cases

- Neither s_j nor s_k increases. In this case, both s_j and s_k become zero.
- Either s_j or s_k increases and the other becomes zero. We denote by P_j^k the probability that only s_k increases when $a_j = 1$. The notation for other cases is defined analogously. The probabilities can be obtained easily using the results in Section 2.3.
- Both s_j and s_k increase. We denote by P_j the probability that both s_j and s_k increase when $a_j = 1$. P_k is defined analogously. The probabilities can be obtained easily using the results in Section 2.3.

We notice that $\delta^{j,k}(x)$ implies the tendency of the base station to choose between the two users. The larger $\delta^{j,k}(x)$ is, the more the base station tends to choose user k. Thus, we investigate the base station's propensity to choose user k when s_k increases but s_j stays the same. We ignore the case where the resulting s_k is zero since it is independent of the increase in s_k. With this in mind, we first notice that $P_k^k \leq P_j^k$. Meanwhile, we can easily verify that $\frac{P_j}{P_k} = \frac{\Gamma_j^{\hat{r}_j}}{\Gamma_k^{\hat{r}_k}}$. When $\Gamma_j^{\hat{r}_j} \leq \Gamma_k^{\hat{r}_k}$, we have $P_j \leq P_k$. Then, there exists a subtle trade-off. More precisely, choosing user k will result in $P_k^k \leq P_j^k$, but at the cost of $P_k \geq P_j$. Hence, in this case, the propensity of the base station is hard to determine. Following the same line, we can show that choosing user j will lead to $P_j^j \leq P_k^j$ and $P_j \leq P_k$. Thus, there exists no such trade-off when we investigate the base station's propensity to choose user j as s_j increases but s_k stays the same.

Leveraging Theorem 1, we can provide some specific structural properties of the optimal policy.

Corollary 1 (Application of Theorem 1). *When $M = 1$, the optimal policy for PP must satisfy the following*

1. *The user i with $\hat{r}_i = p_{e,i}^0 = 0$ or $s_i = 0$ will not be chosen unless it is to break the tie.*
2. *When user j is chosen at state x_1, then for state x_2, such that $\hat{r}_{1,j} \leq \hat{r}_{2,j}$ and $s_{1,i} = s_{2,i}$ for $1 \leq i \leq N$, the optimal choice must be in the set $G = \{j\} \cup \{k : \hat{r}_{1,k} < \hat{r}_{2,k}\}$.*
3. *When $N = 2$, we consider two states, x_1 and x_2, which differ only in the value of s_j. Specifically, $s_{1,j} \leq s_{2,j}$. If user j is chosen at state x_1 and $\Gamma_j^{\hat{r}_{1,j}} \leq \Gamma_k^{\hat{r}_{1,k}}$, the optimal choice at state x_2 will also be user j.*
4. *When $N = 2$, we consider two states, x_1 and x_2, which differ only in the value of s_k. Specifically, $s_{1,k} \geq s_{2,k}$. If user j is chosen at state x_1 and $\Gamma_j^{\hat{r}_{1,j}} \geq \Gamma_k^{\hat{r}_{1,k}}$, the optimal choice at state x_2 will also be user j.*
5. *When all users are statistically identical, the optimal choice at any time slot must be either the user with $x = (s_{max,1}, 1)$ where $s_{max,1} \triangleq \max_{s_i}\{(s_i, 1)\}$ or the user with $x = (s_{max,0}, 0)$ where $s_{max,0} \triangleq \max_{s_i}\{(s_i, 0)\}$. Moreover,*
    - *If $s_{max,1} \geq s_{max,0}$, it is optimal to choose the user with $x = (s_{max,1}, 1)$.*
    - *If $s_{max,1} < s_{max,0}$, the optimal choice will switch from the user with $x = (s_{max,0}, 0)$ to the user with $x = (s_{max,1}, 1)$ when $s_{max,1}$ increases from 0 to $s_{max,0}$ solely.*

Proof. The first property follows directly from Property 1 and Property 3 of Theorem 1. For the second property, leveraging Property 2 of Theorem 1, we have $\delta^{j,k}(x_2) \leq \delta^{j,k}(x_1) \leq 0$ if $\hat{r}_{1,j} \leq \hat{r}_{2,j}$, $\hat{r}_{1,k} \geq \hat{r}_{2,k}$, and $s_{1,i} = s_{2,i}$ for $1 \leq i \leq N$. Thus, the optimal choice will not be user k in this case. Then, we can conclude that the optimal choice must be in the set $G = \{j\} \cup \{k : \hat{r}_{1,k} < \hat{r}_{2,k}\}$.

For the third property, we have proved in Property 4 of Theorem 1 that $\delta^{j,k}(x)$ is non-increasing in s_j if $\Gamma_j^{\hat{r}_j} \leq \Gamma_k^{\hat{r}_k}$. Hence, $\delta^{j,k}(x_2) \leq \delta^{j,k}(x_1) \leq 0$. As we consider the case of $N = 2$, the optimal choice at state x_2 will also be user j. The fourth property can be shown in a similar way by noticing that $\delta^{j,k}(x)$ is non-decreasing in s_k when $\Gamma_j^{\hat{r}_j} \geq \Gamma_k^{\hat{r}_k}$.

For the last property, we recall from Property 5 of Theorem 1 that it is always better to choose the user with a larger s if they are statistically identical and have the same $\hat{r}$. Thus, we can conclude that the optimal choice must be either the user with $x = (s_{max,1}, 1)$ or the user with $x = (s_{max,0}, 0)$. Without a loss of generality, we assume $x_j = (s_{max,1}, 1)$ and $x_k = (s_{max,0}, 0)$. Now, we distinguish between the following cases

- According to Property 5 of Theorem 1, we can conclude that it is optimal to choose user j when $s_{max,1} \geq s_{max,0}$.
- To determine the optimal choice in the case of $s_{max,1} < s_{max,0}$, we recall that the optimal choice will be user k (i.e., $\delta^{j,k}(x) \geq 0$) if $s_j = 0$ and will be user j (i.e., $\delta^{j,k}(x) \leq 0$) if $s_j = s_k$. At the same time, Property 4 of Theorem 1 tells us that $\delta^{j,k}(x)$ is non-increasing in s_j when users j and k are statistically identical. Therefore, we can conclude that the optimal choice will switch from user k to user j when s_j increases from 0 to s_k solely.

$\square$

4. Whittle's Index Policy

Whittle's index policy is a well-known low-complexity heuristic that shows a strong performance in many problems that belong to RMAB [22–24]. In this section, we develop Whittle's index policy for PP. We first present the general procedures we adopt to obtain Whittle's index.

- We first formulate a relaxed version of PP and apply the Lagrangian approach.
- Then, we decouple the problem of minimizing the Lagrangian function into N decoupled problems, each of which only considers a single user. By casting the decoupled problem into an MDP, we investigate the structural properties and performance of the optimal policy.
- Leveraging the results above and under a simple condition, we establish the indexability of the decoupled problem.
- Finally, we obtain the expression of Whittle's index by solving the Bellman equation.

4.1. Relaxed Problem

The first step in obtaining Whittle's index is to formulate the Relaxed Problem (RP). More precisely, instead of requiring the limit on the number of updates allowed per transmission attempt to be met in each time slot, we relax the constraint such that the limit is not violated in an average sense. Then, RP can be formulated as

$$\arg\min_{\phi \in \Phi} \quad \bar{\Delta}_\phi \triangleq \lim_{T \to \infty} \frac{1}{T} \mathbb{E}_\phi \left(\sum_{t=0}^{T-1} \sum_{i=1}^{N} f_i(s_{i,t}) \right) \tag{6a}$$

$$\text{subject to} \quad \bar{\rho}_\phi \triangleq \lim_{T \to \infty} \frac{1}{T} \mathbb{E}_\phi \left(\sum_{t=0}^{T-1} \sum_{i=1}^{N} a_{i,t} \right) \leq M. \tag{6b}$$

As RP is specified, we apply the Lagrangian approach. First of all, we write RP into its Lagrangian form.

$$\mathcal{L}(\lambda, \phi) = \lim_{T \to \infty} \frac{1}{T} \mathbb{E}_\phi \left(\sum_{t=0}^{T-1} \sum_{i=1}^{N} (f_i(s_{i,t}) + \lambda a_{i,t}) \right) - \lambda M,$$

where $\lambda \geq 0$ is the Lagrange multiplier. Then, we investigate the problem of minimizing the Lagrangian function. Since λM is independent of policies, we can ignore it. More precisely, we consider the following minimization problem

$$\min_{\phi \in \Phi} \quad \lim_{T \to \infty} \frac{1}{T} \mathbb{E}_\phi \left(\sum_{t=0}^{T-1} \sum_{i=1}^{N} (f_i(s_{i,t}) + \lambda a_{i,t}) \right). \tag{7}$$

4.2. Decoupled Model

In this section, we formulate the decoupled problem and investigate its optimal policy. The decoupled model associated with each user follows the system model with $N = 1$.

Since all the users share the same structure, we drop the user-dependent subscript i for simplicity. Then, the decoupled problem can be formulated as

$$\underset{\phi \in \Phi'}{\text{minimize}} \quad \lim_{T \to \infty} \frac{1}{T} \mathbb{E}_\phi \left(\sum_{t=0}^{T-1} (f(s_t) + \lambda a_t) \right), \tag{8}$$

where Φ' is the set of all causal policies when $N = 1$. We notice that problem (8) can be cast into the MDP $\mathcal{M}_1(\lambda, -1)$. We define $M = -1$ when there is no restriction on the number of updates allowed per transmission attempt.

We first investigate the structural properties of the optimal policy for $\mathcal{M}_1(\lambda, -1)$ when λ is a given non-negative constant. We start with characterizing the corresponding value function $V(x)$.

Corollary 2 (Extension of Lemma 1). *For $\mathcal{M}_1(\lambda, -1)$, $V(x)$ is non-decreasing in s.*

Proof. The proof follows the same steps as in the proof of Lemma 1. The complete proof can be found in Appendix D. $\quad \square$

Equipped with the above corollary, we can characterize the structural properties of the optimal policy for (8).

Proposition 1 (Optimal policy for decoupled problem). *The optimal policy for the decoupled problem is a threshold policy with the following properties.*

- *The optimal policy can be fully captured by $\mathbf{n} = (n_0, n_1)$. More precisely, when the system is at state $(s, \hat{r})$, it is optimal to make a transmission attempt only when $s \geq n_{\hat{r}}$.*
- *$n_0 \geq n_1 > 0$.*

Proof. We define $\Delta V(x) \triangleq V^1(x) - V^0(x)$, where $V^a(x)$ is the value function resulting from taking action a at state x. Then, the optimal action at state x is $a = 1$ if $\Delta V(x) < 0$, and $a = 0$ is optimal otherwise. We use Corollary 2 to characterize the sign of $\Delta V(x)$. The complete proof can be found in Appendix E. $\quad \square$

In the following, we evaluate the performance of the threshold policy detailed in Proposition 1. More precisely, we calculate the expected AoII $\bar{\Delta}_n$ and the expected transmission rate $\bar{\rho}_n$ resulting from the adoption of threshold policy $\mathbf{n}$. We will see in the following that $\bar{\Delta}_n$ and $\bar{\rho}_n$ are essential for establishing the indexability and obtaining the expression of Whittle's index.

Proposition 2 (Performance). *Under threshold policy $\mathbf{n} = (n_0, n_1)$,*

$$\bar{\Delta}_n = \pi_0 p \left[\sum_{k=1}^{n_1-1} f(k)(1-p)^{k-1} + (1-p)^{n_1-1} \left(\sum_{k=n_1}^{n_0-1} f(k) c_1^{k-n_1} + c_1^{n_0-n_1} \sum_{k=n_0}^{+\infty} f(k) c_2^{k-n_0} \right) \right],$$

$$\bar{\rho}_n = \pi_0 p (1-p)^{n_1-1} \left[\frac{\gamma}{1-c_1} + c_1^{n_0-n_1} \left(\frac{1}{1-c_2} - \frac{\gamma}{1-c_1} \right) \right],$$

where

$$\pi_0 = \frac{1}{2 + p(1-p)^{n_1-1} \left[\frac{1}{1-c_1} - \frac{1}{p} + c_1^{n_0-n_1} \left(\frac{1}{1-c_2} - \frac{1}{1-c_1} \right) \right]},$$

$c_1 = (1-\gamma)(1-p) + \gamma\alpha$, *and* $c_2 = (1-\gamma)\beta + \gamma\alpha$.

Proof. We notice that the dynamic of AoII under the threshold policy can be fully captured by a Discrete-Time Markov Chain (DTMC). Then, combined with the fact that $\hat{r}$ is an independent Bernoulli random variable, we can obtain the desired results from the stationary distribution of the induced DTMC. The complete proof can be found in Appendix F. $\quad \square$

As $f(\cdot)$ can be any non-decreasing function, $\bar{\Delta}$ can grow indefinitely. Thus, it is necessary to require that there exists at least one threshold policy that causes a finite $\bar{\Delta}$. By noting that $1 - p \geq c_1 \geq c_2$, we have

$$\bar{\Delta} \geq \pi_0 p \left[\sum_{k=1}^{n_1-1} f(k)c_2^{k-1} + c_2^{n_1-1} \left(\sum_{k=n_1}^{n_0-1} f(k)c_2^{k-n_1} + c_2^{n_0-n_1} \sum_{k=n_0}^{+\infty} f(k)c_2^{k-n_0} \right) \right]$$

$$= \pi_0 p \left(\sum_{k=1}^{+\infty} f(k)c_2^{k-1} \right).$$

The equality is achieved when $n_0 = n_1 = 1$. Then, we can conclude that it is sufficient to require $\sum_{k=1}^{+\infty} f(k)c_2^{k-1} < +\infty$. This will be the underlying assumption throughout the rest of this paper.

4.3. Indexability

In this section, we establish the indexability of the decoupled problem, which ensures the existence of Whittle's index. We start with the definition of indexability.

Definition 2 (Indexability). *The decoupled problem is indexable if the set of states in which $a = 0$ is the optimal action increases with λ, that is,*

$$\lambda' < \lambda \implies D(\lambda') \subseteq D(\lambda),$$

where $D(\lambda)$ is the set of states in which $a = 0$ is optimal when Lagrange multiplier λ is adopted.

The Lagrange multiplier λ can be viewed as a cost associated with each transmission attempt. Intuitively, as λ increases, the base station should stay idle (i.e., $a = 0$) for a longer time until s becomes large enough to offset the cost. Although it is intuitively correct that the decoupled problem is indexable, the indexability is hard to establish as the optimal policy is characterized by two thresholds. Thus, Whittle's index does not necessarily exist. However, the indexability can be established when the following condition is satisfied

$$p_{e,i}^0 = 0 \quad for\ 1 \leq i \leq N. \tag{9}$$

Remark 3. *Problem* (9) *only requires the estimate $\hat{r}_i$ to be perfect when $\hat{r}_i = 0$. In the case of $\hat{r}_i = 1$, we still allow the estimate to be inaccurate.*

When (9) is satisfied, Propositions 1 and 2 reduce to the following

Corollary 3 (Consequences of (9)). *When* (9) *is satisfied, the optimal policy for the decoupled problem* (8) *is the threshold policy $n = (+\infty, n)$. The corresponding $\bar{\Delta}_n$ and $\bar{\rho}_n$ are*

$$\bar{\Delta}_n = \pi_0 p \left(\sum_{k=1}^{n-1} f(k)(1-p)^{k-1} + (1-p)^{n-1} \sum_{k=n}^{+\infty} f(k)c_1^{k-n} \right),$$

$$\bar{\rho}_n = \pi_0 p (1-p)^{n-1} \left(\frac{\gamma}{1-c_1} \right),$$

where

$$\pi_0 = \frac{1}{2 + p(1-p)^{n-1} \left(\dfrac{1}{1-c_1} - \dfrac{1}{p} \right)}.$$

Proof. We continue with the same notations as in the proof of Propositions 1 and 2. It is sufficient to show that $n_0 = +\infty$. To this end, we consider the state $x = (s, 0)$. By following the same steps as in the proof of Proposition 1, we have

$$\Delta V(s,0) = \lambda \geq 0.$$

Therefore, it is optimal to stay idle (i.e., $a = 0$) at state $x = (s,0)$ for any $s \geq 0$. Equivalently, $n_0 = +\infty$. Then, the corresponding $\tilde{\Delta}_n$ and $\bar{\rho}_n$ can be calculated as a special case of Proposition 2 where $n_0 = +\infty$, $n_1 = n$, and $p_e^0 = 0$. $\quad\square$

Leveraging Corollary 3, we can establish the indexability of the decoupled problem.

Proposition 3 (Indexability of decoupled problem). *The decoupled problem is indexable when* (9) *is satisfied.*

Proof. According to Proposition 2.2 of [25], we only need to verify that the expected transmission rate $\bar{\rho}_n$ is strictly decreasing in n. From Corollary 3, we have

$$\bar{\rho}_n = \frac{\gamma\left(\dfrac{p}{1-c_1}\right)}{\dfrac{2}{(1-p)^{n-1}} + \left(\dfrac{p}{1-c_1} - 1\right)}.$$

As $\frac{1}{2} < 1 - p < 1$, we can easily verify that $\bar{\rho}_n$ is strictly decreasing in n. Thus, the decoupled problem is indexable when (9) is satisfied. $\quad\square$

4.4. Whittle's Index Policy

In this section, we proceed with finding the expression of Whittle's index and defining Whittle's index policy. First of all, we give the definition of Whittle's index.

Definition 3 (Whittle's index). *When the decoupled problem is indexable, Whittle's index at state x is defined as the infimum λ, such that both actions are equally desirable. Equivalently, Whittle's index at state x is defined as the infimum λ such that $V^0(x) = V^1(x)$.*

Let us denote by W_x the Whittle's index at state x. Then, the expression of Whittle's index is given by the following Proposition.

Proposition 4 (Whittle's index). *When* (9) *is satisfied, Whittle's index is*

$$W_x = \begin{cases} 0 & \text{when } x = (0,\hat{r}) \text{ or } x = (s,0), \\[2em] \dfrac{(1-c_1)\displaystyle\sum_{k=s+1}^{+\infty} f(k)c_1^{k-s-1} - \tilde{\Delta}_s}{\dfrac{(1-c_1)(1-p) - \gamma(1-p-\alpha)}{c_1(1-p-\alpha)} + \bar{\rho}_s} & \text{when } x = (s,1), \end{cases}$$

where $s > 0$ and $c_1 = (1-\gamma)(1-p) + \gamma\alpha$. $\tilde{\Delta}_s$ and $\bar{\rho}_s$ are the expected AoII and the expected transmission rate when threshold policy $n = (+\infty, s)$ is adopted, respectively. At the same time, W_x is non-negative and is non-decreasing in s.

Proof. Whittle's indexes at state $x = (0,\hat{r})$ and $x = (s,0)$ are obtained easily from the proof of Proposition 1. For state $x = (s,1)$, we first use backward induction to calculate the expressions of some value functions. Then, the expression of Whittle's index can be obtained from its definition. The complete proof can be found in Appendix G. $\quad\square$

Definition 4 (Whittle's index policy). *At any state $x = (x_1, x_2, \ldots, x_N)$, the base station will transmit the updates from M users with the largest W_{x_i}. The ties are broken arbitrarily. W_{x_i} is calculated using Proposition 4 with the parameters of user i.*

Remark 4. *Whittle's index policy possesses the structural properties detailed in Corollary 1.*

- *The first two properties can be verified by noting that $W_{x_i} \geq 0$ and the equality holds when $\hat{r}_i = 0$ or $s_i = 0$. At the same time, W_{x_i} is non-decreasing in $\hat{r}_i$.*
- *The third and fourth properties can be verified by noting that W_{x_i} is non-decreasing in s_i.*
- *For the last property, we first notice that $W_{x_j} = W_{x_k}$ when users j and k are statistically identical and $x_j = x_k$. Then, the property can be verified by noting that W_{x_i} is non-decreasing in both s_i and $\hat{r}_i$.*

5. Optimal Policy for Relaxed Problem

In this section, we provide an efficient algorithm to obtain the optimal policy for RP, based on which we will develop another scheduling policy for PP in the next section that is free from indexability. At the same time, the performance of the optimal policy for RP forms a universal lower bound because the following ordering holds

$$\bar{\Delta}_{AoII}^{RP} \leq \bar{\Delta}_{AoII}^{PP},$$

where $\bar{\Delta}_{AoII}^{RP}$ and $\bar{\Delta}_{AoII}^{PP}$ are the minimal expected AoII of RP and PP, respectively.

Remark 5. *Note that the optimal policy for RP may not necessarily be a valid policy for PP, as the transmitter may transmit more than M updates in one transmission attempt under RP-optimal policy.*

To solve RP, we follow the discussion in Section 4.1. More precisely, we take the Lagrangian approach and consider the problem reported in (7). We will see in the following discussion that the optimal policy for RP can be characterized by the optimal policies for problem (7). Therefore, we first cast problem (7) into the MDP $\mathcal{M}_N(\lambda, -1)$. However, the optimal policy for $\mathcal{M}_N(\lambda, -1)$ is difficult to obtain because the state space is infinite. Even though we can make the state space finite by imposing an upper limit on the value of s, the state space and the action space grow exponentially with the number of users in the system. To overcome the difficulty, we investigate the optimal policy for $\mathcal{M}_1^i(\lambda, -1)$ where $1 \leq i \leq N$. The superscript i means that the only user in the system is user i. We will show later that the optimal policy for $\mathcal{M}_N(\lambda, -1)$ can be fully characterized by the optimal policies for $\mathcal{M}_1^i(\lambda, -1)$ where $1 \leq i \leq N$.

5.1. Optimal Policy for Single User

In this section, we tackle the problem of finding the optimal policy for $\mathcal{M}_1^i(\lambda, -1)$. Since the users share the same structure, we ignore the superscript i for simplicity. To find the optimal policy, we first use the Approximating Sequence Method (ASM) introduced in [26] to make the state space finite. More precisely, we impose $s \leq m$ where m is a predetermined upper limit. The state transition probabilities $P'_{s,s'}(a, \hat{r})$ are modified in the following way

$$P'_{s,s'}(a, \hat{r}) = \begin{cases} P_{s,s'}(a, \hat{r}) & \text{if } s' < m, \\ P_{s,s'}(a, \hat{r}) + \sum_{z>m} P_{s,z}(a, \hat{r}) & \text{if } s' = m. \end{cases} \tag{10}$$

The action space and the instant cost remain unchanged. Then, we can apply Relative Value Iteration (RVI) with convergence criteria ϵ to obtain the optimal policy. We notice that $\mathcal{M}_1(\lambda, -1)$ coincides with the decoupled model studied in Section 4.2. Hence, we can utilize the threshold structure of the optimal policy to improve RVI. To this end, we class a state as active if the optimal action at this state is $a = 1$. Then, the threshold structure detailed in Proposition 1 tells us the following. For any state x, if there exists an active state x_1 with $s_1 \leq s$ and $\hat{r}_1 \leq \hat{r}$, then x must also be active. Hence, we can determine the optimal action at state x immediately instead of comparing all feasible actions. In this way, we can reduce the running time of RVI. The pseudocode for the improved RVI can be found in Algorithm A1 of Appendix M. A similar technique is also presented in [5].

For $\mathcal{M}_1(\lambda, -1)$, when problem (9) is satisfied, Whittle's index exists and can be calculated efficiently using Proposition 4. Therefore, we can obtain the optimal policy using Whittle's index and further reduce the computational complexity. To this end, we denote by n_λ the optimal policy for $\mathcal{M}_1(\lambda, -1)$ and present the following proposition

Proposition 5 (Optimal deterministic policy). *When (9) is satisfied, the optimal policy for $\mathcal{M}_1(\lambda, -1)$ is $n_\lambda = (+\infty, n)$ where n is given by*

$$n = \begin{cases} 1 & \text{if } \lambda = 0, \\ \max\{s \in \mathbb{N}_0 : W_s \leq \lambda\} + 1 & \text{if } \lambda > 0. \end{cases}$$

W_s is the Whittle's index at state $(s, 1)$.

Proof. We first notice that $\mathcal{M}_1(\lambda, -1)$ coincides with the decoupled model studied in Section 4.2. Then, we show the optimal action for each state with $\hat{r} = 1$ using the definition of Whittle's index and the fact that the decoupled problem is indexable when (9) is satisfied. The complete proof can be found in Appendix H. $\square$

In the following, we provide a randomized policy that is also optimal for $\mathcal{M}_1(\lambda, -1)$. We will see later that the randomized policy is the key to obtaining the optimal policy for RP.

Theorem 2 (Optimal randomized policy). *There exist two deterministic policies n_{λ_+} and n_{λ_-}, which are both optimal for $\mathcal{M}_1(\lambda, -1)$. We consider the following randomized policy n_λ: every time the system reaches state $(0, 0)$, the base station will make the choice between n_{λ_-} with probability μ and n_{λ_+} with probability $1 - \mu$. The chosen policy will be followed until the next choice. Then, the randomized policy n_λ is optimal for $\mathcal{M}_1(\lambda, -1)$ under any $\mu \in [0, 1]$.*

Proof. We show that our system verifies the assumptions given in [27]. Then, leveraging the characteristics of our system, we can obtain the optimal randomized policy. The complete proof can be found in Appendix I. $\square$

In practice, we approximate $\lambda_+ \approx \lambda + \zeta$ and $\lambda_- \approx \lambda - \zeta$ where ζ is a small perturbation. Then, the deterministic policies n_{λ_+} and n_{λ_-} can be obtained by following the discussion at the beginning of this subsection. Note that, in most cases, n_{λ_+} and n_{λ_-} are the same.

5.2. Optimal Policy for RP

In this section, we characterize the optimal policy for RP. Let us denote by $V(x)$ and $V^i(x_i)$ the value functions of $\mathcal{M}_N(\lambda, -1)$ and $\mathcal{M}_1^i(\lambda, -1)$, respectively. Then, we can prove the following

Proposition 6 (Separability). *$V(x) = \sum_{i=1}^{N} V^i(x_i)$ where $x = (x_1, \ldots, x_N)$. In other words, the policy, under which each user adopts its own optimal policy, is optimal for $\mathcal{M}_N(\lambda, -1)$.*

Proof. We show $V(x) = \sum_{i=1}^{N} V^i(x_i)$ by comparing the Bellman equations they must satisfy. The complete proof can be found in Appendix J. $\square$

We denote the optimal policy for $\mathcal{M}_N(\lambda, -1)$ as $\phi_\lambda = [n_{\lambda,1}, \ldots, n_{\lambda,N}]$ where $n_{\lambda,i}$ is the optimal policy for $\mathcal{M}_1^i(\lambda, -1)$. For simplicity, we define $\bar{\Delta}(\lambda)$ and $\bar{\rho}(\lambda)$ as the expected AoII and the expected transmission rate associated with ϕ_λ, respectively. $\bar{\Delta}^i(\lambda)$ and $\bar{\rho}^i(\lambda)$ are defined analogously for user i under policy $n_{\lambda,i}$. We also define $\lambda^* \triangleq \inf\{\lambda > 0 : \bar{\rho}(\lambda) \leq M\}$. With Proposition 6 and the above definitions in mind, we proceed with constructing the optimal policy for RP.

Theorem 3 (Optimal policy for RP). *The optimal policy for RP can be characterized by two deterministic policies* $\phi_{\lambda_+^*} = [n_{\lambda_+^*,1}, \ldots, n_{\lambda_+^*,N}]$ *and* $\phi_{\lambda_-^*} = [n_{\lambda_-^*,1}, \ldots, n_{\lambda_-^*,N}]$ *where* $n_{\lambda_+^*,i}$ *and* $n_{\lambda_-^*,i}$ *are both the optimal deterministic policies for* $\mathcal{M}_1^i(\lambda^*, -1)$. *Then, we mix* $\phi_{\lambda_+^*}$ *and* $\phi_{\lambda_-^*}$ *in the following way: for each user* i, *every time the user reaches state* $(0,0)$, *the base station will make the choice between* $n_{\lambda_-^*,i}$ *with probability* μ_i *and* $n_{\lambda_+^*,i}$ *with probability* $1 - \mu_i$. *The chosen policy will be followed by user* i *until the next choice. Where* $1 \leq i \leq N$, *the* μ_i *is chosen in such a way as to satisfy*

$$\sum_{i=1}^N \bar{\rho}^i(\lambda^*) = \sum_{i=1}^N \left(\mu_i \bar{\rho}^i(\lambda_-^*) + (1 - \mu_i)\bar{\rho}^i(\lambda_+^*) \right) = M. \tag{11}$$

Then, the mixed policy, denoted by ϕ_{λ^*}, *is optimal for RP.*

Proof. According to Lemma 3.10 of [27], a policy is optimal for RP if

1. It is optimal for $\mathcal{M}_N(\lambda^*, -1)$;
2. The resulting expected transmission rate is equal to M.

Then, we construct such a policy using Theorem 2 and Proposition 6. The complete proof can be found in Appendix K. $\square$

Since we approximate $\lambda_+^* \approx \lambda^* + \xi$ and $\lambda_-^* \approx \lambda^* - \xi$ in practice, $\bar{\rho}^i(\lambda_+^*) \leq \bar{\rho}^i(\lambda_-^*)$ for all i according to the monotonicity given by Lemma 3.4 of [27]. Combining with the definition of λ^*, we must have $\bar{\rho}(\lambda_+^*) \leq M < \bar{\rho}(\lambda_-^*)$. Therefore, we can always find μ_i's that realize (11). In this paper, we choose

$$\mu_i = \mu = \frac{M - \bar{\rho}(\lambda_+^*)}{\bar{\rho}(\lambda_-^*) - \bar{\rho}(\lambda_+^*)}, \quad for \ 1 \leq i \leq N. \tag{12}$$

Then, we describe the algorithm used to obtain the optimal policy for RP. As detailed in Theorem 3, it is essential to find λ^*. To this end, we recall that, for any user i under given λ, the optimal deterministic policy $n_{\lambda,i}$ can be obtained using the results in Section 5.1 and the resulting expected transmission rate $\bar{\rho}^i(\lambda)$ is given by Proposition 2. Since $\bar{\rho}^i(\lambda)$ is non-increasing in λ for all i according to Lemma 3.4 of [27], $\bar{\rho}(\lambda) = \sum_{i=1}^N \bar{\rho}^i(\lambda)$ is also non-increasing in λ. Hence, we can regard $\bar{\rho}(\lambda)$ as a non-increasing function of λ. Then, according to the definition of λ^*, we can use the Bisection search to obtain λ^* efficiently. The main steps can be summarized as follows.

1. Initialize $\lambda_- = 0$ and $\lambda_+ = 1$.
2. Do $\lambda_- = \lambda_+$ and $\lambda_+ = 2\lambda_+$ until $\bar{\rho}(\lambda_+) < M$.
3. Run Bisection search on the interval $[\lambda_-, \lambda_+]$ until the tolerance 2ξ is met.

Then, λ_-^* and λ_+^* can simply be the boundaries of the final interval. The pseudocode for the Bisection search can be found in Algorithm A2 of Appendix M. After obtaining λ_-^* and λ_+^*, the optimal policy ϕ_{λ^*} is detailed in Theorem 3 and the mixing probabilities μ_i's are given by (12).

Remark 6. *We recall that the optimal deterministic policy for each user can be characterized by two positive thresholds (i.e., $n_0, n_1 > 0$). Consequently, under RP-optimal policy, the base station will never choose the user at state $(0, \hat{r})$. Then, when M increases, the expected transmission rate achieved by RP-optimal policy will saturate before M reaches N. When the expected transmission rate saturates, the RP-optimal policy is* $\phi^* = [n_1, \ldots, n_N]$ *where* $n_i = (1, 1)$ *for* $1 \leq i \leq N$. *The saturation happens when M is larger than or equal to the expected transmission rate achieved by* ϕ^*.

6. Indexed Priority Policy

Although the performance of Whittle's index policy is known to be good, it requires indexability, which is usually difficult to establish. In this section, based on the primal-dual heuristic introduced in [28], we develop a policy that does not require indexability and has comparable performance to Whittle's index policy. We start with presenting the primal-dual heuristic.

6.1. Primal-Dual Heuristic

The heuristic is based on the optimal primal and dual solution pair to the linear program associated with RP. To introduce the linear program, we define $\pi_{x_i}^{a_i}(\phi) \geq 0$ as the expected time that user i is at state x_i and action a_i is taken according to policy ϕ. Then, for any ϕ, $\pi_{x_i}^{a_i}(\phi)$ must satisfy the following problems

$$\pi_{x_i}^0(\phi) + \pi_{x_i}^1(\phi) = \sum_{x_i'} \sum_{a_i'} P_{x_i',x_i}(a_i') \pi_{x_i'}^{a_i'}(\phi), \quad \forall x_i, i.$$

$$\sum_{x_i} \sum_{a_i} \pi_{x_i}^{a_i}(\phi) = 1, \quad \forall i.$$

The objective function of RP can be rewritten as

$$\underset{\phi \in \Phi}{\text{minimize}} \ \sum_{i=1}^{N} \sum_{x_i, a_i} C(x_i) \pi_{x_i}^{a_i}(\phi),$$

where $C(x_i) = f_i(s_i)$ is the instant cost at state x_i. The constraint on the expected transmission rate can be rewritten as

$$\sum_{i=1}^{N} \sum_{x_i} \pi_{x_i}^1(\phi) \leq M.$$

Thus, the linear program associated with RP can be formulated as the following

$$\underset{\pi_{x_i}^{a_i}}{\text{minimize}} \ \sum_{i=1}^{N} \sum_{x_i, a_i} C(x_i) \pi_{x_i}^{a_i} \tag{13a}$$

$$\text{subject to} \quad \pi_{x_i}^0 + \pi_{x_i}^1 - \sum_{x_i'} \sum_{a_i'} P_{x_i',x_i}(a_i') \pi_{x_i'}^{a_i'} = 0, \quad \forall x_i, i, \tag{13b}$$

$$\sum_{x_i} \sum_{a_i} \pi_{x_i}^{a_i} = 1, \quad \forall i, \tag{13c}$$

$$\sum_{i=1}^{N} \sum_{x_i} \pi_{x_i}^1 \leq M, \tag{13d}$$

$$\pi_{x_i}^{a_i} \geq 0, \quad \forall x_i, a_i, i. \tag{13e}$$

The corresponding dual problem is

$$\underset{\sigma, \sigma_i, \sigma_{x_i}}{\text{maximize}} \ \sum_{i=1}^{N} \sigma_i - M\sigma \tag{14a}$$

$$\text{subject to} \quad \sigma_{x_i} + \sigma_i - \sum_{x_i'} P_{x_i,x_i'}(0)\sigma_{x_i'} \leq C(x_i), \quad \forall x_i, i, \tag{14b}$$

$$\sigma_{x_i} + \sigma_i - \sum_{x_i'} P_{x_i,x_i'}(1)\sigma_{x_i'} - \sigma \leq C(x_i), \quad \forall x_i, i, \tag{14c}$$

$$\sigma \geq 0. \tag{14d}$$

Let $\{\pi_{x_i}^{a_i}\}$ and $\{\bar{\sigma}, \bar{\sigma}_i, \bar{\sigma}_{x_i}\}$ be the optimal primal and dual solution pair to the problems reported in (13) and (14). We define

$$\bar{\psi}_{x_i}^0 = \sum_{x_i'} P_{x_i,x_i'}(0)\bar{\sigma}_{x_i'} + C(x_i) - \bar{\sigma}_i - \bar{\sigma}_{x_i} \geq 0,$$

$$\bar{\psi}_{x_i}^1 = \sum_{x_i'} P_{x_i,x_i'}(1)\bar{\sigma}_{x_i'} + \bar{\sigma} + C(x_i) - \bar{\sigma}_i - \bar{\sigma}_{x_i} \geq 0.$$

For any state $x = (x_1, \ldots, x_N)$, let $h(x) = \sum_{i=1}^{N} \mathbb{1}_{\{\pi_{x_i}^1 > 0\}}$. Then, the heuristic operates in the following way

- If $h(x) \geq M$, the base station will choose the M users with the largest $\bar{\psi}_{x_i}^0$ among the $h(x)$ users.
- If $h(x) < M$, these $h(x)$ users are chosen by the base station. The base station will choose $M - h(x)$ additional users with the smallest $\bar{\psi}_{x_i}^1$.

However, Linear Programming (LP) is a very general technique and does not appear to take advantage of the special structure of the problem. Although there are algorithms for solving rational LP that take time polynomial in the number of variables and constraints, they run extremely slowly in practice [29]. For our problem, we notice that the users have separate activity areas that are linked through a common resource constraint. Therefore, the primal problem can be solved using Dantzig-Wolfe decomposition. Even so, the problem is still computationally demanding when the system scales up. We recall that we solved the exact problem efficiently using MDP-specific algorithms in Section 5. It is more efficient because of the following reasons

- According to Proposition 6, we can decompose the problem into N subproblems.
- For each subproblem, the threshold structure of the optimal policy is utilized to reduce the running time of RVI.
- As we will see later, the developed policy can be obtained directly from the result of RVI in practice.

In the following, we will translate the results in Section 5 into the optimal primal and dual solution pair and propose Indexed priority policy.

6.2. Indexed Priority Policy

We first define the Lagrangian function associated with (13).

$$\mathcal{L}(\pi_{x_i}^{a_i}, \sigma, \sigma_i, \sigma_{x_i}, \psi_{x_i}^{a_i}) = \left(\sum_{i=1}^{N} \sum_{x_i, a_i} C(x_i) \pi_{x_i}^{a_i} \right) + \sum_{i, x_i} \sigma_{x_i} \left(\sum_{x_i'} \sum_{a_i'} P_{x_i', x_i}(a_i') \pi_{x_i'}^{a_i'} - \pi_{x_i}^0 - \pi_{x_i}^1 \right) +$$

$$\sum_{i=1}^{N} \sigma_i \left(1 - \sum_{x_i} \sum_{a_i} \pi_{x_i}^{a_i} \right) + \sigma \left(\sum_{i=1}^{N} \sum_{x_i} \pi_{x_i}^1 - M \right) - \sum_{i, x_i, a_i} \psi_{x_i}^{a_i} \pi_{x_i}^{a_i}.$$

Then, the corresponding Lagrangian dual function is

$$g(\sigma, \sigma_i, \sigma_{x_i}, \psi_{x_i}^{a_i}) = \inf_{\pi_{x_i}^{a_i}} \mathcal{L}(\pi_{x_i}^{a_i}, \sigma, \sigma_i, \sigma_{x_i}, \psi_{x_i}^{a_i}).$$

Let π_{x_i} be the expected time that user i is at state x_i caused by the adoption of ϕ_{λ^*}, where ϕ_{λ^*} is the optimal policy detailed in Theorem 3. Then, we define $\{\pi_{x_i}^{a_i}\}$ as follows

- State x_i is where randomization happens (randomization happens when the actions suggested by the two optimal deterministic policies are different), and it has a value of $\pi_{x_i}^0 = a_{n_{\lambda^*,i}}(x_i)(1 - \mu_i)\pi_{x_i} + a_{n_{\lambda^*_+,i}}(x_i)\mu_i\pi_{x_i}$ and $\pi_{x_i}^1 = \pi_{x_i} - \pi_{x_i}^0$ where μ_i is given by (12) and $a_{n_{\lambda,i}}(x_i)$ is the action suggested by $n_{\lambda,i}$ at state x_i.
- For other values of x_i, we have $\pi_{x_i}^0 = (1 - a_{n_{\lambda^*,i}}(x_i))\pi_{x_i}$ and $\pi_{x_i}^1 = \pi_{x_i} - \pi_{x_i}^0$.

We also define $\sigma = \lambda^*$, $\sigma_i = \theta_i$, and $\sigma_{x_i} = V^i(x_i)$ where λ^* is specified in Section 5.2, θ_i is the optimal value of $\mathcal{M}_1^i(\lambda^*, -1)$, and $V^i(x_i)$ is the value function associated with $\mathcal{M}_1^i(\lambda^*, -1)$. Lastly, we define $\{\psi_{x_i}^{a_i}\}$ as follows

$$\psi_{x_i}^0 = \sum_{x_i'} P_{x_i, x_i'}(0)\sigma_{x_i'} + C(x_i) - \sigma_i - \sigma_{x_i},$$

$$\psi_{x_i}^1 = \sum_{x_i'} P_{x_i, x_i'}(1)\sigma_{x_i'} + \sigma + C(x_i) - \sigma_i - \sigma_{x_i}.$$

Then, we can prove the following proposition.

Proposition 7 (Optimal solution pair). *$\{\pi_{x_i}^{a_i}\}$ and $\{\sigma, \sigma_i, \sigma_{x_i}, \psi_{x_i}^{a_i}\}$ are primal and dual solutions to* (13), *respectively.*

Proof. Since (13) is linear and strictly feasible, it is sufficient to show that $\{\pi_{x_i}^{a_i}\}$ and $\{\sigma, \sigma_i, \sigma_{x_i}, \psi_{x_i}^{a_i}\}$ verify the KKT conditions, which can be expressed as the following four conditions.

1. Primal feasibility: the constraints in (13) are satisfied.
2. Dual feasibility: $\sigma \geq 0$ and $\psi_{x_i}^{a_i} \geq 0$ for all x_i, a_i, and i.
3. Complementary slackness: $\sigma\left(\sum_{i=1}^{N} \sum_{x_i} \pi_{x_i}^1 - M\right) = 0$ and $\psi_{x_i}^{a_i} \pi_{x_i}^{a_i} = 0$ for all x_i, a_i, and i.
4. Stationarity: the gradient of $\mathcal{L}(\pi_{x_i}^{a_i}, \sigma, \sigma_i, \sigma_{x_i}, \psi_{x_i}^{a_i})$ with respect to $\{\pi_{x_i}^{a_i}\}$ vanishes.

Apparently, the first condition is satisfied by $\{\pi_{x_i}^{a_i}\}$. For the second condition, $\sigma \geq 0$ since $\sigma = \lambda^* \geq 0$ by definition. For $\psi_{x_i}^{a_i}$, we can verify that $\psi_{x_i}^{a_i} = V^{i,a_i}(x_i) - V^i(x_i)$ where $V^{i,a_i}(x_i)$ is the value function resulting from taking action a_i at state x_i. Then, the non-negativity is guaranteed by the Bellman equation. For the third condition, the first term is zero because we choose the μ_i's given by (12). For the second term, we recall that $\psi_{x_i}^{a_i} = V^{i,a_i}(x_i) - V^i(x_i)$. According to the definition of $\pi_{x_i}^{a_i}$, we know $V^i(x_i) = V^{i,a_i}(x_i)$ if $\pi_{x_i}^{a_i} > 0$. Combined together, we can conclude that $\psi_{x_i}^{a_i} = 0$ when $\pi_{x_i}^{a_i} > 0$. Thus, the third condition is satisfied. For the last condition, setting the gradient equal to zero yields a system of linear equations. More precisely, for each x_i and $1 \leq i \leq N$

$$
\begin{cases}
\sum_{x_i'} P_{x_i, x_i'}(0)\sigma_{x_i'} + C(x_i) = \sigma_{x_i} + \sigma_i + \psi_{x_i}^0. \\[2mm]
\sum_{x_i'} P_{x_i, x_i'}(1)\sigma_{x_i'} + \sigma + C(x_i) = \sigma_{x_i} + \sigma_i + \psi_{x_i}^1.
\end{cases}
$$

Then, $\{\sigma, \sigma_i, \sigma_{x_i}, \psi_{x_i}^{a_i}\}$ verifies the system of linear equations by definition. Since all four conditions are satisfied, we can conclude our proof. $\square$

According to Proposition 7, we know that $\{\pi_{x_i}^{a_i}\}$ and $\{\sigma, \sigma_i, \sigma_{x_i}\}$ defined above are the optimal solutions to problems (13) and (14), respectively. As the optimal solutions are obtained, we can adopt the heuristic detailed in Section 6.1.

The heuristic can be expressed equivalently as an index policy. To this end, we define the index I_{x_i} for state x_i as

$$
I_{x_i} \triangleq \bar{\psi}_{x_i}^0 - \bar{\psi}_{x_i}^1.
$$

According to the complementary slackness, I_{x_i} can be reduced to the following.

- For state x_i such that $\bar{\pi}_{x_i}^1 > 0$ and $\bar{\pi}_{x_i}^0 = 0$, we have $\bar{\psi}_{x_i}^1 = 0$. Therefore, $I_{x_i} = \bar{\psi}_{x_i}^0 \geq 0$.
- For state x_i such that $\bar{\pi}_{x_i}^1 > 0$ and $\bar{\pi}_{x_i}^0 > 0$, we have $\bar{\psi}_{x_i}^1 = \bar{\psi}_{x_i}^0 = 0$. Therefore, $I_{x_i} = 0$.
- For state x_i such that $\bar{\pi}_{x_i}^1 = 0$ and $\bar{\pi}_{x_i}^0 > 0$, we have $\bar{\psi}_{x_i}^0 = 0$. Therefore, $I_{x_i} = -\bar{\psi}_{x_i}^1 \leq 0$.

We can show that I_{x_i} possesses the following properties.

Proposition 8 (Properties of I_{x_i}). *For $1 \leq i \leq N$, $I_{x_i} \geq -\lambda^*$ for any x_i. The equality holds when $\hat{r}_i = p_{e,i}^0 = 0$ or $s_i = 0$. At the same time, I_{x_i} is non-decreasing in both s_i and $\hat{r}_i$.*

Proof. We notice that I_{x_i} can be expressed as a function of $V^i(x_i)$ and λ^*. Meanwhile, $\mathcal{M}_1^i(\lambda^*, -1)$ coincides with the decoupled model studied in Section 4.2. Then, we can verify the properties of I_{x_i} using the results in Section 4.2. The complete proof can be found in Appendix L. $\square$

Comparing with the heuristic detailed in Section 6.1, we can define the Indexed priority policy.

Definition 5 (Indexed priority policy). *At any state $x = (x_1, x_2, \ldots, x_N)$, the base station will transmit the updates from M users with the largest I_{x_i}. The ties are broken arbitrarily.*

Remark 7. *Indexed priority policy belongs to the class of priority policies introduced in [30]. These priority policies are asymptotically optimal when certain conditions are satisfied.*

Remark 8. *Indexed priority policy possesses the structural properties detailed in Corollary 1.*
- *The first two properties can be verified by noting that $I_{x_i} \geq -\lambda^*$ and the equality holds when $\hat{r}_i = p^0_{e,i} = 0$ or $s_i = 0$. At the same time, I_{x_i} is non-decreasing in $\hat{r}_i$.*
- *The third and fourth properties can be verified by noting that I_{x_i} is non-decreasing in s_i.*
- *For the last property, we first notice that $I_{x_j} = I_{x_k}$ when users j and k are statistically identical and $x_j = x_k$. Then, the property can be verified by noting that I_{x_i} is non-decreasing in both s_i and $\hat{r}_i$.*

We notice that θ_i's and $C(x_i)$'s are canceled out by the definition of I_{x_i}. Therefore, I_{x_i} can be calculated using λ^* and the value function of $\mathcal{M}^i_1(\lambda^*, -1)$. In practice, we can use either λ^*_- or λ^*_+ to approximate λ^*, and the value function can be approximated by the result of the RVI detailed in Section 5.1. Since the state space is infinite, we only calculate a finite number of $V^i(x_i)$, the number of which depends on the truncation parameter m of ASM. Meanwhile, the probabilities $P_{x_i, x'_i}(a_i)$ in I_{x_i} are modified according to (10).

7. Numerical Results

In this section, we provide numerical results to showcase the performance of the developed scheduling policies. To eliminate the effect of N, we plot the expected average AoII. In particular, we provide the expected average AoII achieved by the Indexed priority policy and Whittle's index policy when $M = 1$. The policies are calculated using the results detailed in Sections 4–6. When obtaining the Indexed priority policy, we set the tolerance in the Bisection search to $\xi = 0.005$. Meanwhile, we choose the truncation parameter in ASM $m = 800$ and the convergence criteria in RVI $\epsilon = 0.01$. We notice that the calculation of Whittle's index involves an infinite sum. In practice, we approximate the result by replacing $+\infty$ with a large enough number k_{max}. Here, we choose $k_{max} = 800$. For both scheduling policies, the resulting expected average AoII is obtained via simulations. Each data point is the average of 15 runs with 15,000 time slots considered in each run.

We also compare the developed policies with the optimal policy for RP, which can be calculated by following the discussion in Section 5.2. We adopt the same choices of parameters as we used to obtain the developed policies. The corresponding performance is calculated using Proposition 2. Like before, the infinite sum is approximated by replacing $+\infty$ with $k_{max} = 800$. We also provide the expected average AoII achieved by the Greedy policy to show the performance advantages of the developed policies. When the Greedy policy is adopted, the base station always chooses the user with the largest AoII. The resulting expected average AoII is obtained via the same simulations as applied to the developed policies.

Figures 3 and 4 illustrate the performance when the source processes have different dynamics and when each user's communication goal is different, respectively. Figure 3a provides the performance when $p_i = 0.05 + \frac{0.4(i-1)}{N-1}$ for $1 \leq i \leq N$. For other parameters, the users make the same choices. More precisely, $f_i(s) = s$, $\gamma_i = 0.6$, and $p^0_{e,i} = p^1_{e,i} = 0.1$ for $1 \leq i \leq N$. Figure 4a provides the performance when $f_i(s) = s^{0.5 + \frac{i-1}{N-1}}$ for $1 \leq i \leq N$. Same as before, the users make the same choices for other parameters. More precisely, $p_i = 0.3$, $\gamma_i = 0.6$, and $p^0_{e,i} = p^1_{e,i} = 0.1$ for $1 \leq i \leq N$. In Figures 3b and 4b, we force $p^0_{e,i} = 0$ for all users to ensure the existence of Whittle's index. Other choices remain the same as in Figures 3a and 4a. According to Corollary 1, the optimal policy will never choose the user with $\hat{r} = p^0_e = 0$ unless it is to break the tie. Therefore, in Figures 3b and 4b, we also consider the Greedy+ policy where the base station always chooses the user with the largest AoII among the users with $\hat{r} = 1$. The resulting expected average AoII is obtained via the same simulations as applied to the Greedy policy.

Figure 5 shows the performance in systems where the parameters for each user are generated uniformly and randomly within their ranges. In Figure 5a, we consider $N = 5$, $\gamma \in [0, 1]$, $p \in [0.05, 0.45]$, $p_e^{\hat{r}} \in [0, 0.45]$, and $f(s) = s^{\tau}$, where $\tau \in [0.5, 1.5]$. There are a total of 300 different choices and the results are sorted by the performance of RP-optimal policy in ascending order. Figure 5b adopts the same system settings except that we impose $p_{e,i}^0 = 0$ for $1 \leq i \leq N$ to ensure the feasibility of Whittle's index policy. Meanwhile, we ignore the Greedy policy since the Greedy+ policy achieves a better performance, as indicated by Figures 3b and 4b.

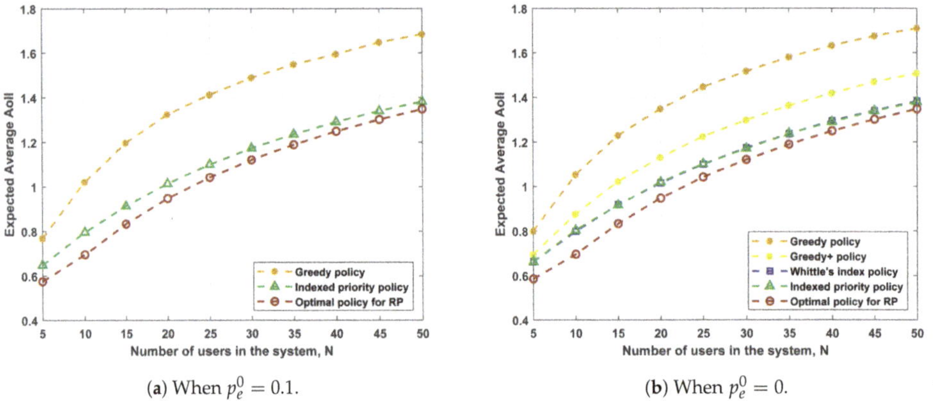

(a) When $p_e^0 = 0.1$.

(b) When $p_e^0 = 0$.

Figure 3. Performance when the source processes vary. We choose $p_i = 0.05 + \frac{0.4(i-1)}{N-1}$, $f_i(s) = s$, $\gamma_i = 0.6$, $p_{e,i}^0 = p_e^0$, and $p_{e,i}^1 = 0.1$ for $1 \leq i \leq N$.

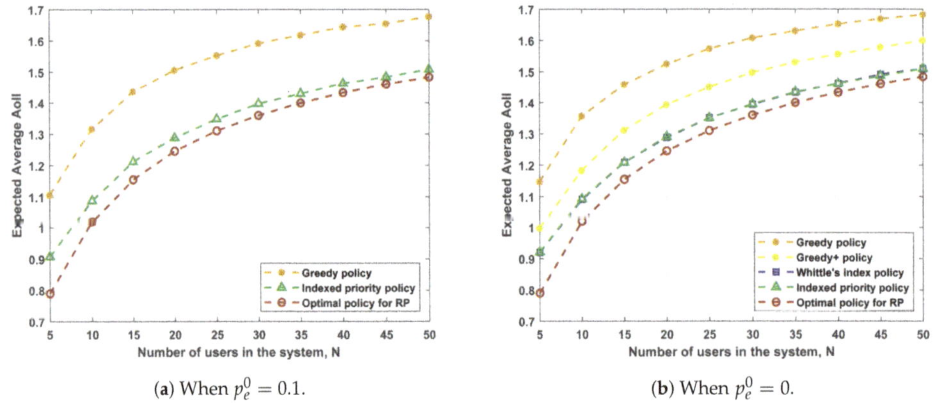

(a) When $p_e^0 = 0.1$.

(b) When $p_e^0 = 0$.

Figure 4. Performance when the communication goals vary. We choose $f_i(s) = s^{0.5 + \frac{i-1}{N-1}}$, $p_i = 0.3$, $\gamma_i = 0.6$, $p_{e,i}^0 = p_e^0$, and $p_{e,i}^1 = 0.1$ for $1 \leq i \leq N$.

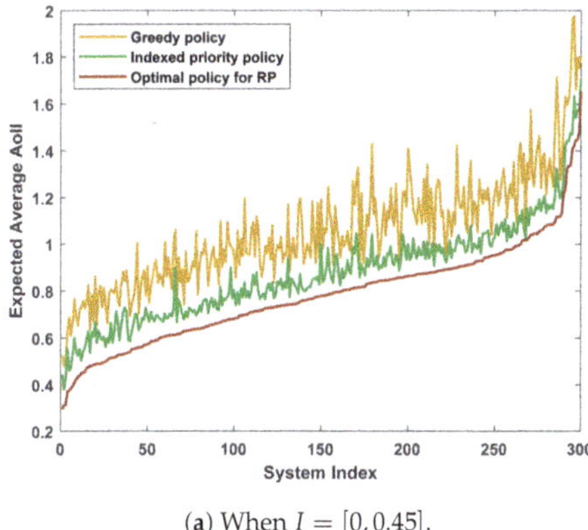

(**a**) When $I = [0, 0.45]$.

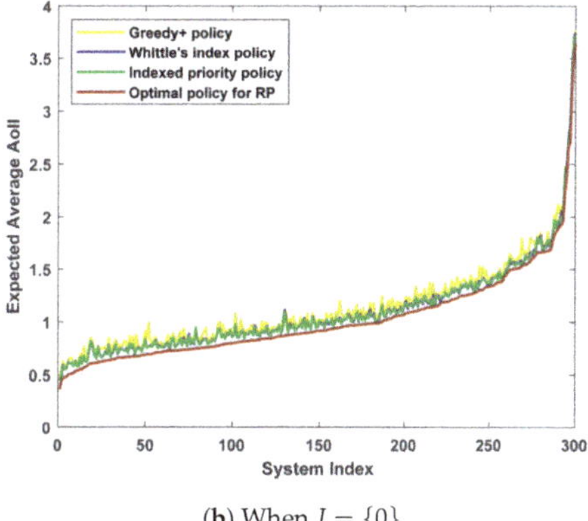

(**b**) When $I = \{0\}$.

Figure 5. Performance in systems with random parameters when $N = 5$. The parameters for each user are chosen randomly within the following intervals: $\gamma \in [0, 1]$, $p \in [0.05, 0.45]$, $p_e^0 \in I$, $p_e^1 \in [0, 0.45]$, and $f(s) = s^\tau$ where $\tau \in [0.5, 1.5]$.

We can make the following observations from the figures.

- The Greedy+ policy yields a smaller expected average AoII than that achieved by the Greedy policy. Recall that we obtained the Greedy+ policy by applying the structural properties detailed in Corollary 1. Therefore, simple applications of the structural properties of the optimal policy can improve the performance of scheduling policies.
- The Indexed priority policy has comparable performance to Whittle's index policy in all the system settings considered. The two policies have their own advantages. The Indexed priority policy has a broader scope of application, while Whittle's index policy has a lower computational complexity.
- The performance of the Indexed priority policy and Whittle's index policy is better than that of the Greedy/Greedy+ policies and is not far from the performance of the RP-optimal policy. Recall that the performance of the RP-optimal policy forms a universal lower bound on the performance of all admissible policies for PP. Hence, we can conclude that both the Indexed priority policy and Whittle's index policy achieve good performances.

8. Conclusions

In this paper, we studied the problem of minimizing the Age of Incorrect Information in a slotted-time system where a base station needs to schedule M users among N available users. Meanwhile, the base station has access to imperfect channel state information in each time slot. The problem is a restless multi-armed bandit problem which is SPACE-hard. However, by casting the problem into a Markov decision process, we obtain the structural properties of the optimal policy. Then, we introduce a relaxed version of the original problem and investigate the decoupled model. Under a simple condition, we establish the indexability of the decoupled problem and obtain the expression of Whittle's index. On this basis, we developed Whittle's index policy. To get rid of the requirement for indexability, we developed the Indexed priority policy based on the optimal policy for the relaxed problem. The characteristics of the relaxed problem are explored to make the calculation of its optimal policy more efficient. Finally, through numerical results, we show that simple applications of the structural properties can improve the performance of scheduling policies. Moreover, Whittle's index policy and the Indexed priority policy achieve good and comparable performances.

Author Contributions: Formal analysis, Y.C.; Investigation, Y.C.; Methodology, Y.C.; Supervision, A.E.; Validation, Y.C.; Writing—original draft, Y.C.; Writing—review & editing, Y.C. and A.E. All authors have read and agreed to the published version of the manuscript.

Funding: This research received no external funding.

Institutional Review Board Statement: Not applicable.

Informed Consent Statement: Not applicable.

Data Availability Statement: Not applicable.

Conflicts of Interest: The authors declare no conflict of interest.

Appendix A. Proof of Lemma 1

We consider two states, x_1 and x_2, that differ only in the value of s_j. Without the loss of generality, we assume $s_{1,j} < s_{2,j}$. Then, it is sufficient to show that, for any $1 \leq j \leq N$, $V(x_1) \leq V(x_2)$. Leveraging the iterative nature of VIA, we use mathematical induction to prove the monotonicity. First of all, the base case (i.e., $\nu = 0$) is true by initialization. We assume the lemma holds at iteration ν. Then, we want to examine whether it holds at iteration $\nu + 1$. The update step reported in problem (5) can be rewritten as follows.

$$V_{\nu+1}(x) = \min_{a \in \mathcal{A}_N(1)} V_{\nu+1}^a(x), \tag{A1}$$

where

$$V_{\nu+1}^a(x) = C(x) - \theta + \sum_{x' - \{x_j'\}} \left\{ \left(\prod_{i \neq j} P_{x_i, x_i'}(a_i) \right) \sum_{\hat{r}_j'} P(\hat{r}_j') U_\nu^j(x, x') \right\},$$

$$U_\nu^j(x, x') = \sum_{s_j'} P_{s_j, s_j'}(a_j, \hat{r}_j) V_\nu(x').$$

To prove the desired results, we distinguish between the following cases.

- We first consider the case of $s_{1,j} = 0 < s_{2,j}$ and $\hat{r}_{1,j} = \hat{r}_{2,j} = 0$. When $a_j = 1$ and for any $x' - \{s_j'\}$, we have

$$U_\nu^j(x_1, x') = p_j V_\nu(x'; s_j' = 1) + (1 - p_j) V_\nu(x'; s_j' = 0),$$

$$U_\nu^j(x_2, x') = \beta_j V_\nu(x'; s_j' = s_{2,j} + 1) + (1 - \beta_j) V_\nu(x'; s_j' = 0),$$

where $V_\nu(x'; s_j' = 0)$ is the estimated value function of the state x' with $s_j' = 0$ at iteration ν (at the risk of abusing the notation, we use $V(x; s_j = s_1)$ and $V(x; s_j = s_2)$ to represent the value functions of two states that differ only in the value of s_j). Then, we get

$$U_\nu^j(x_1, x') - U_\nu^j(x_2, x') \leq (p_j - \beta_j)(V_\nu(x'; s_j' = 1) - V_\nu(x'; s_j' = 0)) \leq 0.$$

The inequalities hold since $\beta_j > p_j$ and Lemma 1 are true at iteration ν by assumption. Therefore, we have $U_\nu^j(x_1, x') \leq U_\nu^j(x_2, x')$ when $a_j = 1$ for any $x' - \{s_j'\}$.
For the case of $a_i = 1$ where $i \neq j$, we notice that $a_j = 0$. Then, for any $x' - \{s_j'\}$, we obtain

$$U_\nu^j(x_1, x') = p_j V_\nu(x'; s_j' = 1) + (1 - p_j) V_\nu(x'; s_j' = 0),$$

$$U_\nu^j(x_2, x') = (1 - p_j) V_\nu(x'; s_j' = s_{2,j} + 1) + p_j V_\nu(x'; s_j' = 0).$$

Therefore, when $a_i = 1$, we have

$$U_\nu^j(x_1, x') - U_\nu^j(x_2, x') \leq (2p_j - 1)(V_\nu(x'; s_j' = 1) - V_\nu(x'; s_j' = 0)) \leq 0.$$

The inequalities hold since $2p_j - 1 < 0$ and Lemma 1 is true at iteration ν by assumption. Combining with the case of $a_j = 1$, $U_\nu^j(x_1, x') \leq U_\nu^j(x_2, x')$ holds for any

$x' - \{s'_j\}$ under any feasible action. Since x_1 and x_2 differ only in the value of s_j and $C(x)$ is non-decreasing in s_i for $1 \leq i \leq N$, we can see that $V^a_{\nu+1}(x_1) \leq V^a_{\nu+1}(x_2)$ for any feasible a. Then, by (A1), we can conclude that the lemma holds at iteration $\nu + 1$ when $s_{1,j} = 0 < s_{2,j}$ and $\hat{r}_{1,j} = \hat{r}_{2,j} = 0$.

- When $s_{1,j} = 0 < s_{2,j}$ and $\hat{r}_{1,j} = \hat{r}_{2,j} = 1$, by replacing the β_j's in the above case with α_j's, we can achieve the same result.
- When $0 < s_{1,j} < s_{2,j}$ and $\hat{r}_{1,j} = \hat{r}_{2,j}$, we notice that

$$P_{s_{1,j},s_{1,j}+1}(a_j, \hat{r}_{1,j}) = P_{s_{2,j},s_{2,j}+1}(a_j, \hat{r}_{2,j}),$$
$$P_{s_{1,j},0}(a_j, \hat{r}_{1,j}) = P_{s_{2,j},0}(a_j, \hat{r}_{2,j}).$$

Then, leveraging the monotonicity of $V_\nu(x)$ and $C(x)$, we can conclude with the same result.

Combining the three cases, we prove that the lemma also holds at iteration $\nu + 1$ of VIA. Therefore, the lemma holds at any iteration ν by mathematical induction. Since the results hold for any $1 \leq j \leq N$ and VIA is guaranteed to converge to the value function when $\nu \to +\infty$, we can conclude our proof.

Appendix B. Proof of Lemma 2

We inherit the notations in the proof of Lemma 1. We still use mathematical induction to obtain the desired results. The base case $\nu = 0$ is true by initialization. We assume the lemma holds at iterative ν and examine whether it still holds at iteration $\nu + 1$. In the case of $M = 1$, we rewrite (5) as

$$V_{\nu+1}(x) = \min_{1 \leq j \leq N} V^j_{\nu+1}(x), \tag{A2}$$

where

$$V^j_{\nu+1}(x) = C(x) - \theta + \sum_{x'} \left\{ \left(\prod_{i \neq j} P^i_{x_i,x'_i}(0) \right) P^j_{x_j,x'_j}(1) V_\nu(x') \right\}, \tag{A3}$$

and $P^i_{x,x'}(a_i)$ is the probability that action a_i will lead to state x' when user i is at state x. To get the desired results, we distinguish between the following cases

- We first show that $V^j_{\nu+1}(x) = V^k_{\nu+1}(\mathscr{P}(x))$. According to (A3), we have

$$V^j_{\nu+1}(x) = C(x) - \theta + \sum_{x'} \left\{ \left(\prod_{i \neq j,k} P^i_{x_i,x'_i}(0) \right) P^k_{x_k,x'_k}(0) P^j_{x_j,x'_j}(1) V_\nu(x') \right\}.$$

$$V^k_{\nu+1}(\mathscr{P}(x)) = C(\mathscr{P}(x)) - \theta +$$
$$\sum_{\mathscr{P}(x)'} \left(\prod_{i \neq j,k} P^i_{\mathscr{P}(x)_i,\mathscr{P}(x)'_i}(0) \right) P^k_{\mathscr{P}(x)_k,\mathscr{P}(x)'_k}(1) P^j_{\mathscr{P}(x)_j,\mathscr{P}(x)'_j}(0) V_\nu(\mathscr{P}(x)').$$

It is obvious that for any $\mathscr{P}(x)'$, there always exists $\mathscr{P}(x'') = \mathscr{P}(x)'$. Then, we obtain

$$V^k_{\nu+1}(\mathscr{P}(x)) = C(\mathscr{P}(x)) - \theta +$$
$$\sum_{\mathscr{P}(x'')} \left(\prod_{i \neq j,k} P^i_{x_i,x''_i}(0) \right) P^k_{x_j,\mathscr{P}(x'')_k}(1) P^j_{x_k,\mathscr{P}(x'')_j}(0) V_\nu(\mathscr{P}(x''))$$
$$= C(\mathscr{P}(x)) - \theta + \sum_{x''} \left(\prod_{i \neq j,k} P^i_{x_i,x''_i}(0) \right) P^k_{x_j,x''_j}(1) P^j_{x_k,x''_k}(0) V_\nu(x'')$$
$$= C(\mathscr{P}(x)) - \theta + \sum_{x'} \left(\prod_{i \neq j,k} P^i_{x_i,x'_i}(0) \right) P^k_{x_j,x'_j}(1) P^j_{x_k,x'_k}(0) V_\nu(x').$$

The second equality follows from the definition of $\mathscr{P}(\cdot)$, the property of summation, and the assumption at iteration ν. The last equality follows from the variable renaming. Then, by the definition of statistically identical, we have $P^k_{x_j,x'_j}(1) = P^j_{x_j,x'_j}(1)$, $P^j_{x_k,x'_k}(0) = P^k_{x_k,x'_k}(0)$, and $C(x) = C(\mathscr{P}(x))$. Therefore, we can conclude that $V^j_{\nu+1}(x) = V^k_{\nu+1}(\mathscr{P}(x))$.

- Along the same lines, we can easily show that $V^k_{\nu+1}(x) = V^j_{\nu+1}(\mathscr{P}(x))$ and $V^i_{\nu+1}(x) = V^i_{\nu+1}(\mathscr{P}(x))$ for $i \neq j,k$.

Combining the above cases with (A2), we prove that $V_{\nu+1}(x) = V_{\nu+1}(\mathscr{P}(x))$. Then, by induction, we have $V_\nu(x) = V_\nu(\mathscr{P}(x))$ at any iteration ν. Since VIA is guaranteed to converge to the value function when $\nu \to +\infty$, we can conclude our proof.

Appendix C. Proof of Theorem 1

For arbitrary j and k

$$\delta^{j,k}(x) = \sum_{x'-\{x'_j,x'_k\}} \left\{ \left(\prod_{i \neq j,k} P_{x_i,x'_i}(0) \right) \sum_{\hat{r}'_j,\hat{r}'_k} P(\hat{r}'_j)P(\hat{r}'_k)R^{j,k}(x,x') \right\}, \tag{A4}$$

where

$$R^{j,k}(x,x') = \sum_{s'_j,s'_k} \left[\left(P_{s_k,s'_k}(0,\hat{r}_k)P_{s_j,s'_j}(1,\hat{r}_j) - P_{s_k,s'_k}(1,\hat{r}_k)P_{s_j,s'_j}(0,\hat{r}_j) \right) V(x') \right]. \tag{A5}$$

With this in mind, we will prove the properties one by one.

Property 1—$\delta^{j,k}(x) \leq 0$ if $\hat{r}_k = p^0_{e,k} = 0$. The equality holds when $s_j = 0$ or $\hat{r}_j = p^0_{e,j} = 0$.

When $\hat{r}_k = p^0_{e,k} = 0$, transmitting the update from user k will necessarily fail. Therefore, $P_{s_k,s'_k}(0,0) = P_{s_k,s'_k}(1,0)$ for any s_k and s'_k. Then, we have

$$R^{j,k}(x,x') = \sum_{s'_k} P_{s_k,s'_k}(0,0) \sum_{s'_j} \left[\left(P_{s_j,s'_j}(1,\hat{r}_j) - P_{s_j,s'_j}(0,\hat{r}_j) \right) V(x') \right].$$

To identify the sign of $R^{j,k}(x,x')$, we distinguish between the following cases

- When $s_j = 0$, we can easily show that $R^{j,k}(x,x') = 0$ for any $x' - \{s'_j,s'_k\}$ by noticing that the two possible actions with respect to user j (i.e., $a_j = 1$ and $a_j = 0$) are equivalent when $s_j = 0$. Since $\delta^{j,k}(x)$ is a linear combination of $R^{j,k}(x,x')$'s with non-negative coefficients, we can conclude that $\delta^{j,k}(x) = 0$ in this case.
- When $s_j > 0$ and $\hat{r}_j = 1$, for any $x' - \{s'_j,s'_k\}$, we have

$$R^{j,k}(x,x') = \sum_{s'_k} P_{s_k,s'_k}(0,0)(\alpha_j + p_j - 1)\left(V(x';s'_j = s_j+1) - V(x';s'_j = 0)\right)$$
$$\tag{A6}$$
$$\leq 0.$$

The inequality holds because of Lemma 1 and the fact that $\alpha_j + p_j < 1$. We recall that $\delta^{j,k}(x)$ is a linear combination of $R^{j,k}(x,x')$'s with non-negative coefficients. Then, we can conclude that $\delta^{j,k}(x) \leq 0$ in this case.

- When $s_j > 0$ and $\hat{r}_j = 0$, by replacing the α_j in (A6) with β_j, we can get the same result. In this case, the equality holds when $\beta_j + p_j = 1$, or, equivalently, $p^0_{e,j} = 0$.

Combining the cases, we prove the first property.

Property 2—$\delta^{j,k}(x)$ is non-increasing in $\hat{r}_j$ and is non-decreasing in $\hat{r}_k$ when $s_j, s_k > 0$. At the same time, $\delta^{j,k}(x)$ is independent of $\hat{r}_i$ for any $i \neq j, k$.

We first prove the monotonicity of $\delta^{j,k}(x)$ with respect to $\hat{r}_j$. To this end, we define x_1 and x_2 as two states that differ only in the value of $\hat{r}_j$. Without a loss of generality, we assume $\hat{r}_{1,j} = 1$ and $\hat{r}_{2,j} = 0$. Then, we investigate the sign of $\delta^{j,k}(x_1) - \delta^{j,k}(x_2)$. We define $x_i \triangleq x_{1,i} = x_{2,i}$ for $i \neq j$. Then, according to (A4), $\delta^{j,k}(x_1) - \delta^{j,k}(x_2)$ can be written as

$$\delta^{j,k}(x_1) - \delta^{j,k}(x_2) = \\ \sum_{x' - \{x_j', x_k'\}} \left\{ \left(\prod_{i \neq j,k} P_{x_i, x_i'}(0) \right) \sum_{\hat{r}_j', \hat{r}_k'} P(\hat{r}_j') P(\hat{r}_k') \left(R^{j,k}(x_1, x') - R^{j,k}(x_2, x') \right) \right\}.$$

Since $x_{1,k} = x_{2,k}$, we have $P_{s_{1,k}, s_k'}(a, \hat{r}_{1,k}) = P_{s_{2,k}, s_k'}(a, \hat{r}_{2,k})$ for any s_k'. We recall that the transition probability is independent of $\hat{r}$ when $a = 0$. Combining with the fact that $s_{1,j} = s_{2,j}$, we also have $P_{s_{1,j}, s_j'}(0, \hat{r}_{1,j}) = P_{s_{2,j}, s_j'}(0, \hat{r}_{2,j})$ for any s_j'. Combining together, we obtain

$$P_{s_{1,k}, s_k'}(1, \hat{r}_{1,k}) P_{s_{1,j}, s_j'}(0, \hat{r}_{1,j}) = P_{s_{2,k}, s_k'}(1, \hat{r}_{2,k}) P_{s_{2,j}, s_j'}(0, \hat{r}_{2,j}),$$

$$P_{s_{1,k}, s_k'}(0, \hat{r}_{1,k}) = P_{s_{2,k}, s_k'}(0, \hat{r}_{2,k}).$$

Leveraging the above two problems, we have

$$R^{j,k}(x_1, x') - R^{j,k}(x_2, x') = \\ \sum_{s_j', s_k'} \left[P_{s_k, s_k'}(0, \hat{r}_k) \left(P_{s_{1,j}, s_j'}(1, \hat{r}_{1,j}) - P_{s_{2,j}, s_j'}(1, \hat{r}_{2,j}) \right) V(x') \right].$$

Consequently, we obtain

$$\delta^{j,k}(x_1) - \delta^{j,k}(x_2) = \\ \sum_{x' - \{x_j'\}} \left\{ \prod_{i \neq j} P_{x_i, x_i'}(0) \left[\sum_{\hat{r}_j'} P(\hat{r}_j') \sum_{s_j'} \left(P_{s_{1,j}, s_j'}(1, 1) - P_{s_{2,j}, s_j'}(1, 0) \right) V(x') \right] \right\}.$$

In the following, we characterize the sign of

$$R_1 \triangleq \sum_{s_j'} \left(P_{s_{1,j}, s_j'}(1, 1) - P_{s_{2,j}, s_j'}(1, 0) \right) V(x').$$

As $s_{1,j} = s_{2,j} > 0$, for any $x' - \{s_j'\}$, we have

$$R_1 = \left((1 - \alpha_j) - (1 - \beta_j) \right) V(x'; s_j' = 0) + (\alpha_j - \beta_j) V(x'; s_j' = s_{1,j} + 1) \leq 0.$$

The inequality follows from Lemma 1 and the fact that $\beta_j > \alpha_j$. Since $\delta^{j,k}(x_1) - \delta^{j,k}(x_2)$ is a linear combination of R_1's with non-negative coefficients, we can conclude that $\delta^{j,k}(x_1) \leq \delta^{j,k}(x_2)$. Since $\hat{r}_{1,j} > \hat{r}_{2,j}$, we can see that $\delta^{j,k}(x)$ is non-increasing in $\hat{r}_j$.

In a very similar way, we can show that $\delta^{j,k}(x)$ is non-decreasing in $\hat{r}_k$. We recall that $\hat{r}_i$ will not affect the system dynamic if $a_i = 0$. Consequently, we can conclude that $\delta^{j,k}(x)$ is independent of $\hat{r}_i$ for any $i \neq j, k$.

Combining together, we prove the second property.

Property 3—$\delta^{j,k}(x) \leq 0$ if $s_k = 0$. The equality holds when $s_j = 0$ or $\hat{r}_j = p_{e,j}^0 = 0$.

Since the probabilities are non-negative, it is sufficient to show that $R^{j,k}(x, x')$ satisfies Property 3 for any $x' - \{s_j', s_k'\}$. More precisely, it is sufficient to show that $R^{j,k}(x, x') \leq 0$

for any $x' - \{s'_j, s'_k\}$ when $s_k = 0$ and the equality holds when $s_j = 0$ or $\hat{r}_j = p^0_{e,j} = 0$. We recall that $P_{s_k,s'_k}(1, \hat{r}_k) = P_{s_k,s'_k}(0, \hat{r}_k)$ for any s'_k when $s_k = 0$. Hence, for any $x' - \{s'_j, s'_k\}$, we have

$$R^{j,k}(x, x') = \sum_{s'_k} \left[P_{s_k,s'_k}(0, \hat{r}_k) \sum_{s'_j} \left(P_{s_j,s'_j}(1, \hat{r}_j) - P_{s_j,s'_j}(0, \hat{r}_j) \right) V(x') \right].$$

Then, we investigate the following quantity for any $x' - \{s'_j\}$

$$R_2 \triangleq \sum_{s'_j} \left(P_{s_j,s'_j}(1, \hat{r}_j) - P_{x_j,x'_j}(0, \hat{r}_j) \right) V(x').$$

To this end, we distinguish between the following cases
- When $s_j = 0$, we have $P_{s_j,s'_j}(1, \hat{r}_j) = P_{s_j,s'_j}(0, \hat{r}_j)$ for any s'_j. Thus, we conclude that $R_2 = 0$ for any $x' - \{s'_j\}$. Consequently, $R^{j,k}(x, x') = 0$ for any $x' - \{s'_j, s'_k\}$.
- When $s_j > 0$ and $\hat{r}_j = 1$, for any $x' - \{s'_j\}$, we have

$$R_2 = (\alpha_j - 1 + p_j)V(x'; s'_j = s_j + 1) + (1 - \alpha_j - p_j)V(x'; s'_j = 0) \leq 0 \qquad (A7)$$

The inequality follows from Lemma 1 and the fact that $\alpha_j + p_j < 1$. Thus, $R^{j,k}(x, x') \leq 0$ for any $x' - \{s'_j, s'_k\}$.
- When $s_j > 0$ and $\hat{r}_j = 0$, by replacing the α_j in (A7) with β_j, we can get the same result. In this case, the equality holds when $\beta_j + p_j = 1$, or, equivalently, $p^0_{e,j} = 0$.

Combined together, we can conclude that Property 3 is true.

Property 4—$\delta^{j,k}(x)$ is non-increasing in s_j if $\Gamma^{\hat{r}_j}_j \leq \Gamma^{\hat{r}_k}_k$ and is non-decreasing in s_k if $\Gamma^{\hat{r}_j}_j \geq \Gamma^{\hat{r}_k}_k$ when $s_j, s_k > 0$. We define $\Gamma^1_i \triangleq \frac{\alpha_i}{1 - p_i}$ and $\Gamma^0_i \triangleq \frac{\beta_i}{1 - p_i}$ for $1 \leq i \leq N$.

Such as we did in the proof of Property 3, it is sufficient to show that $R^{j,k}(x, x')$ satisfies Property 4 for any $x' - \{s'_j, s'_k\}$. We recall that $R^{j,k}(x, x')$ depends on the values of $\hat{r}_j$ and $\hat{r}_k$. Therefore, we distinguish between the following cases
- In the case of $\hat{r}_j = \hat{r}_k = 1$ and $s_j, s_k > 0$, for any $x' - \{s'_j, s'_k\}$, (A5) can be written as

$$R^{j,k}(x, x') = \sum_{s'_j,s'_k} \left[\left(P_{s_k,s'_k}(0,1)P_{s_j,s'_j}(1,1) - P_{s_k,s'_k}(1,1)P_{s_j,s'_j}(0,1) \right) V(x') \right]$$

$$= (p_k\alpha_j - (1 - p_j)(1 - \alpha_k))V(x'; s'_j = s_j + 1; s'_k = 0)$$
$$+ ((1 - p_k)(1 - \alpha_j) - p_j\alpha_k)V(x'; s'_j = 0; s'_k = s_k + 1)$$
$$+ ((1 - p_k)\alpha_j - (1 - p_j)\alpha_k)V(x'; s'_j = s_j + 1; s'_k = s_k + 1)$$
$$+ (p_k(1 - \alpha_j) - p_j(1 - \alpha_k))V(x'; s'_j = 0; s'_k = 0).$$

As we can verify

$$p_k\alpha_j - (1 - p_j)(1 - \alpha_k) < \frac{1}{2}(p_k + p_j - 1) < 0,$$

$$(1 - p_k)(1 - \alpha_j) - p_j\alpha_k > \frac{1}{2}(1 - p_k - p_j) > 0.$$

We define $\Gamma^1_i \triangleq \frac{\alpha_i}{1 - p_i}$ and $\Gamma^0_i \triangleq \frac{\beta_i}{1 - p_i}$ for $1 \leq i \leq N$. Then, we have

$$\Gamma^1_j \lessgtr \Gamma^1_k \implies (1 - p_k)\alpha_j - (1 - p_j)\alpha_k \lessgtr 0.$$

Combining with Lemma 1, we can conclude that, for any $x' - \{s_j', s_k'\}$, $R^{j,k}(x, x')$ is non-increasing in s_j if $\Gamma_j^1 \leq \Gamma_k^1$ and is non-decreasing in s_k if $\Gamma_j^1 \geq \Gamma_k^1$.

- In the case of $\hat{r}_j = \hat{r}_k = 0$ and $s_j, s_k > 0$, by replacing the α's in the above case with β's, we can conclude with the same result.
- In the case of $\hat{r}_j = 1$, $\hat{r}_k = 0$, and $s_j, s_k > 0$, for any $x' - \{s_j', s_k'\}$, (A5) can be written as

$$
R^{j,k}(x, x') = \sum_{s_j', s_k'} \left[\left(P_{s_k, s_k'}(0, 0) P_{s_j, s_j'}(1, 1) - P_{s_k, s_k'}(1, 0) P_{s_j, s_j'}(0, 1) \right) V(x') \right]
$$

$$
\begin{aligned}
&= \left(p_k \alpha_j - (1 - p_j)(1 - \beta_k) \right) V(x'; s_j' = s_j + 1; s_k' = 0) \\
&\quad + \left((1 - p_k)(1 - \alpha_j) - p_j \beta_k \right) V(x'; s_j' = 0; s_k' = s_k + 1) \\
&\quad + \left((1 - p_k)\alpha_j - (1 - p_j)\beta_k \right) V(x'; s_j' = s_j + 1; s_k' = s_k + 1) \\
&\quad + \left(p_k(1 - \alpha_j) - p_j(1 - \beta_k) \right) V(x'; s_j' = 0; s_k' = 0).
\end{aligned}
$$

As we can verify

$$
p_k \alpha_j - (1 - p_j)(1 - \beta_k) < p_k \left(p_j - \frac{1}{2} \right) < 0,
$$

$$
(1 - p_k)(1 - \alpha_j) - p_j \beta_k > (1 - p_k)\left(\frac{1}{2} - p_j \right) > 0.
$$

At the same time

$$
\Gamma_j^1 \lesseqgtr \Gamma_k^0 \implies (1 - p_k)\alpha_j - (1 - p_j)\beta_k \lesseqgtr 0.
$$

Combined with Lemma 1, we can conclude that, for any $x' - \{s_j', s_k'\}$, $R^{j,k}(x, x')$ is non-increasing in s_j if $\Gamma_j^1 \leq \Gamma_k^0$ and is non-decreasing in s_k if $\Gamma_j^1 \geq \Gamma_k^0$.

- In the case of $\hat{r}_j = 0$, $\hat{r}_k = 1$, and $s_j, s_k > 0$, by swapping the α's and β's in the above case, we can conclude with the same result.

Combined together, we conclude that $R^{j,k}(x, x')$ satisfies Property 3 for any $x' - \{s_j', s_k'\}$. Consequently, $\delta^{j,k}(x)$ is non-increasing in s_j if $\Gamma_j^{\hat{r}_j} \leq \Gamma_k^{\hat{r}_k}$ and is non-decreasing in s_k if $\Gamma_j^{\hat{r}_j} \geq \Gamma_k^{\hat{r}_k}$ when $s_j, s_k > 0$.

Property 5—$\delta^{j,k}(x) \leq 0$ if $s_j \geq s_k$, $\hat{r}_j \geq \hat{r}_k$, and users j and k are statistically identical.

According to Property 3, it is sufficient to consider the case where $s_j, s_k > 0$. We notice that the sign of $\delta^{j,k}(x)$ can be captured by the sign of the quantity $Q^{j,k}(x, x') \triangleq \sum_{\hat{r}_j', \hat{r}_k'} P(\hat{r}_j') P(\hat{r}_k') R^{j,k}(x, x')$. Thus, we divide our discussion into the following cases.

- We first consider the case of $s_j \geq s_k > 0$ and $\hat{r}_j = \hat{r}_k = 0$. Leveraging the definition of statistically identical, for any $x' - \{x_j', x_k'\}$, we have

$$
\begin{aligned}
Q^{j,k}(x, x') = \sum_{\hat{r}_j', \hat{r}_k'} P(\hat{r}_j') P(\hat{r}_k') \kappa_1 \Big(&V(x'; x_j' = (0, \hat{r}_j'); x_k' = (s_k + 1, \hat{r}_k')) - \\
&V(x'; x_j' = (s_j + 1, \hat{r}_j'); x_k' = (0, \hat{r}_k')) \Big),
\end{aligned}
$$

where $\kappa_1 = 1 - p_j - \beta_j \geq 0$. Then, by substituting the values of $P(\hat{r})$ and using Lemma 2, we obtain

$$Q^{j,k}(\boldsymbol{x},\boldsymbol{x}') = \gamma_j \gamma_k \kappa_1 V(\boldsymbol{x}'; x'_j = (s_k+1,1); x'_k = (0,1)) -$$
$$\gamma_j \gamma_k \kappa_1 V(\boldsymbol{x}'; x'_j = (s_j+1,1); x'_k = (0,1)) +$$
$$(1-\gamma_j)(1-\gamma_k)\kappa_1 V(\boldsymbol{x}'; x'_j = (s_k+1,0); x'_k = (0,0)) -$$
$$(1-\gamma_j)(1-\gamma_k)\kappa_1 V(\boldsymbol{x}'; x'_j = (s_j+1,0); x'_k = (0,0)) +$$
$$\gamma_k(1-\gamma_j)\kappa_1 V(\boldsymbol{x}'; x'_j = (s_k+1,1); x'_k = (0,0)) -$$
$$\gamma_k(1-\gamma_j)\kappa_1 V(\boldsymbol{x}'; x'_j = (s_j+1,0); x'_k = (0,1)) +$$
$$\gamma_j(1-\gamma_k)\kappa_1 V(\boldsymbol{x}'; x'_j = (s_k+1,0); x'_k = (0,1)) -$$
$$\gamma_j(1-\gamma_k)\kappa_1 V(\boldsymbol{x}'; x'_j = (s_j+1,1); x'_k = (0,0)).$$

Since users j and k are statistically identical, we have $\gamma_j = \gamma_k$. Then, by Lemma 1, we have $Q^{j,k}(\boldsymbol{x},\boldsymbol{x}') \le 0$ for any $\boldsymbol{x}' - \{x'_j, x'_k\}$. Since $\delta^{j,k}(\boldsymbol{x})$ is a linear combination of $Q^{j,k}(\boldsymbol{x},\boldsymbol{x}')$'s with non-negative coefficients, we can conclude that $\delta^{j,k}(\boldsymbol{x}) \le 0$.

- For the case of $s_j \ge s_k > 0$ and $\hat{r}_j = \hat{r}_k = 1$, by replacing β_j in κ_1 with α_j, we can conclude with the same result.
- Then, we consider the case of $s_j \ge s_k > 0$, $\hat{r}_j = 1$, and $\hat{r}_k = 0$. We first notice that, for any $\boldsymbol{x}' - \{s'_j, s'_k\}$

$$R^{j,k}(\boldsymbol{x},\boldsymbol{x}') = \big(p_k \alpha_j - (1-p_j)(1-\beta_k)\big) V(\boldsymbol{x}'; s'_j = s_j+1; s'_k = 0) +$$
$$\big((1-p_k)(1-\alpha_j) - p_j \beta_k\big) V(\boldsymbol{x}'; s'_j = 0; s'_k = s_k+1) +$$
$$\big((1-p_k)\alpha_j - (1-p_j)\beta_k\big) V(\boldsymbol{x}'; s'_j = s_j+1; s'_k = s_k+1) +$$
$$\big(p_k(1-\alpha_j) - p_j(1-\beta_k)\big) V(\boldsymbol{x}'; s'_j = 0; s'_k = 0).$$

As users j and k are statistically identical, we have $p_j = p_k$ and $\alpha_j < \beta_k$. Leveraging Lemma 1, we have

$$R^{j,k}(\boldsymbol{x},\boldsymbol{x}') \le (\alpha_j + p_j - 1)\Big(V(\boldsymbol{x}'; s'_j = s_j+1; s'_k = 0) -$$
$$V(\boldsymbol{x}'; s'_j = 0; s'_k = s_k+1)\Big).$$

Then, for any $\boldsymbol{x}' - \{x'_j, x'_k\}$

$$Q^{j,k}(\boldsymbol{x},\boldsymbol{x}') \le \sum_{\hat{r}'_j, \hat{r}'_k} P(\hat{r}'_j) P(\hat{r}'_k) \kappa_2 \Big(V(\boldsymbol{x}'; x'_j = (0,\hat{r}'_j); x'_k = (s_k+1,\hat{r}'_k)) -$$
$$V(\boldsymbol{x}'; x'_j = (s_j+1,\hat{r}'_j); x'_k = (0,\hat{r}'_k)) \Big),$$

where $\kappa_2 = 1 - p_j - \alpha_j > 0$. Such as we did in the previous cases, we can leverage Lemmas 1 and 2 to conclude that $Q^{j,k}(\boldsymbol{x},\boldsymbol{x}') \le 0$ for any $\boldsymbol{x}' - \{x'_j, x'_k\}$. Consequently, $\delta^{j,k}(\boldsymbol{x}) \le 0$ in this case. The details are omitted for the sake of space.

Combined together, we conclude the proof of Property 5.

Appendix D. Proof of Corollary 2

We follow the same steps as in the proof of Lemma 1. To prove the corollary, it is sufficient to show that $V(x_1) \le V(x_2)$ when $s_1 < s_2$ and $\hat{r}_1 = \hat{r}_2$. We use mathematical induction to prove the monotonicity. First of all, the base case (i.e., $v = 0$) is true by initialization. We assume the lemma holds at iteration v. Then, we want to examine whether it holds at iteration $\nu + 1$. For the system with a single user, the update step reported in problem (5) can be simplified and rewritten as follows

$$V_{\nu+1}(x) = \min_{a \in \{0,1\}} V^a_{\nu+1}(x),\tag{A8}$$

where

$$V^a_{\nu+1}(x) = C(x,a) - \theta + \sum_{\hat{r}'} P(\hat{r}') \sum_{s'} P_{s,s'}(a,\hat{r}) V_{\nu}(x'),$$

and θ is the optimal value for $\mathcal{M}_1(\lambda, -1)$. To prove the desired results, we distinguish between the following cases

- We first consider the case of $s_1 = 0 < s_2$ and $\hat{r}_1 = \hat{r}_2 = 0$. When $a = 1$, we have

$$V^1_{\nu+1}(x_1) = C(x_1,1) - \theta + \sum_{\hat{r}'} P(\hat{r}') \left(p V_{\nu}(1,\hat{r}') + (1-p) V_{\nu}(0,\hat{r}') \right),$$

$$V^1_{\nu+1}(x_2) = C(x_2,1) - \theta + \sum_{\hat{r}'} P(\hat{r}') \left(\beta V_{\nu}(s_2+1,\hat{r}') + (1-\beta) V_{\nu}(0,\hat{r}') \right).$$

Subtracting the two expressions yields

$$V^1_{\nu+1}(x_1) - V^1_{\nu+1}(x_2)$$
$$\leq C(x_1,1) - C(x_2,1) + \sum_{\hat{r}'} P(\hat{r}') \left[(p-\beta)\left(V_{\nu}(1,\hat{r}') - V_{\nu}(0,\hat{r}')\right) \right] \leq 0.$$

The inequalities hold since $\beta > p$, $C(x,a)$ is non-decreasing in s, and Corollary 2 is true at iteration ν by assumption.
For the case of $a = 0$, we obtain

$$V^0_{\nu+1}(x_1) = C(x_1,0) - \theta + \sum_{\hat{r}'} P(\hat{r}') \left(p V_{\nu}(1,\hat{r}') + (1-p) V_{\nu}(0,\hat{r}') \right),$$

$$V^0_{\nu+1}(x_2) = C(x_2,0) - \theta + \sum_{\hat{r}'} P(\hat{r}') \left((1-p) V_{\nu}(s_2+1,\hat{r}') + p V_{\nu}(0,\hat{r}') \right).$$

Therefore, when $a = 0$, we have

$$V^0_{\nu+1}(x_1) - V^0_{\nu+1}(x_2)$$
$$\leq C(x_1,0) - C(x_2,0) + \sum_{\hat{r}'} P(\hat{r}') \left[(2p-1)\left(V_{\nu}(1,\hat{r}') - V_{\nu}(0,\hat{r}')\right) \right] \leq 0.$$

The inequalities hold since $2p - 1 < 0$, $C(x,a)$ is non-decreasing in s, and Corollary 2 is true at iteration ν by assumption. Combined together, we can see that $V^a_{\nu+1}(x_1) \leq V^a_{\nu+1}(x_2)$ for any feasible a. Then, by problem (A8), we can conclude that the lemma holds at iteration $\nu + 1$ when $s_1 = 0 < s_2$ and $\hat{r}_1 = \hat{r}_2 = 0$.
- When $s_1 = 0 < s_2$ and $\hat{r}_1 = \hat{r}_2 = 1$, by replacing the β's in the above case with α's, we can achieve the same result.
- When $0 < s_1 < s_2$ and $\hat{r}_1 = \hat{r}_2$, we notice that $P_{s_1,s_1+1}(a,\hat{r}_1) = P_{s_2,s_2+1}(a,\hat{r}_2)$ and $P_{s_1,0}(a,\hat{r}_1) = P_{s_2,0}(a,\hat{r}_2)$. Then, leveraging the monotonicity of $V_{\nu}(x)$ and $C(x,a)$, we can conclude with the same result.

Combining the three cases, we prove that the lemma holds at iteration $\nu + 1$ of VIA. Therefore, the lemma holds at any iteration ν by mathematical induction. Since VIA is guaranteed to converge to the value function when $\nu \to +\infty$, we can conclude our proof.

Appendix E. Proof of Proposition 1

We define $\Delta V(x) \triangleq V^1(x) - V^0(x)$ where $V^a(x)$ is the value function resulting from taking action a at state x. Then, $V^a(x)$ can be calculated as follows

$$V^a(x) = C(x,a) - \theta + \sum_{x' \in \mathcal{X}} P_{x,x'}(a) V(x'), \tag{A9}$$

where θ is the optimal value for $\mathcal{M}_1(\lambda, -1)$. Hence, the optimal action at state x can be fully characterized by the sign of $\Delta V(x)$. More precisely, the optimal action at state x is $a = 1$ if $\Delta V(x) < 0$, and $a = 0$ is optimal otherwise. To determine the sign of $\Delta V(x)$ for each state, we distinguish between the following cases

- We first consider the state $x = (0, \hat{r})$. Applying the results in Section 2.3 to problem (A9), we obtain

$$V^0(0,\hat{r}) = -\theta + (1-\gamma)(1-p)V(0,0) + (1-\gamma)pV(1,0)+$$
$$\gamma(1-p)V(0,1) + \gamma p V(1,1),$$

$$V^1(0,\hat{r}) = \lambda + V^0(0,\hat{r}). \tag{A10}$$

Therefore, $\Delta V(0,\hat{r}) = \lambda \geq 0$. Thus, the optimal action at state $(0,\hat{r})$ is $a = 0$.

- Then, we consider the state $x = (s,0)$ where $s > 0$. Applying the results in Section 2.3 to Equation (A9), we obtain

$$V^0(s,0) = f(s) - \theta + (1-\gamma)pV(0,0) + (1-\gamma)(1-p)V(s+1,0)+$$
$$\gamma p V(0,1) + \gamma(1-p)V(s+1,1),$$

$$V^1(s,0) = f(s) + \lambda - \theta + (1-\gamma)(1-\beta)V(0,0) + (1-\gamma)\beta V(s+1,0)+$$
$$\gamma(1-\beta)V(0,1) + \gamma\beta V(s+1,1).$$

Then,
$$\Delta V(s,0) = \lambda + p_e^0(1-2p)\omega, \tag{A11}$$

where $\omega = (1-\gamma)[V(0,0) - V(s+1,0)] + \gamma[V(0,1) - V(s+1,1)] \leq 0$.

- Finally, we consider the state $x = (s,1)$ where $s > 0$. Following the same trajectory, we have

$$\Delta V(s,1) = \lambda + (1 - p_e^1)(1-2p)\omega.$$

According to Corollary 2 and the fact that $p < 0.5$, we can see that $\Delta V(s,0)$ and $\Delta V(s,1)$ are both a constant λ plus a term that is non-increasing in s. As the time penalty function is unbounded, the value function must also be unbounded. Then, combining the three cases, we can conclude the following. For fixed $\hat{r}$, there always exists a threshold $n_{\hat{r}} > 0$ such that the optimal action at state $(s,\hat{r})$ where $s \geq n_{\hat{r}}$ is $a = 1$, otherwise $a = 0$ is optimal. Since $\hat{r} \in \{0,1\}$, the optimal policy can be fully captured by the pair (n_0, n_1).

In the following, we determine the relationship between n_0 and n_1. We have

$$\Delta V(s,1) - \Delta V(s,0) = (1 - p_e^1 - p_e^0)(1-2p)\omega \leq 0.$$

At the same time, for the threshold n_0, we know $\Delta V(n_0,0) < 0$. Then, we have $\Delta V(n_0,1) \leq \Delta V(n_0,0) < 0$. Combined with the fact that $\Delta V(s,\hat{r})$ is non-increasing in s, we can conclude that the ordering $n_0 \geq n_1$ is true.

Appendix F. Proof of Proposition 2

We notice that the dynamic of AoII under threshold policy can be fully captured by a Discrete-Time Markov Chain (DTMC). Then, the expected AoII $\bar{\Delta}_n$ and the expected transmission rate $\bar{\rho}_n$ under threshold policy $n = (n_0, n_1)$ can be obtained from the stationary distribution of the induced DTMC. Let the states of the induced DTMC be the values of s. We recall that $\hat{r}$ is an independent Bernoulli random variable with parameter γ. Combined with the results in Section 2.3, we can easily obtain the state transition probabilities of the induced DTMC, which are shown in Figure A1.

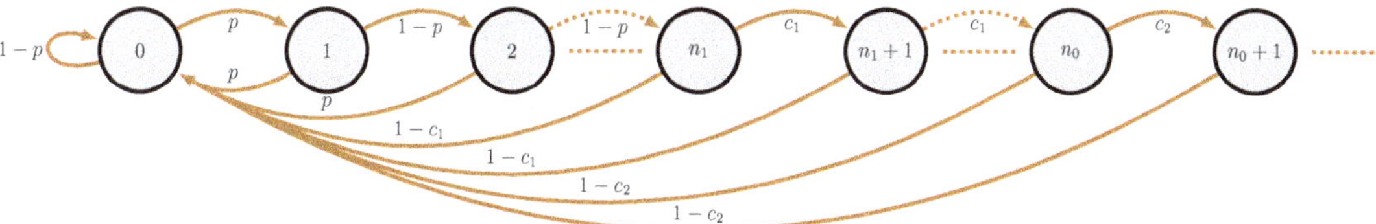

Figure A1. DTMC induced by the threshold policy $n = (n_0, n_1)$. In the figure, $c_1 = (1 - \gamma)(1 - p) + \gamma\alpha$ and $c_2 = (1 - \gamma)\beta + \gamma\alpha$.

The balance equations of the induced DTMC are the following

$$(1 - p)\pi_0 + p \sum_{k=1}^{n_1 - 1} \pi_k + (1 - c_1) \sum_{k=n_1}^{n_0 - 1} \pi_k + (1 - c_2) \sum_{k=n_0}^{+\infty} \pi_k = \pi_0.$$

$$p\pi_0 = \pi_1.$$

$$(1 - p)\pi_{k-1} = \pi_k \ \ for \ 2 \leq k \leq n_1.$$

$$c_1\pi_{k-1} = \pi_k \ \ for \ n_1 + 1 \leq k \leq n_0.$$

$$c_2\pi_{k-1} = \pi_k \ \ for \ n_0 + 1 \leq k.$$

$$\sum_{k=0}^{+\infty} \pi_k = 1.$$

Then, we can easily solve the above system of linear equations. After some algebraic manipulation, we obtain the following

$$\pi_0 = \frac{1}{2 + p(1 - p)^{n_1 - 1}\left[\dfrac{1}{1 - c_1} - \dfrac{1}{p} + c_1^{n_0 - n_1}\left(\dfrac{1}{1 - c_2} - \dfrac{1}{1 - c_1}\right)\right]}.$$

$$\pi_k = p(1 - p)^{k-1}\pi_0 \ \ for \ 1 \leq k \leq n_1.$$

$$\pi_k = p(1 - p)^{n_1 - 1}c_1^{k - n_1}\pi_0 \ \ for \ n_1 + 1 \leq k \leq n_0.$$

$$\pi_k = p(1 - p)^{n_1 - 1}c_1^{n_0 - n_1}c_2^{k - n_0}\pi_0 \ \ for \ n_0 + 1 \leq k.$$

Equipped with the above results, we proceed with calculating $\bar{\Delta}_n$ and $\bar{\rho}_n$. According to problem (6a), the expected AoII is:

$$\bar{\Delta}_n = \sum_{k=0}^{+\infty} f(k)\pi_k.$$

Substituting the expressions of π_k's, we can get the expression of $\bar{\Delta}_n$. Proposition 1 tells us the following.

- For state $(s, \hat{r})$ where $s < n_1$, it is optimal to stay idle (i.e., $a = 0$).
- For state $(s, \hat{r})$ where $n_1 \leq s < n_0$, it is optimal to make a transmission attempt only when $\hat{r} = 1$. We recall that $\hat{r}$ is an independent Bernoulli random variable with parameter γ. Therefore, the expected proportion of time that the system is at state $(s, 1)$ is $\gamma\pi_s$.

- For state $(s, \hat{r})$ where $s \geq n_0$, it is optimal to make transmission attempt regardless of $\hat{r}$.

Combined with problem (6b), we have

$$\bar{\rho}_n = \gamma \sum_{k=n_1}^{n_0-1} \pi_k + \sum_{k=n_0}^{+\infty} \pi_k.$$

Substituting the expressions of π_k's, we can obtain the closed-form expression of $\bar{\rho}_n$.

Appendix G. Proof of Proposition 4

We first tackle the Whittle's indexes at state $(0, \hat{r})$ and $(s, 0)$ where $s > 0$. To this end, we distinguish between the following cases

- We first consider the state $x = (0, \hat{r})$. By definition, Whittle's index is the infimum λ such that $V^0(x) = V^1(x)$. According to (A10), we can conclude that $W_x = 0$ when $x = (0, \hat{r})$.
- Then, we consider the state $x = (s, 0)$ where $s > 0$. We recall that $p_e^0 = 0$. Then, we can conclude, from (A11), that $W_x = 0$ for all $x = (s, 0)$ where $s > 0$.

Now, we tackle the Whittle's index at state $x = (s, 1)$ where $s > 0$. For convenience, we denote by W_n the Whittle's index at state $x = (n, 1)$. According to the monotonicity of $\Delta V(x)$ shown in the proof of Proposition 1, we can conclude that threshold policy $n = (+\infty, n+1)$ is optimal when $V^0(n, 1) = V^1(n, 1)$. Then, we can prove the following

Lemma A1. *When (9) is satisfied and* $V^0(n, 1) = V^1(n, 1)$, $V(s, 1) = V(s, 0) \triangleq V(s)$ *for* $0 \leq s \leq n$.

Proof. Since the value function satisfies the Bellman equation, it is sufficient to show that $V(s, 1)$ and $V(s, 0)$ satisfy the same Bellman equation. We recall that the Bellman equation for $V(x)$ is given by

$$V(x) = \min_{a \in \{0,1\}} V^a(x),$$

where

$$V^a(x) = C(x, a) - \theta + \sum_{x'} P_{x,x'}(a) V(x'), \tag{A12}$$

and θ is the optimal value of the decoupled problem. We recall, from Corollary 3, that the optimal action at state $(s, 0)$ is staying idle (i.e., $a = 0$) for any s. We also know that threshold policy $n = (+\infty, n+1)$ is optimal when $V^0(n, 1) = V^1(n, 1)$. Therefore, the optimal actions at states $(s, 0)$ and $(s, 1)$ where $s \leq n$ are the same (i.e., $a = 0$). Equivalently, we have

$$V(s, \hat{r}) = V^0(s, \hat{r}), \quad for\ s \leq n. \tag{A13}$$

According to the system dynamic reported in Section 2.3, we know that the state transition probabilities are independent of $\hat{r}$ when $a = 0$. Meanwhile, $\hat{r}$ does not affect the instant cost. Let $x_1 = (s, 1)$ and $x_2 = (s, 0)$. Then, for any x', we have

$$P_{x_1, x'}(0) = P_{x_2, x'}(0).$$

$$C(x_1, 0) = C(x_2, 0).$$

Hence, according to (A12), we can see that $V^0(s, 0) = V^0(s, 1)$ for any $s \leq n$. Combined with problem (A13), we can conclude that $V(s, 0) = V(s, 1)$ for any $0 \leq s \leq n$. $\square$

By definition, Whittle's index W_n is the infimum λ such that $V^0(n, 1) = V^1(n, 1)$. In this case, according to Lemma A1, $V(0, 1) = V(0, 0) = V(0)$. Then, $V^0(n, 1)$ and $V^1(n, 1)$ can be written as

$$V^0(n, 1) = f(n) - \theta + pV(0) + (1-p)[(1-\gamma)V(n+1, 0) + \gamma V(n+1, 1)]. \tag{A14}$$

$$V^1(n,1) = f(n) + W_n - \theta + (1-\alpha)V(0) + \alpha[(1-\gamma)V(n+1,0) + \gamma V(n+1,1)].$$

Without a loss of generality, we assume $V(0) = 0$. Then, equating the two expressions yields

$$W_n = (1 - p - \alpha)(\gamma V(n+1,1) + (1-\gamma)V(n+1,0)). \tag{A15}$$

Combining problems (A14) and (A15), we conclude that W_n is

$$W_n = \frac{(1 - p - \alpha)(V^0(n,1) + \theta - f(n))}{1 - p}.$$

Since the optimal action at state $(n,1)$ is $a = 0$, we have $V^0(n,1) = V(n,1) = V(n)$. Finally, we obtain

$$W_n = \frac{(1 - p - \alpha)(V(n) + \theta - f(n))}{1 - p}. \tag{A16}$$

Now, we tackle the expression of $V(n)$. When $V^0(n,1) = V^1(n,1)$, the optimal action at state $(s,\hat{r})$ where $0 \le s < n$ is staying idle. Then, leveraging Lemma A1, value function $V(s)$ where $0 \le s < n$ satisfies the following

$$V(s) = \begin{cases} -\theta + f(0) + pV(1) & \text{when } s = 0, \\ -\theta + f(s) + (1-p)V(s+1) & \text{when } 0 < s < n. \end{cases} \tag{A17}$$

By backward induction, we end up with the following equation for $0 < s < n$.

$$V(s) = \frac{-\theta(1 - (1-p)^{n-s})}{p} + \sum_{k=1}^{n-s} f(n-k)(1-p)^{n-s-k} + (1-p)^{n-s}V(n).$$

Letting $s = 1$ yields

$$V(1) = \frac{-\theta(1 - (1-p)^{n-1})}{p} + \sum_{k=1}^{n-1} f(n-k)(1-p)^{n-1-k} + (1-p)^{n-1}V(n).$$

From problem (A17), $V(1)$ also satisfies the following

$$V(1) = \frac{\theta - f(0)}{p}.$$

Equating the two expressions of $V(1)$, we obtain

$$V(n) = \frac{-f(0)}{p(1-p)^{n-1}} + \theta\left(\frac{2}{p(1-p)^{n-1}} - \frac{1}{p}\right) - \sum_{k=1}^{n-1} f(n-k)(1-p)^{-k}. \tag{A18}$$

We recall that, when $V^0(n,1) = V^1(n,1)$, threshold policy $n = (+\infty, n+1)$ is optimal and both actions at state $x = (n,1)$ are equally desirable. Thus, threshold policy $n = (+\infty, n)$ is also optimal. Then, we know

$$\theta = \bar{\Delta}_n + W_n \bar{\rho}_n, \tag{A19}$$

where $\bar{\Delta}_n$ and $\bar{\rho}_n$ are the expected AoII and the expected transmission rate under threshold policy $n = (+\infty, n)$, respectively. Finally, combining problems (A16), (A18) and (A19), we obtain

$$W_n = \frac{\dfrac{-f(0)}{p(1-p)^n} + \bar{\Delta}_n \dfrac{2 - (1-p)^n}{p(1-p)^n} - (1-p)^{-n}\left(\sum_{k=1}^{n} f(k)(1-p)^{k-1}\right)}{\dfrac{1}{1-p-\alpha} - \bar{\rho}_n \dfrac{2 - (1-p)^n}{p(1-p)^n}}.$$

After some algebraic manipulation, we have

$$W_n = \frac{(1-c_1)\sum_{k=n+1}^{+\infty} f(k)c_1^{k-n-1} - \bar{\Delta}_n}{\dfrac{(1-c_1)(1-p) - \gamma(1-p-\alpha)}{c_1(1-p-\alpha)}} + \bar{\rho}_n,$$

where $c_1 = (1-\gamma)(1-p) + \gamma\alpha$.

In the following, we investigate some properties of Whittle's index. First of all, W_n is non-negative since $1 - p - \alpha$ and $V(n+1, \hat{r})$ in (A15) are all non-negative. Meanwhile, combining (A15) with the fact that $V(n, \hat{r})$ is non-decreasing in n, we can verify that W_n is non-decreasing in n. Combined with the Whittle's indexes in two other cases (i.e., $x = (0, \hat{r})$ and $x = (s, 0)$ where $s > 0$), we can easily obtain the properties of W_x as detailed in Proposition 4.

Appendix H. Proof of Proposition 5

We notice that $\mathcal{M}_1(\lambda, -1)$ coincides with the decoupled model studied in Section 4.2. When problem (9) is satisfied, the decoupled problem is indexable, and, according to Corollary 3, we only need to show that n is the optimal threshold for the states with $\hat{r} = 1$. We first tackle the case of $\lambda > 0$. To this end, we divide our discussion into the following cases

- For state $(s, 1)$ where $s < n$, $W_s \leq \lambda$ by definition. As the problem is indexable, we have $D(W_s) \subseteq D(\lambda)$. We recall that $W_s \triangleq \min\{\lambda' \geq 0 : V^0(s, 1) = V^1(s, 1)\}$. Equivalently, $W_s \triangleq \min\{\lambda' \geq 0 : (s, 1) \in D(\lambda')\}$. Then, we know $(s, 1) \in D(W_s)$. Combined together, we conclude that $(s, 1) \in D(\lambda)$. In other words, the optimal action at state $(s, 1)$ where $s < n$ is to stay idle (i.e., $a = 0$).
- For state $(s, 1)$ where $s \geq n$, we first recall that $W_s = \min\{\lambda' \geq 0 : (s, 1) \in D(\lambda')\}$. Consequently, for any $\lambda' < W_s$, we know $(s, 1) \notin D(\lambda')$. Meanwhile, we have $W_s \geq W_n > \lambda$ by the monotonicity of Whittle's index and the definition of n. Hence, we can conclude that $(s, 1) \notin D(\lambda)$. In other words, the optimal action at state $(s, 1)$ where $s \geq n$ is to make the transmission attempt.

Then, we conclude that n is the optimal threshold for the states with $\hat{r} = 1$ when $\lambda > 0$. In the case of $\lambda = 0$, according to the proof of Proposition 1, we can easily verify that the optimal threshold is 1.

Appendix I. Proof of Theorem 2

We first make the following definitions. When $\mathcal{M}_1(\lambda, -1)$ is at state x and action a is taken, cost $C_1(x, a) \triangleq f(s)$ and $C_2(x, a) \triangleq \lambda a$ are incurred. We denote the expected C_1-cost and the expected C_2-cost under policy ϕ as $\bar{C}_1(\phi)$ and $\bar{C}_2(\phi)$, respectively. Let G be a non-empty set of states. For the given state i, we define $\mathcal{R}^*(i, G)$ as the class of policies ϕ, for which the following hold

- The probability $P_\phi(x_n \in G \text{ for some } n \geq 1 \mid x_0 = i) = 1$ where x_n is the state of $\mathcal{M}_1(\lambda, -1)$ at time n.
- The expected time $m_{iG}(\phi)$ of a first passage from i to G under ϕ is finite.
- The expected C_1-cost $\bar{C}_1^{i,G}(\phi)$ and the expected C_2-cost $\bar{C}_2^{i,G}(\phi)$ of a first passage form i to G under ϕ are finite.

With the definitions in mind, we proceed with verifying the assumptions given in [27].

1. *For all $d > 0$, the set $A(d) = \{x \mid \text{there exists an action } a \text{ such that } C_1(x, a) + C_2(x, a) \leq d\}$ is finite*: For any state x, the cost satisfies $C_1(x, a) + C_2(x, a) = f(s) + \lambda a \geq f(s)$. The equality holds when $a = 0$. Then, the states in $A(d)$ must satisfy $f(s) \leq d$. Combined with the fact that $f(s)$ is a non-decreasing and unbounded function when $s \in \mathbb{N}_0$, we can conclude that $A(d)$ is finite.
2. *There exists a stationary policy e such that the induced Markov chain has the following properties: the state space $\mathcal{S}$ consists of a single (non-empty) positive recurrent class R and a*

set U of transient states such that $e \in \mathcal{R}^*(i,R)$ for $i \in U$. Moreover, both $\bar{C}_1(e)$ and $\bar{C}_2(e)$ on R are finite: We consider the policy under which the base station makes a transmission attempt at every time slot. According to the system dynamic detailed in Section 2.3, we can see that all the states communicate with state $(0,0)$ and $(0,0)$ communicates with all other states. Thus, the state space $\mathcal{S}$ consists of a single (non-empty) positive recurrent class and the set of transient states can simply be an empty set. $\bar{C}_1(e)$ and $\bar{C}_2(e)$ are trivially finite as we can verify using Proposition 2.

3. *Given any two state $x \neq y$, there exists a policy ϕ such that $\phi \in \mathcal{R}^*(x,y)$*: We notice that, under any policy, the maximum increase of s between two consecutive time slots is 1. Meanwhile, when s decreases, it decreases to zero. Combined with the fact that $\hat{r}$ is an independent Bernoulli random variable, we can conclude that there always exists a path between any x and y with positive probability. $m_{xy}(\phi)$, $\bar{C}_1^{x,y}(\phi)$, and $\bar{C}_2^{x,y}(\phi)$ are trivially finite.

4. *If a stationary policy ϕ has at least one positive recurrent state, then it has a single positive recurrent class R. Moreover, if $x = (0,0) \notin R$, then $\phi \in \mathcal{R}^*(x,R)$*: Given that $\hat{r}$ is an independent Bernoulli random variable, we can easily conclude from the system dynamic that all the states communicate with state $(0,0)$ and $(0,0)$ communicates with all other states under any stationary policy. Therefore, any positive recurrent class must contain state $(0,0)$. Thus, there must have only one positive recurrent class which is $R = \mathcal{S}$.

5. *There exists a policy ϕ such that $\bar{C}_1(\phi) < \infty$ and $\bar{C}_2(\phi) < K$ where $K \in (0,1]$*: We notice that $\bar{C}_1(\phi)$ and $\bar{C}_2(\phi)$ are nothing but the expected AoII and the expected transmission rate achieved by ϕ, respectively. Then, we can easily verify that such policy exists using Proposition 2.

As the assumptions are verified, we proceed with introducing the optimal randomized policy for given λ. We say a policy is λ-optimal if the policy is optimal for $\mathcal{M}_1(\lambda, -1)$. We consider two monotone sequences $\lambda_+^n \downarrow \lambda$ and $\lambda_-^n \uparrow \lambda$. Then, there exist subsequences of λ_+^n and λ_-^n such that the corresponding sequences of optimal policies converge. Then, according to Lemma 3.7 of [27], the limit points, denoted by n_{λ_+} and n_{λ_-}, are both λ-optimal. By Proposition 3.2 of [27], the Markov chains induced by n_{λ_+} and n_{λ_-} both contain a single non-empty positive recurrent class and state $(0,0)$ is positive recurrent in both induced Markov chains. Hence, the base station can choose which policy to follow each time the system reaches state $(0,0)$ while keeping the resulting randomized policy λ-optimal as suggested by Lemma 3.9 of [27]. More precisely, we consider the following randomized policy: each time the system reaches state $(0,0)$, the base station will choose n_{λ_-} with probability μ and n_{λ_+} with probability $1 - \mu$. The chosen policy will be followed until the next choice. We denote such policy as n_λ and conclude that n_λ is λ-optimal under any $\mu \in [0,1]$.

Appendix J. Proof of Proposition 6

The value function $V(x)$ and $V^i(x_i)$ must satisfy their own Bellman equations. More precisely

$$V(x) + \theta = \min_{a \in \mathcal{A}_N(-1)} \left\{ C(x,a) + \sum_{x'} Pr(x' \mid x, a) V(x') \right\},$$

$$V^i(x_i) + \theta_i = \min_{a_i \in \{0,1\}} \left\{ C(x_i, a_i) + \sum_{x_i'} Pr(x_i' \mid x_i, a_i) V^i(x_i') \right\}, \qquad (A20)$$

where θ and θ_i are the optimal values of $\mathcal{M}_N(\lambda, -1)$ and $\mathcal{M}_1^i(\lambda, -1)$, respectively. We recall from Section 2.3 that the users are independent when action a and current state x are given. Thus

$$Pr(x' \mid x, a) = \prod_{i=1}^{N} Pr(x_i' \mid x, a),$$

where $x' = (x_1', \ldots, x_N')$. Then, we have

$$\sum_{\mathbf{x}'-\{x_i'\}} Pr(\mathbf{x}' - \{x_i'\} \mid \mathbf{x}, \mathbf{a}) = \sum_{\mathbf{x}'-\{x_i'\}} \prod_{j \neq i} Pr(x_j' \mid \mathbf{x}, \mathbf{a}) = 1.$$

We also recall from Section 2.3 that the state of user i depends only on its previous state and the action with respect to user i. Thus

$$Pr(x_i' \mid \mathbf{x}, \mathbf{a}) = Pr(x_i' \mid x_i, a_i).$$

Combined together, we obtain

$$\begin{aligned}
\sum_{i=1}^{N} \sum_{x_i'} Pr(x_i' \mid x_i, a_i) V^i(x_i') &= \sum_{i=1}^{N} \sum_{x_i'} \left[\sum_{\mathbf{x}'-\{x_i'\}} \prod_{j \neq i} Pr(x_j' \mid \mathbf{x}, \mathbf{a}) \right] Pr(x_i' \mid x_i, a_i) V^i(x_i') \\
&= \sum_{i=1}^{N} \sum_{x_i'} \left(\sum_{\mathbf{x}'-\{x_i'\}} \prod_{i=1}^{N} Pr(x_i' \mid \mathbf{x}, \mathbf{a}) V^i(x_i') \right) \qquad \text{(A21)} \\
&= \sum_{\mathbf{x}'} Pr(\mathbf{x}' \mid \mathbf{x}, \mathbf{a}) \left(\sum_{i=1}^{N} V^i(x_i') \right).
\end{aligned}$$

Then, we sum problem (A20) over all users which yields

$$\sum_{i=1}^{N} (V^i(x_i) + \theta_i) = \min_{\mathbf{a}} \left\{ \sum_{i=1}^{N} \left(C(x_i, a_i) + \sum_{x_i'} Pr(x_i' \mid x_i, a_i) V^i(x_i') \right) \right\}.$$

We recall that $C(\mathbf{x}, \mathbf{a}) = \sum_{i=1}^{N} C(x_i, a_i)$ by definition. Then, leveraging problem (A21), we obtain

$$\sum_{i=1}^{N} V^i(x_i) + \sum_{i=1}^{N} \theta_i = \min_{\mathbf{a} \in \mathcal{A}_N(-1)} \left\{ C(\mathbf{x}, \mathbf{a}) + \sum_{\mathbf{x}'} Pr(\mathbf{x}' \mid \mathbf{x}, \mathbf{a}) \left(\sum_{i=1}^{N} V^i(x_i') \right) \right\}.$$

Since the solution to the Bellman equation is unique [21], we must have $\sum_{i=1}^{N} V^i(x_i) = V(\mathbf{x})$ and $\sum_{i=1}^{N} \theta_i = \theta$. Then, we can conclude that it is optimal for $\mathcal{M}_N(\lambda, -1)$ if each user adopts its own optimal policy.

Appendix K. Proof of Theorem 3

In this proof, we class a policy as λ^*-optimal if it is optimal for $\mathcal{M}_N(\lambda^*, -1)$. In Section 4.2, we ensure that, for each user, there exists at least one threshold policy that yields a finite expected AoII. Therefore, we can conclude that, for RP, there exists at least one policy that causes the expected AoII and the expected transmission rate to be both finite. Then, according to Lemma 3.10 of [27], a policy is optimal for RP if

1. It is λ^*-optimal;
2. The resulting expected transmission rate is equal to M.

We first construct a policy ϕ_{λ^*} that is λ^*-optimal. We recall from Proposition 6 that a policy is λ^*-optimal if it consists of the optimal policies for each $\mathcal{M}_1^i(\lambda^*, -1)$ where $1 \leq i \leq N$. According to Theorem 2, for any i, there exist $n_{\lambda_-^*, i}$ and $n_{\lambda_+^*, i}$ that are both optimal for $\mathcal{M}_1^i(\lambda^*, -1)$. Then, we can construct the policy ϕ_{λ^*} in the following way.

- For user i with $n_{\lambda_-^*, i} = n_{\lambda_+^*, i} \triangleq n_{\lambda^*, i}$, the threshold policy $n_{\lambda^*, i}$ is used. Then, the deterministic policy $n_{\lambda^*, i}$ is optimal for $\mathcal{M}_1^i(\lambda^*, -1)$ and

$$\bar{\rho}^i(\lambda^*) = \bar{\rho}^i(\lambda_-^*) = \bar{\rho}^i(\lambda_+^*).$$

  In this case, the choice of μ_i makes no difference.
- For user i with $n_{\lambda_-^*, i} \neq n_{\lambda_+^*, i}$, the randomized policy $n_{\lambda^*, i}$ as detailed in Theorem 2 is used. Then, for any $\mu_i \in [0, 1]$, the randomized policy $n_{\lambda^*, i}$ is optimal for $\mathcal{M}_1^i(\lambda^*, -1)$ and

$$\bar{\rho}^i(\lambda^*) = \mu_i \bar{\rho}^i(\lambda_-^*) + (1 - \mu_i)\bar{\rho}^i(\lambda_+^*).$$

Combing the two cases, we conclude that $\phi_{\lambda^*} = [n_{\lambda^*,1}, \ldots, n_{\lambda^*,N}]$ is λ^*-optimal under any $\mu_i \in [0,1]$. Hence, as long as the chosen μ_i's realize $\sum_{i=1}^{N} \bar{\rho}^i(\lambda^*) = M$, we can conclude that the randomized policy ϕ_{λ^*} is optimal for RP.

Appendix L. Proof of Proposition 8

We notice that $\mathcal{M}_1^i(\lambda^*, -1)$ coincides with the decoupled model studied in Section 4.2. Therefore, we can use the results in Section 4.2 to prove the properties. Since the users share the same structure, we ignore the user index i for simplicity. According to the definition of I_x, we have

$$I_x = \sum_{x'} P_{x,x'}(0)V(x') - \sum_{x'} P_{x,x'}(1)V(x') - \lambda^*$$
$$= -\Delta V(x).$$

Leveraging the results in the proof of Proposition 1, we have the following

- For state $x = (0, \hat{r})$, $I_x = -\lambda^*$.
- For state $x = (s, 0)$ where $s > 0$, $I_x = -\lambda^* - p_e^0(1 - 2p)\omega$ where $\omega = (1 - \gamma)[V(0,0) - V(s+1,0)] + \gamma[V(0,1) - V(s+1,1)] \leq 0$.
- For state $x = (s, 1)$ where $s > 0$, $I_x = -\lambda^* - (1 - p_e^1)(1 - 2p)\omega$.

From the above three cases, we can easily conclude that $I_x \geq -\lambda^*$ and the equality holds when $\hat{r} = p_e^0 = 0$ or $s = 0$. As is proven in Corollary 2, $V(x)$ is non-decreasing in s. Hence, we can conclude that I_x is also non-decreasing in s. To show that I_x is monotone in $\hat{r}$, we consider two states $x_1 = (s, 1)$ and $x_2 = (s, 0)$. Then, we have

$$I_{x_2} - I_{x_1} = \Delta V(s,1) - \Delta V(s,0) = (1 - p_e^1 - p_e^0)(1 - 2p)\omega \leq 0.$$

Therefore, we can conclude that I_x is non-decreasing in $\hat{r}$.

Appendix M

Algorithm A1 Improved Relative Value Iteration

Require:
    MDP $\mathcal{M} = (\mathcal{X}, \mathcal{P}, \mathcal{A}, \mathcal{C})$
    Convergence Criteria ϵ
1: **procedure** RELATIVEVALUEITERATION($\mathcal{M}, \epsilon$)
2:     Initialize $V_0(x) = 0$; $v = 0$
3:     Choose $x^{ref} \in \mathcal{X}$ arbitrarily
4:     **while** V_v is not converged (RVI converges when the maximum difference between the results of two consecutive iterations is less than ϵ) **do**
5:         **for** $x = (s, \hat{r}) \in \mathcal{X}$ **do**
6:         **if** $\exists$ active state $(s_1, \hat{r}_1)$ s.t. $s_1 \leq s$ and $\hat{r}_1 \leq \hat{r}$ **then**
7:            $a^*(x) = 1$
8:            $Q_{v+1}(x) = C(x,1) + \sum_{x'} P_{xx'}(1)V_v(x')$
9:         **else**
10:           **for** $a \in \mathcal{A}$ **do**
11:              $H_{x,a} = C(x,a) + \sum_{x'} P_{xx'}(a)V_v(x')$
12:            $a^*(x) = \arg\min_a\{H_{x,a}\}$
13:            $Q_{v+1}(x) = H_{x,a^*}$
14:         **for** $x \in \mathcal{X}$ **do**
15:            $V_{v+1}(x) = Q_{v+1}(x) - Q_{v+1}(x^{ref})$
16:         $v = v + 1$
        return $n \leftarrow a^*(x)$

Algorithm A2 Bisection Search

Require:
    Maximum updates per transmission attempt M
    MDP $\mathcal{M}_N(\lambda, -1) = (\mathcal{X}_N, \mathcal{A}_N(-1), \mathcal{P}_N, \mathcal{C}_N(\lambda))$
    Tolerance ζ
    Convergence criteria ϵ

1: **procedure** BISECTIONSEARCH($\mathcal{M}_N(\lambda, -1), M, \zeta, \epsilon$)
2:     Initialize $\lambda_- = 0; \lambda_+ = 1$
3:     $\phi_{\lambda_+} \leftarrow (\mathcal{M}_N(\lambda_+, -1), \epsilon)$ using Section 5.1 and Proposition 6
4:     $\bar{\rho}(\lambda_+) \leftarrow \phi_{\lambda_+}$ using Proposition 2
5:     **while** $\bar{\rho}(\lambda_+) \geq M$ **do**
6:         $\lambda_- = \lambda_+; \lambda_+ = 2\lambda_+$
7:         $\phi_{\lambda_+} \leftarrow (\mathcal{M}_N(\lambda_+, -1), \epsilon)$ using Section 5.1 and Proposition 6
8:         $\bar{\rho}(\lambda_+) \leftarrow \phi_{\lambda_+}$ using Proposition 2
9:     **while** $\lambda_+ - \lambda_- \geq 2\zeta$ **do**
10:        $\lambda = \frac{\lambda_+ + \lambda_-}{2}$
11:        $\phi_\lambda \leftarrow (\mathcal{M}_N(\lambda, -1), \epsilon)$ using Section 5.1 and Proposition 6
12:        $\bar{\rho}(\lambda) \leftarrow \phi_\lambda$ using Proposition 2
13:        **if** $\bar{\rho}(\lambda) > M$ **then**
14:           $\lambda_- = \lambda$
15:        **else**
16:           $\lambda_+ = \lambda$
       return $(\lambda_+^*, \lambda_-^*) \leftarrow (\lambda_+, \lambda_-)$

References

1. Maatouk, A.; Kriouile, S.; Assaad, M.; Ephremides, A. The age of incorrect information: A new performance metric for status updates. *IEEE/ACM Trans. Netw.* **2020**, *28*, 2215–2228. [CrossRef]
2. Uysal, E.; Kaya, O.; Ephremides, A.; Gross, J.; Codreanu, M.; Popovski, P.; Assaad, M.; Liva, G.; Munari, A.; Soleymani, T.; et al. Semantic communications in networked systems. *arXiv* **2021**, arXiv:2103.05391.
3. Kam, C.; Kompella, S.; Ephremides, A. Age of incorrect information for remote estimation of a binary markov source. In Proceedings of the IEEE INFOCOM 2020-IEEE Conference on Computer Communications Workshops (INFOCOM WKSHPS), Toronto, ON, Canada, 6–9 July 2020; pp. 1–6.
4. Maatouk, A.; Assaad, M.; Ephremides, A. The age of incorrect information: An enabler of semantics-empowered communication. *arXiv* **2020**, arXiv:2012.13214.
5. Chen, Y.; Ephremides, A. Minimizing Age of Incorrect Information for Unreliable Channel with Power Constraint. *arXiv* **2021**, arXiv:2101.08908.
6. Kriouile, S.; Assaad, M. Minimizing the Age of Incorrect Information for Real-time Tracking of Markov Remote Sources. *arXiv* **2021**, arXiv:2102.03245.
7. Kadota, I.; Sinha, A.; Uysal-Biyikoglu, E.; Singh, R.; Modiano, E. Scheduling policies for minimizing age of information in broadcast wireless networks. *IEEE/ACM Trans. Netw.* **2018**, *26*, 2637–2650. [CrossRef]
8. Hsu, Y.P. Age of information: Whittle index for scheduling stochastic arrivals. In Proceedings of the 2018 IEEE International Symposium on Information Theory (ISIT), Vail, CO, USA, 17–22 June 2018; pp. 2634–2638.
9. Tripathi, V.; Modiano, E. A whittle index approach to minimizing functions of age of information. In Proceedings of the 2019 57th Annual Allerton Conference on Communication, Control, and Computing (Allerton), Monticello, IL, USA, 24–27 September 2019; pp. 1160–1167.
10. Maatouk, A.; Kriouile, S.; Assad, M.; Ephremides, A. On the optimality of the Whittle's index policy for minimizing the age of information. *IEEE Trans. Wirel. Commun.* **2020**, *20*, 1263–1277. [CrossRef]
11. Sun, J.; Jiang, Z.; Krishnamachari, B.; Zhou, S.; Niu, Z. Closed-form Whittle's index-enabled random access for timely status update. *IEEE Trans. Commun.* **2019**, *68*, 1538–1551. [CrossRef]
12. Nguyen, G.D.; Kompella, S.; Kam, C.; Wieselthier, J.E. Information freshness over a Markov channel: The effect of channel state information. *Ad Hoc Networks* **2019**, *86*, 63–71. [CrossRef]
13. Talak, R.; Karaman, S.; Modiano, E. Optimizing age of information in wireless networks with perfect channel state information. In Proceedings of the 2018 16th International Symposium on Modeling and Optimization in Mobile, Ad Hoc, and Wireless Networks (WiOpt), Shanghai, China, 7–11 May 2018; pp. 1–8.
14. Shi, L.; Cheng, P.; Chen, J. Optimal periodic sensor scheduling with limited resources. *IEEE Trans. Autom. Control* **2011**, *56*, 2190–2195. [CrossRef]
15. Leong, A.S.; Dey, S.; Quevedo, D.E. Sensor scheduling in variance based event triggered estimation with packet drops. *IEEE Trans. Autom. Control* **2016**, *62*, 1880–1895. [CrossRef]

16. Mo, Y.; Garone, E.; Casavola, A.; Sinopoli, B. Stochastic sensor scheduling for energy constrained estimation in multi-hop wireless sensor networks. *IEEE Trans. Autom. Control* **2011**, *56*, 2489–2495. [CrossRef]
17. Kaul, S.; Yates, R.; Gruteser, M. Real-time status: How often should one update? In Proceedings of the 2012 Proceedings IEEE INFOCOM, Orlando, FL, USA, 25–30 March 2012; pp. 2731–2735.
18. Leong, A.S.; Ramaswamy, A.; Quevedo, D.E.; Karl, H.; Shi, L. Deep reinforcement learning for wireless sensor scheduling in cyber–physical systems. *Automatica* **2020**, *113*, 108759. [CrossRef]
19. Wang, J.; Ren, X.; Mo, Y.; Shi, L. Whittle index policy for dynamic multichannel allocation in remote state estimation. *IEEE Trans. Autom. Control* **2019**, *65*, 591–603. [CrossRef]
20. Gittins, J.; Glazebrook, K.; Weber, R. *Multi-Armed Bandit Allocation Indices*; John Wiley & Sons: Hoboken, NJ, USA, 2011.
21. Russell, S.; Norvig, P. *Artificial Intelligence: A Modern Approach*, 3rd ed.; Prentice Hall Press: Hoboken, NJ, USA, 2009.
22. Whittle, P. Restless bandits: Activity allocation in a changing world. *J. Appl. Probab.* **1988**, *25*, 287–298. [CrossRef]
23. Weber, R.R.; Weiss, G. On an index policy for restless bandits. *J. Appl. Probab.* **1990**, *27*, 637–648. [CrossRef]
24. Glazebrook, K.D.; Ruiz-Hernandez, D.; Kirkbride, C. Some indexable families of restless bandit problems. *Adv. Appl. Probab.* **2006**, *38*, 643–672. [CrossRef]
25. Larrañaga, M. Dynamic Control of Stochastic and Fluid Resource-Sharing Systems. Ph.D. Thesis, Université de Toulouse, Toulouse, France, 2015.
26. Sennott, L.I. On computing average cost optimal policies with application to routing to parallel queues. *Math. Methods Oper. Res.* **1997**, *45*, 45–62. [CrossRef]
27. Sennott, L.I. Constrained average cost Markov decision chains. *Probab. Eng. Inf. Sci.* **1993**, *7*, 69–83. [CrossRef]
28. Bertsimas, D.; Niño-Mora, J. Restless bandits, linear programming relaxations, and a primal-dual index heuristic. *Oper. Res.* **2000**, *48*, 80–90. [CrossRef]
29. Littman, M.L.; Dean, T.L.; Kaelbling, L.P. On the complexity of solving Markov decision problems. *arXiv* **2013**, arXiv:1302.4971.
30. Verloop, I.M. Asymptotically optimal priority policies for indexable and nonindexable restless bandits. *Ann. Appl. Probab.* **2016**, *26*, 1947–1995. [CrossRef]

Article

Distribution of the Age of Gossip in Networks

Mohamed A. Abd-Elmagid * and Harpreet S. Dhillon

Wireless@VT, Bradley Department of Electrical and Computer Engineering, Virginia Tech,
Blacksburg, VA 24061, USA
* Correspondence: maelaziz@vt.edu

Abstract: We study a general setting of gossip networks in which a source node forwards its measurements (in the form of status updates) about some observed physical process to a set of monitoring nodes according to independent Poisson processes. Furthermore, each monitoring node sends status updates about its information status (about the process observed by the source) to the other monitoring nodes according to independent Poisson processes. We quantify the freshness of the information available at each monitoring node in terms of Age of Information (AoI). While this setting has been analyzed in a handful of prior works, the focus has been on characterizing the average (i.e., marginal first moment) of each age process. In contrast, we aim to develop methods that allow the characterization of higher-order marginal or joint moments of the age processes in this setting. In particular, we first use the stochastic hybrid system (SHS) framework to develop methods that allow the characterization of the stationary marginal and joint moment generating functions (MGFs) of age processes in the network. These methods are then applied to derive the stationary marginal and joint MGFs in three different topologies of gossip networks, with which we derive closed-form expressions for marginal or joint high-order statistics of age processes, such as the variance of each age process and the correlation coefficients between all possible pairwise combinations of age processes. Our analytical results demonstrate the importance of incorporating the higher-order moments of age processes in the implementation and optimization of age-aware gossip networks rather than just relying on their average values.

Keywords: Age of Information; information freshness; gossip networks; stochastic hybrid systems

Citation: Abd-Elmagid, M.A.;
Dhillon, H.S. Distribution of the Age
of Gossip in Networks. *Entropy* **2023**,
25, 364. https://doi.org/10.3390/
e25020364

Academic Editors: Anthony
Ephremides and Yin Sun

Received: 16 January 2023
Revised: 12 February 2023
Accepted: 14 February 2023
Published: 16 February 2023

1. Introduction

Timely delivery of status updates is crucial for enabling the operation of many emerging Internet of Things (IoT)-based real-time status updating systems [1]. The concept of AoI was introduced in [2] to quantify the freshness of information available at some node about a physical process as a result of status update receptions over time. In particular, for a single source of information queueing theoretic model in which status updates about a single physical process are generated randomly at a *transmitter node* and are then sent to a *destination node* through a single server, the AoI at the destination was defined in [2] as the following random process: $x(t) = t - u(t)$, where $u(t)$ is the generation time instant of the latest status update received at the destination by a time t. Assuming that the AoI process is ergodic, in [2], the stationary average value of the AoI under the first-come-first-serve (FCFS) queueing discipline was derived by leveraging the properties of the AoI's sample functions and applying appropriate geometric arguments. Although this geometric approach has been considered in a series of subsequent prior works [3–13] to analyze the marginal distributional properties of AoI or peak AoI (an AoI-related metric introduced in [3] to capture the peak values of AoI over time) for adaptations of the queueing model studied in [2], it often requires tedious calculations of joint moments that limit its tractability in analyzing more sophisticated queueing models or disciplines.

Motivated by the above limitations of the geometric approach to AoI analysis, the authors of [14,15] developed an SHS-based framework to allow the analysis of the marginal distributional properties of each AoI process (in a network with multiple AoI processes) through the characterization of its stationary marginal moments and MGF. Furthermore, by using the notion of tensors, the authors of [16] generalized the analysis in [14,15] and developed an SHS-based general framework that facilitates the analysis of the joint distributional properties of an arbitrary set of AoI processes in a network through the characterization of their stationary joint moments and MGFs. In the piecewise linear SHS model with linear reset maps considered in the analyses in [14–16], the discrete state of the system $q(t)$ is modeled as a finite-state, continuous-time Markov chain, and the continuous state of the system is modeled using the vector $\mathbf{x}(t)$, which contains the AoI or age processes at different nodes in the network. When a transition l occurs in $q(t)$ (as a result of status update generation or reception at one of the nodes in the network), the continuous state is updated according to the following linear mapping of $\mathbf{x}(t)$: $\mathbf{x}'(t) = \mathbf{x}(t)\mathbf{A}_l$, where $\mathbf{x}'(t)$ is the updated version of $\mathbf{x}(t)$ and $\mathbf{A}_l$ is the reset mapping matrix associated with a transition l. Additionally, in the absence of a transition in $q(t)$, the age processes in $\mathbf{x}(t)$ grow at a unit rate with time, which yields piecewise linear age processes over time. Based on this description of the piecewise linear SHS model with linear reset maps, one can realize that the frameworks in [14–16] are not applicable to age analysis in classes of status-updating systems where it is not possible for every transition l in $q(t)$ to express the updated value of each age process in the network as a linear combination of the age processes in $\mathbf{x}(t)$. A popular class of such systems is the gossip-based status-updating system, where each node in the network randomly shares its information status over time with the other nodes [17,18]. Here, when there is a transition caused by a status update reception at node j from node i, the updated value of the age process at node j is given by the minimum between the values of the age processes at nodes i and j. As a result, there have been a handful of recent efforts for developing new SHS-based methods that are suitable for age analysis in such gossip networks [19,20]. However, the methods developed thus far have been limited to the characterization of the stationary marginal first moment (average value) of each age process in the network. In this paper, we develop new SHS-based methods that allow the evaluation of the stationary marginal or joint high-order moments of the age processes in gossip networks through the characterization of their stationary marginal or joint MGFs.

1.1. Related Work

The literature relevant to this paper can be categorized into the following two categories: (1) prior analyses of AoI applying the SHS approach with linear reset maps and (2) prior analyses of AoI in gossip networks. We now discuss the relevant prior work in these two directions.

Analyses of AoI applying the SHS approach with linear reset maps. The SHS approach with linear reset maps developed in [14,15] has been applied to characterize the marginal distributional properties of AoI under a variety of system settings or queueing disciplines [21–33]. In particular, the average AoI was characterized for single-source systems in [21,22] and multi-source systems in [23–27], whereas the MGF of AoI was derived for single-source systems in [28,29], two-source systems in [30], and multi-source systems in [31–33]. Note that a multi-source system refers to the set-up where a transmitter has multiple sources of information generating status updates about multiple physical processes. The authors of [21] derived the average AoI under the last-come-first-serve (LCFS) with preemption in service queueing discipline when the transmitter contained multiple parallel servers. Furthermore, the authors of [22] derived the average AoI under the LCFS with preemption in service queueing discipline when the transmitter contained multiple servers in series or there existed a series of nodes between the transmitter and destination nodes. In [23], the average AoI was characterized under the priority LCFS with preemption in service or waiting queueing model. The authors of [24] derived the average AoI in the presence of packet delivery errors under stationary randomized and round-robin scheduling policies. In [25], the average AoI was characterized under the LCFS with preemption in service queueing discipline when the transmitter contained multiple parallel servers. The authors of [26] analyzed the average AoI for a network in which multiple transmitter-destination pairs contended for the channel using the carrier sense multiple access scheme. In [27] (in [30]), the average AoI (the MGF of AoI) was derived under several source-aware packet management scheduling policies at the transmitter. For the case where the transmitter was powered by energy harvesting (EH), the authors of [28,31] derived the MGF of AoI under several queueing disciplines, including the LCFS with and without preemption in service or waiting strategies. On the other hand, the authors of [16,34] applied their SHS-based framework (developed to allow the analysis of the joint distributional properties of AoI processes in networks) to characterize the joint MGF of an arbitrary set of AoI processes in a multi-source updating system under non-preemptive and source-agnostic or source-aware preemptive-in-service queueing disciplines.

Analyses of AoI in gossip networks. There are only a handful of recent works focusing on the analysis or optimization of AoI and its variants in gossip networks [19,20,35–41]. For a general setting of gossip networks, the author of [19,20] first developed SHS-based methods for the evaluation of the average AoI and the average version age at each node in the network. Note that the version age is a discrete form of AoI defined as the number of versions where the current status of information at a node is out of date compared with the current status of the original source of information. The authors of [35] applied the results of [20] to derive the average version age at each node in several topologies of clustered gossip networks and characterized the average version age scaling as a function of the network size. The authors of [36] extended the SHS-based method developed in [19] for the evaluation of the average AoI in the setting where a timestomping adversary is present and then obtained the average AoI scaling for several network topologies. In [37], each node was assumed to have the ability to estimate the information at the source by applying the majority rule to the information received from the other nodes, and an error metric was introduced to quantify the average percentage of nodes that could accurately obtain the most up-to-date information. The authors of [38–40] developed gossip protocols with the objective of improving the average version age scaling. In [41], the problem of optimizing the average version age was formulated as a Markov decision process for a setting where an energy harvesting (EH)-powered sensor was sending status updates to an aggregator with caching capabilities (which served the requests of a gossip network), and the structural properties of the optimal policy were analytically characterized. Different from the analyses in [19,20,35–41], which were focused on characterizing or optimizing the stationary marginal first moment of AoI or some other AoI-related metrics, this paper is the first to develop SHS-based methods that allow the characterization of the stationary marginal or joint MGFs of AoI processes in gossip networks.

Before delving into more detail about our contributions, it is worth noting that aside from the above queueing theory-based analyses of AoI, there have been efforts to evaluate and optimize AoI or some other AoI-related metrics in a variety of communication systems that deal with time-sensitive information (see [42] for a comprehensive book and [43] for a recent survey). For instance, AoI has been studied in the context of age-optimal transmission scheduling policies [44–52], multi-hop networks [53–55], broadcast networks [56,57], ultra-reliable low-latency vehicular networks [58], unmanned aerial vehicle (UAV)-assisted communication systems [59–61], Internet of Underwater Things networks [62], reconfigurable intelligent surface (RIS)-assisted communication systems [63,64], EH systems [65–74], large-scale analysis of IoT networks [75–77], remote estimation [78,79], information-theoretic analysis [80–83], timely source coding [84,85], cache updating systems [86–88], economic systems [89], and timely communication in federated learning [90,91].

1.2. Contributions

A general setting of gossip networks is analyzed in this paper, where a source node forwards its measurements (in the form of status updates) about some observed physical process to a set of monitoring nodes according to independent Poisson processes. Furthermore, each monitoring node sends status updates about its information status (about the process observed by the source) to the other monitoring nodes according to independent Poisson processes. We quantify the freshness of the information available at each monitoring node in terms of AoI. The continuous state of the system is then formed by the AoI or age processes at different monitoring nodes. For this set-up, our main contributions are listed below.

Developing SHS-based methods for the evaluation of the MGF of age of gossip in networks. For the general setting of gossip networks described above, we use the SHS framework to characterize (1) the stationary marginal MGF of each age process in the network and (2) the stationary joint MGF of any two arbitrarily selected age processes in the network. In particular, we first construct two classes of test functions (functions whose expected values are quantities of interest) that are suitable for analyzing the marginal or joint MGF. By applying Dynkin's formula to each test function, we derive two systems of first-order ordinary differential equations characterizing the temporal evolution of the marginal and joint MGFs, from which the stationary marginal and joint MGFs are evaluated. To the best of our knowledge, this paper makes the first attempt at developing SHS-based methods for the characterization of the marginal or joint MGF of age of gossip in networks.

Analysis of the stationary marginal or joint MGF of age of gossip in three different network topologies. We apply our developed SHS-based methods to study the marginal or joint distributional properties of age processes in the following three network topologies: (1) a serially-connected topology, (2) a parallelly-connected topology, and (3) a clustered topology. For each of these topologies, we derive close-form expressions for (1) the stationary marginal MGF of the age process at each node and (2) the stationary joint MGFs of all possible pairwise combinations of the age processes.

System design insights. Using the MGF expressions derived for each network topology considered in this paper, we obtain closed-form expressions for the following quantities: (1) the stationary marginal first and second moments of each age process, (2) the variance of each age process, and (3) the correlation coefficients between all possible pairwise combinations of the age processes. For these derived quantities, we characterize their structural properties in terms of their convexity and monotonic nature with respect to the status updating rates and further provide asymptotic results showing their behaviors when each of the status updating rates becomes small or large. A key insight drawn from our analysis is that it is crucial to incorporate the higher-order moments of age processes in the implementation or optimization of age-aware gossip networks rather than just relying on the average values of the age processes (as has been performed in the existing literature thus far). This insight promotes the importance of the SHS-based methods developed in this paper for the characterization of the marginal or joint MGFs of different age processes in a general setting of gossip networks.

1.3. Organization

The rest of this paper is organized as follows. Section 2 presents the system model and the problem statement. Afterward, in Section 3, we develop the SHS-based methods that allow the evaluation of the stationary marginal or joint high-order moments of the age processes in gossip networks through the characterization of their stationary marginal or joint MGFs. Section 4 applies the SHS-based methods developed in Section 3 to derive the marginal or joint MGFs of age processes at different nodes in three different connected network settings. For each considered connected network setting, we further use the derived MGF expressions to obtain the marginal or joint high-order statistics of age processes such as the variance of each age process and the correlation coefficients between all possible pairwise combinations of the age processes. Finally, Section 5 concludes the paper.

2. System Model and Problem Statement

We consider a general setting of gossip networks where a source node (referred to as node 0) provides its measurements about some observed physical process for a set of nodes $\mathcal{N} = \{1, 2, \cdots, N\}$ in the form of status updates. In particular, all the nodes in $\mathcal{N}$ are tracking the age of the process observed by the source, and the status updates sent by node 0 to node $j \in \mathcal{N}$ are assumed to follow an independent Poisson process with a rate λ_{0j}. Aside from that, node $i \in \mathcal{N}$ sends updates about its information status (about the process observed by the source) to each node $j \in \mathcal{N} \setminus \{i\}$ according to an independent Poisson process with a rate λ_{ij}. When $\lambda_{ij} > 0$, we say that nodes i and j are connected to each other. Since we allow each λ_{ij} ($i \in \{0\} \cup \mathcal{N}$ and $j \in \mathcal{N}$) to take a value in $[0, \infty]$, we refer to the above setting as an arbitrarily connected gossip network. Note that this gossip network setting is of interest in many practical networks, such as low-latency vehicular networks and UAV-assisted communication networks. The freshness of status of the information available at each node is quantified in terms of AoI. Let $x_i(t)$ denote the AoI process (or equivalently the age process) at node $i \in \mathcal{N}$. Assuming that node 0 always maintains a fresh status of information about the observed physical process, the age or AoI at node $j \in \mathcal{N}$ is reset to zero whenever it receives a status update from node 0. Furthermore, when node $j \in \mathcal{N}$ receives a status update from node $i \in \mathcal{N} \setminus \{j\}$ at time t, its age $x_j(t)$ is reset to the age of node i $x_i(t)$ only if $x_i(t)$ is smaller than $x_j(t)$. To summarize, when node $j \in \mathcal{N}$ receives a status update from node $i \in \{0\} \cup \mathcal{N}$, the age at node $k \in \mathcal{N}$ is updated as follows:

$$
x'_k(t) = \begin{cases} 0, & \text{if } i = 0 \text{ and } k = j, \\ \min\big[x_j(t), x_i(t)\big], & \text{if } i \in \mathcal{N} \text{ and } k = j, \\ x_k(t), & \text{otherwise.} \end{cases} \tag{1}
$$

For an arbitrary set $S \subseteq \mathcal{N}$, define $x_S(t) = \min\limits_{i \in S} x_i(t)$ as the age or AoI process associated with S (or simply the age or AoI of S). For the above gossip network setting, the method developed in [19] has been limited to the characterization of the stationary marginal first moment of $x_S(t)$ (i.e., the stationary average value of $x_S(t)$). In this paper, our prime objective is to develop a method that allows characterizing (1) the stationary marginal higher-order moments of $x_S(t)$ and (2) the stationary joint high-order moments of the two age processes associated with two arbitrary sets S_1 and S_2 (i.e., $x_{S_1}(t)$ and $x_{S_2}(t)$, respectively). Note that we do not place any restrictions on the construction of S_1 or S_2. For instance, they could even have common elements. Formally, we aim at characterizing the stationary marginal MGF of $x_S(t)$ and the stationary joint MGF of $x_{S_1}(t)$ and $x_{S_2}(t)$, which are of the following forms: $\lim\limits_{t \to \infty} \mathbb{E}[\exp[n x_S(t)]]$ and $\lim\limits_{t \to \infty} \mathbb{E}\big[\exp\big[n_1 x_{S_1}(t) + n_2 x_{S_2}(t)\big]\big]$, respectively, where $n, n_1, n_2 \in \mathbb{R}$ and $S, S_1, S_2 \subseteq \mathcal{N}$. As will be evident from the technical sections shortly, the characterization of such MGFs allows one to derive the marginal or joint high-order statistics of the AoI processes at different nodes in the network, such as the variance of each AoI process and the correlation coefficients between all possible pairwise combinations of the AoI processes. Given the generality of the system setting considered in this paper, the importance of our method lies in the fact that it is applicable to the marginal or joint analysis of AoI processes for an arbitrary structured gossip network setting.

3. MGF Analysis of Age in Arbitrarily Connected Gossip Networks

In this section, we first formulate the problem at hand as an SHS. We then use the SHS framework to characterize (1) the stationary marginal MGF of the age process associated with an arbitrary set $S \subseteq \mathcal{N}$ (i.e., $x_S(t)$) and (2) the stationary joint MGF of the two age processes associated with two arbitrary sets $S_1 \subseteq \mathcal{N}$ and $S_2 \subseteq \mathcal{N}$ (i.e., $x_{S_1}(t)$ and $x_{S_2}(t)$, respectively) for the arbitrarily connected gossip network setting described in Section 2.

The SHS framework is used to analyze hybrid queueing systems that can be modeled by a combination of discrete and continuous state parameters. For the gossip network setting considered in this paper, the continuous state of the system is modeled using the row vector $\mathbf{x}(t) = [x_1(t) \, x_2(t) \, \cdots \, x_N(t)]$ containing the AoI or age processes at different nodes in the network. Furthermore, since the status updates sent by each node in the network to the other nodes are assumed to follow independent Poisson processes, it is sufficient to model the discrete state of the system as a singleton set. To complete the description of an SHS, one needs to define a set of transitions $\mathcal{L}$ along with the continuous and discrete states of the system. This set $\mathcal{L}$ refers to changes in either the continuous state or the discrete state. Since the discrete state of the SHS under consideration is a singleton set, the set $\mathcal{L}$ corresponds to only the changes in the continuous state of the system. In our system setting, a change in the continuous state of the system occurs when there is a status update reception at some node in the network. Furthermore, as long as there is no status update reception at any of the nodes, the AoI or age at each node grows linearly with time (which yields piecewise linear age processes over time); in other words, $\dot{\mathbf{x}}(t) = \mathbf{1}_N$, where $\mathbf{1}_N$ is the row vector $[1 \ \cdots \ 1] \in \mathbb{R}^{1 \times N}$. By inspecting the age updating rule in (1), the set $\mathcal{L}$ can be defined as follows:

$$\mathcal{L} = \{(0,j) : j \in \mathcal{N}\} \cup \{(i,j) : i,j \in \mathcal{N}\}. \tag{2}$$

For the above SHS-based formulation, we derive two systems of linear equations for evaluating the stationary marginal MGF $\lim_{t\to\infty} \mathbb{E}[\exp[n x_S(t)]]$ and the stationary joint MGF $\lim_{t\to\infty} \mathbb{E}\left[\exp\left[n_1 x_{S_1}(t) + n_2 x_{S_2}(t)\right]\right]$. The description of these systems of equations and the presentation of the subsequent results require defining the following quantities:

$$v_S^{(n)}(t) = \mathbb{E}[\exp[n x_S(t)]], \ \bar{v}_S^{(n)} = \lim_{t\to\infty} v_S^{(n)}, \ \forall S \subseteq \mathcal{N}, \tag{3}$$

$$v_{S_1,S_2}^{(n_1,n_2)}(t) = \mathbb{E}\left[\exp\left[n_1 x_{S_1}(t) + n_2 x_{S_2}(t)\right]\right], \ \bar{v}_{S_1,S_2}^{(n_1,n_2)} = \lim_{t\to\infty} v_{S_1,S_2}^{(n_1,n_2)}, \ \forall S_1, S_2 \subseteq \mathcal{N}, \tag{4}$$

$$v_S^{(m)}(t) = \mathbb{E}[x_S^m(t)], \ \bar{v}_S^{(m)} = \lim_{t\to\infty} v_S^{(m)}, \ \forall S \subseteq \mathcal{N}, \tag{5}$$

$$v_{S_1,S_2}^{(m_1,m_2)}(t) = \mathbb{E}\left[x_{S_1}^{m_1}(t) x_{S_2}^{m_2}(t)\right], \ \bar{v}_{S_1,S_2}^{(m_1,m_2)} = \lim_{t\to\infty} v_{S_1,S_2}^{(m_1,m_2)}, \ \forall S_1, S_2 \subseteq \mathcal{N}, \tag{6}$$

where $v_S^{(m)}$ is the marginal mth moment of the age process $x_S(t)$ and $v_{S_1,S_2}^{(m_1,m_2)}$ is the joint moment of the two age processes $x_{S_1}(t)$ and $x_{S_2}(t)$. From (3) and (5), $v_S^{(1)}(t)$ may generally refer to $v_S^{(n)}(t)|_{n=1}$ or $v_S^{(m)}(t)|_{m=1}$. To eliminate this conflict, the convention that $v_S^{(i)}(t)$ for an integer i refers to $v_S^{(m)}(t)$ at $m = i$ is maintained here. The previous argument also applies to $v_{S_1,S_2}^{(n_1,n_2)}(t)$ and $v_{S_1,S_2}^{(m_1,m_2)}(t)$ in (4) and (6), respectively, where $v_{S_1,S_2}^{(i,j)}(t)$, for integers i and j, refers to $v_{S_1,S_2}^{(m_1,m_2)}(t)$ at $m_1 = i$ and $m_2 = j$. Furthermore, following the notations in [19], we define the update rate of node i into set S and the set of updating neighbors of S as

$$\lambda_i(S) = \begin{cases} \sum_{j \in S} \lambda_{i,j}, & \text{if } i \notin S, \\ 0, & \text{otherwise}, \end{cases} \tag{7}$$

$$N(S) = \{i \in \mathcal{N} : \lambda_i(S) > 0\}. \tag{8}$$

We are now ready to present the two systems of linear equations for the evaluation of $\bar{v}_S^{(n)}$ and $\bar{v}_{S_1,S_2}^{(n_1,n_2)}$ in the following two theorems:

Theorem 1. *For an arbitrarily connected gossip network, there exists a threshold $\delta > 0$ such that for $n \in [0, \delta)$, the stationary marginal MGF of AoI of set $S \subseteq \mathcal{N}$ is given by*

$$\bar{v}_S^{(n)} = \frac{\lambda_0(S) + \sum_{i \in \mathcal{N}(S)} \lambda_i(S) \bar{v}_{S \cup \{i\}}^{(n)}}{\lambda_0(S) + \sum_{i \in \mathcal{N}(S)} \lambda_i(S) - n}. \tag{9}$$

Furthermore, for $m \geq 1$, the stationary marginal m-th moment of AoI of set $S \subseteq \mathcal{N}$ is given by

$$\bar{v}_S^{(m)} = \frac{m \bar{v}_S^{(m-1)} + \sum_{i \in \mathcal{N}(S)} \lambda_i(S) \bar{v}_{S \cup \{i\}}^{(m)}}{\lambda_0(S) + \sum_{i \in \mathcal{N}(S)} \lambda_i(S)}. \tag{10}$$

Proof of Theorem 1. See Appendix A. $\square$

Theorem 2. *For an arbitrarily connected gossip network, there exists a threshold $\delta > 0$ such that for $0 \leq n_1 + n_2 < \delta$, the stationary joint MGF of the two AoI processes associated with the two sets S_1 and S_2 is given by*

$$
\begin{aligned}
\bar{v}_{S_1,S_2}^{(n_1,n_2)} = &\frac{1}{\lambda_0(S_1 \cup S_2) + \sum_{i \in \mathcal{N} \setminus (S_1 \cap S_2)} \lambda_i(S_1 \cap S_2) + \sum_{i \in \mathcal{N} \setminus S_1} \lambda_i(S_1 \setminus S_2) + \sum_{i \in \mathcal{N} \setminus S_2} \lambda_i(S_2 \setminus S_1) - (n_1 + n_2)} \times \Big[\lambda_0(S_1 \cap S_2) \\
&+ \lambda_0(S_1 \setminus S_2) \bar{v}_{S_2}^{(n_2)} + \lambda_0(S_2 \setminus S_1) \bar{v}_{S_1}^{(n_1)} + \sum_{i \in \mathcal{N} \setminus S_1} \lambda_i(S_1 \setminus S_2) \bar{v}_{S_1 \cup \{i\}, S_2}^{(n_1,n_2)} + \sum_{i \in \mathcal{N} \setminus S_2} \lambda_i(S_2 \setminus S_1) \bar{v}_{S_1, S_2 \cup \{i\}}^{(n_1,n_2)} \\
&+ \sum_{i \in \mathcal{N} \setminus (S_1 \cup S_2)} \lambda_i(S_1 \cap S_2) \bar{v}_{S_1 \cup \{i\}, S_2 \cup \{i\}}^{(n_1,n_2)} + \sum_{i \in S_1 \setminus S_2} \lambda_i(S_1 \cap S_2) \bar{v}_{S_1, S_2 \cup \{i\}}^{(n_1,n_2)} + \sum_{i \in S_2 \setminus S_1} \lambda_i(S_1 \cap S_2) \bar{v}_{S_1 \cup \{i\}, S_2}^{(n_1,n_2)} \Big].
\end{aligned}
\tag{11}
$$

Furthermore, for $m_1, m_2 \geq 1$, the stationary joint (m_1, m_2)-th moment of the AoI processes associated with the two sets S_1 and S_2 is given by

$$
\begin{aligned}
\bar{v}_{S_1,S_2}^{(m_1,m_2)} = &\frac{1}{\lambda_0(S_1 \cup S_2) + \sum_{i \in \mathcal{N} \setminus (S_1 \cap S_2)} \lambda_i(S_1 \cap S_2) + \sum_{i \in \mathcal{N} \setminus S_1} \lambda_i(S_1 \setminus S_2) + \sum_{i \in \mathcal{N} \setminus S_2} \lambda_i(S_2 \setminus S_1)} \times \Big[m_1 \bar{v}_{S_1,S_2}^{(m_1-1,m_2)} \\
&+ m_2 \bar{v}_{S_1,S_2}^{(m_1,m_2-1)} + \sum_{i \in \mathcal{N} \setminus S_1} \lambda_i(S_1 \setminus S_2) \bar{v}_{S_1 \cup \{i\}, S_2}^{(m_1,m_2)} + \sum_{i \in \mathcal{N} \setminus S_2} \lambda_i(S_2 \setminus S_1) \bar{v}_{S_1, S_2 \cup \{i\}}^{(m_1,m_2)} + \sum_{i \in \mathcal{N} \setminus (S_1 \cup S_2)} \lambda_i(S_1 \cap S_2) \bar{v}_{S_1 \cup \{i\}, S_2 \cup \{i\}}^{(m_1,m_2)} \\
&+ \sum_{i \in S_1 \setminus S_2} \lambda_i(S_1 \cap S_2) \bar{v}_{S_1, S_2 \cup \{i\}}^{(m_1,m_2)} + \sum_{i \in S_2 \setminus S_1} \lambda_i(S_1 \cap S_2) \bar{v}_{S_1 \cup \{i\}, S_2}^{(m_1,m_2)} \Big].
\end{aligned}
\tag{12}
$$

Proof of Theorem 2. See Appendix B. $\square$

Remark 1. *Note that the stationary marginal MGF of S_1 or S_2 can be obtained from the stationary joint MGF in (11). In particular, when $n_2 = 0$ and $S_2 = \varnothing$, $\bar{v}_{S_1,S_2}^{(n_1,n_2)}$ reduces to*

$$\bar{v}_{S_1,\varnothing}^{(n_1,0)} = \frac{\lambda_0(S_1) + \sum_{i \in \mathcal{N}(S_1)} \lambda_i(S_1) \bar{v}_{S_1 \cup \{i\}}^{(n)}}{\lambda_0(S_1) + \sum_{i \in \mathcal{N}(S_1)} \lambda_i(S_1) - n_1} \overset{(a)}{=} \bar{v}_{S_1}^{(n_1)}, \tag{13}$$

where step (a) follows from (9). Similarly, one can observe that $\bar{v}_{\varnothing,S_2}^{(0,n_2)} = \bar{v}_{S_2}^{(n_2)}$.

Furthermore, when $m = 1$, (10) reduces to ([19] Theorem 1) characterizing the stationary marginal first moment of the AoI of set $S \subseteq \mathcal{N}$.

It is worth highlighting that the generality of Theorems 1 and 2 lies in the fact that they allow one to investigate the stationary marginal or joint MGFs of the age processes at different nodes in an arbitrarily connected gossip network. This opens the door for the application of Theorems 1 and 2 to characterize the marginal or joint high-order moments of age processes for different configurations or topologies of gossip networks studied in the literature, which have only been analyzed in terms of the marginal first moments of age processes (i.e., average age values) until now. Furthermore, the expressions in (10) and (12) provide a straightforward way for the numerical evaluation of the stationary marginal or joint high-order moments.

4. Applications of Theorems 1 and 2

In this section, we first apply Theorems 1 and 2 to understand the distributional properties of the age processes in the two canonical settings depicted in Figure 1 (i.e., the the serially and parallelly-connected network settings). We then aim to analyze a more complicated network setting, which was chosen to be the clustered gossip network topology depicted in Figure 2. Our choice for the clustered gossip network setting was inspired by the recent interest in analyzing its different topologies in terms of the marginal first moment of each age process (average age) in the network [35].

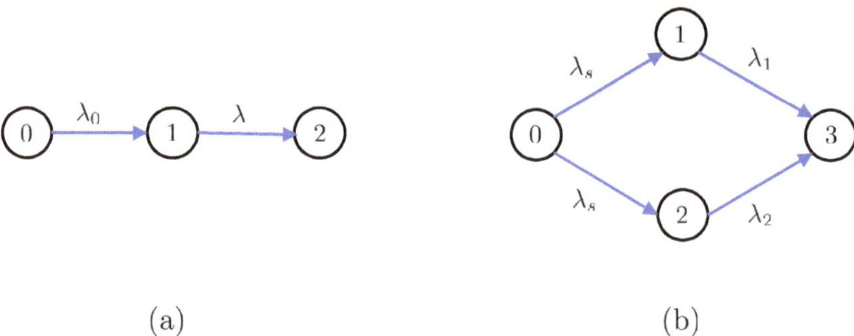

Figure 1. (**a**) A serially-connected network setting. (**b**) A parallelly-connected network setting.

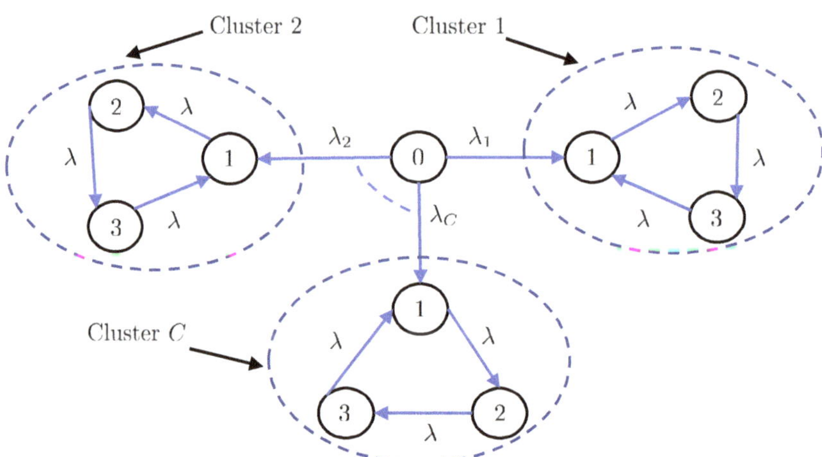

Figure 2. A clustered gossip network topology consisting of C clusters such that the status updating rate from node 0 to the c-th cluster is λ_c.

4.1. Serially-Connected Networks

Theorem 3. *For the serially-connected network in Figure 1a, the stationary marginal MGFs of the AoI processes at nodes 1 and 2 are respectively given by*

$$\bar{v}^{(n)}_{\{1\}} = \frac{\lambda_0}{\lambda_0 - n},$$

(14)

$$\bar{v}^{(n)}_{\{2\}} = \frac{\lambda_0 \lambda}{(\lambda_0 - n)(\lambda - n)}. \tag{15}$$

Additionally, the stationary joint MGF of the two AoI processes at nodes 1 and 2 is given by

$$\bar{v}^{(n_1,n_2)}_{\{2\},\{1\}} = \frac{\lambda_0 \lambda}{\lambda_0 + \lambda - (n_1 + n_2)} \left(\frac{\lambda_0}{(\lambda_0 - n_1)(\lambda - n_1)} + \frac{1}{\lambda_0 - (n_1 + n_2)} \right). \tag{16}$$

Proof of Theorem 3. See Appendix C. $\square$

Proposition 1. *For the serially-connected network in Figure 1a, the first moment, second moment, and variance of the AoI process at each node are given by*

$$\bar{v}^{(1)}_{\{1\}} = \lambda_0^{-1}, \ \ \bar{v}^{(2)}_{\{1\}} = 2\lambda_0^{-2}, \ \ \mathrm{var}[x_1(t)] = \lambda_0^{-2}, \tag{17}$$

$$\bar{v}^{(1)}_{\{2\}} = \frac{1}{\lambda_0} + \frac{1}{\lambda}, \ \ \bar{v}^{(2)}_{\{2\}} = 2\left(\frac{1}{\lambda_0^2} + \frac{1}{\lambda_0 \lambda} + \frac{1}{\lambda^2} \right), \ \ \mathrm{var}[x_2(t)] = \frac{1}{\lambda_0^2} + \frac{1}{\lambda^2}. \tag{18}$$

Furthermore, the correlation coefficient between the AoI processes at nodes 1 and 2 can be expressed as

$$\mathrm{cor}[x_1(t), x_2(t)] = \frac{\lambda^2}{(\lambda_0 + \lambda)\sqrt{\lambda_0^2 + \lambda^2}}. \tag{19}$$

Proof of Proposition 1. See Appendix D. $\square$

Remark 2. *Note that the expressions of the stationary marginal MGFs in Theorem 3 and the stationary marginal moments in Proposition 1 match their corresponding ones for the preemptive line networks analyzed in [15].*

Remark 3. *Note that the stationary moments and variance of the age process at node 1 in (17) are univariate functions of λ_0. This happens because node 1 is directly connected to node 0. This argument will also apply to: (i) the expressions derived for the age processes at nodes 1 and 2 in the parallelly-connected network in Figure 1b, and (ii) the expressions derived for the age process at node 1 inside each cluster of the clustered gossip network in Figure 2.*

Remark 4. *Note that the stationary moments and variance of the age process at node 2 in (18) are invariant to exchanging λ and λ_0. These quantities are also jointly convex functions in (λ_0, λ), where the minimum value (zero) of each function is achieved at $\lambda_0 = \lambda = \infty$. Furthermore, for a given λ or λ_0, each quantity in (18) is a monotonically non-increasing function with respect to λ_0 or λ. This can also be observed in Figure 3.*

Remark 5. *For a given λ, $\mathrm{cor}[x_1(t), x_2(t)]$ in (19) monotonically decreases as a function of λ_0 in the form $\lim\limits_{\lambda_0 \to 0} \mathrm{cor}[x_1(t), x_2(t)] = 1$ until it approaches $\lim\limits_{\lambda_0 \to \infty} \mathrm{cor}[x_1(t), x_2(t)] = 0$. On the other hand, for a given λ_0, $\mathrm{cor}[x_1(t), x_2(t)]$ monotonically increases as a function of λ in the form $\lim\limits_{\lambda \to 0} \mathrm{cor}[x_1(t), x_2(t)] = 0$ until it approaches $\lim\limits_{\lambda \to \infty} \mathrm{cor}[x_1(t), x_2(t)] = 1$. This can also be observed in Figure 4.*

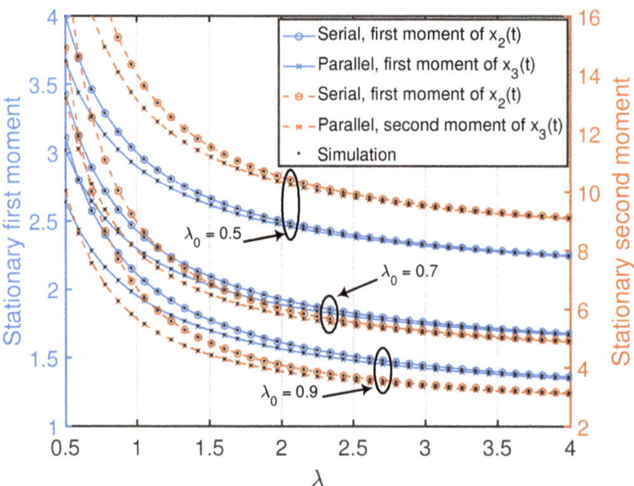

Figure 3. Stationary first and second moments of age processes in the serially and parallelly-connected network settings. We set $\lambda_s = 0.5\lambda_0$ and $\lambda = \lambda_1 = \lambda_2$. The simulated curves are obtained from the numerical evaluation of the stationary marginal moments using (10) in Theorem 1.

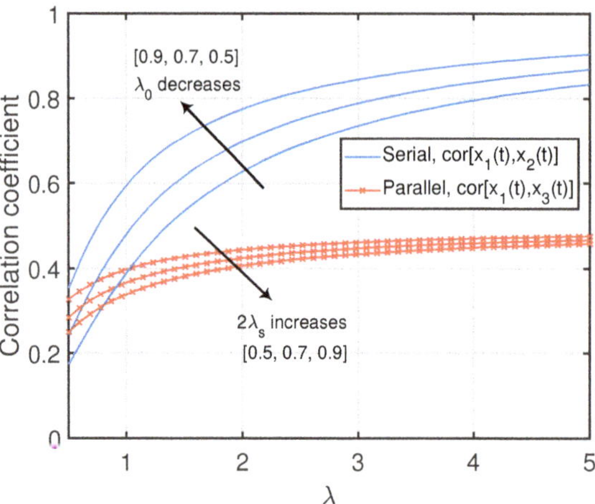

Figure 4. Correlation coefficients between age processes in the serially and parallelly-connected network settings.

4.2. Parallelly-Connected Networks

Theorem 4. *For the parallelly-connected network in Figure 1b, the stationary marginal MGFs of the AoI processes at nodes 1, 2, and 3 are given by*

$$\bar{v}_{\{1\}}^{(n)} = \bar{v}_{\{2\}}^{(n)} = \frac{\lambda_s}{\lambda_s - n}, \tag{20}$$

$$\bar{v}_{\{3\}}^{(n)} = \frac{\lambda_s(2\lambda_s - n)\left[\lambda_1(\lambda_s + \lambda_1 - n) + \lambda_2(\lambda_s + \lambda_2 - n)\right] + 2\lambda_s\lambda_1\lambda_2(2\lambda_s + \lambda_1 + \lambda_2 - 2n)}{(2\lambda_s - n)(\lambda_1 + \lambda_2 - n)(\lambda_s + \lambda_1 - n)(\lambda_s + \lambda_2 - n)}. \tag{21}$$

Additionally, the stationary joint MGF of the two AoI processes at nodes 1 and 3 is given by

$$\bar{v}^{(n_1,n_2)}_{\{3\},\{1\}} = \frac{\sum_{i=1}^{4} \alpha_i(n_1,n_2)}{[\lambda_s + \lambda_1 + \lambda_2 - (n_1 + n_2)][2\lambda_s + \lambda_1 - (n_1 + n_2)][2\lambda_s - (n_1 + n_2)][\lambda_s + \lambda_2 - (n_1 + n_2)]}$$
$$\times \frac{1}{(\lambda_s - n_2)(\lambda_1 + \lambda_2 - n_1)(2\lambda_s - n_1)(\lambda_s + \lambda_2 - n_1)(\lambda_s + \lambda_1 - n_1)}, \tag{22}$$

where

$$\alpha_1(n_1,n_2) = \lambda_s^2(\lambda_s - n_2)[\lambda_s + \lambda_2 - (n_1 + n_2)][2\lambda_s + \lambda_1 - (n_1 + n_2)][2\lambda_s - (n_1 + n_2)]$$
$$\times \Big[(2\lambda_2 - n_1)\big[\lambda_1(\lambda_s + \lambda_1 - n_1) + \lambda_2(\lambda_s + \lambda_2 - n_1)\big] + 2\lambda_1\lambda_2(2\lambda_s + \lambda_1 + \lambda_2 - 2n_1)\Big], \tag{23}$$

$$\alpha_2(n_1,n_2) = \lambda_s^2\lambda_2(\lambda_1 + \lambda_2 - n_1)(2\lambda_s - n_1)(\lambda_s + \lambda_1 - n_1)(\lambda_s + \lambda_2 - n_1)[2\lambda_s + \lambda_1 - (n_1 + n_2)][\lambda_s + \lambda_1 + \lambda_2 - (n_1 + n_2)], \tag{24}$$

$$\alpha_3(n_1,n_2) = \lambda_s^2\lambda_2(\lambda_1 + \lambda_2 - n_1)(\lambda_s + \lambda_2 - n_1)(2\lambda_s + 2\lambda_1 - n_1)(\lambda_s - n_2)[\lambda_s + \lambda_2 - (n_1 + n_2)][2\lambda_s - (n_1 + n_2)], \tag{25}$$

$$\alpha_4(n_1,n_2) = \lambda_s\lambda_1(\lambda_s - n_2)(\lambda_1 + \lambda_2 - n_1)(2\lambda_s - n_1)(\lambda_s + \lambda_2 - n_1)(\lambda_s + \lambda_1 - n_1)$$
$$\times \Big[[2\lambda_s + \lambda_1 - (n_1 + n_2)][2\lambda_s + \lambda_2 - (n_1 + n_2)] + \lambda_2[\lambda_s + \lambda_2 - (n_1 + n_2)]\Big]. \tag{26}$$

Proof of Theorem 4. See Appendix E. $\square$

Proposition 2. *For the parallelly-connected network in Figure 1b, the first moment, second moment, and variance of the AoI process at each node are given by*

$$\bar{v}^{(1)}_{\{1\}} = \bar{v}^{(1)}_{\{2\}} = \lambda_s^{-1}, \quad \bar{v}^{(2)}_{\{1\}} = \bar{v}^{(2)}_{\{2\}} = 2\lambda_s^{-2}, \quad \mathrm{var}[x_1(t)] = \mathrm{var}[x_2(t)] = \lambda_s^{-2}, \tag{27}$$

$$\bar{v}^{(1)}_{\{3\}} = \frac{2\lambda_s(\lambda_s + \lambda_1)(\lambda_s + \lambda_2) + \lambda_1(2\lambda_s + \lambda_2)(\lambda_s + \lambda_1) + \lambda_2(2\lambda_s + \lambda_1)(\lambda_s + \lambda_2)}{2\lambda_s(\lambda_s + \lambda_1)(\lambda_s + \lambda_2)(\lambda_1 + \lambda_2)}, \tag{28}$$

$$\bar{v}^{(2)}_{\{3\}} = \frac{\sum_{i=0}^{6} \gamma_i \lambda_s^i}{2\lambda_s^2(\lambda_1 + \lambda_2)^2(\lambda_s + \lambda_1)^2(\lambda_s + \lambda_2)^2}, \tag{29}$$

$$\mathrm{var}[x_3(t)] = \frac{\sum_{i=0}^{6} \eta_i \lambda_s^i}{4\lambda_s^2(\lambda_1 + \lambda_2)^2(\lambda_s + \lambda_1)^2(\lambda_s + \lambda_2)^2}, \tag{30}$$

where

$$\gamma_6 = 4, \ \gamma_5 = 12(\lambda_1 + \lambda_2), \ \gamma_4 = 4\Big[4(\lambda_1 + \lambda_2)^2 + \lambda_1\lambda_2\Big], \ \gamma_3 = 12(\lambda_1 + \lambda_2)^3,$$
$$\gamma_2 = (\lambda_1 + \lambda_2)^2\Big[4(\lambda_1 + \lambda_2)^2 + \lambda_1\lambda_2\Big], \ \gamma_1 = 3\lambda_1\lambda_2(\lambda_1 + \lambda_2)^3, \ \gamma_0 = \lambda_1^2\lambda_2^2(\lambda_1 + \lambda_2)^2,$$
$$\eta_6 = 4, \ \eta_5 = 8(\lambda_1 + \lambda_2), \ \eta_4 = 8\Big[(\lambda_1 + \lambda_2)^2 + \lambda_1\lambda_2\Big], \ \eta_3 = 4(\lambda_1 + \lambda_2)\Big(2\lambda_1^2 + 3\lambda_1\lambda_2 + 2\lambda_2^2\Big),$$
$$\eta_2 = 2(\lambda_1 + \lambda_2)^2\Big(2\lambda_1^2 + \lambda_1\lambda_2 + 2\lambda_2^2\Big), \ \eta_1 = 2\lambda_1\lambda_2(\lambda_1 + \lambda_2)^3, \ \eta_0 = \lambda_1^2\lambda_2^2(\lambda_1 + \lambda_2)^2.$$

Furthermore, the correlation coefficient between the AoI processes at nodes 1 and 3 can be expressed as

$$\text{cor}[x_1(t), x_3(t)] = \frac{\lambda_1(\lambda_1 + \lambda_2)}{2(\lambda_s + \lambda_1 + \lambda_2)(2\lambda_s + \lambda_1)(\lambda_s + \lambda_2)\sqrt{\sum_{i=0}^{6} \delta_i \lambda_s^i}}$$
$$\times \left[8\lambda_s^4 + \lambda_s^3(12\lambda_1 + 7\lambda_2) + 2\lambda_s^2(\lambda_1 + 2\lambda_2)(2\lambda_1 + \lambda_2) + \lambda_s\lambda_2\left(3\lambda_1^2 + 5\lambda_1\lambda_2 + \lambda_2^2\right) + \lambda_1\lambda_2^2(\lambda_1 + \lambda_2)\right], \tag{31}$$

where

$$\delta_6 = 4, \ \delta_5 = 8(\lambda_1 + \lambda_2), \ \delta_4 = 8\left[(\lambda_1 + \lambda_2)^2 + \lambda_1\lambda_2\right], \ \delta_3 = 4(\lambda_1 + \lambda_2)\left(2\lambda_1^2 + 3\lambda_1\lambda_2 + 2\lambda_2^2\right),$$
$$\delta_2 = 2(\lambda_1 + \lambda_2)^2\left(2\lambda_1^2 + \lambda_1\lambda_2 + 2\lambda_2^2\right), \ \delta_1 = 2\lambda_1\lambda_2(\lambda_1 + \lambda_2)^3, \ \delta_0 = \lambda_1^2\lambda_2^2(\lambda_1 + \lambda_2)^2.$$

Proof of Proposition 2. See Appendix F. □

Remark 6. *When λ_1 or λ_2 is zero, the parallelly-connected network reduces to a serially-connected network with a single path from node 0 to node 3. Thus, in this case, the stationary moments and variance of the age process at node 3 reduce to the corresponding expressions associated with the age process at node 2 in the serially-connected network such that λ_0 and λ are replaced by λ_s and λ_1 or λ_2. On the other hand, when λ_1 and λ_2 approach ∞, we have $\lim_{\lambda_1 \to \infty, \lambda_2 \to \infty} \bar{v}_{\{3\}}^{(1)} = \frac{1}{2\lambda_s}$, $\lim_{\lambda_1 \to \infty, \lambda_2 \to \infty} \bar{v}_{\{3\}}^{(2)} = \frac{1}{2\lambda_s^2}$, and $\lim_{\lambda_1 \to \infty, \lambda_2 \to \infty} \text{var}[x_3(t)] = \frac{1}{4\lambda_s^2}$. Note that the stationary moments and variance of $x_3(t)$ reduce to the ones associated with $x_{\{1,2\}}(t)$.*

Remark 7. *Note that the stationary moments and variance of the age process at node 3 in (28)–(30) are invariant to exchanging λ_1 and λ_2. Furthermore, for a given (λ_s, λ_2), (λ_s, λ_1), or (λ_1, λ_2), each quantity in (28)–(30) is a monotonically non-increasing function with respect to λ_1, λ_2, or λ_s. This can also be observed in Figure 3.*

Remark 8. *For the same status updating rate from node 0 (i.e., $\lambda_0 = 2\lambda_s$) and $\lambda = \lambda_1 = \lambda_2$, one can compare the achievable age performance at node 3 in the parallelly-connected network with the achievable age performance at node 2 in the serially-connected network using Propositions 1 and 2 as follows:*

$$\bar{v}_{\{2\}}^{(1)} - \bar{v}_{\{3\}}^{(1)} = \frac{\lambda_0}{2\lambda(\lambda_0 + 2\lambda)}, \tag{32}$$

$$\bar{v}_{\{2\}}^{(2)} - \bar{v}_{\{3\}}^{(2)} = \frac{3\lambda_0^2 + 4\left(\lambda^2 + 2\lambda_0\lambda\right)}{2\lambda^2(\lambda_0 + 2\lambda)^2}, \tag{33}$$

$$\text{var}[x_2(t)] - \text{var}[x_3(t)] = \frac{3\lambda_0(\lambda_0 + 4\lambda)}{4\lambda^2(\lambda_0 + 2\lambda)^2}. \tag{34}$$

By inspecting (32)–(34), one can see that these are positive quantities for any choice of values of (λ_0, λ). This certainly indicates that node 3 in the parallelly-connected network achieved a better age performance than the one achievable by node 2 in the serially-connected network. The improvement in the age performance at node 3 resulted from the existence of two status-updating paths from node 0 to node 3, as opposed to only a single path from node 0 to node 2 in the serially-connected network. Furthermore, each quantity in (32)–(34) is a monotonically decreasing function of λ for a given λ_0 such that its value approaches zero as $\lambda \to \infty$. This can also be observed in Figure 3.

Remark 9. *Due to the symmetry in the configuration of the parallelly-connected network, note that the correlation coefficient between $x_2(t)$ and $x_3(t)$ (i.e., $\mathrm{cor}[x_2(t), x_3(t)]$) can be obtained by replacing λ_1 and λ_2 with λ_2 and λ_1, respectively, in (31). Furthermore, for a given (λ_1, λ_2), $\mathrm{cor}[x_1(t), x_3(t)]$ monotonically decreases as a function of λ_s from $\lim_{\lambda_s \to 0} \mathrm{cor}[x_1(t), x_3(t)] = \frac{1}{2}$ until it approaches $\lim_{\lambda_s \to \infty} \mathrm{cor}[x_1(t), x_3(t)] = 0$. On the other hand, for a given (λ_s, λ_2), $\mathrm{cor}[x_1(t), x_3(t)]$ monotonically increases as a function of λ_1 from $\lim_{\lambda_1 \to 0} \mathrm{cor}[x_1(t), x_3(t)] = 0$ until it approaches $\lim_{\lambda_1 \to \infty} \mathrm{cor}[x_1(t), x_3(t)] = \frac{4\lambda_s^2 + 3\lambda_s\lambda_2 + \lambda_2^2}{2(\lambda_s + \lambda_2)\sqrt{4\lambda_s^2 + 2\lambda_s\lambda_2 + \lambda_2^2}}$. Finally, for a given (λ_s, λ_1), one can deduce the following asymptotic results: $\lim_{\lambda_2 \to 0} \mathrm{cor}[x_1(t), x_3(t)] = \frac{\lambda_1^2}{(\lambda_s + \lambda_1)\sqrt{\lambda_s^2 + \lambda_1^2}}$ and $\lim_{\lambda_2 \to \infty} \mathrm{cor}[x_1(t), x_3(t)] = \frac{\lambda_1(\lambda_s + \lambda_1)}{2(2\lambda_s + \lambda_1)\sqrt{4\lambda_s^2 + 2\lambda_s\lambda_1 + \lambda_1^2}}$. Clearly, when $\lambda_2 = 0$, there will only be a single status-updating path from node 0 to node 3 (through node 1), and hence we observe that $\mathrm{cor}[x_1(t), x_3(t)]$ reduced to the same expression as $\mathrm{cor}[x_1(t), x_2(t)]$ in (19) for the serially-connected network after replacing λ_0 and λ with λ_s and λ_1, respectively. Some of the above insights can also be seen in Figure 4.*

4.3. Clustered Gossip Networks

Theorem 5. *For the clustered gossip network in Figure 2, the stationary marginal MGFs of the AoI processes at nodes 1, 2, and 3 in the c-th cluster are respectively given by*

$$\bar{v}_{\{1\}}^{(n)} = \frac{\lambda_c}{\lambda_c - n}, \tag{35}$$

$$\bar{v}_{\{2\}}^{(n)} = \frac{\lambda_c \lambda}{(\lambda_c - n)(\lambda - n)}, \tag{36}$$

$$\bar{v}_{\{3\}}^{(n)} = \frac{\lambda_c \lambda^2}{(\lambda_c - n)(\lambda - n)^2}. \tag{37}$$

Additionally, the stationary joint MGF of each pair of AoI processes at nodes 1, 2, and 3 is given by

$$\bar{v}_{\{1\},\{2\}}^{(n_1,n_2)} = \frac{\lambda_c \lambda[(\lambda_c + \lambda - n_2)(\lambda_c - n_2) - \lambda_c n_1]}{(\lambda_c - n_2)(\lambda - n_2)[\lambda_c + \lambda - (n_1 + n_2)][\lambda_c - (n_1 + n_2)]}, \tag{38}$$

$$\bar{v}_{\{1\},\{3\}}^{(n_1,n_2)} = \frac{\lambda_c \lambda^2[\lambda_c + 2\lambda - (n_1 + n_2)]^3 \left[\lambda_c[\lambda_c - (n_1 + n_2)][\lambda_c + 2\lambda - n_1 - 2n_2] + (\lambda_c - n_2)(\lambda - n_2)^2\right]}{(\lambda_c - n_2)(\lambda - n_2)^2[\lambda_c - (n_1 + n_2)][\lambda_c + \lambda - (n_1 + n_2)]^2[\lambda_c + 2\lambda - (n_1 + n_2)]^3}, \tag{39}$$

$$\bar{v}_{\{2\},\{3\}}^{(n_1,n_2)} = \frac{\lambda_c \lambda^2 \sum_{i=1}^{4} \beta_i(n_1, n_2)}{(\lambda_c - n_2)(\lambda - n_2)^2[\lambda_c - (n_1 + n_2)][\lambda - (n_1 + n_2)][2\lambda - (n_1 + n_2)][\lambda_c + \lambda - (n_1 + n_2)]^2[\lambda_c + 2\lambda - (n_1 + n_2)]^2}, \tag{40}$$

where

$$\beta_1(n_1, n_2) = (\lambda_c - n_2)(\lambda - n_2)^2[\lambda_c + \lambda - (n_1 + n_2)]^2[\lambda_c + 2\lambda - (n_1 + n_2)]^2, \tag{41}$$

$$\beta_2(n_1, n_2) = \lambda^2(\lambda_c - n_2)(\lambda - n_2)^2[\lambda - (n_1 + n_2)][3\lambda_c + 4\lambda - 3(n_1 + n_2)], \tag{42}$$

$$\beta_3(n_1, n_2) = \lambda(\lambda_c - n_2)(\lambda - n_2)^2[\lambda - (n_1 + n_2)][\lambda_c - (n_1 + n_2)][\lambda_c + \lambda - (n_1 + n_2)], \tag{43}$$

$$\beta_4(n_1, n_2) = \lambda\lambda_c[\lambda - (n_1 + n_2)][\lambda_c - (n_1 + n_2)]\Big[[\lambda_c + 2\lambda - (n_1 + n_2)]^2[\lambda_c + \lambda - (n_1 + n_2)] + (\lambda - n_2)[\lambda_c + \lambda - (n_1 + n_2)]^2$$

$$+ \lambda(\lambda - n_2)[2\lambda_c + 3\lambda - 2(n_1 + n_2)]\Big]. \tag{44}$$

Proof of Theorem 5. See Appendix G. $\square$

Proposition 3. *For the clustered gossip network in Figure 2, the first moment, second moment, and variance of the AoI process at each node in the c-th cluster are given by*

$$\bar{v}_{\{1\}}^{(1)} = \lambda_c^{-1}, \quad \bar{v}_{\{1\}}^{(2)} = 2\lambda_c^{-2}, \quad \mathrm{var}[x_1(t)] = \lambda_c^{-2}, \tag{45}$$

$$\bar{v}_{\{2\}}^{(1)} = \lambda_c^{-1} + \lambda^{-1}, \quad \bar{v}_{\{2\}}^{(2)} = 2\left(\lambda_c^{-2} + \lambda_c^{-1}\lambda^{-1} + \lambda^{-2}\right), \quad \mathrm{var}[x_2(t)] = \lambda_c^{-2} + \lambda^{-2}, \tag{46}$$

$$\bar{v}_{\{3\}}^{(1)} = \lambda_c^{-1} + 2\lambda^{-1}, \quad \bar{v}_{\{3\}}^{(2)} = 2\left(\lambda_c^{-2} + 2\lambda_c^{-1}\lambda^{-1} + 3\lambda^{-2}\right), \quad \mathrm{var}[x_3(t)] = \lambda_c^{-2} + 2\lambda^{-2}. \tag{47}$$

Furthermore, the correlation coefficient between each pair of nodes can be expressed as

$$\mathrm{cor}[x_1(t), x_2(t)] = \frac{\lambda^2}{(\lambda_c + \lambda)\sqrt{\lambda_c^2 + \lambda^2}}, \tag{48}$$

$$\mathrm{cor}[x_1(t), x_3(t)] = \frac{\lambda^3}{(\lambda_c + \lambda)^2\sqrt{2\lambda_c^2 + \lambda^2}}, \tag{49}$$

$$\mathrm{cor}[x_2(t), x_3(t)] = \frac{\lambda_c^4 + 2\lambda_c^3\lambda + 2\lambda_c^2\lambda^2 + 2\lambda_c\lambda^3 + 2\lambda^4}{2(\lambda_c + \lambda)^2\sqrt{(\lambda_c^2 + \lambda^2)(2\lambda_c^2 + \lambda^2)}}. \tag{50}$$

Proof of Proposition 3. See Appendix H. $\square$

Proposition 4. *Let $\mathcal{N}_c$ denote the set of nodes inside cluster c. For $i, j \in \{1, 2, \cdots, C\}$, the two age processes $x_{\mathcal{N}_i}(t)$ and $x_{\mathcal{N}_j}(t)$ are not correlated.*

Proof of Proposition 4. See Appendix I. $\square$

Remark 10. *From Proposition 3, one can deduce that $\bar{v}_{\{1\}}^{(1)} \leq \bar{v}_{\{2\}}^{(1)} \leq \bar{v}_{\{3\}}^{(1)}$, $\bar{v}_{\{1\}}^{(2)} \leq \bar{v}_{\{2\}}^{(2)} \leq \bar{v}_{\{3\}}^{(2)}$, and $\mathrm{var}[x_1(t)] \leq \mathrm{var}[x_2(t)] \leq \mathrm{var}[x_3(t)]$ for any choice of values of λ_c and λ. This follows from the fact that the configuration of each cluster in the clustered gossip network under consideration is a uni-directional ring, where each node has a single status-updating path from node 0 passing through its preceding node in the cluster.*

Remark 11. *Similar to Remark 4, note that the quantities in (46) and (47) associated with the age processes at nodes 2 and 3 are jointly convex functions in (λ_c, λ), where the minimum value (zero) of each function is achieved at $\lambda_c = \lambda = \infty$. Furthermore, for a given λ or λ_c, each quantity in (46) and (47) is a monotonically non-increasing function with respect to λ_c or λ. This can also be observed in Figure 5.*

Remark 12. *Note that the correlation coefficients in (48)–(50) are monotonically non-increasing functions of λ_c for a given λ, whereas they are monotonically non-decreasing functions of λ for a given λ_c. In particular, $\mathrm{cor}[x_1(t), x_2(t)]$ and $\mathrm{cor}[x_1(t), x_3(t)]$ monotonically increase as functions of λ from $\lim_{\lambda \to 0}\mathrm{cor}[x_1(t), x_2(t)] = \lim_{\lambda \to 0}\mathrm{cor}[x_1(t), x_3(t)] = 0$ until they approach*

$\lim_{\lambda \to \infty} \mathrm{cor}[x_1(t), x_2(t)] = \lim_{\lambda \to \infty} \mathrm{cor}[x_1(t), x_3(t)] = 1$ *and monotonically decrease as functions of* λ_c *from* $\lim_{\lambda_c \to 0} \mathrm{cor}[x_1(t), x_2(t)] = \lim_{\lambda_c \to 0} \mathrm{cor}[x_1(t), x_3(t)] = 1$ *until they approach* $\lim_{\lambda_c \to \infty} \mathrm{cor}[x_1(t),$ $x_2(t)] = \lim_{\lambda_c \to \infty} \mathrm{cor}[x_1(t), x_3(t)] = 0.$ *Additionally,* $\mathrm{cor}[x_2(t), x_3(t)]$ *monotonically increases as a function of* λ *from* $\lim_{\lambda \to 0} \mathrm{cor}[x_2(t), x_3(t)] = \frac{1}{2\sqrt{2}}$ *until it approaches* $\lim_{\lambda \to \infty} \mathrm{cor}[x_2(t), x_3(t)] = 1$ *and monotonically decreases as a function of* λ_c *from* $\lim_{\lambda_c \to 0} \mathrm{cor}[x_2(t), x_3(t)] = 1$ *until it approaches* $\lim_{\lambda_c \to \infty} \mathrm{cor}[x_2(t), x_3(t)] = \frac{1}{2\sqrt{2}}.$ *These insights can also be seen in Figure* 6.

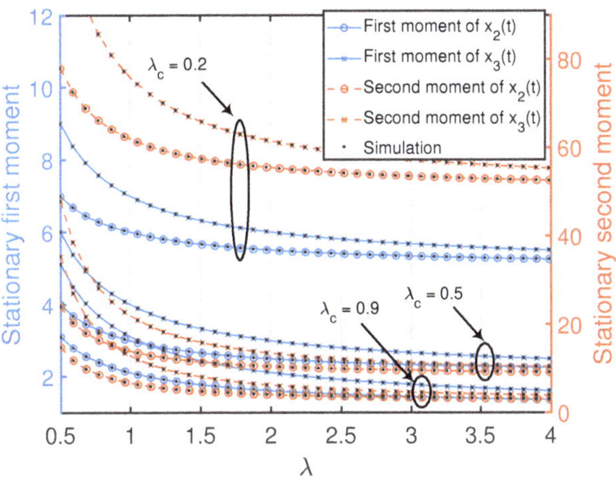

Figure 5. Stationary first and second moments of age processes at the nodes inside the *c*th cluster of the clustered gossip network topology. The simulated curves were obtained from the numerical evaluation of the stationary marginal moments using (10) in Theorem 1.

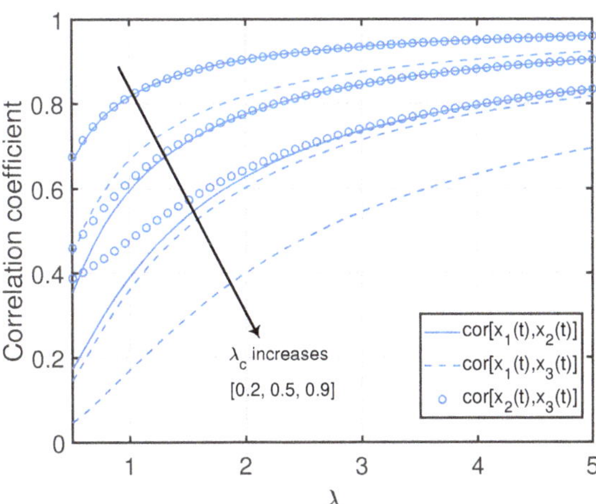

Figure 6. Correlation coefficients between age processes in the clustered gossip network topology.

Remark 13. *Note that the result of Proposition* 4 *agrees with the intuition. In particular, since the nodes in each cluster are disconnected from the nodes in the other clusters, the two age processes associated with any two arbitrary clusters in the network are uncorrelated.*

Remark 14. *From Propositions 1–3, one can see that the standard deviation of $x_1(t)$ (i.e., $\sqrt{\mathrm{var}[x_1(t)]}$) was equal to its average value $\bar{v}_{\{1\}}^{(1)}$. Additionally, the standard deviations of the age processes at the other nodes were relatively large with respect to their average values (which is also demonstrated numerically in Figures 7–10). This key insight promotes the importance of incorporating the higher-order moments of age processes in the implementation or optimization of age-aware gossip networks rather than just relying on the average values of the age processes (as has been performed in the existing literature thus far). This insight also demonstrates the need for the development of Theorems 1 and 2 in this paper, which allow the characterization of the marginal or joint MGFs of different age processes in the network that can then be used to evaluate the marginal or joint higher-order moments.*

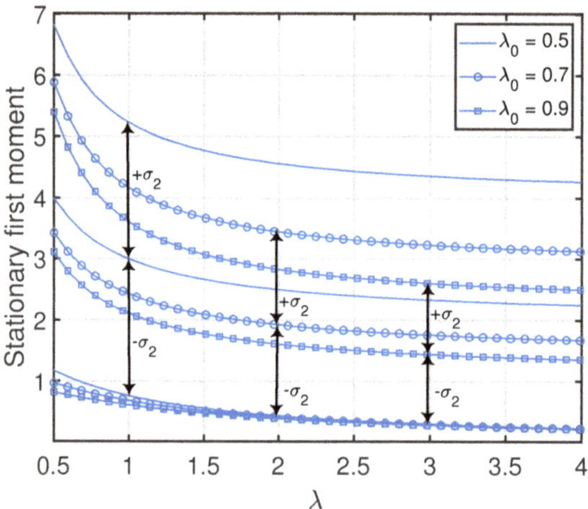

Figure 7. Variance of $x_2(t)$ in the serially-connected network setting. We denote the standard deviation of $x_2(t)$ as σ_2.

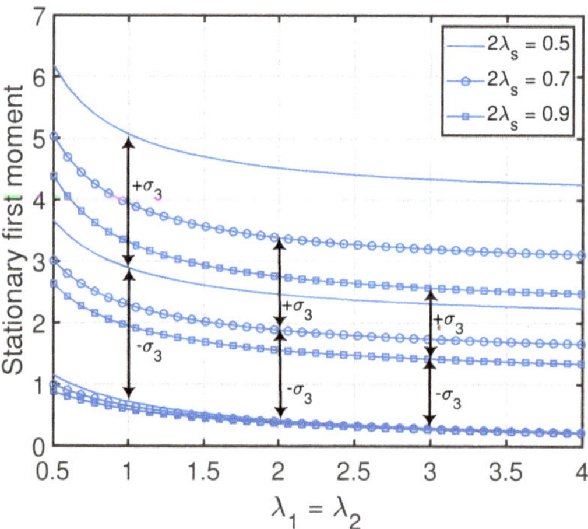

Figure 8. Variance of $x_3(t)$ in the parallelly-connected network setting. We denote the standard deviation of $x_3(t)$ as σ_3.

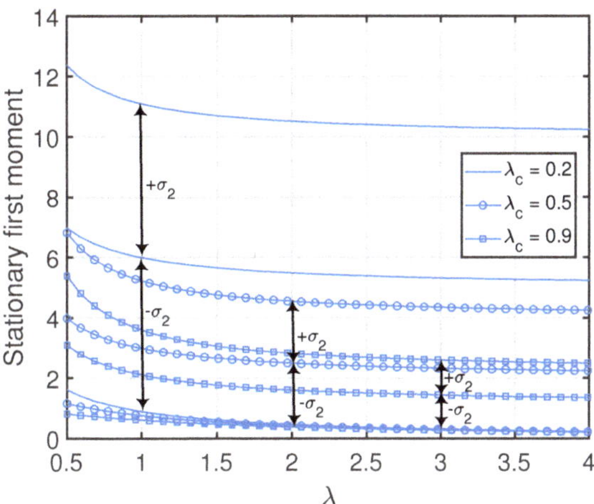

Figure 9. Variance of $x_2(t)$ in the clustered gossip network topology. We denote the standard deviation of $x_2(t)$ as σ_2.

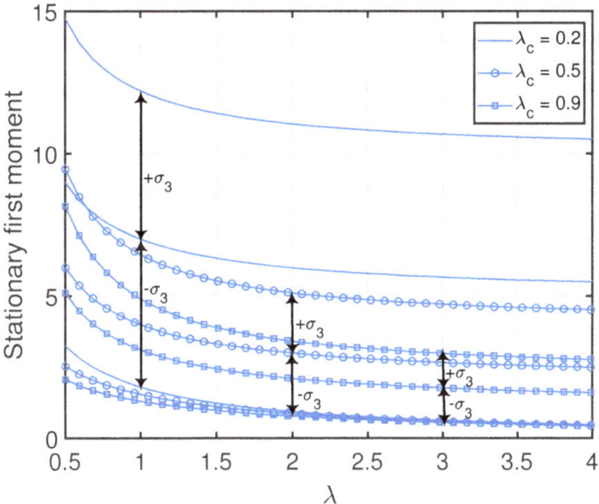

Figure 10. Variance of $x_3(t)$ in the clustered gossip network topology. We denote the standard deviation of $x_3(t)$ as σ_3.

5. Conclusions

In this paper, we developed SHS-based methods that allow the characterization of the stationary marginal and joint MGFs of age processes in a general setting of gossip networks. In particular, we used the SHS framework to derive two systems of first-order ordinary differential equations characterizing the temporal evolution of the marginal and joint MGFs, from which the stationary marginal and joint MGFs were evaluated. Afterward, these methods were applied to derive the stationary marginal and joint MGFs in the following three network topologies: (1) a serially-connected topology, (2) a parallelly-connected topology, and (3) a clustered topology. Using the MGF expressions derived for each network topology, we obtained closed-form expressions for the following quantities: (1) the stationary marginal first and second moments of each age process, (2) the variance of each age process, and (3) the correlation coefficients between all possible pairwise combinations of the age processes. We further characterized the structural properties of these quantities in terms of their convexity and monotonic nature with respect to the status updating rates and provided asymptotic results showing their behaviors when each of the status updating rates became small or large. Our analytical findings demonstrated that the standard deviations of the age processes in each network topology considered in this paper were relatively large with respect to their average values. This key insight promotes the importance of incorporating the higher-order moments of age processes in the implementation and optimization of age-aware gossip networks rather than just relying on the average values of the age processes (as has been performed in the existing literature thus far).

Given the generality of the setting of gossip networks analyzed in this paper, our developed methods can be applied to understand the marginal or joint distributional properties of age processes in any arbitrary gossip network topology. This opens the door for the use of these methods in the future to characterize the stationary marginal or joint moments and MGFs of the age processes in gossip network topologies that have only been analyzed in terms of the stationary first moment of each age process in the network until now. It would also be interesting to investigate how the stationary marginal or joint moments scale as functions of the network size.

Author Contributions: Conceptualization, M.A.A.-E. and H.S.D.; formal analysis, M.A.A.-E.; writing—original draft preparation, M.A.A.-E.; writing—review and editing, M.A.A.-E. and H.S.D.; funding acquisition, H.S.D. All authors have read and agreed to the published version of the manuscript.

Funding: This work was supported in part by the U.S. NSF (Grants CNS-1814477 and CNS-1923807). The publication charges of this article were covered in part by Virginia Tech's Open Access Subvention Fund.

Data Availability Statement: Not applicable.

Conflicts of Interest: The authors declare no conflict of interest.

Appendix A. Proof of Theorem 1

We derive this result by first using the SHS framework to obtain a system of differential equations characterizing the temporal evolution of the marginal MGFs of the age processes associated with all sets $S \subseteq \mathcal{N}$. We then obtain the stationary marginal MGFs as the fixed point of this system of equations (i.e., when $t \to \infty$). To derive the system of differential equations, we follow a similar approach to that in [15,92], where the idea is to define the test functions $\{\psi(\mathbf{x}(t))\}$ whose expected values $\{\mathbb{E}[\psi(\mathbf{x}(t))]\}$ are quantities of interest. Since we are interested here in the characterization of the marginal MGFs, we define the following class of test functions that is appropriate for this analysis:

$$\psi_S^{(n)}(\mathbf{x}(t)) = \exp[n x_S(t)], \ \forall S \subset \mathcal{N}, \tag{A1}$$

where the expected value $\mathbb{E}\left[\psi_S^{(n)}(\mathbf{x}(t))\right]$ is $v_S^{(n)}(t)$. We apply the SHS mapping $\psi(\mathbf{x}(t)) \to L\psi(\mathbf{x}(t))$ (known as the extended generator) to every test function in (A1). Since the test functions defined above are time-invariant, it follows from [92] Theorem 1 that the extended generator of a test function $\psi(\mathbf{x}(t))$ under the considered piecewise linear SHS is given by

$$L\psi(\mathbf{x}(t)) = \frac{\mathrm{d}\psi(\mathbf{x}(t))}{\mathrm{d}\mathbf{x}(t)}\mathbf{1}_N^{\mathrm{T}} + \sum_{l=(i,j)\in\mathcal{L}} \lambda_{ij}\left[\psi(\mathbf{x}'(t)) - \psi(\mathbf{x}(t))\right], \tag{A2}$$

where $\mathbf{x}'(t) = \begin{bmatrix} x_1'(t) & x_2'(t) & \cdots & x_N'(t) \end{bmatrix}$ such that the updated age at node k, $x_k'(t)$ resulting from the transition (i,j) is given by (1). In addition, note that $x_S'(t) = \min_{i\in S} x_i'(t)$. Now, we proceed to evaluate $L\psi_S^{(n)}(\mathbf{x}(t))$. From the age updating rule in (1), the set of transitions $\mathcal{L}$ in (2), and the structure of $\psi_S^{(n)}(\mathbf{x}(t))$ in (A1), we have

$$\frac{\mathrm{d}\psi_S^{(n)}(\mathbf{x}(t))}{\mathrm{d}\mathbf{x}(t)}\mathbf{1}_N^{\mathsf{T}} = n\psi_S^{(n)}(\mathbf{x}(t)),\tag{A3}$$

$$\psi_S^{(n)}(\mathbf{x}'(t)) = \exp\left[nx_S'(t)\right] = \begin{cases} \exp[n \times 0] = 1, & l = (0,j), j \in S, \\ \exp\left[nx_{S\cup\{i\}}(t)\right] = \psi_{S\cup\{i\}}^{(n)}(\mathbf{x}(t)), & l = (i,j), j \in S, i \in \mathcal{N}\setminus S, \\ \exp[nx_S(t)] = \psi_S^{(n)}(\mathbf{x}(t)), & \text{otherwise.} \end{cases}\tag{A4}$$

Substituting (A3) and (A4) into (A2) gives

$$L\psi_S^{(n)}(\mathbf{x}(t)) = n\psi_S^{(n)}(\mathbf{x}(t)) + \sum_{j\in S}\lambda_{0j}\left[1 - \psi_S^{(n)}(\mathbf{x}(t))\right] + \sum_{i\in\mathcal{N}\setminus S}\sum_{j\in S}\lambda_{ij}\left[\psi_{S\cup\{i\}}^{(n)}(\mathbf{x}(t)) - \psi_S^{(n)}(\mathbf{x}(t))\right],\tag{A5}$$

The system of differential equations characterizing the temporal evolution of the marginal MGFs $\{v_S^{(n)}(t)\}_{S\subseteq\mathcal{N}}$ can be derived by applying Dynkin's formula [92] to each test function and its associated extended generator. In particular, for a test function $\psi(\mathbf{x}(t))$, the Dynkin's formula can be expressed as

$$\frac{\mathrm{d}\mathbb{E}[\psi(\mathbf{x}(t))]}{\mathrm{d}t} = \mathbb{E}[L\psi(\mathbf{x}(t))].\tag{A6}$$

Plugging $\psi_S^{(n)}(\mathbf{x}(t))$ and $L\psi_S^{(n)}(\mathbf{x}(t))$ into (A6) gives

$$\dot{v}_S^{(n)}(t) = nv_S^{(n)}(t) + \sum_{j\in S}\lambda_{0j}\left[1 - v_S^{(n)}(t)\right] + \sum_{i\in\mathcal{N}\setminus S}\sum_{j\in S}\lambda_{ij}\left[v_{S\cup\{i\}}^{(n)}(t) - v_S^{(n)}(t)\right]$$

$$\overset{(a)}{=} \lambda_0(S) + v_S^{(n)}(t)\left[n - \lambda_0(S) - \sum_{i\in N(S)}\lambda_i(S)\right] + \sum_{i\in N(S)}\lambda_i(S)v_{S\cup\{i\}}^{(n)}(t),\tag{A7}$$

where step (a) directly follows from the definitions of $\lambda_i(S)$ and $N(S)$ in (7) and (8), respectively. Note that there exists a range of n values for which the differential equation in (A7) is asymptotically stable for any arbitrary set $S \subseteq \mathcal{N}$. To see this, let us first express $\dot{v}_\mathcal{N}^{(n)}(t)$ using (A7) as follows:

$$\dot{v}_\mathcal{N}^{(n)}(t) = \lambda_0(\mathcal{N}) + v_\mathcal{N}^{(n)}(t)[n - \lambda_0(\mathcal{N})].\tag{A8}$$

For $0 \leq n < \lambda_0(\mathcal{N})$, (A8) is asymptotically stable, and the stationary marginal MGF $\bar{v}_\mathcal{N}^{(n)}$ can be obtained by setting $\dot{v}_\mathcal{N}^{(n)}(t)$ to zero and replacing $v_\mathcal{N}^{(n)}(t)$ with $\bar{v}_\mathcal{N}^{(n)}$. Now, when $S = \mathcal{N}\setminus\{k\}$, (A7) is given by

$$\dot{v}_S^{(n)}(t) = \lambda_0(S) + v_S^{(n)}(t)[n - \lambda_0(S) - \lambda_k(S)] + \lambda_k(S)v_\mathcal{N}^{(n)}(t).\tag{A9}$$

For $0 \leq n < \min[\lambda_0(S) + \lambda_k(S), \lambda_0(\mathcal{N})]$ and $S = \mathcal{N}\setminus\{k\}$, $v_S^{(n)}(t)$ converges as $t \to \infty$, and the differential equation in (A9) is asymptotically stable. The stationary marginal MGF $\bar{v}_{\mathcal{N}\setminus\{k\}}^{(n)}$ is then the fixed point of (A9), which can be obtained after setting the derivative to zero. Afterward, when $S = \mathcal{N}\setminus\{k_1, k_2\}$, one can follow the above procedure to obtain the range of n values under which (A7) is asymptotically stable. Generally, for an arbitrary set S, there exists a threshold δ such that for $n \in [0, \delta)$, the stationary marginal MGF $\bar{v}_S^{(n)}$ is the fixed point of (A7).

Finally, the stationary marginal mth moment $\bar{v}_S^{(m)}$ in (10) can be obtained by substituting $\psi(\mathbf{x}(t))$ in (A2) with $\psi_S^{(m)}(\mathbf{x}(t)) = x_S^m(t)$ and following similar steps to those in (A2)–(A9). This completes the proof.

Appendix B. Proof of Theorem 2

The flow of this proof is similar to that for Theorem 1 in Appendix A. In particular, we start by constructing a class of test functions that is appropriate for the joint MGF analysis. We then use (A2) to derive the extended generator for each test function, which is then plugged into Dynkin's formula in (A6) to obtain the system of differential equations characterizing the temporal evolution of the joint MGFs $\{v_{S_1,S_2}^{(n_1,n_2)}\}_{S_1,S_2 \subseteq \mathcal{N}}$. The class of test functions we define here for the joint MGF analysis is given by

$$\psi_{S_1,S_2}^{(n_1,n_2)}(\mathbf{x}(t)) = \exp\left[n_1 x_{S_1}(t) + n_2 x_{S_2}(t)\right], \ \forall S_1, S_2 \subseteq \mathcal{N}, \tag{A10}$$

such that the expected value $\mathbb{E}\left[\psi_{S_1,S_2}^{(n_1,n_2)}(\mathbf{x}(t))\right]$ is $v_{S_1,S_2}^{(n_1,n_2)}(t)$. For such a structure of test functions, we have

$$\frac{d\psi_{S_1,S_2}^{(n_1,n_2)}(\mathbf{x}(t))}{d\mathbf{x}(t)}\mathbf{1}_N^{\mathrm{T}} = (n_1 + n_2)\psi_{S_1,S_2}^{(n_1,n_2)}(\mathbf{x}(t)). \tag{A11}$$

Compared with the proof for Theorem 1 in Appendix A, a key challenge in the derivation of the extended generator here is to carefully identify all the possible transitions in $\mathcal{L}$ that result in having $\psi_{S_1,S_2}^{(n_1,n_2)}(\mathbf{x}'(t)) \neq \psi_{S_1,S_2}^{(n_1,n_2)}(\mathbf{x}(t))$. We provide Figure A1 to help one easily visualize the following arguments. For the first subset of transitions $\{(0,j) : j \in \mathcal{N}\} \subset \mathcal{L}$, we have

$$\psi_{S_1,S_2}^{(n_1,n_2)}(\mathbf{x}'(t)) = \exp\left[n_1 x_{S_1}'(t) + n_2 x_{S_2}'(t)\right] = \begin{cases} \psi_{S_2}^{(n_2)}(\mathbf{x}(t)), & l = (0,j), j \in S_1 \setminus S_2, \\ \psi_{S_1}^{(n_1)}(\mathbf{x}(t)), & l = (0,j), j \in S_2 \setminus S_1, \\ 1, & l = (0,j), j \in S_1 \cap S_2, \\ \psi_{S_1,S_2}^{(n_1,n_2)}(\mathbf{x}(t)), & \text{otherwise.} \end{cases} \tag{A12}$$

To help one easily grasp the different cases in (A12), we elaborate more on the construction of the first case, and the other cases can be interpreted similarly. In particular, when $j \in S_1 \setminus S_2$, the transition $(0,j)$ results in resetting the age of S_1 to zero, whereas the age of S_2 will not change. As a result, $\psi_{S_1,S_2}^{(n_1,n_2)}(\mathbf{x}'(t)) = \exp\left[n_1 \times 0 + n_2 \times x_{S_2}(t)\right] \overset{(a)}{=} \psi_{S_2}^{(n_2)}(\mathbf{x}(t))$, where step (a) follows from (A1).

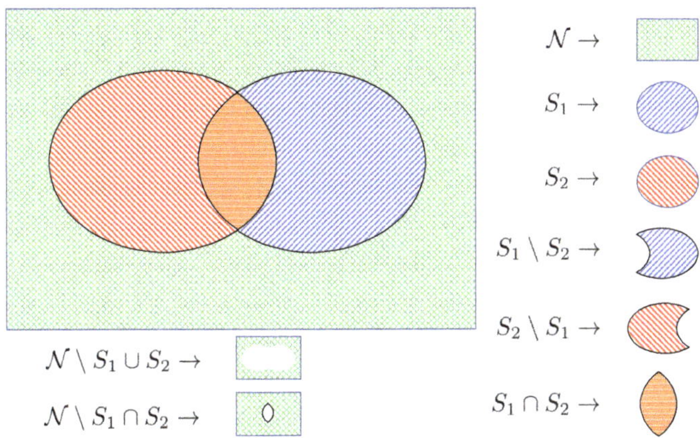

Figure A1. A Venn diagram representation.

For the second subset of transitions $\{(i,j) : i,j \in \mathcal{N}\} \subset \mathcal{L}$, we have

$$
\psi_{S_1,S_2}^{(n_1,n_2)}\big(\mathbf{x}'(t)\big) = \exp\Big[n_1 x_{S_1}'(t) + n_2 x_{S_2}'(t)\Big] =
\begin{cases}
\psi_{S_1 \cup \{i\},S_2}^{(n_1,n_2)}(\mathbf{x}(t)), & l = (i,j), j \in S_1 \setminus S_2, i \in \mathcal{N} \setminus S_1, \\[4pt]
\psi_{S_1,S_2 \cup \{i\}}^{(n_1,n_2)}(\mathbf{x}(t)), & l = (i,j), j \in S_2 \setminus S_1, i \in \mathcal{N} \setminus S_2, \\[4pt]
\psi_{S_1 \cup \{i\},S_2 \cup \{i\}}^{(n_1,n_2)}(\mathbf{x}(t)), & l = (i,j), j \in S_1 \cap S_2, i \in \mathcal{N} \setminus S_1 \cup S_2, \\[4pt]
\psi_{S_1,S_2 \cup \{i\}}^{(n_1,n_2)}(\mathbf{x}(t)), & l = (i,j), j \in S_1 \cap S_2, i \in S_1 \setminus S_2, \\[4pt]
\psi_{S_1 \cup \{i\},S_2}^{(n_1,n_2)}(\mathbf{x}(t)), & l = (i,j), j \in S_1 \cap S_2, i \in S_2 \setminus S_1, \\[4pt]
\psi_{S_1,S_2}^{(n_1,n_2)}(\mathbf{x}(t)), & \text{otherwise.}
\end{cases}
\tag{A13}
$$

Plugging (A11)–(A13) into (A2) gives

$$
L\psi_{S_1,S_2}^{(n_1,n_2)}(\mathbf{x}(t)) = (n_1 + n_2)\psi_{S_1,S_2}^{(n_1,n_2)}(\mathbf{x}(t)) + \sum_{j \in S_1 \setminus S_2} \lambda_{0j}\Big[\psi_{S_2}^{(n_2)}(\mathbf{x}(t)) - \psi_{S_1,S_2}^{(n_1,n_2)}(\mathbf{x}(t))\Big] + \sum_{j \in S_2 \setminus S_1} \lambda_{0j}\Big[\psi_{S_1}^{(n_1)}(\mathbf{x}(t)) - \psi_{S_1,S_2}^{(n_1,n_2)}(\mathbf{x}(t))\Big]
$$

$$
+ \sum_{j \in S_1 \cap S_2} \lambda_{0j}\Big[1 - \psi_{S_1,S_2}^{(n_1,n_2)}(\mathbf{x}(t))\Big] + \sum_{i \in \mathcal{N} \setminus S_1} \sum_{j \in S_1 \setminus S_2} \lambda_{ij}\Big[\psi_{S_1 \cup \{i\},S_2}^{(n_1,n_2)}(\mathbf{x}(t)) - \psi_{S_1,S_2}^{(n_1,n_2)}(\mathbf{x}(t))\Big]
$$

$$
+ \sum_{i \in \mathcal{N} \setminus S_2} \sum_{j \in S_2 \setminus S_1} \lambda_{ij}\Big[\psi_{S_1,S_2 \cup \{i\}}^{(n_1,n_2)}(\mathbf{x}(t)) - \psi_{S_1,S_2}^{(n_1,n_2)}(\mathbf{x}(t))\Big] + \sum_{i \in \mathcal{N} \setminus S_1 \cup S_2} \sum_{j \in S_1 \cap S_1} \lambda_{ij}\Big[\psi_{S_1 \cup \{i\},S_2 \cup \{i\}}^{(n_1,n_2)}(\mathbf{x}(t)) - \psi_{S_1,S_2}^{(n_1,n_2)}(\mathbf{x}(t))\Big]
$$

$$
+ \sum_{i \in S_1 \setminus S_2} \sum_{j \in S_1 \cap S_2} \lambda_{ij}\Big[\psi_{S_1,S_2 \cup \{i\}}^{(n_1,n_2)}(\mathbf{x}(t)) - \psi_{S_1,S_2}^{(n_1,n_2)}(\mathbf{x}(t))\Big] + \sum_{i \in S_2 \setminus S_1} \sum_{j \in S_1 \cap S_2} \lambda_{ij}\Big[\psi_{S_1 \cup \{i\},S_2}^{(n_1,n_2)}(\mathbf{x}(t)) - \psi_{S_1,S_2}^{(n_1,n_2)}(\mathbf{x}(t))\Big].
\tag{A14}
$$

By applying Dynkin's formula in (A6) to $\psi_{S_1,S_2}^{(n_1,n_2)}(\mathbf{x}(t))$ and $L\psi_{S_1,S_2}^{(n_1,n_2)}(\mathbf{x}(t))$, we have

$$\dot{v}_{S_1,S_2}^{(n_1,n_2)}(t) = (n_1+n_2)v_{S_1,S_2}^{(n_1,n_2)}(t) + \sum_{j\in S_1\setminus S_2}\lambda_{0j}\left[v_{S_2}^{(n_2)}(t) - v_{S_1,S_2}^{(n_1,n_2)}(t)\right] + \sum_{j\in S_2\setminus S_1}\lambda_{0j}\left[v_{S_1}^{(n_1)}(t) - v_{S_1,S_2}^{(n_1,n_2)}(t)\right]$$

$$+ \sum_{j\in S_1\cap S_2}\lambda_{0j}\left[1 - v_{S_1,S_2}^{(n_1,n_2)}(t)\right] + \sum_{i\in\mathcal{N}\setminus S_1}\sum_{j\in S_1\setminus S_2}\lambda_{ij}\left[v_{S_1\cup\{i\},S_2}^{(n_1,n_2)}(t) - v_{S_1,S_2}^{(n_1,n_2)}(t)\right]$$

$$+ \sum_{i\in\mathcal{N}\setminus S_2}\sum_{j\in S_2\setminus S_1}\lambda_{ij}\left[v_{S_1,S_2\cup\{i\}}^{(n_1,n_2)}(t) - v_{S_1,S_2}^{(n_1,n_2)}(t)\right] + \sum_{i\in\mathcal{N}\setminus S_1\cup S_2}\sum_{j\in S_1\cap S_2}\lambda_{ij}\left[v_{S_1\cup\{i\},S_2\cup\{i\}}^{(n_1,n_2)}(t) - v_{S_1,S_2}^{(n_1,n_2)}(t)\right]$$

$$+ \sum_{i\in S_1\setminus S_2}\sum_{j\in S_1\cap S_2}\lambda_{ij}\left[v_{S_1,S_2\cup\{i\}}^{(n_1,n_2)}(t) - v_{S_1,S_2}^{(n_1,n_2)}(t)\right] + \sum_{i\in S_2\setminus S_1}\sum_{j\in S_1\cap S_2}\lambda_{ij}\left[v_{S_1\cup\{i\},S_2}^{(n_1,n_2)}(t) - v_{S_1,S_2}^{(n_1,n_2)}(t)\right]$$

$$\overset{(a)}{=} v_{S_1,S_2}^{(n_1,n_2)}(t)\left[(n_1+n_2) - \lambda_0(S_1\cup S_2) - \sum_{i\in\mathcal{N}\setminus(S_1\cap S_2)}\lambda_i(S_1\cap S_2) - \sum_{i\in\mathcal{N}\setminus S_1}\lambda_i(S_1\setminus S_2) - \sum_{i\in\mathcal{N}\setminus S_2}\lambda_i(S_2\setminus S_1)\right]$$

$$+ \lambda_0(S_1\cap S_2) + \lambda_0(S_1\setminus S_2)v_{S_2}^{(n_2)}(t) + \lambda_0(S_2\setminus S_1)v_{S_1}^{(n_1)}(t) + \sum_{i\in\mathcal{N}\setminus S_1}\lambda_i(S_1\setminus S_2)v_{S_1\cup\{i\},S_2}^{(n_1,n_2)}(t)$$

$$+ \sum_{i\in\mathcal{N}\setminus S_2}\lambda_i(S_2\setminus S_1)v_{S_1,S_2\cup\{i\}}^{(n_1,n_2)}(t) + \sum_{i\in\mathcal{N}\setminus(S_1\cup S_2)}\lambda_i(S_1\cap S_2)v_{S_1\cup\{i\},S_2\cup\{i\}}^{(n_1,n_2)}(t) + \sum_{i\in S_1\setminus S_2}\lambda_i(S_1\cap S_2)v_{S_1,S_2\cup\{i\}}^{(n_1,n_2)}(t)$$

$$+ \sum_{i\in S_2\setminus S_1}\lambda_i(S_1\cap S_2)v_{S_1\cup\{i\},S_2}^{(n_1,n_2)}(t), \tag{A15}$$

where step (a) follows from applying the definition of $\lambda_i(S)$ in (7), followed by some algebraic simplifications. Now, following a similar procedure to that in (A8) and (A9) in Appendix A, one can show that for any two arbitrary sets S_1 and S_2, there exists a threshold δ (such that $0 \leq n_1 + n_2 < \delta$) under which the differential equation in (A15) is asymptotically stable. Thus, the final expression of the stationary joint MGF $\bar{v}_{S_1,S_2}^{(n_1,n_2)}$ in (11) can be obtained by taking the limit as $t\to\infty$ in (A15) (i.e., setting $\dot{v}_{S_1,S_2}^{(n_1,n_2)}(t)$ to zero and replacing $v_{S_1,S_2}^{(n_1,n_2)}(t)$ with $\bar{v}_{S_1,S_2}^{(n_1,n_2)}$).

Finally, the stationary joint (m_1,m_2)th moment $\bar{v}_{S_1,S_2}^{(m_1,m_2)}$ in (12) can be obtained by substituting $\psi(\mathbf{x}(t))$ in (A2) with $\psi_{S_1,S_2}^{(m_1,m_2)}(\mathbf{x}(t)) = x_{S_1}^{m_1}(t)x_{S_2}^{m_2}(t)$ and following similar steps to those in (A11)–(A15). This completes the proof.

Appendix C. Proof of Theorem 3

We start the proof by showing how one can use Theorem 1 to obtain the stationary marginal MGF of the AoI or age process at each node in the network. In particular, by observing the set of transitions in Figure 1a, repeated application of (9) gives

$$\bar{v}_{\{1\}}^{(n)} = \frac{\lambda_0}{\lambda_0 - n}, \tag{A16}$$

$$\bar{v}_{\{2\}}^{(n)} = \frac{\lambda\bar{v}_{\{1,2\}}^{(n)}}{\lambda - n}, \tag{A17}$$

$$\bar{v}_{\{1,2\}}^{(n)} = \frac{\lambda_0}{\lambda_0 - n}. \tag{A18}$$

By substituting (A18) into (A17), we obtain

$$\bar{v}_{\{2\}}^{(n)} = \frac{\lambda_0\lambda}{(\lambda_0 - n)(\lambda - n)}. \tag{A19}$$

Now, we proceed to the evaluation of the stationary joint MGF $\bar{v}^{(n_1,n_2)}_{\{2\},\{1\}}$ using Theorem 2. In particular, by applying (11) twice (the first time for $S_1 = \{2\}$ and $S_2 = \{1\}$ and the second time for $S_1 = \{1,2\}$ and $S_2 = \{1\}$), we obtain

$$\bar{v}^{(n_1,n_2)}_{\{2\},\{1\}}[\lambda_0 + \lambda - (n_1 + n_2)] = \lambda_0 \bar{v}^{(n_1)}_{\{2\}} + \lambda \bar{v}^{(n_1,n_2)}_{\{1,2\},\{1\}}, \tag{A20}$$

$$\bar{v}^{(n_1,n_2)}_{\{1,2\},\{1\}}[\lambda_0 - (n_1 + n_2)] = \lambda_0. \tag{A21}$$

The final expression of $\bar{v}^{(n_1,n_2)}_{\{2\},\{1\}}$ in (16) can be obtained by substituting $\bar{v}^{(n_1)}_{\{2\}}$ and $\bar{v}^{(n_1,n_2)}_{\{1,2\},\{1\}}$ from (A17) and (A21), respectively, into (A20).

Appendix D. Proof of Proposition 1

The stationary marginal mth moment of the age process at node $i \in \mathcal{N} = \{1,2\}$ (i.e., $\bar{v}^{(m)}_{\{i\}}$) is given by

$$\bar{v}^{(m)}_{\{i\}} = \frac{d^m \left[\bar{v}^{(n)}_{\{i\}} \right]}{dn^m} \Bigg|_{n=0}. \tag{A22}$$

Furthermore, for $i,j \in \mathcal{N}$, the stationary joint moment $\bar{v}^{(m_1,m_2)}_{\{i\},\{j\}}$ of the two age processes at nodes i and j is given by

$$\bar{v}^{(m_1,m_2)}_{\{i\},\{j\}} = \frac{\partial^{m_1+m_2} \left[\bar{v}^{(n_1,n_2)}_{\{i\},\{j\}} \right]}{\partial n_1^{m_1} \partial n_2^{m_2}} \Bigg|_{n_1=0, n_2=0}. \tag{A23}$$

The marginal first and second moments of the age process at each node in the serially-connected network (in (A22) and (A23)) can be obtained by plugging the marginal MGF expressions derived in Theorem 3 into (A22). Furthermore, the variance of the age process at node i is given by

$$\mathrm{var}[x_i(t)] = \bar{v}^{(2)}_{\{i\}} - \left(\bar{v}^{(1)}_{\{i\}} \right)^2. \tag{A24}$$

Finally, for nodes $i,j \in \mathcal{N}$, the correlation coefficient can be evaluated as follows:

$$\mathrm{cor}[x_i(t), x_j(t)] = \frac{\bar{v}^{(1,1)}_{\{i\},\{j\}} - \bar{v}^{(1)}_{\{i\}} \bar{v}^{(1)}_{\{j\}}}{\sqrt{\mathrm{var}[x_i(t)]} \sqrt{\mathrm{var}[x_j(t)]}}, \tag{A25}$$

In order to obtain $\mathrm{cor}[x_1(t), x_2(t)]$, what remains is only to evaluate $\bar{v}^{(1,1)}_{\{2\},\{1\}}$ from (A23) (using the joint MGF expression in (16)) as

$$\bar{v}^{(1,1)}_{\{2\},\{1\}} = \frac{\lambda_0^2 + 2\lambda_0\lambda + 2\lambda^2}{\lambda\lambda_0^2(\lambda_0 + \lambda)}. \tag{A26}$$

By noting that

$$\bar{v}^{(1,1)}_{\{2\},\{1\}} - \bar{v}^{(1)}_{\{2\}} \bar{v}^{(1)}_{\{1\}} = \frac{\lambda_0^2 + 2\lambda_0\lambda + 2\lambda^2}{\lambda\lambda_0^2(\lambda_0 + \lambda)} - \frac{\lambda_0 + \lambda}{\lambda_0^2\lambda} = \frac{\lambda}{\lambda_0^2(\lambda_0 + \lambda)}, \tag{A27}$$

then the final expression of $\mathrm{cor}[x_1(t), x_2(t)]$ in (19) can be obtained, which completes the proof.

Appendix E. Proof of Theorem 4

For the parallelly-connected network in Figure 1b, the stationary marginal MGF of the age process at each node $i \in \mathcal{N} = \{1, 2, 3\}$ can be derived by repeatedly applying (9) as follows:

$$\bar{v}_{\{3\}}^{(n)} = \frac{\lambda_1 \bar{v}_{\{1,3\}}^{(n)} + \lambda_2 \bar{v}_{\{2,3\}}^{(n)}}{\lambda_1 + \lambda_2 - n}, \tag{A28}$$

$$\bar{v}_{\{1,3\}}^{(n)} = \frac{\lambda_s + \lambda_2 \bar{v}_{\{1,2,3\}}^{(n)}}{\lambda_s + \lambda_2 - n}, \tag{A29}$$

$$\bar{v}_{\{2,3\}}^{(n)} = \frac{\lambda_s + \lambda_1 \bar{v}_{\{1,2,3\}}^{(n)}}{\lambda_s + \lambda_1 - n}, \tag{A30}$$

$$\bar{v}_{\{1,2,3\}}^{(n)} = \frac{2\lambda_s}{2\lambda_s - n}, \tag{A31}$$

$$\bar{v}_{\{1\}}^{(n)} = \bar{v}_{\{2\}}^{(n)} = \frac{\lambda_s}{\lambda_s - n}. \tag{A32}$$

The final expression of $\bar{v}_{\{3\}}^{(n)}$ in (21) can be obtained by substituting $\bar{v}_{\{1,3\}}^{(n)}$ and $\bar{v}_{\{2,3\}}^{(n)}$ from (A29)–(A31) into (A28).

We now derive the stationary joint MGF $\bar{v}_{\{3\},\{1\}}^{(n_1,n_2)}$ by repeatedly applying (11) as follows:

$$\bar{v}_{\{3\},\{1\}}^{(n_1,n_2)} [\lambda_s + \lambda_1 + \lambda_2 - (n_1 + n_2)] = \lambda_s \bar{v}_{\{3\}}^{(n_1)} + \lambda_1 \bar{v}_{\{1,3\},\{1\}}^{(n_1,n_2)} + \lambda_2 \bar{v}_{\{2,3\},\{1\}}^{(n_1,n_2)}, \tag{A33}$$

$$\bar{v}_{\{1,3\},\{1\}}^{(n_1,n_2)} [\lambda_s + \lambda_2 - (n_1 + n_2)] = \lambda_s + \lambda_2 \bar{v}_{\{1,2,3\},\{1\}}^{(n_1,n_2)}, \tag{A34}$$

$$\bar{v}_{\{2,3\},\{1\}}^{(n_1,n_2)} [2\lambda_s + \lambda_1 - (n_1 + n_2)] = \lambda_s \left(\bar{v}_{\{1\}}^{(n_2)} + \bar{v}_{\{2,3\}}^{(n_1)} \right) + \lambda_1 \bar{v}_{\{1,2,3\},\{1\}}^{(n_1,n_2)}, \tag{A35}$$

$$\bar{v}_{\{1,2,3\},\{1\}}^{(n_1,n_2)} [2\lambda_s - (n_1 + n_2)] = \lambda_s + \lambda_s \bar{v}_{\{1\}}^{(n_2)}. \tag{A36}$$

By substituting (A34)–(A36) into (A33), $\bar{v}_{\{3\},\{1\}}^{(n_1,n_2)}$ can expressed as

$$\bar{v}_{\{3\},\{1\}}^{(n_1,n_2)} = \frac{1}{[\lambda_s + \lambda_1 + \lambda_2 - (n_1 + n_2)][2\lambda_s + \lambda_1 - (n_1 + n_2)][2\lambda_s - (n_1 + n_2)][\lambda_s + \lambda_2 - (n_1 + n_2)]} \times$$
$$\left[\lambda_s[2\lambda_s + \lambda_1 - (n_1 + n_2)][2\lambda_s - (n_1 + n_2)][\lambda_s + \lambda_2 - (n_1 + n_2)]\bar{v}_{\{3\}}^{(n_1)} + \lambda_s\lambda_2[2\lambda_s + \lambda_1 - (n_1 + n_2)] \right.$$
$$\times [\lambda_s + \lambda_1 + \lambda_2 - (n_1 + n_2)]\bar{v}_{\{1\}}^{(n_2)} + \lambda_s\lambda_2[\lambda_s + \lambda_2 - (n_1 + n_2)][2\lambda_s - (n_1 + n_2)]\bar{v}_{\{2,3\}}^{(n_1)} + \lambda_s\lambda_1\lambda_2$$
$$\left. \times [\lambda_s + \lambda_2 - (n_1 + n_2)] + \lambda_s\lambda_1[2\lambda_s + \lambda_1 - (n_1 + n_2)][2\lambda_s + \lambda_2 - (n_1 + n_2)] \right]. \tag{A37}$$

The final expression of $\bar{v}_{\{3\},\{1\}}^{(n_1,n_2)}$ in (22) can be obtained by substituting $\bar{v}_{\{3\}}^{(n_1)}$, $\bar{v}_{\{1\}}^{(n_2)}$ and $\bar{v}_{\{2,3\}}^{(n_1)}$ from (21), (A32) and (A30), respectively, into (A37).

Appendix F. Proof of Proposition 2

The expressions in (27)–(30) of the first moment, second moment, and variance of the age process at each node $i \in \mathcal{N} = \{1,2,3\}$ can be derived by plugging the marginal MGF expressions in Theorem 2 into (A22). Furthermore, by plugging the joint MGF $\bar{v}_{\{3\},\{1\}}^{(n_1,n_2)}$ in (22) into (A23), one can obtain $\bar{v}_{\{3\},\{1\}}^{(1,1)}$ as follows:

$$
\begin{aligned}
\bar{v}_{\{3\},\{1\}}^{(1,1)} = {} & \frac{1}{4\lambda_s^2(\lambda_s + \lambda_1 + \lambda_2)(\lambda_s + \lambda_2)^2(2\lambda_s + \lambda_1)(\lambda_1 + \lambda_2)(\lambda_s + \lambda_1)} \times \Big[8\lambda_s^6 + 4\lambda_s^5(7\lambda_1 + 8\lambda_2) + 4\lambda_s^4\big(11\lambda_1^2 + 24\lambda_1\lambda_2 + 12\lambda_2^2\big) \\
& + \lambda_s^3(\lambda_1 + \lambda_2)\big(32\lambda_1^2 + 75\lambda_1\lambda_2 + 32\lambda_2^2\big) + 4\lambda_s^2(\lambda_1 + \lambda_2)^2\big(2\lambda_1^2 + 9\lambda_1\lambda_2 + 2\lambda_2^2\big) + 3\lambda_s\lambda_1\lambda_2\big(3\lambda_1^2 + 7\lambda_1\lambda_2 + 3\lambda_2^2\big) \\
& \times (\lambda_1 + \lambda_2) + 3\lambda_1^2\lambda_2^2(\lambda_1 + \lambda_2)^2 \Big],
\end{aligned}
\tag{A38}
$$

The final expression of $\mathrm{cor}[x_1(t), x_3(t)]$ in (31) can be obtained from (A25) while noting that we have

$$
\bar{v}_{\{3\},\{1\}}^{(1,1)} - \bar{v}_{\{3\}}^{(1)}\bar{v}_{\{1\}}^{(1)} = \frac{\lambda_1\big[8\lambda_s^4 + \lambda_s^3(12\lambda_1 + 7\lambda_2) + 2\lambda_s^2(\lambda_1 + 2\lambda_2)(\lambda_2 + 2\lambda_1) + \lambda_s\lambda_2(3\lambda_1^2 + 5\lambda_1\lambda_2 + \lambda_2^2) + \lambda_1\lambda_2^2(\lambda_1 + \lambda_2)\big]}{4\lambda_s^2(\lambda_s + \lambda_1 + \lambda_2)(\lambda_s + \lambda_2)^2(2\lambda_s + \lambda_1)(\lambda_s + \lambda_1)},
$$

This completes the proof.

Appendix G. Proof of Theorem 5

Repeated application of (9) gives

$$
\bar{v}_{\{1\}}^{(n)} = \frac{\lambda_c + \lambda\bar{v}_{\{1,3\}}^{(n)}}{\lambda_c + \lambda - n},
\tag{A39}
$$

$$
\bar{v}_{\{1,3\}}^{(n)} = \frac{\lambda_c + \lambda\bar{v}_{\{1,2,3\}}^{(n)}}{\lambda_c + \lambda - n} \overset{(a)}{=} \frac{\lambda_c}{\lambda_c - n},
\tag{A40}
$$

$$
\bar{v}_{\{1,2,3\}}^{(n)} = \frac{\lambda_c}{\lambda_c - n},
\tag{A41}
$$

$$
\bar{v}_{\{2\}}^{(n)} = \frac{\lambda\bar{v}_{\{1,2\}}^{(n)}}{\lambda - n},
\tag{A42}
$$

$$
\bar{v}_{\{1,2\}}^{(n)} = \frac{\lambda_c + \lambda\bar{v}_{\{1,2,3\}}^{(n)}}{\lambda_c + \lambda - n} \overset{(a)}{=} \frac{\lambda_c}{\lambda_c - n},
\tag{A43}
$$

$$
\bar{v}_{\{3\}}^{(n)} = \frac{\lambda\bar{v}_{\{2,3\}}^{(n)}}{\lambda - n},
\tag{A44}
$$

$$
\bar{v}_{\{2,3\}}^{(n)} = \frac{\lambda\bar{v}_{\{1,2,3\}}^{(n)}}{\lambda - n} \overset{(a)}{=} \frac{\lambda_c\lambda}{(\lambda_c - n)(\lambda - n)},
\tag{A45}
$$

where step (a) in (A40), (A43), and (A45) follows from substituting $\bar{v}^{(n)}_{\{1,2,3\}}$ from (A41). The expressions of $\{\bar{v}^{(n)}_{\{i\}}\}_{i\in\{1,2,3\}}$ in (35)–(37) are obtained from substituting (1) $\bar{v}^{(n)}_{\{1,3\}}$ from (A40) into (A39), (2) $\bar{v}^{(n)}_{\{1,2\}}$ from (A43) into (A42), and (3) $\bar{v}^{(n)}_{\{2,3\}}$ from (A45) into (A44).

Regarding the evaluation of the stationary joint MGF expressions, we start by deriving $\bar{v}^{(n_1,n_2)}_{\{1\},\{2\}}$. Repeated application of (11) gives

$$\bar{v}^{(n_1,n_2)}_{\{1\},\{2\}}[\lambda_c + 2\lambda - (n_1 + n_2)] = \lambda_c \bar{v}^{(n_2)}_{\{2\}} + \lambda \bar{v}^{(n_1,n_2)}_{\{1,3\},\{2\}} + \lambda \bar{v}^{(n_1,n_2)}_{\{1\},\{1,2\}}, \tag{A46}$$

$$\bar{v}^{(n_1,n_2)}_{\{1,3\},\{2\}}[\lambda_c + 2\lambda - (n_1 + n_2)] = \lambda_c \bar{v}^{(n_2)}_{\{2\}} + \lambda \bar{v}^{(n_1,n_2)}_{\{1,2,3\},\{2\}} + \lambda \bar{v}^{(n_1,n_2)}_{\{1,3\},\{1,2\}}, \tag{A47}$$

$$\bar{v}^{(n_1,n_2)}_{\{1\},\{1,2\}}[\lambda_c + \lambda - (n_1 + n_2)] = \lambda_c + \lambda \bar{v}^{(n_1,n_2)}_{\{1,3\},\{1,2\}}, \tag{A48}$$

$$\bar{v}^{(n_1,n_2)}_{\{1,2,3\},\{2\}}[\lambda_c + \lambda - (n_1 + n_2)] = \lambda_c \bar{v}^{(n_2)}_{\{2\}} + \lambda \bar{v}^{(n_1,n_2)}_{\{1,2,3\},\{1,2\}}, \tag{A49}$$

$$\bar{v}^{(n_1,n_2)}_{\{1,3\},\{1,2\}}[\lambda_c + 2\lambda - (n_1 + n_2)] = \lambda_c + \lambda \bar{v}^{(n_1,n_2)}_{\{1,2,3\},\{1,2\}} + \lambda \bar{v}^{(n_1,n_2)}_{\{1,3\},\{1,2,3\}}, \tag{A50}$$

$$\bar{v}^{(n_1,n_2)}_{\{1,2,3\},\{1,2\}}[\lambda_c + \lambda - (n_1 + n_2)] = \lambda_c + \lambda \bar{v}^{(n_1,n_2)}_{\{1,2,3\},\{1,2,3\}}, \tag{A51}$$

$$\bar{v}^{(n_1,n_2)}_{\{1,3\},\{1,2,3\}}[\lambda_c + \lambda - (n_1 + n_2)] = \lambda_c + \lambda \bar{v}^{(n_1,n_2)}_{\{1,2,3\},\{1,2,3\}}, \tag{A52}$$

$$\bar{v}^{(n_1,n_2)}_{\{1,2,3\},\{1,2,3\}} = \frac{\lambda_c}{\lambda_c - (n_1 + n_2)}. \tag{A53}$$

By substituting $\bar{v}^{(n_1,n_2)}_{\{1,2,3\},\{1,2,3\}}$ from (A53) into (A48) and (A50)–(A52), we obtain

$$\bar{v}^{(n_1,n_2)}_{\{1,2,3\},\{1,2\}} = \bar{v}^{(n_1,n_2)}_{\{1,3\},\{1,2,3\}} = \bar{v}^{(n_1,n_2)}_{\{1,3\},\{1,2\}} = \bar{v}^{(n_1,n_2)}_{\{1\},\{1,2\}} = \frac{\lambda_c}{\lambda_c - (n_1 + n_2)}, \tag{A54}$$

Furthermore, from (A47), (A49), (A50) and (A54), $\bar{v}^{(n_1,n_2)}_{\{1,3\},\{2\}}$ can be expressed as

$$\bar{v}^{(n_1,n_2)}_{\{1,3\},\{2\}} = \frac{\lambda_c[\lambda_c - (n_1 + n_2)]\bar{v}^{(n_2)}_{\{2\}} + \lambda\lambda_c}{[\lambda_c + \lambda - (n_1 + n_2)][\lambda_c - (n_1 + n_2)]}. \tag{A55}$$

The final expression of $\bar{v}^{(n_1,n_2)}_{\{1\},\{2\}}$ in (38) can be obtained by substituting $\bar{v}^{(n_1,n_2)}_{\{1,3\},\{2\}}$, $\bar{v}^{(n_1,n_2)}_{\{1\},\{1,2\}}$ and $\bar{v}^{(n_2)}_{\{2\}}$ from (A55), (A54) and (36), respectively, into (A46). Now, we proceed with the evaluation of $\bar{v}^{(n_1,n_2)}_{\{1\},\{3\}}$. Repeated application of (11) gives

$$\bar{v}^{(n_1,n_2)}_{\{1\},\{3\}}[\lambda_c + 2\lambda - (n_1 + n_2)] = \lambda_c \bar{v}^{(n_2)}_{\{3\}} + \lambda \bar{v}^{(n_1,n_2)}_{\{1,3\},\{3\}} + \lambda \bar{v}^{(n_1,n_2)}_{\{1\},\{2,3\}}, \tag{A56}$$

$$\bar{v}^{(n_1,n_2)}_{\{1,3\},\{3\}}[\lambda_c + \lambda - (n_1 + n_2)] = \lambda_c \bar{v}^{(n_2)}_{\{3\}} + \lambda \bar{v}^{(n_1,n_2)}_{\{1,2,3\},\{2,3\}}, \tag{A57}$$

$$\bar{v}^{(n_1,n_2)}_{\{1\},\{2,3\}}[\lambda_c + 2\lambda - (n_1 + n_2)] = \lambda_c \bar{v}^{(n_2)}_{\{2,3\}} + \lambda \bar{v}^{(n_1,n_2)}_{\{1,3\},\{2,3\}} + \lambda \bar{v}^{(n_1,n_2)}_{\{1\},\{1,2,3\}}, \tag{A58}$$

$$\bar{v}^{(n_1,n_2)}_{\{1,2,3\},\{2,3\}}[\lambda_c + \lambda - (n_1 + n_2)] = \lambda_c \bar{v}^{(n_2)}_{\{2,3\}} + \lambda \bar{v}^{(n_1,n_2)}_{\{1,2,3\},\{1,2,3\}}, \tag{A59}$$

$$\bar{v}^{(n_1,n_2)}_{\{1,3\},\{2,3\}}[\lambda_c + 2\lambda - (n_1 + n_2)] = \lambda_c \bar{v}^{(n_2)}_{\{2,3\}} + \lambda \bar{v}^{(n_1,n_2)}_{\{1,3\},\{1,2,3\}} + \lambda \bar{v}^{(n_1,n_2)}_{\{1,2,3\},\{2,3\}}, \tag{A60}$$

$$\bar{v}^{(n_1,n_2)}_{\{1\},\{1,2,3\}}[\lambda_c + \lambda - (n_1 + n_2)] = \lambda_c + \lambda \bar{v}^{(n_1,n_2)}_{\{1,3\},\{1,2,3\}}, \tag{A61}$$

where $\bar{v}^{(n_1,n_2)}_{\{1,3\},\{1,2,3\}} = \bar{v}^{(n_1,n_2)}_{\{1,2,3\},\{1,2,3\}} = \frac{\lambda_c}{\lambda_c - (n_1 + n_2)}$. By substituting $\bar{v}^{(n_1,n_2)}_{\{1,3\},\{1,2,3\}}$ into (A61), we obtain

$$\bar{v}^{(n_1,n_2)}_{\{1\},\{1,2,3\}} = \frac{\lambda_c}{\lambda_c - (n_1 + n_2)}. \tag{A62}$$

Furthermore, from (A57)–(A62), $\bar{v}^{(n_1,n_2)}_{\{1,3\},\{3\}}$ and $\bar{v}^{(n_1,n_2)}_{\{1\},\{2,3\}}$ can be respectively expressed as

$$\bar{v}^{(n_1,n_2)}_{\{1,3\},\{3\}} = \frac{\lambda_c[\lambda_c + \lambda - (n_1 + n_2)]\bar{v}^{(n_2)}_{\{3\}} + \lambda\left(\lambda_c \bar{v}^{(n_2)}_{\{2,3\}} + \lambda \bar{v}^{(n_1,n_2)}_{\{1,2,3\},\{1,2,3\}}\right)}{[\lambda_c + \lambda - (n_1 + n_2)]^2}, \tag{A63}$$

$$\bar{v}^{(n_1,n_2)}_{\{1\},\{2,3\}} = \frac{\lambda_c \bar{v}^{(n_2)}_{\{2,3\}} + \lambda \bar{v}^{(n_1,n_2)}_{\{1\},\{1,2,3\}}}{\lambda_c + \lambda - (n_1 + n_2)}. \tag{A64}$$

The final expression of $\bar{v}^{(n_1,n_2)}_{\{1\},\{3\}}$ in (39) can be obtained from substituting (A63) and (A64) into (A56), followed by some algebraic simplifications. Finally, to derive $\bar{v}^{(n_1,n_2)}_{\{2\},\{3\}}$, we first repeatedly use (11) as follows:

$$\bar{v}^{(n_1,n_2)}_{\{2\},\{3\}}[2\lambda - (n_1 + n_2)] = \lambda \bar{v}^{(n_1,n_2)}_{\{1,2\},\{3\}} + \lambda \bar{v}^{(n_1,n_2)}_{\{2\},\{2,3\}}, \tag{A65}$$

$$\bar{v}^{(n_1,n_2)}_{\{1,2\},\{3\}}[\lambda_c + 2\lambda - (n_1 + n_2)] = \lambda_c \bar{v}^{(n_2)}_{\{3\}} + \lambda \bar{v}^{(n_1,n_2)}_{\{1,2,3\},\{3\}} + \lambda \bar{v}^{(n_1,n_2)}_{\{1,2\},\{2,3\}}, \tag{A66}$$

$$\bar{v}^{(n_1,n_2)}_{\{2\},\{2,3\}}[\lambda - (n_1 + n_2)] = \lambda \bar{v}^{(n_1,n_2)}_{\{1,2\},\{1,2,3\}}, \tag{A67}$$

$$\bar{v}^{(n_1,n_2)}_{\{1,2,3\},\{3\}}[\lambda_c + \lambda - (n_1 + n_2)] = \lambda_c \bar{v}^{(n_2)}_{\{3\}} + \lambda \bar{v}^{(n_1,n_2)}_{\{1,2,3\},\{2,3\}}, \tag{A68}$$

$$\bar{v}^{(n_1,n_2)}_{\{1,2\},\{2,3\}}[\lambda_c + 2\lambda - (n_1 + n_2)] = \lambda_c \bar{v}^{(n_2)}_{\{2,3\}} + \lambda \bar{v}^{(n_1,n_2)}_{\{1,2,3\},\{2,3\}} + \lambda \bar{v}^{(n_1,n_2)}_{\{1,2\},\{1,2,3\}}. \tag{A69}$$

From (A66)–(A69), $\bar{v}^{(n_1,n_2)}_{\{1,2\},\{3\}}$ and $\bar{v}^{(n_1,n_2)}_{\{2\},\{2,3\}}$ can be respectively expressed as

$$\bar{v}^{(n_1,n_2)}_{\{1,2\},\{3\}} = \frac{1}{[\lambda_c + \lambda - (n_1 + n_2)][\lambda_c + 2\lambda - (n_1 + n_2)]^2} \times \Big[\lambda_c[\lambda_c + 2\lambda - (n_1 + n_2)]^2 \bar{v}^{(n_2)}_{\{3\}}$$
$$+ \lambda[\lambda_c + \lambda - (n_1 + n_2)]\left(\lambda_c \bar{v}^{(n_2)}_{\{2,3\}} + \lambda \bar{v}^{(n_1,n_2)}_{\{1,2\},\{1,2,3\}}\right) + \lambda^2[2\lambda_c + 3\lambda - 2(n_1 + n_2)]\bar{v}^{(n_1,n_2)}_{\{1,2,3\},\{2,3\}}\Big], \tag{A70}$$

$$\bar{v}^{(n_1,n_2)}_{\{2\},\{2,3\}} = \frac{\lambda_c \lambda}{[\lambda_c - (n_1 + n_2)][\lambda - (n_1 + n_2)]}. \tag{A71}$$

The final expression of $\bar{v}^{(n_1,n_2)}_{\{2\},\{3\}}$ in (40) can be obtained from plugging (A70) and (A71) into (A65), followed by substituting (1) $\bar{v}^{(n_2)}_{\{3\}}$ from (37), (2) $\bar{v}^{(n_2)}_{\{2,3\}}$ from (A45), (3) $\bar{v}^{(n_1,n_2)}_{\{1,2,3\},\{2,3\}}$ from (A59), and (4) $\bar{v}^{(n_1,n_2)}_{\{1,2\},\{1,2,3\}}$ as $\frac{\lambda_c}{\lambda_c - (n_1 + n_2)}$.

Appendix H. Proof of Proposition 3

The results of this proposition can be derived by following similar steps to those in Appendices D and F while noting that

$$\bar{v}^{(1,1)}_{\{1\},\{2\}} = \frac{\lambda^2 + (\lambda_c + \lambda)^2}{\lambda \lambda_c^2 (\lambda_c + \lambda)}, \tag{A72}$$

$$\bar{v}^{(1,1)}_{\{1\},\{2\}} - \bar{v}^{(1)}_{\{1\}} \bar{v}^{(1)}_{\{2\}} = \frac{\lambda}{\lambda_c^2 (\lambda_c + \lambda)}, \tag{A73}$$

$$\bar{v}^{(1,1)}_{\{1\},\{3\}} = \frac{2\lambda_c^3 + 5\lambda_c^2 \lambda + 4\lambda_c \lambda^2 + 2\lambda^3}{\lambda \lambda_c^2 (\lambda_c + \lambda)^2}, \tag{A74}$$

$$\bar{v}^{(1,1)}_{\{1\},\{3\}} - \bar{v}^{(1)}_{\{1\}} \bar{v}^{(1)}_{\{3\}} = \frac{\lambda^2}{\lambda_c^2 (\lambda_c + \lambda)^2}, \tag{A75}$$

$$\bar{v}^{(1,1)}_{\{2\},\{3\}} = \frac{5\lambda_c^4 + 16\lambda_c^3 \lambda + 20\lambda_c^2 \lambda^2 + 12\lambda_c \lambda^3 + 4\lambda^4}{2\lambda_c^2 \lambda^2 (\lambda_c + \lambda)^2}, \tag{A76}$$

$$\bar{v}^{(1,1)}_{\{2\},\{3\}} - \bar{v}^{(1)}_{\{2\}} \bar{v}^{(1)}_{\{3\}} = \frac{\lambda_c^4 + 2\lambda_c^3 \lambda + 2\lambda_c^2 \lambda^2 + 2\lambda_c \lambda^3 + 2\lambda^4}{2\lambda_c^2 \lambda^2 (\lambda_c + \lambda)^2}. \tag{A77}$$

Appendix I. Proof of Proposition 4

We first apply (11) to obtain $\bar{v}^{(n_1,n_2)}_{\mathcal{N}_i,\mathcal{N}_j}$ as

$$\bar{v}^{(n_1,n_2)}_{\mathcal{N}_i,\mathcal{N}_j} = \frac{\lambda_i \bar{v}^{(n_2)}_{\mathcal{N}_j} + \lambda_j \bar{v}^{(n_1)}_{\mathcal{N}_i}}{\lambda_i + \lambda_j - (n_1 + n_2)} \overset{(a)}{=} \frac{\lambda_i \lambda_j}{(\lambda_i - n_1)(\lambda_j - n_2)}, \tag{A78}$$

where step (a) follows from substituting $\bar{v}^{(n_1)}_{\mathcal{N}_i}$ and $\bar{v}^{(n_2)}_{\mathcal{N}_j}$ from (A41) as $\frac{\lambda_i}{\lambda_i - n_1}$ and $\frac{\lambda_j}{\lambda_j - n_2}$, respectively. We then obtain $\dfrac{\partial^2 \bar{v}^{(n_1,n_2)}_{\mathcal{N}_i,\mathcal{N}_j}}{\partial n_2 \partial n_1}$ as

$$\frac{\partial^2 \bar{v}^{(n_1,n_2)}_{\mathcal{N}_i,\mathcal{N}_j}}{\partial n_2 \partial n_1} = \frac{\lambda_i \lambda_j}{(\lambda_i - n_1)^2 (\lambda_j - n_2)^2}. \tag{A79}$$

Thus, from (A79), we have

$$\bar{v}_{\mathcal{N}_i,\mathcal{N}_j}^{(1,1)} = \frac{1}{\lambda_i \lambda_j}. \tag{A80}$$

The conclusion that the two age processes $x_{\mathcal{N}_i}(t)$ and $x_{\mathcal{N}_j}(t)$ are uncorrelated follows from noting that $\bar{v}_{\mathcal{N}_i,\mathcal{N}_j}^{(1,1)} - \bar{v}_{\mathcal{N}_i}^{(1)} \bar{v}_{\mathcal{N}_j}^{(1)} = 0$, and hence the correlation coefficient between $x_{\mathcal{N}_i}(t)$ and $x_{\mathcal{N}_j}(t)$ is zero.

References

1. Abd-Elmagid, M.A.; Pappas, N.; Dhillon, H.S. On the Role of Age of Information in the Internet of Things. *IEEE Commun. Mag.* **2019**, *57*, 72–77. [CrossRef]
2. Kaul, S.; Yates, R.; Gruteser, M. Real-time status: How often should one update? In Proceedings of the IEEE INFOCOM, Orlando, FL, USA, 25–30 March 2012.
3. Costa, M.; Codreanu, M.; Ephremides, A. On the age of information in status update systems with packet management. *IEEE Trans. Inf. Theory* **2016**, *62*, 1897–1910. [CrossRef]
4. Soysal, A.; Ulukus, S. Age of Information in G/G/1/1 Systems: Age Expressions, Bounds, Special Cases, and Optimization. *IEEE Trans. Inf. Theory* **2021**, *67*, 7477–7489. [CrossRef]
5. Kam, C.; Kompella, S.; Nguyen, G.D.; Wieselthier, J.E.; Ephremides, A. On the age of information with packet deadlines. *IEEE Trans. Inf. Theory* **2018**, *64*, 6419–6428. [CrossRef]
6. Zou, P.; Ozel, O.; Subramaniam, S. Waiting before serving: A companion to packet management in status update systems. *IEEE Trans. Inf. Theory* **2019**, *66*, 3864–3877. [CrossRef]
7. Inoue, Y.; Masuyama, H.; Takine, T.; Tanaka, T. A General Formula for the Stationary Distribution of the Age of Information and Its Application to Single-Server Queues. *IEEE Trans. Inf. Theory* **2019**, *65*, 8305–8324. [CrossRef]
8. Kosta, A.; Pappas, N.; Ephremides, A.; Angelakis, V. The Age of Information in a Discrete Time Queue: Stationary Distribution and Non-Linear Age Mean Analysis. *IEEE J. Sel. Areas Commun.* **2021**, *39*, 1352–1364. [CrossRef]
9. Champati, J.P.; Al-Zubaidy, H.; Gross, J. On the Distribution of AoI for the GI/GI/1/1 and GI/GI/1/2* Systems: Exact Expressions and Bounds. In Proceedings of the IEEE INFOCOM, Paris, France, 29 April–2 May 2019.
10. Chiariotti, F.; Vikhrova, O.; Soret, B.; Popovski, P. Peak Age of Information Distribution for Edge Computing With Wireless Links. *IEEE Trans. Commun.* **2021**, *69*, 3176–3191. [CrossRef]
11. Moltafet, M.; Leinonen, M.; Codreanu, M. On the Age of Information in Multi-Source Queueing Models. *IEEE Trans. Commun.* **2020**, *68*, 5003–5017. [CrossRef]
12. Xu, J.; Gautam, N. Peak Age of Information in Priority Queuing Systems. *IEEE Trans. Inf. Theory* **2021**, *67*, 373–390. [CrossRef]
13. Dogan, O.; Akar, N. The Multi-Source Probabilistically Preemptive M/PH/1/1 Queue With Packet Errors. *IEEE Trans. Commun.* **2021**, *69*, 7297–7308. [CrossRef]
14. Yates, R.D.; Kaul, S.K. The age of information: Real-time status updating by multiple sources. *IEEE Trans. Inf. Theory* **2019**, *65*, 1807–1827. [CrossRef]
15. Yates, R.D. The age of information in networks: Moments, distributions, and sampling. *IEEE Trans. Inf. Theory* **2020**, *66*, 5712–5728. [CrossRef]
16. Abd-Elmagid, M.A.; Dhillon, H.S. Joint distribution of ages of information in networks. *arXiv* **2022**, arXiv:2205.07448v3.
17. Boyd, S.; Ghosh, A.; Prabhakar, B.; Shah, D. Randomized gossip algorithms. *IEEE Trans. Inf. Theory* **2006**, *52*, 2508–2530. [CrossRef]
18. Shah, D. *Gossip Algorithms*; Now Publishers Inc.: Norwell, MA, USA, 2009.
19. Yates, R.D. Timely Gossip. In Proceedings of the IEEE Workshop on Signal Processing Advances in Wireless Communications (SPAWC), Oulu, Finland 27–30 September 2021.
20. Yates, R.D. The age of gossip in networks. In Proceedings of the IEEE International Symposium on Information Theory (ISIT), Melbourne, VIC, Australia, 12–20 July 2021.
21. Yates, R.D. Status Updates through Networks of Parallel Servers. In Proceedings of the IEEE International Symposium on Information Theory (ISIT), Vail, CO, USA, 17–22 June 2018.
22. Yates, R.D. Age of information in a network of preemptive servers. In Proceedings of the IEEE INFOCOM Workshops, Honolulu, HI, USA, 15–19 April 2018.
23. Kaul, S.K.; Yates, R.D. Age of Information: Updates with Priority. In Proceedings of the IEEE International Symposium on Information Theory (ISIT), Vail, CO, USA, 17–22 June 2018.
24. Farazi, S.; Klein, A.G.; Brown, D.R. Average Age of Information in Update Systems With Active Sources and Packet Delivery Errors. *IEEE Wirel. Commun. Lett.* **2020**, *9*, 1164–1168. [CrossRef]
25. Javani, A.; Zorgui, M.; Wang, Z. Age of Information for Multiple-Source Multiple-Server Networks. *arXiv* **2021**, arXiv:2106.07247v1.

26.	Maatouk, A.; Assaad, M.; Ephremides, A. On the Age of Information in a CSMA Environment. *IEEE/ACM Trans. Netw.* **2020**, *28*, 818–831. [CrossRef]

27.	Moltafet, M.; Leinonen, M.; Codreanu, M. Average AoI in Multi-Source Systems With Source-Aware Packet Management. *IEEE Trans. Commun.* **2021**, *69*, 1121–1133. [CrossRef]

28.	Abd-Elmagid, M.A.; Dhillon, H.S. Closed-Form Characterization of the MGF of AoI in Energy Harvesting Status Update Systems. *IEEE Trans. Inf. Theory* **2022**, *68*, 3896–3919. [CrossRef]

29.	Abd-Elmagid, M.A.; Dhillon, H.S. Distributional Properties of Age of Information in Energy Harvesting Status Update Systems. In Proceedings of the IEEE International Symposium on Modeling and Optimization in Mobile, Ad Hoc, and Wireless Networks (WiOpt), Philadelphia, PA, USA, 18–21 October 2021.

30.	Moltafet, M.; Leinonen, M.; Codreanu, M. Moment Generating Function of the AoI in a Two-Source System With Packet Management. *IEEE Wirel. Commun. Lett.* **2021**, *10*, 882–886. [CrossRef]

31.	Abd-Elmagid, M.A.; Dhillon, H.S. Age of Information in Multi-source Updating Systems Powered by Energy Harvesting. *IEEE J. Sel. Areas Inf. Theory* **2022**, *3*, 98–112. [CrossRef]

32.	Abd-Elmagid, M.A.; Dhillon, H.S. Distribution of AoI in EH-powered Multi-source Systems with Source-aware Packet Management. In Proceedings of the IEEE International Conference on Communications, Seoul, Republic of Korea, 16–20 May 2022.

33.	Abd-Elmagid, M.A.; Dhillon, H.S. Distribution of AoI in EH-powered Multi-source Systems under Non-preemptive and Preemptive Policies. In Proceedings of the IEEE INFOCOM Workshops, Online, 2–5 May 2022.

34.	Abd-Elmagid, M.A.; Dhillon, H.S. A Stochastic Hybrid Systems Approach to the Joint Distribution of Ages of Information in Networks. In Proceedings of the IEEE International Symposium on Modeling and Optimization in Mobile, Ad Hoc, and Wireless Networks (WiOpt), Torino, Italy, 19–23 September 2022.

35.	Buyukates, B.; Bastopcu, M.; Ulukus, S. Version Age of Information in Clustered Gossip Networks. *IEEE J. Sel. Areas Info. Theory* **2022**, *3*, 85–97. [CrossRef]

36.	Kaswan, P.; Ulukus, S. Timestomping Vulnerability of Age-Sensitive Gossip Networks. *arXiv* **2022**, arXiv:2212.14421v1.

37.	Bastopcu, M.; Etesami, S.R.; Basar, T. The Role of Gossiping for Information Dissemination over Networked Agents. *arXiv* **2022**, arXiv:2201.08365v1.

38.	Kaswan, P.; Ulukus, S. Timely gossiping with file slicing and network coding. *arXiv* **2022**, arXiv:2202.00649v1.

39.	Mitra, P.; Ulukus, S. ASUMAN: Age sense updating multiple access in networks. In Proceedings of the Annual Allerton Conference on Communication, Control, and Computing (Allerton), Monticello, IL, USA, 28–30 September 2022.

40.	Mitra, P.; Ulukus, S. Timely Opportunistic Gossiping in Dense Networks. *arXiv* **2023**, arXiv:2301.00798v1.

41.	Delfani, E.; Pappas, N. Version Age-Optimal Cached Status Updates in a Gossiping Network with Energy Harvesting Sensor. *TechRxiv* **2022**. . [CrossRef]

42.	Pappas, N.; Abd-Elmagid, M.A.; Zhou, B.; Saad, W.; Dhillon, H.S. *Age of Information: Foundations and Applications*; Cambridge University Press: Cambridge, UK, 2023.

43.	Yates, R.D.; Sun, Y.; Brown, D.R.; Kaul, S.K.; Modiano, E.; Ulukus, S. Age of Information: An Introduction and Survey. *IEEE J. Sel. Areas Commun.* **2021**, *39*, 1183–1210. [CrossRef]

44.	Sun, Y.; Uysal-Biyikoglu, E.; Yates, R.D.; Koksal, C.E.; Shroff, N.B. Update or wait: How to keep your data fresh. *IEEE Trans. Inf. Theory* **2017**, *63*, 7492–7508. [CrossRef]

45.	Bedewy, A.M.; Sun, Y.; Shroff, N.B. Optimizing data freshness, throughput, and delay in multi-server information-update systems. In Proceedings of the IEEE International Symposium on Information Theory, Barcelona, Spain, 10–15 July 2016.

46.	He, Q.; Yuan, D.; Ephremides, A. Optimal Link Scheduling for Age Minimization in Wireless Systems. *IEEE Trans. Inf. Theory* **2018**, *64*, 5381–5394. [CrossRef]

47.	Lu, N.; Ji, B.; Li, B. Age-based scheduling: Improving data freshness for wireless real-time traffic. In Proceedings of the ACM International Symposium on Mobile Ad Hoc Networking and Computing, Los Angeles, CA, USA, 26–29 June 2018.

48.	Jiang, Z.; Krishnamachari, B.; Zheng, X.; Zhou, S.; Niu, Z. Timely status update in wireless uplinks: Analytical solutions with asymptotic optimality. *IEEE Internet Things J.* **2019**, *6*, 3885–3898. [CrossRef]

49.	Huang, H.; Qiao, D.; Gursoy, M.C. Age-energy tradeoff in fading channels with packet-based transmissions. In Proceedings of the IEEE INFOCOM Workshops, Toino, Italy, 6–9 July 2020.

50.	Dong, Y.; Chen, Z.; Liu, S.; Fan, P.; Letaief, K.B. Age-Upon-Decisions Minimizing Scheduling in Internet of Things: To Be Random or To Be Deterministic? *IEEE Internet Things J.* **2020**, *7*, 1081–1097. [CrossRef]

51.	Saurav, K.; Vaze, R. Minimizing the Sum of Age of Information and Transmission Cost under Stochastic Arrival Model. In Proceedings of the IEEE INFOCOM, Vancouver, BC, Canada, 10–13 May 2021.

52.	Han, B.; Zhu, Y.; Jiang, Z.; Sun, M.; Schotten, H.D. Fairness for Freshness: Optimal Age of Information Based OFDMA Scheduling With Minimal Knowledge. *IEEE Trans. Wirel. Commun.* **2021**, *20*, 7903–7919. [CrossRef]

53.	Talak, R.; Karaman, S.; Modiano, E. Minimizing age-of-information in multi-hop wireless networks. In Proceedings of the Annual Allerton Conference on Communication, Control, and Computing, Monticello, IL, USA, 3–6 October 2017.

54.	Bedewy, A.M.; Sun, Y.; Shroff, N.B. Age-optimal information updates in multihop networks. In Proceedings of the IEEE International Symposium on Information Theory (ISIT), Aachen, Germany, 25–30 June 2017.

55.	Buyukates, B.; Soysal, A.; Ulukus, S. Age of Information in Two-Hop Multicast Networks. In Proceedings of the IEEE Asilomar, Pacific Grove, CA, USA, 28–31 October 2018.

56. Kadota, I.; Uysal-Biyikoglu, E.; Singh, R.; Modiano, E. Minimizing the age of information in broadcast wireless networks. In Proceedings of the Annual Allerton Conference on Communication, Control, and Computing, Monticello, IL, USA, 27–30 September 2016.
57. Hsu, Y.P.; Modiano, E.; Duan, L. Scheduling Algorithms for Minimizing Age of Information in Wireless Broadcast Networks with Random Arrivals. *IEEE Trans. Mobile Comput.* **2020**, *19*, 2903–2915. [CrossRef]
58. Abdel-Aziz, M.K.; Liu, C.F.; Samarakoon, S.; Bennis, M.; Saad, W. Ultra-reliable low-latency vehicular networks: Taming the age of information tail. In Proceedings of the IEEE Globecom, Abu Dhabi, United Arab Emirates, 9–13 December 2018.
59. Abd-Elmagid, M.A.; Dhillon, H.S. Average Peak Age-of-Information Minimization in UAV-Assisted IoT Networks. *IEEE Trans. Veh. Technol.* **2019**, *68*, 2003–2008. [CrossRef]
60. Abd-Elmagid, M.A.; Ferdowsi, A.; Dhillon, H.S.; Saad, W. Deep Reinforcement Learning for Minimizing Age-of-Information in UAV-assisted Networks. In Proceedings of the IEEE Globecom, Big Island, HI, USA, 9–13 December 2019.
61. Ferdowsi, A.; Abd-Elmagid, M.A.; Saad, W.; Dhillon, H.S. Neural Combinatorial Deep Reinforcement Learning for Age-Optimal Joint Trajectory and Scheduling Design in UAV-Assisted Networks. *IEEE J. Sel. Areas Commun.* **2021**, *39*, 1250–1265. [CrossRef]
62. Fang, Z.; Wang, J.; Du, J.; Hou, X.; Ren, Y.; Han, Z. Stochastic Optimization-Aided Energy-Efficient Information Collection in Internet of Underwater Things Networks. *IEEE Internet Things J.* **2022**, *9*, 1775–1789. [CrossRef]
63. Samir, M.; Elhattab, M.; Assi, C.; Sharafeddine, S.; Ghrayeb, A. Optimizing Age of Information Through Aerial Reconfigurable Intelligent Surfaces: A Deep Reinforcement Learning Approach. *IEEE Trans. Veh. Technol.* **2021**, *70*, 3978–3983. [CrossRef]
64. Muhammad, A.; Elhattab, M.; Arfaoui, M.A.; Al-Hilo, A.; Assi, C. Age of Information Optimization in a RIS-Assisted Wireless Network. *arXiv* **2021**, arXiv:2103.06405v3.
65. Bacinoglu, B.T.; Ceran, E.T.; Uysal-Biyikoglu, E. Age of information under energy replenishment constraints. In Proceedings of the Information Theory and its Applications (ITA), San Diego, CA, USA, 1–6 February 2015.
66. Wu, X.; Yang, J.; Wu, J. Optimal status update for age of information minimization with an energy harvesting source. *IEEE Trans. Green Commun. Netw.* **2018**, *2*, 193–204. [CrossRef]
67. Arafa, A.; Yang, J.; Ulukus, S.; Poor, H.V. Age-minimal transmission for energy harvesting sensors with finite batteries: Online policies. *IEEE Trans. Inf. Theory* **2020**, *66*, 534–556. [CrossRef]
68. Abd-Elmagid, M.A.; Dhillon, H.S.; Pappas, N. Online Age-minimal Sampling Policy for RF-powered IoT Networks. In Proceedings of the IEEE Globecom, Big Island, HI, USA, 9–13 December 2019.
69. Hatami, M.; Jahandideh, M.; Leinonen, M.; Codreanu, M. Age-aware status update control for energy harvesting IoT sensors via reinforcement learning. In Proceedings of the IEEE PIMRC, London, UK, 31 August–3 September 2020.
70. Abd-Elmagid, M.A.; Dhillon, H.S.; Pappas, N. A Reinforcement Learning Framework for Optimizing Age of Information in RF-powered Communication Systems. *IEEE Trans. Commun.* **2020**, *68*, 4747–4760. [CrossRef]
71. Abd-Elmagid, M.A.; Dhillon, H.S.; Pappas, N. AoI-Optimal Joint Sampling and Updating for Wireless Powered Communication Systems. *IEEE Trans. Veh. Technol.* **2020**, *69*, 14110–14115. [CrossRef]
72. Gindullina, E.; Badia, L.; Gündüz, D. Age-of-Information With Information Source Diversity in an Energy Harvesting System. *IEEE Trans. Green Commun. Netw.* **2021**, *5*, 1529–1540. [CrossRef]
73. Khorsandmanesh, Y.; Emadi, M.J.; Krikidis, I. Average Peak Age of Information Analysis for Wireless Powered Cooperative Networks. *IEEE Trans. Cogn. Commun. Netw.* **2021**, *7*, 1291–1303. [CrossRef]
74. Ren, Q.; Chan, T.T.; Pan, H.; Ho, K.H.; Du, Z. Information Freshness and Energy Harvesting Tradeoff in Network-Coded Broadcasting. *IEEE Wirel. Commun. Lett.* **2022**, *11*, 2061–2065. [CrossRef]
75. Emara, M.; Elsawy, H.; Bauch, G. A Spatiotemporal Model for Peak AoI in Uplink IoT Networks: Time Versus Event-Triggered Traffic. *IEEE Internet Things J.* **2020**, *7*, 6762–6777. [CrossRef]
76. Mankar, P.D.; Abd-Elmagid, M.A.; Dhillon, H.S. Spatial Distribution of the Mean Peak Age of Information in Wireless Networks. *IEEE Trans. Wirel. Commun.* **2021**, *20*, 4465–4479. [CrossRef]
77. Mankar, P.D.; Chen, Z.; Abd-Elmagid, M.A.; Pappas, N.; Dhillon, H.S. Throughput and Age of Information in a Cellular-Based IoT Network. *IEEE Trans. Wirel. Commun.* **2021**, *20*, 8248–8263. [CrossRef]
78. Ornee, T.Z.; Sun, Y. Sampling for remote estimation through queues: Age of information and beyond. In Proceedings of the Modeling and Optimization in Mobile, Ad Hoc and Wireless Networks, Avignon, France, 27–31 May 2019.
79. Tsai, C.H.; Wang, C.C. Unifying AoI Minimization and Remote Estimation-Optimal Sensor/Controller Coordination With Random Two-Way Delay. *IEEE/Acm Trans. Netw.* **2022**, *30*, 229–242. [CrossRef]
80. Sun, Y.; Cyr, B. Information Aging Through Queues: A Mutual Information Perspective. In Proceedings of the IEEE International Workshop on Signal Processing Advances in Wireless Communications (SPAWC), Kalamata, Greece, 25–28 June 2018.
81. Bastopcu, M.; Ulukus, S. Partial updates: Losing information for freshness. In Proceedings of the IEEE International Symposium on Information Theory, Los Angeles, CA, USA, 21–26 June 2020.
82. Wang, Z.; Badiu, M.A.; Coon, J.P. A Framework for Characterizing the Value of Information in Hidden Markov Models. *IEEE Trans. Inf. Theory* **2022**, *68*, 5203–5216. [CrossRef]
83. Shisher, M.; Qin, H.; Yang, L.; Yan, F.; Sun, Y. The age of correlated features in supervised learning based forecasting. *arXiv* **2021**, arXiv:2103.00092v2.
84. Zhong, J.; Yates, R.D. Timeliness in lossless block coding. In Proceedings of the IEEE Data Compression Confrtrnvr (DCC), Snowbird, UT, USA, 29 March–1 April 2016.

85. Feng, S.; Yang, J. Age-optimal transmission of rateless codes in an erasure channel. In Proceedings of the IEEE International Conference on Communications (ICC), Pudong, China, 20–24 May 2019.
86. Tang, H.; Ciblat, P.; Wang, J.; Wigger, M.; Yates, R. Age of information aware cache updating with file-and age-dependent update durations. In Proceedings of the Modeling and Optimization in Mobile, Ad Hoc and Wireless Networks, Volos, Greece, 15–19 June 2020.
87. Ma, M.; Wong, V.W. Age of information driven cache content update scheduling for dynamic contents in heterogeneous networks. *IEEE Trans. Wirel. Commun.* **2020**, *19*, 8427–8441. [CrossRef]
88. Bastopcu, M.; Ulukus, S. Information Freshness in Cache Updating Systems. *IEEE Trans. Wirel. Commun.* **2021**, *20*, 1861–1874. [CrossRef]
89. Zhang, M.; Arafa, A.; Huang, J.; Poor, H.V. How to price fresh data. In Proceedings of the Modeling and Optimization in Mobile, Ad Hoc and Wireless Networks, Avignon, France, 27–31 May 2019.
90. Yang, H.H.; Arafa, A.; Quek, T.Q.; Poor, H.V. Age-based scheduling policy for federated learning in mobile edge networks. In Proceedings of the IEEE International Conference on Acoustics, Speech and Signal Processing (ICASSP), Barcelona, Spain, 4–8 May 2020.
91. Buyukates, B.; Ulukus, S. Timely communication in federated learning. In Proceedings of the IEEE INFOCOM Workshops, Online, 10–13 May 2021.
92. Hespanha, J.P. Modelling and analysis of stochastic hybrid systems. *IEE Proc. Control Theory Appl.* **2006**, *153*, 520–535. [CrossRef]

Article

Optimizing Urgency of Information through Resource Constrained Joint Sensing and Transmission

Zhuoxuan Ju, Parisa Rafiee and Omur Ozel *

Department of Electrical and Computer Engineering, George Washington University, Washington, DC 20052, USA

* Correspondence: ozel@gwu.edu

Abstract: Applications requiring services from modern wireless networks, such as those involving remote control and supervision, call for maintaining the timeliness of information flows. Current research and development efforts for 5G, Internet of things, and artificial intelligence technologies will benefit from new notions of timeliness in designing novel sensing, computing, and transmission strategies. The age of information (AoI) metric and a recent related urgency of information (UoI) metric enable promising frameworks in this direction. In this paper, we consider UoI optimization in an interactive point-to-point system when the updating terminal is resource constrained to send updates and receive/sense the feedback of the status information at the receiver. We first propose a new system model that involves Gaussian distributed time increments at the receiving end to design interactive transmission and feedback sensing functions and develop a new notion of UoI suitable for this system. We then formulate the UoI optimization with a new objective function involving a weighted combination of urgency levels at the transmitting and receiving ends. By using a Lyapunov optimization framework, we obtain a decision strategy under energy resource constraints at both transmission and receiving/sensing and show that it can get arbitrarily close to the optimal solution. We numerically study performance comparisons and observe significant improvements with respect to benchmarks.

Keywords: urgency of information; information freshness; resource constraints; Lyapunov optimization

Citation: Ju, Z.; Rafiee, P.; Ozel, O. Optimizing Urgency of Information through Resource Constrained Joint Sensing and Transmission. *Entropy* **2022**, *24*, 1624. https://doi.org/10.3390/e24111624

Academic Editors: Anthony Ephremides and Yin Sun

Received: 7 October 2022
Accepted: 7 November 2022
Published: 9 November 2022

Publisher's Note: MDPI stays neutral with regard to jurisdictional claims in published maps and institutional affiliations.

1. Introduction

As demand from wireless networks exponentially increases to enable emerging technologies, the timeliness of data delivery and adaptation to the context of information becomes essential for improved quality of service and experience in time-sensitive applications. To this end, the measurement and improvement of the timeliness of data delivery and the effective adaptation to the context of delivered data have been fundamental challenges that researchers and practitioners have worked on actively in recent years. The age of information (AoI) is a well-known metric to measure the timeliness of data from the perspective of the nodes receiving or consuming data [1] and is expressed as the time elapsed since the generation of the latest received data. Although AoI has received much interest as a metric representing the freshness of information, new metrics are needed to address nonlinearity in the aging of data and time-varying value or context associated with flowing data. As a matter of fact, context-based applications (e.g., automatic driving and artificial intelligence) and nonlinear age [2–4] (as in many IoT applications) require a departure from AoI definition and analysis. Toward this end, the references [5,6] recently proposed an urgency of information (UoI) framework by combining the timeliness and context associated with information updates. In these papers, UoI was formally defined as the product of context-aware weight and the cost resulting from real-time estimation error in a Gaussian dynamical system, the latter being a well-known nonlinear function of AoI. UoI expression can be expressed in mathematical form as follows:

$$F(t) = w_t \delta(C(t)), \tag{1}$$

where w_t is the nonnegative coefficient representing the context or value at a specific time t, $\delta(.)$ is the cost function, and $C(t)$ is the instantaneous cost measuring the urgency. This formulation subsumes the typical definition of AoI. If $C(t)$ increases by one each time an update is not received, then the common AoI problem can be formulated as $w_t = 1$ and $\delta(Q(t)) = U(t)C(t)$, where $U(t)$ is an indicator that shows whether the information is updated or not. In our current paper, we will pursue a similar metric whereby the urgency level is represented by a coefficient w_t, which will be set as an independent, identically distributed random process that shows how crucial the status information is at a specific moment t. In addition, we will pursue a quadratic cost function. This formulation enables us to analyze error increments and connect the proposed framework to the classical AoI problem.

The UoI framework in this paper will be designed to measure the expected performance degradation as a weighted sum of expected staleness or informativeness of the latest sensed Gaussian process at the receiving end with respect to the transmitter and the lack of synchrony between them, maintained by status updating from the transmitting end to the receiving end. Our goal is to build a systematic understanding on the interaction of feedback sensing and update transmission to maintain improved UoI levels measuring the synchrony and informativeness of information at one side about the other side when both actions are resource constrained. We will employ Lyapunov optimization tools to address this crucial problem.

Lyapunov optimization methods and tools have been well-known to various research communities to control queues and more generally dynamical systems in a near-optimal sense. In the context of queuing theory, the state of a system at a particular time is the vector of realizations of error variables which can easily be brought in queue forms by lower bounding it by zero and studied for upper bounding the optimal cost. Typically, the cost function is defined to take smaller value when the system moves toward the desirable states. System stability is achieved by taking control actions that make the Lyapunov drift in the negative direction toward zero. The key requirement is that all the queues and virtual queues in the system are mean rate stable [7,8]. In addition, the target function is achieved by taking control actions that minimize the Lyapunov penalty. However, because of the system stability awareness, the solution always has a gap with the optimal solution. Due to its general applicability in queuing theory, Lyapunov optimization is also used in AoI analysis and optimization. Ref. [9] used Lyapunov optimization to identify the tradeoff between AoI, accuracy, and completeness with the constrained throughput optimization problem. Ref. [10] used Lyapunov optimization to jointly minimize the average cost of sampling and transmitting status updates by users over a wireless channel subject to average AoI constraints.

Our work's motivation is rooted in AoI research that was presented in the recent past. We next aim to cover some of the literature that relates to the proposed research in this paper. The references [11,12] address varying source update frequency and [13,14] address service rate in various queuing models. In the wireless network scenario, the scheduling algorithms for optimizing AoI is studied extensively, such as those considering the channel state [15,16], throughput [17–19], energy harvesting [20–22], and average resource constraints [23,24], multiple sources [25–28], and multiple channels [29–32] to name a few. Ref. [33] studied the calculation and iterative process of AoI in combination with queuing theory and gave the analytic formula of average AoI under the random scheduling strategy. Ref. [34] explored the impact of service rate on average AoI under fixed deadline constraints and random exponential deadline constraints. Regarding link scheduling in wireless networks, ref. [35] studied the link-scheduling problem for every time slot under periodic data updates, and proposed random, greedy, Lyapunov optimization, Whittle Index, and other strategies for link scheduling to optimize the average AoI of the network. Ref. [36] proposed offline and online scheduling algorithms based on the Markov decision process for the random data arrival scenario.

Feedback is also an essential factor in wireless communication scenarios and can influence the AoI performance significantly. In particular, it is well known that the feedback may help maintain expedient processing, non-repetitive transmission, and hence, energy efficiency in wireless transmission. For the case of battery-based non-energy harvesting devices, it is also vital to schedule appropriate transmission and sensing strategies to prolong the device's life. As a result, the role of feedback and energy cost in AoI analysis and optimization has received much interest from the research community (see e.g., [37–41]). Additionally, ref. [42] proves that the AoI and energy-harvesting scheduling strongly differ with or without the feedback. Refs. [43,44] minimized the AoI when the sensor uses ON/OFF schemes with energy harvesting nodes. Ref. [45] focused on the extreme cases of one unit battery and infinite battery situations to minimize the average peak AoI with energy constraints. Most recently, the paper [46] provided an analysis of feedback cost in AoI optimization over a point-to-point channel and determined specific conditions when feedback may or may not be useful for AoI optimization.

Decisions to sense/receive updates under energy constraints have also been of interest to AoI researchers. In particular, energy constraints can limit the chance of sensing new data and hence cause AoI to increase. In this context, refs. [47,48] proposed the joint scheduling of sense and transmission schemes to optimize the average peak AoI in an energy-harvesting system. In this paper, we will combine the concept of feedback and sensing, which means that the system will decide whether to sense the feedback information as input. As other related research, refs. [49,50] studied the value of information (VoI) in status update systems, and compared the performance of VoI with AoI. We also refer the reader to the related paper [51]. Based on the idea that AoI is only important when the receiver performs a query, refs. [52,53] proposed the age of information at query (QAoI) and optimized the QAoI.

In this paper, we will extend the UoI optimization framework in [6] to an interactive scenario by considering sensing/receiving costs at the updating terminal under energy resource constraint by using a Lyapunov optimization framework. Resource constraints in receiving/sensing the feedback can be interpreted as a limitation due to processing or energy to make it available for decision making on update transmission. Our motivation can be compared to that of [46] as well, which assumes the cost of feedback is incurred at the receiving end. This new problem calls for coordinated decisions to sense the feedback from the receiver and transmit the update to the receiver. Additionally, we need to account for relativity with respect to the transmitter and receiver sides and measure urgency by using a weight representing their importance under resource constraints. Our framework will address these new issues.

As the main contributions of this paper, we extend the UoI optimization framework by using a new definition that addresses the interactive nature of the setting when transmitting and receiving/sensing information is costly and average resource constraints are present on both actions. Constructing the objective function by assigning different weights to the urgency levels at the transmitting and receiving terminals, we determine jointly optimal scheduling of transmission and receiving/sensing the feedback by using a Lyapunov optimization framework. We obtain the Lyapunov gap and show that the result can be made arbitrarily close to the optimal solution. Our simulation results show that the proposed algorithm performs significantly better than two benchmark schemes, namely the greedy and AoI optimal algorithms.

The rest of the paper is organized as follows. In Section 2, we present the system model of the UoI problem. In Section 3, we formulate the UoI problem and analyze it. In Section 4, we offer numerical results to show the behavior of the solution. Finally, we conclude this paper in Section 5 by summarizing our contributions and discussing future directions.

2. System Model

We consider the system model in Figure 1. Here, the time is slotted: $t = 1, 2, \ldots, T$. The information-carrying signal in the service center and terminal, A_t and Q_t, are as follows:

$$A_{t+1} = (1 - U_2(t))A_t + K_t \tag{2}$$

$$Q_{t+1} = (1 - S_t U_1(t))Q_t + U_2(t)A_t. \tag{3}$$

The variable $K_t \sim \mathcal{N}(0, \sigma^2)$ represents the increments added to the information-carrying signal A_t and is a Gaussian random variable independent over time and other variables. For convenience, we take the variable $A_1, Q_1 \sim \mathcal{N}(0, \sigma^2)$; however, the initial conditions are assumed given and do not determine the outcome as long as they come from a well-behaving distribution that makes the expectations well defined (c.f. Lemma 3 below). $U_1(t), U_2(t) \in \{0, 1\}$ are decision variables to determine whether to transmit an update and sense the feedback, respectively. Equation (2) represents the evolution of information at the receiver with respect to the sensing at the transmitter. When $U_2(t) = 1$, the sensing action is activated and the information at both ends are synchronized except an additive Gaussian noise due to causality and one time slot difference. The Equation (3) represents the evolution of the information at the transmitter with respect to the receiver side. These two equations represent the interaction between the transmitter and the receiver. Note that if the transmission or sensing does not happen, i.e., if $U_1(t)S_t = 0$ or $U_2(t) = 0$, then Q_t or A_t, respectively, will become noisier. This is at the heart of the urgency of information notion we pursue in this paper. When a transmission does not happen (due to not transmitting or a channel erasure), the synchrony between the two sides, represented by Q_t, is not affected as long as a new sensing action is not taken. At the beginning of the tth time slot, the terminal first decides $U_1(t) \in \{0, 1\}$ to determine whether to transmit the information-carrying variable Q_t to the service center or not. The transmission takes one time slot and goes through an erasure-type wireless channel represented by S_t with a fixed failure transmission rate p. In particular, $S_t = 1$ if the transmission is successful and $S_t = 0$ otherwise. At the same time, the service center feeds back A_t to the terminal, which also takes one time slot with no failure rate. At the end of the tth time slot, the feedback arrives at the terminal, and the terminal will decide $U_2(t) \in \{0, 1\}$ to determine whether to sense the feedback or not. We can, in principle, let A_t and Q_t evolve as $\max\{(1 - U_2(t))A_t + K_t, 0\}$ and $\max\{(1 - S_t U_1(t))Q_t + U_2(t)A_t, 0\}$ with nonnegative initial values. These versions bring these system states to the form of queues with potentially dependent arrivals and departures. Our Lyapunov drift plus penalty-based analysis will be applicable for both versions. We therefore prefer to keep them as in (2) and (3) in the ensuing analysis.

Now we can elicit our optimization problem P_1 to minimize an upper bound of average UoI:

$$\min_{\pi_t} \lim_{T \to \infty} \sup \frac{1}{T} \sum_{t=0}^{T-1} E\left[w_t(Q_t^2 + MA_t^2)\right] \tag{4}$$

$$\text{s.t} \quad \lim_{T \to \infty} \sup \frac{1}{T} \sum_{t=0}^{T-1} E[U_1(t)] \leq \varphi_1 \tag{5}$$

$$\lim_{T \to \infty} \sup \frac{1}{T} \sum_{t=0}^{T-1} E[U_2(t)] \leq \varphi_2, \tag{6}$$

where π_t is the set of sequence of decisions $\pi_t = \{U_1(t), U_2(t)\}$, w_t is the nonnegative weight of urgency modeled as an i.i.d. random variable, M is the weight of the relative error of the variable A_t at the transmitter side, φ_1 is the energy (or frequency) constraint on transmission, and φ_2 is the energy (or frequency) constraint on sensing. In order to satisfy the average transmission/sensing frequency constraints (5) and (6), we define the virtual queues H_t and G_t as follows, which are both initialized at 0:

$$H_{t+1} = \max\{H_t - \varphi_1 + U_1(t), 0\} \tag{7}$$

$$G_{t+1} = \max\{G_t - \varphi_2 + U_2(t), 0\}. \tag{8}$$

Next, let us consider the evolution of the transmission virtual queue: If the terminal decides to transmit at time slot t, the transmission virtual queue H_t will increase by $1 - \varphi_1$. Otherwise, it will decrease by φ_1. As a result, the longer the virtual queue, the more transfers will be performed. The virtual queue of sensing G_t evolves similarly. Therefore, these two virtual queues can appropriately express the usage of the historical transmission/sensing frequency.

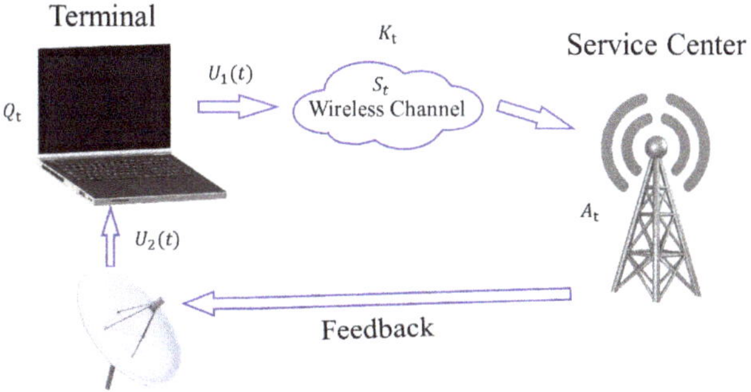

Figure 1. System model with joint transmission and feedback reception.

3. Optimizing the Urgency of Information in P_1

In this section, we will systematically develop a Lyapunov optimization framework for optimizing an upper bound for the solution of P_1. We summarize the notations we use throughout the rest of the paper in Table 1.

Table 1. Definitions of Variables.

Symbol	Description
A_t	Error in Service Center
Q_t	Error in Terminal
H_t	Transmission Frequency Virtual Queue
G_t	Sensing Frequency Virtual Queue
n_t	Number of Time Slots since Last Sense
S_t	Channel Situation
$U_1(t), U_2(t)$	Transmission/Sensing Decision
π_t	Set of Decisions
φ_1, φ_2	Transmission/Sensing Frequency Constraints
w_t	Weight of Urgency
$\tilde{w}$	Average Weight of Urgency
V, Z, θ, β	Weight of $H_t^2, G_t^2, Q_t^2, A_t^2$ in Drift Function
M	Weight of A_t^2 in Target Function
R	Weight of Penalty Compared with Drift
L_t	Summation of all the Queues
Δ_t	Lyapunov Drift
f_t	Lyapunov Penalty
Y_t	Set of Given Parameters in t_{th} Time Slot

3.1. Lyapunov Function Definitions

In order to use the Lyapunov optimization framework, we will first define the Lyapunov drift function Δ_t by using the quadratic sum of system states:

$$L_t = \frac{1}{2}VH_t^2 + \frac{1}{2}ZG_t^2 + \frac{1}{2}\theta Q_t^2 + \frac{1}{2}\beta A_t^2, \tag{9}$$

where V, Z, θ and β are the weights for different variables, which represent different importance levels of the stability of the queues or system states H_t, G_t, Q_t and A_t, respectively. In our analysis, we use the terms "queue" or "system state" interchangeably. Although the evolution of A_t and Q_t in (2) and (3) can take negative values, we can redefine them by lower bounding their evolution by zero and make their definitions suitable as a queue with arrivals and departures potentially depending on the control actions. However, none of the analysis steps we take in this paper will be affected by this redefinition, as the Lyapunov analysis we present essentially optimizes a bound on the system performance. We therefore continue using the original definitions (2) and (3). The Lyapunov drift function for this system can be expressed as

$$\Delta_t = E[L_{t+1} - L_t | Q_t, n_t, H_t, G_t, w_{t+1}], \tag{10}$$

where n_t is the number of time slots since the last time we decide to sense the feedback. It is obvious that in tth time slot, the terminal has a knowledge of H_t, G_t, Q_t. However, the terminal cannot access the specific value of A_t because the latest estimation error arrived at the service center at the end of $(t-1)$st time slot. Nevertheless, the terminal is aware of the number of time slot since the last time it decides to sense n_t, which can be expressed as:

$$n_{t+1} = (1 - U_2(t))n_t + 1. \tag{11}$$

As a result, the terminal will decide whether to sense based on the number of time slot since the last time it decided to sense n_t rather than the error in the service center A_t.

Lemma 1. *In each time slot t, given the error in terminal Q_t, urgency weight at the next time slot w_{t+1}, the number of time slots since the last time terminal decides to sense n_t, virtual queue length H_t and G_t, set $Y_t = \{Q_t, n_t, H_t, G_t, w_{t+1}\}$, we can obtain an upper bound on the Lyapunov drift Δ_t as*

$$\Delta_t \leq \frac{1}{2}(V + Z) + \frac{1}{2}\beta\sigma^2 - V\varphi_1 H_t - Z\varphi_2 G_t + (VH_t - \frac{1}{2}\theta p Q_t^2)E[U_1(t)|Y_t]$$
$$+ (ZG_t + \frac{1}{2}\theta n_t\sigma^2 - \frac{1}{2}\beta n_t\sigma^2)E[U_2(t)|Y_t]. \tag{12}$$

Proof. See Appendix A. $\square$

Denote the penalty in the tth time slot by f_t. Because of causality, $U_1(t), U_2(t)$ will affect UoI in $(t+1)$st time slot. Therefore, we let $f_t = Rw_{t+1}(Q_{t+1}^2 + MA_{t+1}^2)$, where R is the weight of the UoI compared with system stability and the remaining terms represent UoI at $t+1$.

Lemma 2. *If we set the penalty in the tth time slot as $f_t = Rw_{t+1}(Q_{t+1}^2 + MA_{t+1}^2)$, and the average of the weight of the urgency as $\tilde{w}$. The Lyapunov drift plus penalty function is upper bounded as:*

$$\Delta_t + E[f_t|Y_t] \le \frac{1}{2}(V+Z) + \frac{1}{2}\beta\sigma^2 + R\tilde{w}(Q_t^2 + M\sigma^2 + Mn_t\sigma^2) - V\varphi_1 H_t - Z\varphi_2 G_t$$
$$+ (VH_t - \frac{1}{2}\theta p Q_t^2 - R\tilde{w}pQ_t^2)E[U_1(t)|Y_t]$$
$$+ (ZG_t + \frac{1}{2}\theta n_t\sigma^2 - \frac{1}{2}\beta n_t\sigma^2 + (1-M)R\tilde{w}n_t\sigma^2)E[U_2(t)|Y_t]. \tag{13}$$

Proof. See Appendix B. $\square$

Lemma 3. *If $E[L_0] < \infty$, and $\Delta_t + E[f_t] \le C$, where C is a constant, then all the queues and virtual queues in the system are mean rate stable.*

Proof. See Appendix C. $\square$

3.2. Finding Appropriate Weights for the System

Next, we are going to find the optimal value of the weight parameters θ and β to minimize the right hand side of (13) to the extent possible. Note that it is feasible to use a stationary randomized scheme that independently transmits and senses with probability φ_1 and φ_2 at each time slot, which translates to $E[U_1(t)] = \varphi_1$ and $E[U_2(t)] = \varphi_2$. As a result, we reorder (13) to get

$$E[L_{t+1} - L_t + f_t|Y_t] \le (R\tilde{w}M + \frac{1}{2}\beta)\sigma^2 + \frac{1}{2}(V+Z) + (-\frac{1}{2}\theta p\varphi_1 - R\tilde{w}p\varphi_1 + R\tilde{w})Q_t^2$$
$$+ (\frac{1}{2}(\theta - \beta)n_t\sigma^2 + R\tilde{w}M + (1-M)R\tilde{w}\varphi_2)n_t\sigma^2. \tag{14}$$

To make the right hand side of (14) no larger than a constant, we want the coefficients of Q_t^2 and $n_t\sigma^2$ no larger than 0. For the coefficient of Q_t^2,

$$-\frac{1}{2}\theta p\varphi_1 - R\tilde{w}p\varphi_1 + R\tilde{w} \le 0$$
$$\theta \ge \frac{2}{p\varphi_1}(1 - p\varphi_1)R\tilde{w}. \tag{15}$$

For the coefficient of $n_t\sigma^2$,

$$\frac{1}{2}(\theta - \beta)n_t\sigma^2 + R\tilde{w}M + (1-M)R\tilde{w}\varphi_2 \le 0$$
$$\beta \ge \theta + 2(\frac{1}{\varphi_2} - 1)R\tilde{w}M + 2R\tilde{w}. \tag{16}$$

As a result, we take the value of the parameters θ and β as

$$\theta = \frac{2}{p\varphi_1}(1 - p\varphi_1)R\tilde{w} \tag{17}$$

$$\beta = \frac{2}{p\varphi_1}R\tilde{w} + 2(\frac{1}{\varphi_2} - 1)R\tilde{w}M. \tag{18}$$

Put the value of the parameters θ and β back to (14), then we can get the upper bound of $E[L_{t+1} - L_t + f_t|Y_t]$ as

$$E[L_{t+1} - L_t + f_t|Y_t] \le (\frac{1}{p\varphi_1} + \frac{M}{\varphi_2})R\tilde{w}\sigma^2 + \frac{1}{2}(V+Z). \tag{19}$$

Note that the right hand of (19) is a constant, which means that all the queues and virtual queues in the system are mean rate stable under above derived conditions.

3.3. Deriving Lyapunov Optimal Decisions

We now minimize the upper bound in the RHS of (13), which is actually in the following form:

$$\min_{\pi_t} \ (VH_t - \frac{1}{2}\theta p Q_t^2 - Rw_{t+1}pQ_t^2)U_1(t)$$
$$+ (ZG_t + \frac{1}{2}(\theta - \beta)n_t\sigma^2 + (1-M)Rw_{t+1}n_t\sigma^2)U_2(t). \tag{20}$$

We next show the scheduling scheme for each time slot. Putting the value of the parameters θ and β back to (20), we get the following:

$$\min_{\pi_t} \ [VH_t - (w_{t+1} - \tilde{w} + \frac{\tilde{w}}{p\varphi_1})RpQ_t^2]U_1(t)$$
$$+ [ZG_t + ((M-1)\left(\tilde{w} - w_{t+1}\right) - \frac{\tilde{w}M}{\varphi_2})Rn_t\sigma^2]U_2(t). \tag{21}$$

Set the update index $a_t = VH_t - (w_{t+1} - \tilde{w} + \frac{\tilde{w}}{p\varphi_1})RpQ_t^2$ and update index $b_t = ZG_t + ((M-1)\left(\tilde{w} - w_{t+1}\right) - \frac{\tilde{w}M}{\varphi_2})Rn_t\sigma^2$, and then the solution to the scheme (20) can be achieved:

$$U_1(t) = \begin{cases} 1 & , a_t < 0 \\ 0 & , \text{otherwise} \end{cases} \tag{22a} \tag{22b}$$

$$U_2(t) = \begin{cases} 1 & , b_t < 0 \\ 0 & , \text{otherwise.} \end{cases} \tag{23a} \tag{23b}$$

We summarize below the resulting Lyapunov optimal Algorithm 1.

Algorithm 1 Decisions scheduling scheme based on Lyapunov optimization

---

Require: $A_0, Q_0, H_0, G_0, n_0, S_t, K_t, \varphi_1, \varphi_2, w_t, \tilde{w}, V, Z, M, R$
1: **for** each time slot t **do**
2:    Calculate $a_t = VH_t - (w_{t+1} - \tilde{w} + \frac{\tilde{w}}{p\varphi_1})RpQ_t^2$;
3:    Calculate $b_t = ZG_t + ((M-1)\left(\tilde{w} - w_{t+1}\right) - \frac{\tilde{w}M}{\varphi_2})Rn_t\sigma^2$;
4:    **if** $a_t < 0$ **then**
5:      $U_1(t) = 1$
6:    **else**
7:      $U_1(t) = 0$;
8:    **end if**
9:    **if** $b_t < 0$ **then**
10:      $U_2(t) = 1$;
11:    **else**
12:      $U_2(t) = 0$;
13:    **end if**
14:    Calculate $A_{t+1} = (1 - U_2(t))A_t + K_t$;
15:    Calculate $Q_{t+1} = (1 - S_t U_1(t))Q_t + U_2(t)A_t$;
16:    Calculate $H_{t+1} = \max\{H_t - \varphi_1 + U_1(t), 0\}$;
17:    Calculate $G_{t+1} = \max\{G_t - \varphi_2 + U_2(t), 0\}$;
18:    Calculate $n_{t+1} = (1 - U_2(t))n_t + 1$;
19: **end for**

Based on the algorithm, we can make decisions by scheduling every time slot to minimize the value of UoI and maintain the virtual queue stability simultaneously. From the algorithm, it is apparent that we can successfully decouple the joint decisions into two independent threshold schemes, which makes the implementation desirably simple.

3.4. Solving for the Target Function and Lyapunov Gap

In this section, we will solve for the target function and achieve the expression of the gap between the optimal solution and the result obtained by the Lyapunov optimization algorithm. We will also prove that the result gained by the Lyapunov optimization algorithm can be infinitely close to the optimal solution. Now make the summation of the total T-time slot on both sides of (19), and we can get

$$E\left[L_T - L_0 + \sum_{t=0}^{T-1} f_t\right] \le T\left[(\frac{1}{p\varphi_1} + \frac{M}{\varphi_2})R\tilde{w}\sigma^2 + \frac{1}{2}(V+Z)\right]. \tag{24}$$

Note that $L_T \ge 0$ and $\frac{L_0}{T} = 0$, and then divide T on both sides of (24) to get the time-averaged result

$$\frac{1}{T}E\left[\sum_{t=0}^{T-1} f_t\right] \le (\frac{1}{p\varphi_1} + \frac{M}{\varphi_2})R\tilde{w}\sigma^2 + \frac{1}{2}(V+Z). \tag{25}$$

Theorem 1. *Set the problem of* (20) *as* $P_2(\pi_t)$*, and then the solution of* $P_2(\pi_t)$ *will satisfy the following gap:*

$$\frac{1}{T}\sum_{t=0}^{T-1} E\left[w_t(Q_t^2 + MA_t^2)\right] \le (\frac{1}{p\varphi_1} + \frac{M}{\varphi_2})\tilde{w}\sigma^2 + \frac{(V+Z)}{2R} \tag{26}$$

That is, the solution of $P_2(\pi_t)$ *can be approximated by the solution of* $P_1(\pi_t)$*, and the gap between them is* $\frac{(V+Z)}{2R}$*.*

Proof. See Appendix D. □

To be precise, the proof of this gap result in Appendix D requires A_t and Q_t in (2) and (3) to be lower bounded by zero. Nevertheless, our numerical results show consistence with this gap even when they are non-negative. Note that as the value of R is taken as large as possible, and the result obtained by the Lyapunov optimization algorithm $P_2(\pi_t)$ can be made arbitrarily close to the optimal result $P_1^*(\pi_t)$. $\frac{(V+Z)}{2R}$ can also seem to be the ratio of the weight of the energy constraints and UoI, which shows the tradeoff between the UoI and the energy constraints.

4. Numerical Results

In this section, we present extensive numerical results to explore the behaviour of the optimal scheme under various constraints and scenarios. At the beginning of each time slot, the terminal first decides whether to transmit the error packets to the service center or not. The transmission takes 1 ms and goes through a wireless channel with a fixed failure transmission rate. At the same time, the service center transmits the estimation error (feedback) to the terminal, which also takes 1 ms with no failure rate. At the end of each time slot, the feedback arrives at the terminal, and the terminal will decide whether to sense this feedback. Meanwhile, the service channel receives the error packets and the latest estimation of Gaussian noise. The service center will immediately calculate the error difference between the transmitted status information and the received status information and add that new error into the error packet.

4.1. Response to Urgency Levels

To demonstrate the system's response to a new urgency, for every 5000 time slots, we set W = 100 in the 50 consecutive time slots and W = 1 in the rest of the time slots. The transmission/sense energy constraints are set as $\varphi_1 = 0.25$ and $\varphi_2 = 0.5$. The channel error rate is $p = 0.8$, the weight of the UoI is set as $M = 2.5$ and $R = 2$, and the weight of the system states is set as $V = Z = 1$. Additionally, the Gaussian noise variance will be set to unity. Figures 2–4 show a sample evolution of the squared of errors $MA_t^2 + Q_t^2$ and two virtual queue length H_t, G_t. Observing Figures 2–4, we understand that when the urgency level rises, the square of errors will drop significantly, and the virtual queues will keep increasing because update transmissions are ramped up. However, due to the energy constraints, the terminal's probability of transmitting and sensing are affected. This is the reason why the square of errors will increase, and the transmission virtual queue will decrease after the urgency. These show that the system can swiftly respond to urgency levels while keeping the error variance portion of UoI (i.e., $Q_t^2 + MA_t^2$) at a reasonable level at all times.

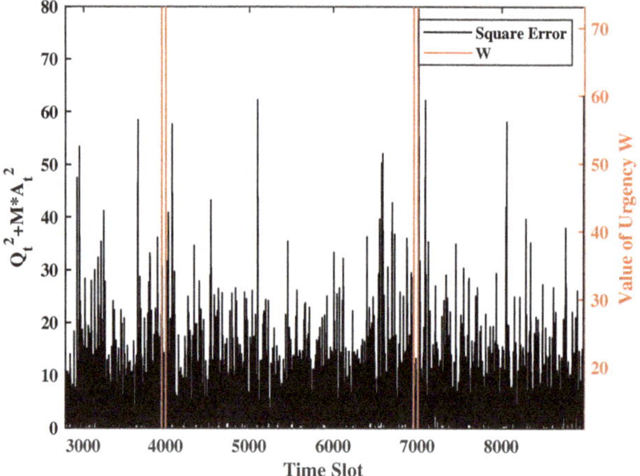

Figure 2. UoI sequence obtained by the proposed Lyapunov algorithm under a specific realization of weights w_t.

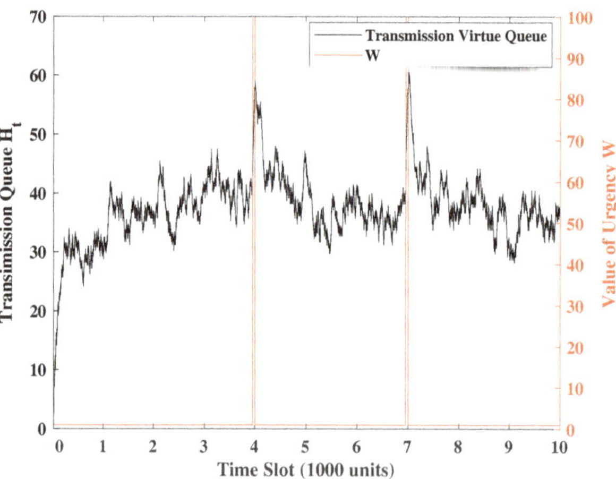

Figure 3. Transmission virtual queue under the same realization of weight w_t in Figure 2.

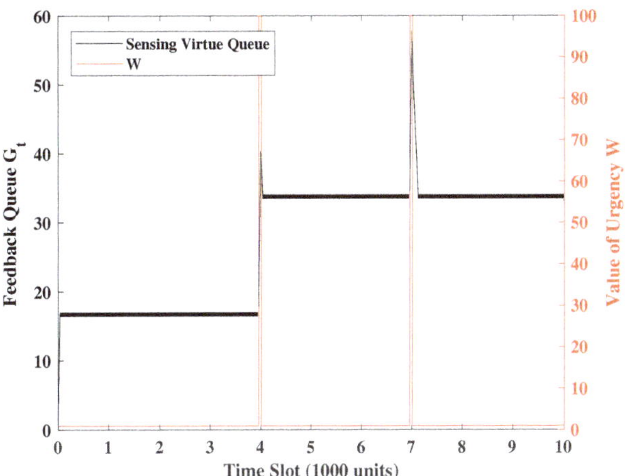

Figure 4. Sensing virtual queue under the same realization of weight w_t in Figure 2.

4.2. Tradeoff between UoI and System Parameters

In this section, we compare how the relationship between different variables will affect the UoI in the system. Unless otherwise specified, we set the energy constraint of transmission/sensing as $\varphi_1 = \varphi_2 = 0.8$, the weight of the system stability as $V = Z = 1$, the weight of totally UoI and the UoI in service center as $R = M = 2$, the channel error rate as $p = 0.8$, and the weight of urgency at each time slot is i.i.d. with probability 0.99 being 1 and probability 0.01 being 100.

Figures 5 and 6 present the relationship between UoI and transmission/sense energy constraint. They also show the effect of system stability weights on UoI. In Figure 5, the energy constraint of transmission ranges from 0.1 to 1.0, and the weight of the queue stability (i.e., the virtual queue levels) in the transmission part will be set as $V = 1, 10, 100$, and 1000. Similarly, in Figure 6, we set the energy constraint of sensing from 0.1 to 1.0 and $Z = 1, 10, 100$, and 1000. We observe that when average energy is less constrained, the UoI decreases. However, the UoI will not change much when the transmission frequency reaches 0.5. This is due to the fact that the frequency constraint becomes inactive after a certain level depending on the sensing activity. As sensing and transmission are in tandem, the higher frequency drives the overall performance. Moreover, when the weight of the stability V and Z are small, e.g., $V = 1$ or $Z = 1$, we pay more attention to the value of UoI than the frequency of transmission levels, yielding a virtual queue significantly above the set constraint. On the other hand, if we set the weight of the stability V and Z at a high level, e.g., $V = 1000$ or $Z = 1000$, the virtual queue stability becomes much more important, which compromises UoI performance.

In Figure 7, the energy constraint of transmission will be set from 0.1 to 1.0, and the failure probability of transmission will be set as $p = 0.2, 0.4, 0.6, 0.8$, and 1.0. We observe that the higher p is, the lower the average UoI is. This is because we need to decide to transmit more frequently to achieve the optimal average of UoI when the success rate is lower. In Figure 8, we observe that the average UoI decreases no matter whether φ_1 or φ_2 increases because we have more chances to transmit or sense when the energy is sufficient. Additionally, as φ_2 gets smaller, the curve will converge earlier because the error packets in the service center are the input of the terminal. When we have less probability of sensing the feedback, the transmission frequency will also not be large because of the input limitation, even if the transmission energy is sufficient.

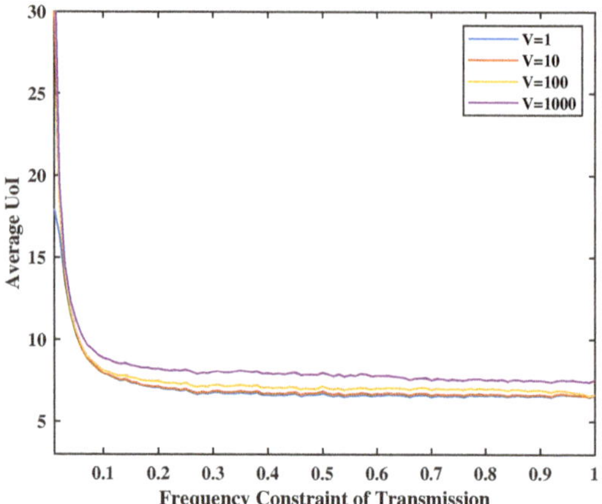

Figure 5. Tradeoff between transmission energy constraint, V and UoI.

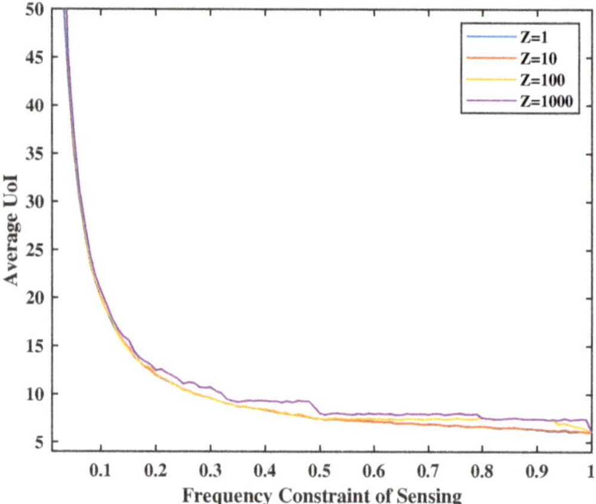

Figure 6. Tradeoff between sense energy constraint, Z and UoI.

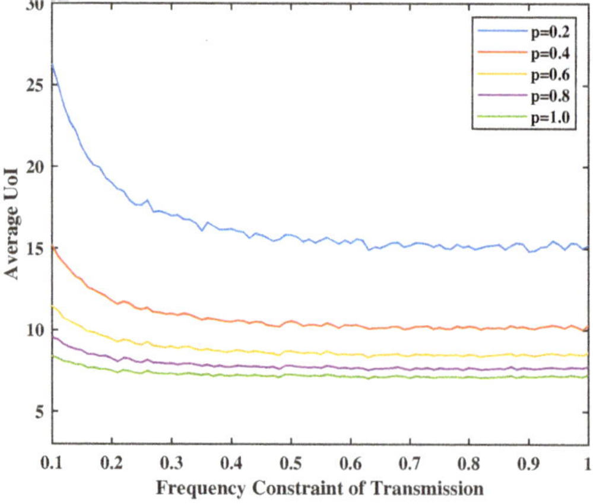

Figure 7. Tradeoff between transmission energy constraint, channel failure rate p, and UoI.

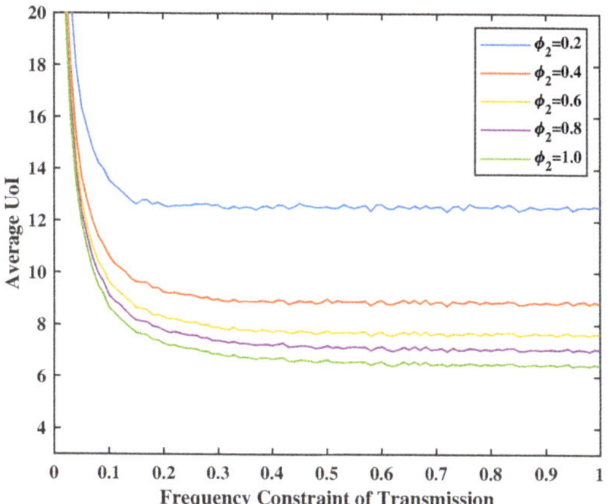

Figure 8. Tradeoff between transmission energy constraint, sensing energy constraint, and UoI.

In Figure 9, we set the weight of total UoI as $R = 1, 8, 16$ and 64, and the weight of two virtual queues as $V = Z = 20$. As expected, the larger the weight of the total UoI is, the smaller the average UoI will be. This is because the system will consider the UoI more important and will take more chances to transmit and sense. Moreover, the Lyapunov gap, i.e., $\frac{(V+Z)}{2R}$, will diminish as R increases.

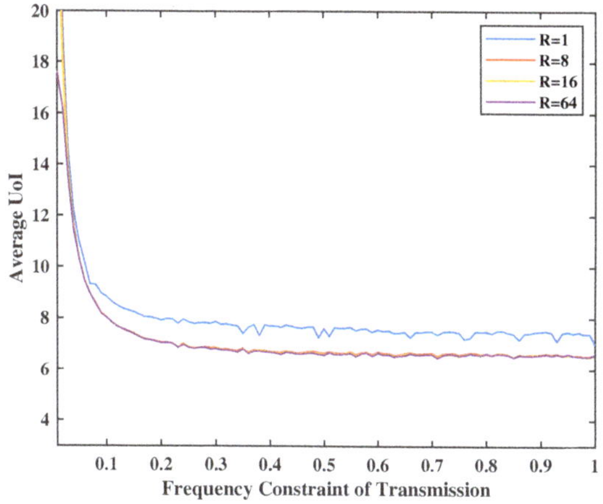

Figure 9. Tradeoff between transmission energy constraint, R and UoI.

4.3. Tradeoff between UoI and System Stability

The tradeoff between the target function and the system stability is always an exciting and crucial question in the Lyapunov optimization framework. This section will show examples of how different weights can affect the system stability and UoI. We set $T = 10,000$ and channel error rate as $p = 0.8$. The urgency weight w_t is determined as an i.i.d. random process with probability 0.99 being 1 and probability 0.01 being 1000. We will observe the number of update transmissions and senses (i.e. the energies spent for update transmission and sensing throughout $T = 10,000$ slots) to represent system stability.

In Figures 10 and 11, we set the weight of the system stability as $V = Z = 10$, and the weight of UoI as $R = M = 2$. As the energy is sufficient, we can have more chances to transmit and sense. In addition, the number of transmissions is always smaller or equal to the number of senses. This makes sense because the input error in the terminal comes from the service center and will be sent together in one transmission. In addition, even if there is no energy constraint for the transmission, e.g., $\varphi_1 = 1.0$, the number of transmissions will not reach the value of constraints. This is due to the fact that the frequency constraint becomes inactive after a certain level. However, when $\varphi_2 \leq 0.2$, the energy spent for sensing goes above the set energy constraints. The reason is that the weight of UoI is much larger than the weight of stability. This means that the system will sacrifice stability for better UoI.

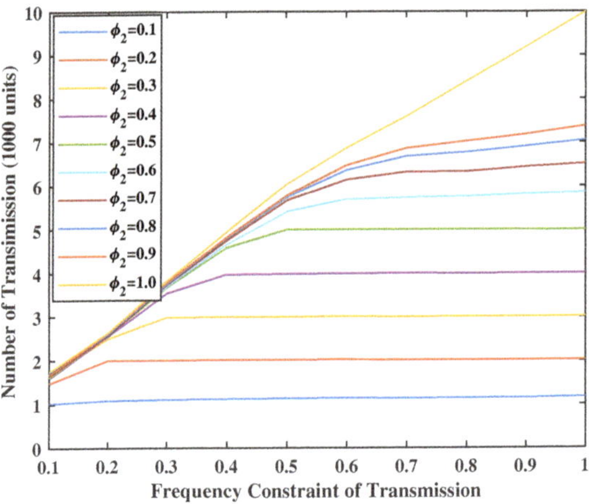

Figure 10. Energy spent for update transmission when V = Z = 10.

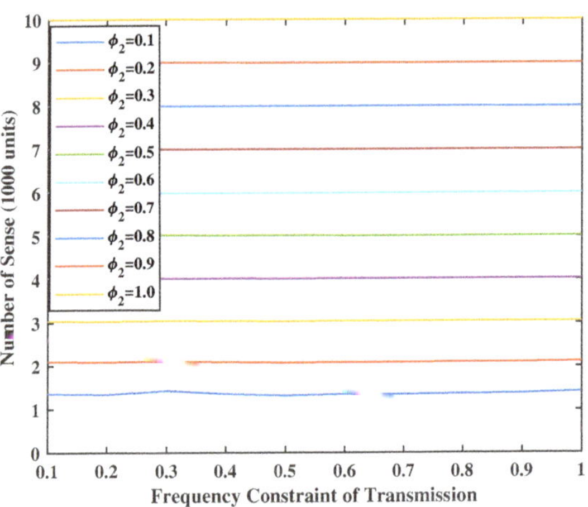

Figure 11. Energy spent for sensing when V = Z = 10.

In Figures 12 and 13, we set the weight of the system stability as $V = Z = 80$, which is larger than the weight of UoI. We see that both the transmission and sensing constraints are not binding. Comparing with the Figures 5 and 6, we observe that the UoI with $V, Z = 100$ is close to the UoI with $V, Z = 1$. Hence, by sacrificing a small amount of UoI, a very stable system can be guaranteed.

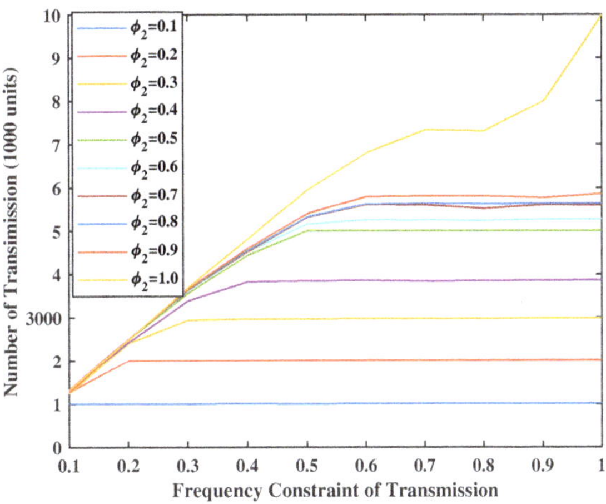

Figure 12. Energy spent for update transmission when V = Z = 80.

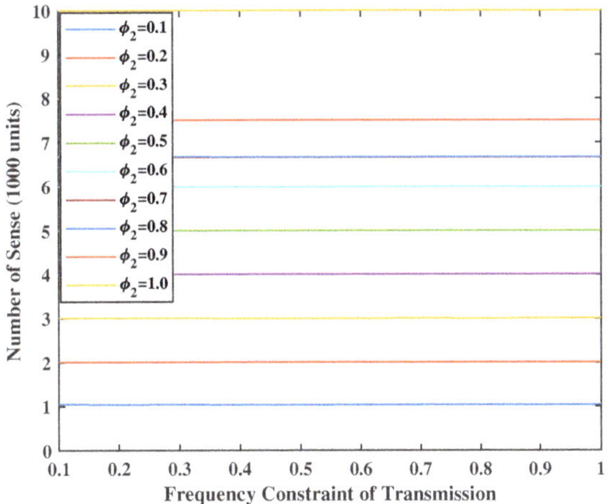

Figure 13. Energy spent for sensing when V = Z = 80.

In Figure 14, we set the weight of the UoI in service center as $M \in [1, 10]$, the weights $Z = 8, 16, 32, 64$, and 128, and $V = 5$, $\varphi_1 = 0.5$ and $\varphi_2 = 0.8$. The virtual queue G_t is small when its weight is large, and the energy constraints are tight. In addition, when the weight of UoI in the service center M increases, the sensing time will keep increasing because the UoI in the service center is much more important than the virtual queue stability and the information in the terminal. This also exemplifies that our framework can accommodate different cases flexibly by using different weights.

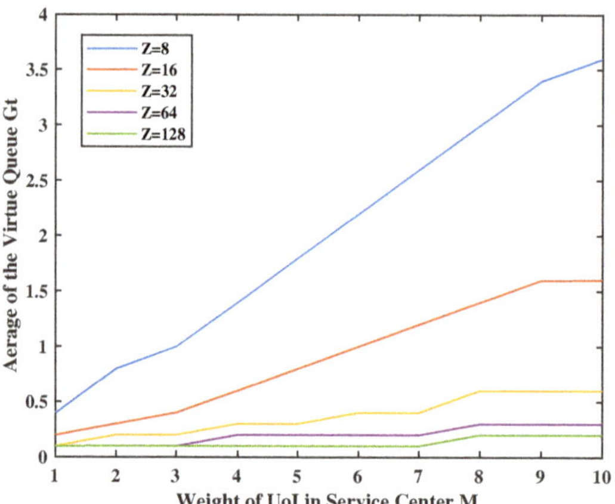

Figure 14. Tradeoff between M, Z and sensing frequency.

4.4. Comparison of Lyapunov Optimal Performance with Other Algorithms

The greedy and probabilistic algorithms are also very suitable naive algorithms to solve this problem. The main idea of the greedy algorithm is that the terminal will decide to transmit/sense at time t if the instantaneous transmission/sensing frequency at time t has not reached the corresponding set limits. Moreover, for the probabilistic algorithm, in each time slot the terminal will transmit/sense with probability equal to the value of frequency constraints. For Figure 15, we set the weight of system stability as $V = Z = 30$. Channel success rate is set as $p = 0.6$, sense energy constraint is set as $\varphi_2 = 0.8$, and the weight of urgency at each time slot is the same as before. Because the greedy algorithm takes action independent of urgency, we will compare the average UoI with $w_t = 1$. From the figure, the average error portion of UoI (i.e., $Q_t^2 + MA_t^2$) obtained by Lyapunov optimization is always lower than the other two algorithms, especially when the energy is insufficient and the gap closes with increasing energy availability.

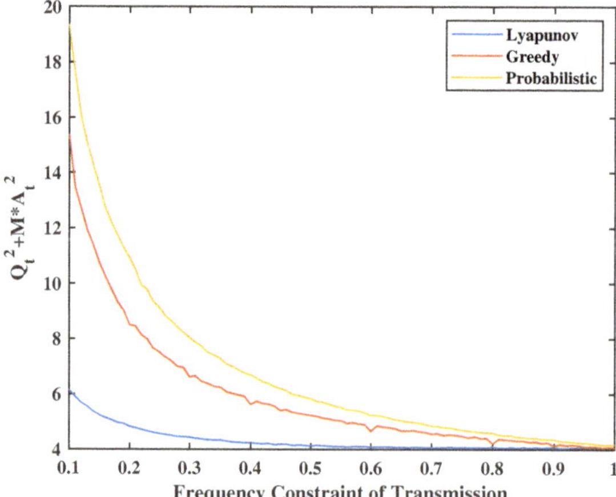

Figure 15. Lyapunov optimal algorithm and greedy algorithm.

Recall that UoI can subsume various AoI problems. For instance, if we set the cost function $\delta(.)$ as a linear function with the unit parameter and the urgency weight $w_t = 1$, then we can express the AoI in terminal $\tilde{Q}_t$ and the AoI in service center $\tilde{A}_t$ as

$$\tilde{A}_{t+1} = (1 - U_2(t))\tilde{A}_t + 1 \tag{27}$$

$$\tilde{Q}_{t+1} = (1 - S(t)U_1(t))\tilde{Q}_t + U_2(t)\tilde{A}_t. \tag{28}$$

Let us use the same Lyapunov optimization algorithm described earlier along with the same weights for system state variables and target function for a fair comparison. In Figure 16, we set the weight of virtual queues as $V = Z = 20$, the weight of UoI as $R = M = 2$, the weight of system states θ, β in AoI optimal will be the same as the value of UoI optimal and will be calculated each round. Additionally, set the probability of fail transmission as $p = 0.6$, sensing frequency limitation as $\varphi_2 = 0.6$. We can deduce that the average UoI obtained by UoI optimal is much better than the value obtained by AoI optimal. In addition, the value of average UoI by UoI optimal is smaller than that of the average weighted AoI by AoI optimal. This is because, in the AoI model, the increment K_t will always be 1; however, the UoI model yields a lower expectation.

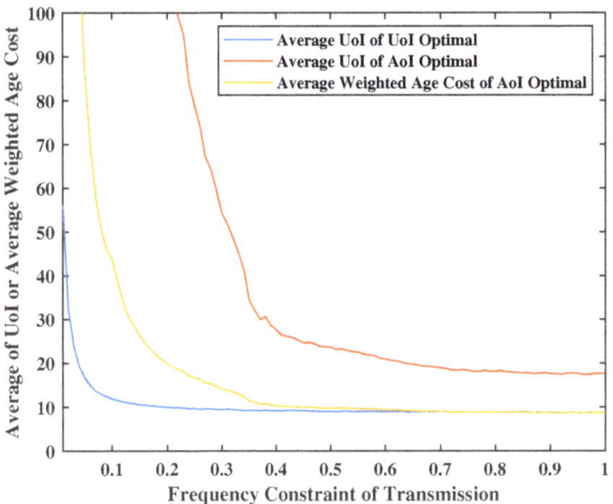

Figure 16. Lyapunov optimization vs. AoI optimal.

5. Conclusions

This paper focused on urgency of information (UoI) optimization through joint sensing and transmission. We proposed a new interactive status updating problem over a point-to-point channel in which transmission and sensing actions are determined to minimize UoI as a combination of the staleness of sensed data and synchronization between two ends under resource constraints, and we used a Lyapunov optimization framework for its optimization. We obtained the gap between the optimal solution and the result gained by the Lyapunov optimal algorithm, and proved that the gap between them can be made arbitrarily small. We presented an extensive numerical study that illustrates various features of the model and resulting algorithm, and potential performance improvements with respect to several schemes. In our future work, we plan to extend this work in multiple directions such as the case of multiple terminals in series or parallel, on demand UoI definition and optimization as well as the cases of computation transmission tradeoffs and dynamical energy constraints.

Author Contributions: Conceptualization, Z.J., P.R. and O.O.; methodology, Z.J., P.R. and O.O.; software, Z.J.; validation, Z.J., P.R. and O.O.; formal analysis, Z.J., P.R. and O.O.; investigation, Z.J., P.R. and O.O.; writing—original draft preparation, Z.J. and P.R.; writing—review and editing, Z.J., P.R. and O.O.; visualization, Z.J. and P.R.; supervision, P.R. and O.O.; project administration, O.O. All authors have read and agreed to the published version of the manuscript.

Funding: This work was supported by National Science Foundation under grant CNS 2219180.

Data Availability Statement: Not applicable.

Conflicts of Interest: The authors declare no conflict of interest.

Appendix A

Based on (7) we can get the following sequence of steps:

$$
E\left[H_{t+1}^2 - H_t^2 | Y_t\right]
$$
$$
\leq E\left[(H_t - \varphi_1 + U_1(t))^2 - H_t^2 | Y_t\right]
$$
$$
= E\left[H_t^2 + \varphi_1^2 + U_1(t)^2 - 2\varphi_1 H_t + 2H_t U_1(t) - 2\varphi_1 U_1(t) - H_t^2 | Y_t\right]
$$
$$
= E\left[(\varphi_1 - U_1(t))^2 + 2(-\varphi_1 + U_1(t))H_t | Y_t\right]
$$
$$
\leq 1 + 2(-\varphi_1 + E[U_1(t)|Y_t])H_t, \tag{A1}
$$

where the first inequality follows from the definition of H_t in (7) used in the identity that for any $X = \max\{a + b - c, 0\}$, $X^2 \leq (a + b - c)^2$, the following equalities follow from rearranging terms and the final inequality follow from $\varphi_1 - U_1(t) \leq 1$. Based on Equation (8), and using the same method as that for obtaining (A1), we get the following inequality:

$$
E\left[G_{t+1}^2 - G_t^2 | Y_t\right] \leq 1 + 2(-\varphi_2 + E[U_2(t)|Y_t])G_t. \tag{A2}
$$

Based on (2), we have

$$
E\left[A_{t+1}^2 - A_t^2 | Y_t\right]
$$
$$
= E\left[(1 - U_2(t))^2 A_t^2 + 2A_t K_t(1 - U_2(t)) + K_t^2 - A_t^2 | Y_t\right]. \tag{A3}
$$

Recall that $K_t \sim (0, \sigma^2)$ follows i.i.d Gaussian distributions. This is due to the fact that the queue A_t is the summation of K_t; it is obvious that the summation of the Gaussian distribution is still a Gaussian distribution. As a result, the error in the service center also follows a Gaussian distribution $A_t \sim (0, n\sigma^2)$. In addition, as $U_2(t) \in \{0, 1\}$, we can simplified (A3) by $U_2(t)^2 = U_2(t)$ and $(1 - U_2(t))^2 = 1 - U_2(t)$. As a result, we have

$$
E\left[A_{t+1}^2 - A_t^2 | Y_t\right] = -n_t \sigma^2 E[U_2(t)|Y_t] + \sigma^2. \tag{A4}
$$

Based on (3) and the fact that $(1 - U_1(t)S_t)^2 = (1 - U_1(t)S_t)$, we have

$$
E\left[Q_{t+1}^2 - Q_t^2 | Y_t\right]
$$
$$
= E\left[(1 - U_1(t)S_t)^2 Q_t^2 + 2Q_t A_t U_2(t)(1 - U_1(t)S_t) + A_t^2 U_2(t)^2 - Q_t^2 | Y_t\right]
$$
$$
= -Q_t^2 p E[U_1(t)|Y_t] + n_t \sigma^2 E[U_2(t)|Y_t]. \tag{A5}
$$

Based on (A1)–(A5), we have

$$\Delta_t = E[L_{t+1} - L_t | Y_t]$$

$$= E\left[\frac{1}{2}V(H_{t+1}^2 - H_t^2) + \frac{1}{2}Z(G_{t+1}^2 - G_t^2) + \frac{1}{2}\theta(Q_{t+1}^2 - Q_t^2) + \frac{1}{2}\beta(A_{t+1}^2 - A_t^2) | Y_t\right]$$

$$\leq \frac{1}{2}(V + Z) + \frac{1}{2}\beta\sigma^2 - V\varphi_1 H_t - Z\varphi_2 G_t + (VH_t - \frac{1}{2}\theta p Q_t^2)E[U_1(t)|Y_t]$$

$$+ (ZG_t + \frac{1}{2}\theta n_t\sigma^2 - \frac{1}{2}\beta n_t\sigma^2)E[U_2(t)|Y_t]. \tag{A6}$$

Appendix B

Set the Lyapunov penalty function as $f_t = Rw_{t+1}(Q_{t+1}^2 + MA_{t+1}^2)$, where $w_t \geq 0$. Based on (A4) and (A5), we can get the Lyapunov penalty function as follows:

$$E[f_t|Y_t] = RE\left[w_{t+1}(-Q_t^2 S_t U_1(t) + 2Q_t A_t U_2(t)(1 - S_t U_1(t)) + A_t^2 U_2(t) + Q_t^2)|Y_t\right]$$

$$+ RME\left[w_{t+1}(-A_t^2 U_2(t) + K_t^2 + A_t^2|Y_t\right]$$

$$= R\tilde{w}(-Q_t^2 pE[U_1(t)|Y_t] + Q_t^2 + M\sigma^2 + Mn_t\sigma^2 + (1 - M)n_t\sigma^2 E[U_2(t)|Y_t]). \tag{A7}$$

Combining (A6) and (A7), the Lyapunov drift plus penalty function satisfies the following inequality:

$$\Delta_t + E[f_t|Y_t] \leq \frac{1}{2}(V + Z) + \frac{1}{2}\beta\sigma^2 + R\tilde{w}(Q_t^2 + M\sigma^2 + Mn_t\sigma^2) - V\varphi_1 H_t - Z\varphi_2 G_t$$

$$+ (VH_t - \frac{1}{2}\theta p Q_t^2 - R\tilde{w}p Q_t^2)E[U_1(t)|Y_t]$$

$$+ (ZG_t + \frac{1}{2}\theta n_t\sigma^2 - \frac{1}{2}\beta n_t\sigma^2 + (1 - M)R\tilde{w}n_t\sigma^2)E[U_2(t)|Y_t]. \tag{A8}$$

Appendix C

We start by assuming that the initial values satisfy $E[L_0] < \infty$. If $\Delta_t + E[f_t] \leq C$, where C is a constant, then take the summation over T time slots to get

$$E\left[L_T - L_0 + \sum_{t=0}^{T-1} f_t\right] \leq TC. \tag{A9}$$

Based on (9) we have

$$E[L_T] \geq \frac{1}{2}VE\left[(H_t^2)\right]. \tag{A10}$$

From the definition of the virtual queue H_t (A1), it is obviously that $E\left[(H_t^2)\right] \geq (E[(H_t)])^2$, and also because the penalty function f_t is always non-negative, we can change (A9) into

$$\frac{1}{2}VE[(H_t)])^2 \leq TC + L_0$$

$$E[(H_t)] \leq \frac{\sqrt{2(TC + L_0)}}{V}$$

$$\frac{E[(H_t)]}{T} \leq \frac{\sqrt{2(TC + L_0)}}{VT}. \tag{A11}$$

Since $T \to \infty$, the right hand side of (A11) is equal to 0. As a result,

$$\frac{E[(H_t)]}{T} \to 0. \tag{A12}$$

As a result, the virtual queue H_t is mean rate stable. The other system states Q_t, A_t and the virtual queue Z_t can be proven as mean rate stable with the same method above. Therefore, the expressions of the queues in the system are appropriate, and the Lyapunov optimization algorithm is applicable. We also recall that the evolution of Q_t, A_t, although not originally in a queue form, can be easily redefined to be bounded below by 0, and the analysis in our paper will be valid without any changes.

Appendix D

First, let us assume that P_1 has an optimal solution, which is to take the best decision for every time slot and get the optimal result of the target function (4). Because this optimal solution does not use the Lyapunov algorithm, the decision has no relationship with the queues and virtual queues in the system. Below $\pi_t = \{U_1(t), U_2(t)\}$ will be used to represent the decision policy of the Lyapunov optimization algorithm and $\pi_t^* = \{U_1(t)^*, U_2^*(t)\}$ is used to represent the decisions of the optimal solution. Based on Equations (A1)–(A5), we have

$$
\begin{aligned}
L_{t+1} - L_t + f_t(\pi_t) \leq{}& \frac{1}{2}(V+Z) + (-\varphi_1 + U_1(t))H_t + (-\varphi_2 + U_2(t))G_t \\
&+ \frac{1}{2}\theta\left(-Q_t{}^2 U_1(t)S_t + 2Q_t A_t U_2(t)(1 - U_1(t)S_t) + A_t{}^2 U_2(t)\right) \\
&+ \frac{1}{2}\beta\left(-U_2(t)A_t{}^2 + 2A_t K_t(1 - U_2(t)) + K_t{}^2\right) + f_t(\pi_t).
\end{aligned} \tag{A13}
$$

Because the optimal solution is a solution of the problem, it should also obey (A13)

$$
\begin{aligned}
L_{t+1} - L_t + f_t(\pi_t) \leq{}& \frac{1}{2}(V+Z) + (-\varphi_1 + U_1^*(t))H_t + (-\varphi_2 + U_2^*(t))G_t \\
&+ \frac{1}{2}\theta\left(-Q_t{}^2 U_1^*(t)S_t + 2Q_t A_t U_2^*(t)(1 - U_1^*(t)S_t) + A_t{}^2 U_2^*(t)\right) \\
&+ \frac{1}{2}\beta\left(-U_2^*(t)A_t{}^2 + 2A_t K_t(1 - U_2^*(t)) + K_t{}^2\right) + f_t(\pi_t^*).
\end{aligned} \tag{A14}
$$

Then take the expectation on both sides of (A14)

$$
\begin{aligned}
E[L_{t+1} - L_t + f_t(\pi_t)|Y_t] \leq{}& \frac{1}{2}(V+Z) + E[(-\varphi_1 + U_1^*(t))H_t] + E[(-\varphi_2 + U_2^*(t))G_t] \\
&+ \frac{1}{2}\theta E\left[-Q_t{}^2 U_1^*(t)S_t + 2Q_t A_t U_2^*(t)(1 - U_1^*(t)S_t) + A_t{}^2 U_2^*(t)\right] \\
&+ \frac{1}{2}\beta E\left[-U_2^*(t)A_t{}^2 + 2A_t K_t(1 - U_2^*(t)) + K_t{}^2\right] + E[f_t(\pi_t^*)].
\end{aligned} \tag{A15}
$$

As is well known in the literature [7,8], there exists a w-optimal decision rule that makes decision randomly and independent of the variables in the system. In the analysis below, we assume such an optimal policy and denote it as $(U_1^*(t), U_2^*(t))$:

$$
\begin{aligned}
E[L_{t+1} - L_t + f_t(\pi_t)|Y_t] \leq{}& \frac{1}{2}(V+Z) + (-\varphi_1 + E[U_1^*(t)])E[H_t] + (-\varphi_2 + E[U_2^*(t)])E[G_t] \\
&+ \frac{1}{2}\theta(-E\left[Q_t{}^2\right]E[U_1^*(t)]p + 2E[Q_t]E[A_t]E[U_2^*(t)](1 - E[U_1^*(t)]p) + E\left[A_t{}^2\right]E[U_2^*(t)]) \\
&+ \frac{1}{2}\beta(-E[U_2^*(t)]E\left[A_t{}^2\right] + 2E[A_t]E[K_t](1 - E[U_2^*(t)]) + E\left[K_t{}^2\right]) + E[f_t(\pi_t^*)].
\end{aligned} \tag{A16}
$$

Placing $E[K_t] = 0$, $E\left[K_t{}^2\right] = \sigma^2$, $E[A_t] = 0$, as well as φ_1 into $E\left[U_1^*(t)\right]$, and φ_2 into $E[U_2^*(t)]$, we get the following. It is worth noting that placing the time-average constraint on $E\left[U_1^*(t)\right]$ and $E\left[U_2^*(t)\right]$ with equality in the Lyapunov drift analysis can be justified easily by observing that the constraints must be active almost always over the T time horizon $O(T)$ time instants:

$$E[L_{t+1} - L_t + f_t(\pi_t)|Y_t] \leq \frac{1}{2}(V + Z) + \frac{1}{2}\theta(-E\left[Q_t^2\right]p\varphi_1 + E[A_t]\varphi_2)$$

$$+ \frac{1}{2}\beta(-\varphi_2 E[A_t] + \sigma^2) + E[f_t(\pi_t^*)]. \tag{A17}$$

Recalling that queues A_t and Q_t are mean rate stable, we have, $E\left[Q_t^2\right] = E\left[Q_{t+1}^2\right]$ and $E\left[A_t^2\right] = E\left[A_{t+1}^2\right]$. From (A3)–(A5), we can get the expectation of A_t^2 and Q_t^2 as

$$E\left[A_t^2\right] = \frac{\sigma^2}{\varphi_2} \tag{A18}$$

$$E\left[Q_t^2\right] = \frac{\sigma^2}{p\varphi_1}. \tag{A19}$$

As a result, (A17) can be simplified as follows:

$$E[L_{t+1} - L_t + Rf_t(\pi_t)|Y_t] \leq \frac{1}{2}(V + Z) + \frac{1}{2}\theta(-\frac{\sigma^2}{p\varphi_1}p\varphi_1 + \frac{\sigma^2}{\varphi_2}\varphi_2)$$

$$+ \frac{1}{2}\beta(-\varphi_2\frac{\sigma^2}{\varphi_2} + \sigma^2) + E[f_t(\pi_t^*)]$$

$$= \frac{1}{2}(V + Z) + E[f_t(\pi_t^*)]. \tag{A20}$$

Now take the summation of the total T-time slot on both sides of (A20) and we have

$$E\left[L_T - L_0 + \sum_{t=0}^{T-1} f_t(\pi_t)|Y_t\right] \leq \frac{1}{2}(V + Z)T + \sum_{t=0}^{T-1} E[f_t(\pi_t^*)]. \tag{A21}$$

Note that $L_T \geq 0$ and $\frac{L_0}{T} = 0$; we then divide T on both sides of (A21) to get the time averaged result

$$\frac{1}{T}E\left[\sum_{t=0}^{T-1} f_t(\pi_t)|Y_t\right] \leq \frac{1}{2}(V + Z) + \frac{1}{T}\sum_{t=0}^{T-1} E[f_t(\pi_t^*)]. \tag{A22}$$

Finally, divide R on both side of (A22) to convert f_t into target function

$$P_1^*(\pi_t) \leq P_2(\pi_t) \leq \frac{V + Z}{2R} + P_1^*(\pi_t). \tag{A23}$$

References

1. Kaul, S.; Yates, R.; Gruteser, M. Real-time status: How often should one update? In Proceedings of the IEEE INFOCOM, Orlando, FL, USA, 10 May 2012; pp. 2731–2735.
2. Miridakis, N.; Tsiftsis, T.; Yang, G. Non-Linear Age of Information: An Energy Efficient Receiver-Centric Approach. *IEEE Wirel. Commun. Lett.* **2022**, *11*, 655–659. [CrossRef]
3. Abd-Elmagid, M.; Pappas, N.; Dhillon, H. On the role of age of information in the Internet of Things. *IEEE Commun. Mag.* **2019**, *57*, 72–77. [CrossRef]
4. Kosta, A.; Pappas, N.; Ephremides, A.; Angelakis, V. The age of information in a discrete time queue: Stationary distribution and non-linear age mean analysis. *IEEE J. Sel. Areas Commun.* **2021**, *39*, 1352–1364. [CrossRef]
5. Zheng, X.; Zhou, S.; Niu, Z. Context-Aware Information Lapse for Timely Status Updates in Remote Control Systems. In Proceedings of the 2019 IEEE Global Communications Conference (GLOBECOM), Waikoloa, HI, USA, 9–13 December 2019; pp. 1–6. [CrossRef]

6. Zheng, X.; Zhou, S.; Niu, Z. Urgency of information for context-aware timely status updates in remote control systems. *IEEE Trans. Wirel. Commun.* **2020**, *19*, 7237–7250. [CrossRef]

7. Neely, M.J. Stochastic network optimization with application to communication and queueing systems. *Synth. Lect. Commun. Netw.* **2010**, *3*.

8. Georgiadis, L.; Neely, M.J.; Tassiulas, L. Resource allocation and cross-layer control in wireless networks. *Found. Trends Netw.* **2006**, *1*, 1–144. [CrossRef]

9. Kam, C.; Kompella, S.; Ephremides, A. The Role of AoI in a Cognitive Radio Network: Lyapunov Optimization and Tradeoffs. In Proceedings of the IEEE MILCOM, San Diego, CA, USA, 29 November–2 December 2021; pp. 303–308.

10. Fountoulakis, E.; Codreanu, M.; Ephremides, A.; Pappas, N. Joint sampling and transmission policies for minimizing cost under aoi constraints. *arXiv* **2021**, arXiv:2103.15450.

11. Zhong, J.; Yates, R.; Soljanin, E. Two freshness metrics for local cache refresh. In Proceedings of the IEEE ISIT, Vail, CO, USA, 17–22 June 2018; pp. 1924–1928.

12. Yates, R.; Zhong, J.; Zhang, W. Updates with multiple service classes. In Proceedings of the IEEE ISIT, Paris, France, 7–12 July 2019; pp. 1017–1021.

13. Champati, J.; Al-Zubaidy, H.; Gross, J. On the Distribution of AoI for the GI/GI/1/1 and GI/GI/1/2 Systems: Exact Expressions and Bounds. In Proceedings of the IEEE INFOCOM, Paris, France, 29 April–2 May 2019; pp. 37–45.

14. Hu, L.; Chen, Z.; Dong, Y.; Jia, Y.; Liang, L.; Wang, M. Status update in IoT networks: Age-of-information violation probability and optimal update rate. *IEEE Internet Things J.* **2021**, *8*, 11329–11344. [CrossRef]

15. Talak, R.; Kadota, I.; Karaman, S.; Modiano, E. Scheduling policies for age minimization in wireless networks with unknown channel state. In Proceedings of the IEEE ISIT, Vail, CO, USA, 17–22 June 2018; pp. 2564–2568.

16. Talak, R.; Karaman, S.; Modiano, E. Optimizing age of information in wireless networks with perfect channel state information. In Proceedings of the IEEE WiOpt, Shanghai, China, 7–11 May 2018.

17. Sun, J.; Jiang, Z.; Zhou, S.; Niu, Z. Age-Optimal Scheduling for Heterogeneous Traffic with Timely-Throughput Constraint. In Proceedings of the IEEE INFOCOM AoI Workshop, Beijing, China, 27 April 2020; pp. 317–322.

18. Kadota, I.; Sinha, A.; Modiano, E. Optimizing age of information in wireless networks with throughput constraints. In Proceedings of the IEEE INFOCOM, Honolulu, HI, USA, 15–19 April 2018; pp. 1844–1852.

19. Kadota, I.; Sinha, A.; Modiano, E. Scheduling algorithms for optimizing age of information in wireless networks with throughput constraints. *IEEE/ACM Trans. Netw.* **2019**, *27*, 1359–1372. [CrossRef]

20. Bacinoglu, B.; Ceran, E.; Uysal-Biyikoglu, E. Age of information under energy replenishment constraints. In Proceedings of the IEEE ITA, Wrexham, UK, 8–11 September 2015; pp. 25–31.

21. Bacinoglu, B.; Uysal-Biyikoglu, E. Scheduling status updates to minimize age of information with an energy harvesting sensor. In Proceedings of the IEEE ISIT, Aachen, Germany, 25–30 June 2017; pp. 1122–1126.

22. Yates, R.D. Lazy is timely: Status updates by an energy harvesting source. In Proceedings of the IEEE ISIT, Hong Kong, China, 14–19 June 2015; pp. 3008–3012.

23. Ceran, E.; Gündüz, D.; György, A. Average age of information with hybrid ARQ under a resource constraint. *IEEE Trans. Wireless Comm.* **2019**, *18*, 1900–1913. [CrossRef]

24. Zhou, B.; Saad, W. Joint status sampling and updating for minimizing age of information in the Internet of Things. *IEEE Trans. Commun.* **2019**, *67*, 7468–7482. [CrossRef]

25. Yates, R.; Kaul, S. Real-time status updating: Multiple sources. In Proceedings of the IEEE ISIT, Cambridge, MA, USA, 1–6 July 2012; pp. 2666–2670.

26. Yates, R.; Kaul, S.K. The age of information: Real-time status updating by multiple sources. *IEEE Trans. Inf. Theory* **2018**, *65*, 1807–1827.

27. Moltafet, M.; Leinonen, M.; Codreanu, M. On the age of information in multi-source queueing models. *IEEE Trans. Commun.* **2020**, *68*, 5003–5017. [CrossRef]

28. Abd-Elmagid, M.; Dhillon, H. Distributional properties of age of information in energy harvesting status update systems. In Proceedings of the IEEE WiOpt, Virtual, 18–21 October 2021.

29. Hirosawa, N.; Iimori, H.; de Abreu, G.; Ishibashi, K. Age-of-Information Minimization in Two-User Multiple Access Channel with Energy Harvesting. In Proceedings of the IEEE CAMSAP, Guadeloupe, France, 15–18 December 2019; pp. 361–365.

30. Chen, Z.; Pappas, N.; Björnson, E.; Larsson, E. Age of information in a multiple access channel with heterogeneous traffic and an energy harvesting node. In Proceedings of the IEEE INFOCOM AoI Workshop, Paris, France, 29 April 2019; pp. 662–667.

31. Yates, R.; Kaul, S. Status updates over unreliable multiaccess channels. In Proceedings of the IEEE ISIT, Aachen, Germany, 25–30 June 2017; pp. 331–335.

32. Tang, H.; Wang, J.; Song, L.; Song, J. Minimizing age of information with power constraints: Multi-user opportunistic scheduling in multi-state time-varying channels. *IEEE J. Sel. Areas Commun.* **2020**, *38*, 854–868. [CrossRef]

33. Champati, J.; Mamduhi, M.; Johansson, K.; Gross, J. Performance characterization using aoi in a single-loop networked control system. In Proceedings of the IEEE INFOCOM AoI Workshop, Paris, France, 29 April 2019; pp. 197–203.

34. Li, J.; Zhou, Y.; Chen, H. Age of information for multicast transmission with fixed and random deadlines in IoT systems. *IEEE Internet Things J.* **2020**, *7*, 8178–8191. [CrossRef]

35. Kadota, I.; Sinha, A.; Uysal-Biyikoglu, E.; Singh, R.; Modiano, E. Scheduling policies for minimizing age of information in broadcast wireless networks. *IEEE/ACM Trans. Netw.* **2018**, *26*, 2637–2650. [CrossRef]
36. Hsu, Y.P.; Modiano, E.; Duan, L. Age of information: Design and analysis of optimal scheduling algorithms. In Proceedings of the IEEE ISIT, Aachen, Germany, 25–30 June 2017; pp. 561–565.
37. Sun, Y.; Polyanskiy, Y.; Uysal-Biyikoglu, E. Remote estimation of the Wiener process over a channel with random delay. In Proceedings of the IEEE ISIT, Aachen, Germany, 25–30 June 2017; pp. 321–325.
38. Ornee, T.; Sun, Y. Sampling for remote estimation through queues: Age of information and beyond. In Proceedings of the IEEE WiOPT, Avignon, France, 3–7 June 2019.
39. Arafa, A.; Banawan, K.; Seddik, K.; Poor, H. Timely estimation using coded quantized samples. In Proceedings of the IEEE ISIT, Los Angeles, CA, USA, 21–26 June 2020; pp. 1812–1817.
40. Tsai, C.H.; Wang, C.C. Unifying AoI minimization and remote estimation—Optimal sensor/controller coordination with random two-way delay. *IEEE/ACM Trans. Netw.* **2021**, *30*, 229–242. [CrossRef]
41. Moltafet, M.; Leinonen, M.; Codreanu, M.; Yates, R. Status Update Control and Analysis under Two-Way Delay. *arXiv* **2022**, arXiv:2208.06177.
42. Feng, S.; Yang, J. Age of information minimization for an energy harvesting source with updating erasures: Without and with feedback. *IEEE Trans. Commun.* **2021**, *69*, 5091–5105. [CrossRef]
43. Rafiee, P.; Ozel, O. Active Status Update Packet Drop Control in an Energy Harvesting Node. In Proceedings of the IEEE SPAWC, Atlanta, GA, USA, 26–29 May 2020.
44. Rafiee, P.; Zou, P.; Ozel, O.; Subramaniam, S. Maintaining Information Freshness in Power-Efficient Status Update Systems. In Proceedings of the IEEE INFOCOM AoI Workshop, Virtual, 6–9 July 2020; pp. 31–36.
45. Rafiee, P.; Oktay, M.; Ozel, O. Intermittent Status Updating with Random Update Arrivals. In Proceedings of the IEEE ISIT, Melbourne, Australia, 12–20 July 2021; pp. 3121–3126.
46. Munari, A.; Badia, L. The Role of Feedback in AoI Optimization Under Limited Transmission Opportunities. *arXiv* **2022**, arXiv:2208.14128.
47. Ozel, O. Timely Status Updating Through Intermittent Sensing and Transmission. In Proceedings of the IEEE ISIT, Los Angeles, CA, USA, 21–26 June 2020; pp. 1788–1793.
48. Ozel, O.; Rafiee, P. Intermittent Status Updating Through Joint Scheduling of Sensing and Retransmissions. In Proceedings of the IEEE INFOCOM AoI Workshop, Virtual, 10 May 2021.
49. Ayan, O.; Vilgelm, M.; Klügel, M.; Hirche, S.; Kellerer, W. Age-of-information vs. value-of-information scheduling for cellular networked control systems. In Proceedings of the ACM/IEEE International Conference on Cyber-Physical Systems, Montreal, QC, Canada, 16–18 April 2019; pp. 109–117.
50. Zou, P.; Ozel, O.; Subramaniam, S. On age and value of information in status update systems. In Proceedings of the IEEE WCNC, Seoul, Korea, 25–28 May 2020.
51. Lin, W.; Wang, X.; Sun, X.; Chen, X. Average age of changed information in the Internet of Things. In Proceedings of the IEEE WCNC, virtual, 25–28 May 2020.
52. Ildiz, M.E.; Yavascan, O.T.; Uysal, E.; Kartal, O.T. Pull or wait: How to optimize query age of information. *arXiv* **2021**, arXiv:2111.02309.
53. Chiariotti, F.; Holm, J.; Kalør, A.E.; Soret, B.; Jensen, S.; Pedersen, T.; Popovski, P. Query age of information: Freshness in pull-based communication. *IEEE Trans. Commun.* **2022**, *70*, 1606–1622. [CrossRef]

Article

Optimal Information Update for Energy Harvesting Sensor with Reliable Backup Energy

Lixin Wang [1], Fuzhou Peng [2], Xiang Chen [2] and Shidong Zhou [1,*]

[1] Department of Electronic Engineering, Tsinghua University, Beijing 100084, China; wanglx19@mails.tsinghua.edu.cn
[2] School of Electronics and Information Technology, Sun Yat-sen University, Guangzhou 510006, China; pengfzh@mail2.sysu.edu.cn (F.P.); chenxiang@mail.sysu.edu.cn (X.C.)
* Correspondence: zhousd@tsinghua.edu.cn

Abstract: The timely delivery of status information collected from sensors is critical in many real-time applications, e.g., monitoring and control. In this paper, we consider a scenario where a wireless sensor sends updates to the destination over an erasure channel with the supply of harvested energy and reliable backup energy. We adopt the metric age of information (AoI) to measure the timeliness of the received updates at the destination. We aim to find the optimal information updating policy that minimizes the time-average weighted sum of the AoI and the reliable backup energy cost. First, when all the environmental statistics are assumed to be known, the optimal information updating policy exists and is proved to have a threshold structure. Based on this special structure, an algorithm for efficiently computing the optimal policy is proposed. Then, for the unknown environment, a learning-based algorithm is employed to find a near-optimal policy. The simulation results verify the correctness of the theoretical derivation and the effectiveness of the proposed method.

Keywords: age of information; information update; energy harvesting; reliable backup energy

Citation: Wang, L.; Peng, F.; Chen, X.; Zhou, S. Optimal Information Update for Energy Harvesting Sensor with Reliable Backup Energy. *Entropy* **2022**, 24, 961. https://doi.org/10.3390/e24070961

Academic Editors: Udo Von Toussaint and Chintha Tellambura

Received: 22 February 2022
Accepted: 7 July 2022
Published: 11 July 2022

Publisher's Note: MDPI stays neutral with regard to jurisdictional claims in published maps and institutional affiliations.

1. Introduction

Timely information updates from wireless sensors to destinations are essential for real-time monitoring and control systems. To describe the timeliness of information updates from the receivers' perspective, a new metric called age of information (AoI) is proposed [1–3]. Unlike general performance metrics, such as delay and throughput, AoI refers to the time elapsed since the generation of the latest received information. A lower AoI generally reflects more timely information at the destination. Therefore, the AoI-minimal status updating policies in sensor networks have been widely studied [4–7].

The destinations always desire information updates in as timely a manner as possible, which is typically constrained by sensors' energy. Generally, energy sources include the grid and sensors' own non-rechargeable batteries. We call these sources *reliable energy* since they enable sensors to reliably operate until the power grid is cut off or sensors' batteries are exhausted [8]. Specifically, if sensors consume energy from the grid, they need to pay the electricity bill; if sensors only use the power of their own batteries, the price of sensing and transmitting updates will be the cost of frequent battery replacement. There is clearly a price to pay for using reliable energy to update. Energy harvesting (EH) is a promising technology that can help reduce the consumption of reliable energy for information update [9,10]. It can continuously extract energy from solar power, ambient RF, and thermal energy and store the harvested energy in sensors' rechargeable batteries. The stored energy is renewable and can be used for free. Hence, in this case, the reliable energy can serve as *backup* energy. The design of the coexistence of reliable backup energy and harvested energy has been researched and promoted in academia and industry [8,11–14]. The mixed energy supply mode can enhance the reliability of the system.

However, the irregular arrivals of harvested energy and the limited capacity of rechargeable batteries still motivate us to schedule the energy usage properly to reduce the cost of using reliable backup energy while maintaining the timeliness of information updates (i.e., the average AoI). Intuitively, the average AoI and the cost of using reliable energy cannot be minimized simultaneously. On the one hand, a lower average AoI means that the sensor senses and transmits updates more frequently, which will increase the consumption of reliable backup energy since the harvested energy is limited. On the other hand, to reduce the cost of reliable backup energy, the sensor will only exploit the harvested energy. Due to the uncertainty of the energy harvesting behavior, the average AoI of the system will inevitably increase. Therefore, in this paper, we focus on achieving the best trade-off between the average AoI and the cost of reliable backup energy.

We consider a sensor-based information update system, where an energy harvesting sensor with reliable backup energy sends timely updates to the destination through an erasure channel. Based on our settings, we will minimize the long-term average weighted sum of the AoI and the paid reliable energy cost to find the optimal information updating policy by Markov decision process (MDP) theory [15]. First, we assume that the sensor knows the relevant statistics in advance, such as the success probability of each transmission and the probability of energy arrival, so that the sensor can make the optimal decision at any time. Then we consider a more realistic scenario where the sensor has no knowledge of the environment. In such an unknown environment, learning-based approaches should be adopted to obtain the updating policy.

1.1. Related Work

There have been a series of related works studying AoI minimization in EH communication systems [16–34]. In these systems, each update consumes harvested energy and is constrained by the energy causality.

Refs. [16–23] focus on how to optimize AoI under general energy causality constraints, where different battery model settings are considered. Constrained by the average power available in the infinite-sized battery, ref. [16] shows that a lazy policy which leaves a certain idle period between updates outperforms the greedy policy under random service times. With the same assumption of an infinite-sized battery, ref. [17] focuses on both offline and online policies under energy replenishment constraints with zero service time. While considering fixed service times, the offline results in [17] are extended to a two-hop scenario in [18], and online policy is provided in [19]. In the case of the delay being controlled by transmission energy, ref. [20] also investigated the optimal offline policy. For the error-free and delay-free channel, the optimal updating policies were investigated for different battery settings [21,22]. Ref. [21] derived the asymptotically optimal policies for the infinite-sized, finite-sized, and unit-sized battery by renewal theory. It turned out to be a threshold policy for the unit-sized battery case. More general battery models were considered in [22]. The optimal policy was also proved to be multi-threshold and the energy-dependent thresholds were characterized explicitly. When the battery is finite sized and there is no feedback from the destination, it was shown that the optimal updating policy is of a threshold structure and the threshold is non-increasing with the battery level [23].

Refs. [24–30] studied how to properly utilize the harvested energy to transmit updates over imperfect channels. For the noisy channel, ref. [24] considered an infinite-sized battery model and derived the different optimal policies for updating with and without feedback. Ref. [25] further derived a closed-form expression for the threshold of the unit-sized battery model and extended the threshold-based policies to multiple sources case. To combat the noisy channel, some channel coding schemes for EH communication were investigated in [26,27]. In [28], the HARQ protocol was applied for a single EH sensor to send updates to the destination. The optimal policies were obtained by employing reinforcement learning in both known and unknown environments, but no clear intuition on the policy structure was provided. Considering energy harvesting wireless sensor networks (EH-WSNs), ref. [29] suggested to estimate the channel state of a Rayleigh fading channel before transmitting to improve the AoI, update interval and packet loss performance, despite the associated time and energy costs. Ref. [30] aimed to minimize the average AoI of an EH-aided secondary user(SU) in a cognitive ratio network. The SU has to make sensing and updating decisions subject to random energy arrivals and the available spectrum. The sequential decision problem is formulated as a partially observable Markov decision process (POMDP).

Refs. [31–34] paid attention to other AoI-related metrics in EH communication and even the distributional properties of AoI, not just the average AoI. Different freshness metrics were considered, such as nonlinear AoI [31], urgency-aware AoI (U-AoI) [32], and peak AoI [33] in EH sensor network. To better understand the distributional properties of AoI, ref. [34] further derived closed-form expressions of the moment generating function (MGF) of AoI in an EH-powered queuing system using the stochastic hybrid systems (SHS) framework.

The above works focus on optimizing information freshness under the EH supply. Different from them, energy sources in this paper include both harvested energy and reliable backup energy, and our goal is to achieve the best trade-off between age and reliable energy consumption, instead of merely optimizing AoI. Among the above works, refs. [23,25] are the most related to our paper. The following Table 1 summarizes the detailed differences. It is worth noting that by letting the reliable energy consumption be small enough, our results can be compared with some prior results in [23,25].

Table 1. Comparative summary of the most related works in contrast to our paper.

Feature \ Ref	[23]	[25]	Our
Energy supply	EH	EH	EH + reliable energy
Battery capacity	Finite-sized	Unit-sized	Finite-sized
Wireless channel	Error-free	Error-prone	Error-prone
Optimization objective	AoI	AoI	AoI-reliable energy trade-off

The age–energy trade-off has been widely studied in [35–39]. The age–energy trade-off in the erasure channel was studied in [35], and the fading channel case was investigated in [36]. Ref. [37] adopted a truncated automatic repeat request (TARQ) scheme and characterized the age–energy trade-off for the IoT monitoring system. Optimum energy efficiency and AoI trade-off was considered in a multicast system in [38]. In [39] , the authors investigated the optimal age–energy trade-off, where status sensing and data transmission can be carried out separately. By the MDP analysis similar to [6,15], the optimal policy exists and is proved to have two thresholds. The energy sources are all reliable in these works, which means that the energy cost of the update is easy to track. However, the uncertainty of the energy arrival and mixed energy supplies bring more challenges to the MDP analysis in this paper. To the best of our knowledge, this paper is the first to consider the timeliness of the system under mixed energy supplies. The preliminary results of this paper are presented in [40].

1.2. Main Contributions

The main contributions of this paper are as follows:

- We consider an information update system where the harvested energy and reliable energy coexist. The goal is to find the optimal policy that achieves the best trade-off between age and reliable energy consumption. Compared to the existing works [23,25], our problem is more practical and general, which will provide some insights for future green and durable update system designs.
- For the case that all the statistics such as channel erasure probability and EH probability are known a priori, we formulate an unconstrained infinite space Markov decision process (MDP) problem, and prove the existence of the optimal policy. By revealing the monotonicity and *proportional differential property* of the value function, we find

that the optimal policy is of the threshold-type. Based on this special structure, we propose an efficient algorithm to compute the optimal policy.

- In an unknown environment, we propose an average cost Q-learning algorithm to obtain the updating policy.
- Simulation results show that the optimal policy outperforms other baseline policies when the environmental statistics are known. At the same time, the performance of the policy learned in the unknown environment is very close to the theoretical optimal policy. We also compare the age-reliable energy trade-off curves of the optimal updating policies under different energy supply conditions, which reflects the rationality of mixed energy supplies. The optimal policy can also be particularized to a special case, where the sensor can only utilize the harvested energy and the battery is unit-sized, and its performance coincides with the existing results in [23,25].

1.3. Organization

The rest of this paper is organized as follows. In Section 2, we introduce the model of the information update system and formulate the problem. In Section 3, we analyze the optimal policy when all the statistics are known. In Section 4, we aim to minimize the average cost of updating in an unknown environment. In Section 5, we present the simulation results. Finally, in Section 6, we conclude the paper.

2. System Model and Problem Formulation

2.1. System Model

In this paper, we consider a point-to-point information update system, where a wireless sensor and a destination are connected by an erasure channel, as shown in Figure 1. The channel is assumed to be noisy and time invariant, and each update is corrupted with probability p during transmission (Note $p \in (0,1)$). Both the free harvested energy stored in the rechargeable battery and the reliable backup energy that needs to be paid can be used for real-time environmental status updates.

Without loss of generality, time is slotted with equal length and indexed by $t = 0, 1, 2 \ldots$. At the beginning of each time slot, the sensor decides whether to generate and transmit an update to the destination or stay idle. The decision action at slot t, denoted by $a[t]$, takes value from action set $\mathcal{A} = \{0, 1\}$, where $a[t] = 1$ means that the sensor decides to generate and transmit an update to the destination while $a[t] = 0$ means the sensor is idle. The destination will feed back an instantaneous ACK to the sensor through an error-free channel when it has successfully received an update and a NACK otherwise. We assume the above processes can be completed in one time slot. The destination keeps track of the environment status through the received updates. We apply the metric age of information to measure the freshness of the status information available at the destination.

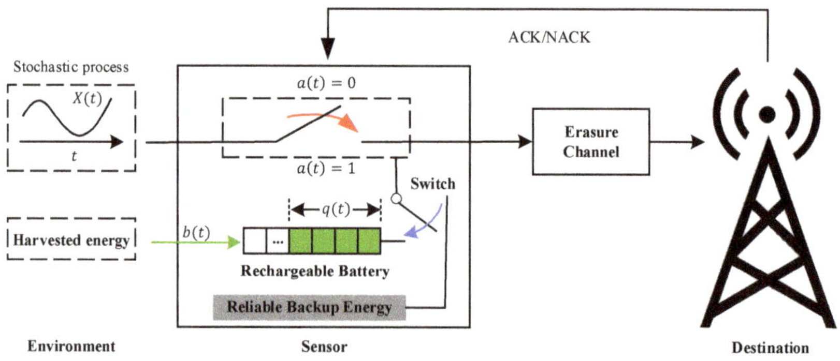

Figure 1. System model.

2.1.1. Age of Information

Age of Information (AoI) is defined as the elapsed time since the generation of the latest successfully received update [1–3]. Denote $\Delta[t]$ as the AoI of destination in time slot t. Then, we have

$$\Delta[t] = t - U[t]. \tag{1}$$

where $U[t]$ denotes the time slot when the most recently received update was generated before time slot t. In particular, the AoI will decrease to one if a new update is successfully received. Otherwise, it will increase by one. The evolution of AoI can be expressed as follows:

$$\Delta[t+1] = \begin{cases} 1, & \text{successful transmission,} \\ \Delta[t] + 1, & \text{otherwise.} \end{cases} \tag{2}$$

A sample path of AoI is depicted in Figure 2.

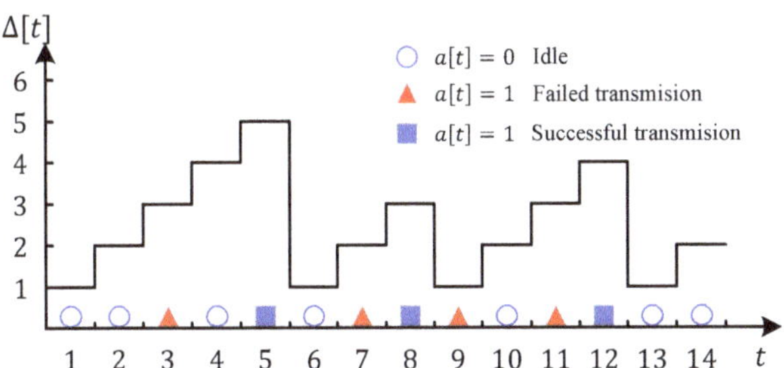

Figure 2. A sample path of AoI with initial age 1.

2.1.2. Description of Energy Supply

We assume that only the sensor's measurement and transmission process will consume energy and ignore other energy consumption. The energy unit is normalized, so the generation and transmission for each update will consume one energy unit. As previously described, the energy sources of the sensor include energy harvested from nature and reliable backup energy.

For the harvested energy, the sensor can store it in a rechargeable battery for later use. The maximum capacity of the rechargeable battery is B units ($B > 1$). Considering the scarcity of energy in nature, the total energy harvested in one time slot may sometimes not reach an energy unit. So we consider using the Bernoulli process with the parameter λ to approximately capture the arrival process of harvested energy, which was also adopted in [41–43]. Let $b[t]$ be the accumulated harvested energy in time slot t. That is, we have $\Pr\{b[t] = 1\} = \lambda$ and $\Pr\{b[t] = 0\} = 1 - \lambda$ in each time slot t (note $\lambda \in (0, 1)$). Here, we assume that the energy arrival at each slot is independently and identically distributed. Time-correlated energy arrival processes, such as Markov process, will be considered in future work.

For reliable backup energy, we assume that it contains much more energy units than the rechargeable battery, so the energy it contains can be viewed as infinite. However, it needs to be used for a fee compared to the free renewable energy stored in the rechargeable battery. Therefore, when the stored renewable energy is not zero, the sensor will prioritize using it for status updates; otherwise, it will automatically switch to the reliable backup energy until the sensor has harvested energy. Defining the power of the rechargeable battery at the beginning of time slot t as the battery state $q[t]$, then the evolution of battery state from time slot t to $t + 1$ can be summarized as follows:

$$q[t+1] = \min\{q[t] + b[t] - a[t]u(q[t]), B\}, \tag{3}$$

where $u(\cdot)$ is unit step function, which is defined as

$$u(x) = \begin{cases} 1, & \text{if } x > 0, \\ 0, & \text{otherwise.} \end{cases} \tag{4}$$

Suppose that under reliable energy supply, the cost of generating and transmitting an update is a non-negative value C_r. Defining $E[t]$ as the paid reliable energy cost at the time slot t, then we have

$$E[t] = C_r a[t](1 - u(q[t])). \tag{5}$$

2.2. Problem Formulation

Let Π denote the set of non-anticipative policies in which scheduling decision $a[t]$ are made based on the action history $\{a[k]\}_{k=0}^{t-1}$, the evolution of AoI $\{\Delta[k]\}_{k=0}^{t}$, the evolution of battery state $\{q[k]\}_{k=0}^{t}$ as well as the system parameters (e.g., p and λ). In order to keep the information freshness at the destination, the sensor needs to send updates. However, due to the randomness of energy arrivals, the battery energy may sometimes be insufficient to support updates, and the sensor has to take energy from reliable backup energy. To balance the information freshness and the paid reliable backup energy cost, we aim to find the optimal information updating policy $\pi \in \Pi$ that achieves the minimum of the time-average weighted sum of the AoI and the paid reliable backup energy cost. The problem is formulated as follows:

$$\min_{\pi \in \Pi} \limsup_{T \to \infty} \frac{1}{T} \mathbb{E}_\pi \left\{ \sum_{t=0}^{T-1} [\Delta[t] + \omega E[t]] \right\}, \tag{6}$$
$$\text{s.t.} \quad (2), (3), (5),$$

where ω is the pre-defined non-negative weighting factor. If $\omega = 0$, the optimal policy is to update in each time slot, i.e., zero-wait policy [4]. Since the effect of energy can be ignored, if the rechargeable battery is not empty, the sensor uses the renewable energy; otherwise, the sensor will use the reliable energy directly. When $\omega > 0$, the optimal policy is non-trivial. So we will focus on the optimal policy for $\omega > 0$ in the rest of the paper. The smaller ω is, the more we attach importance to the system AoI; otherwise, the more emphasis is placed on the cost of reliable energy.

Remark 1. *The optimal trade-off between age and reliable energy consumption can also be formulated as a constrained problem, where the reliable energy consumption serves as a constraint (not exceeding E_m) but not a penalty, and the goal is to minimize the long-term average age. By the Lagrangian method, it can be converted into an unconstrained weighted sum problem, where the Lagrangian multiplier is exactly the weight factor ω. So the solution proposed in this paper can be used. If there exists an ω such that the average reliable energy consumption in the minimum weighted sum is E_m, the optimal policy of the weighted sum problem also minimizes the long-term average age with the E_m constraint. Otherwise, a randomized optimal policy for the constrained problem needs to be considered; see details in [44].*

3. Optimal Policy Analysis In A Known Environment

In this section, we aim to solve the problem (6) in a known environment and obtain the optimal policy. It is difficult to solve the original problem directly due to the random erasures and the temporal dependency in both AoI and battery state evolution. However, since the statistics such as channel erasure probability and EH probability are known, we can reformulate the original problem as a time-average cost MDP with infinite state space and analyze the structure of the optimal policy.

3.1. Markov Decision Process Formulation

According to the system description mentioned in the previous section, the MDP is formulated as follows:

- **State space** . The sensor's state $x[t]$ in slot t is a couple of the current destination AoI and the battery state, i.e., $(\Delta[t], q[t])$. Define $\mathcal{B} = \{0, 1, \dots, B\}$. The state space $\mathcal{S} = \mathbb{Z}^+ \times \mathcal{B}$ is thus infinite countable.
- **Action space**. The sensor's action $a[t]$ in time slot t takes value from the action set $\mathcal{A} = \{0, 1\}$.
- **Transition probabilities**. Denote $\Pr(x[t+1]|x[t], a[t])$ as the transition probability that current state $x[t]$ transits to next state $x[t+1]$ after taking action $a[t]$. Suppose the current state $x[t] = (\Delta, q)$ and action $a[t] = a$, then the transition probability is divided into two following cases conditioned on different values of action.
  Case 1 . $a = 0$,

$$
\begin{cases}
\Pr\{(\Delta+1, q+1)|(\Delta, q), 0\} = \lambda, & \text{if } q < B, \\
\Pr\{(\Delta+1, B)|(\Delta, B), 0\} = 1, & \text{if } q = B, \\
\Pr\{(\Delta+1, q)|(\Delta, q), 0\} = 1 - \lambda, & \text{if } q < B.
\end{cases}
\tag{7}
$$

  Case 2. $a = 1$,

$$
\begin{cases}
\Pr\{(\Delta+1, q)|(\Delta, q), 1\} = p\lambda, & \text{if } q > 0, \\
\Pr\{(1, q)|(\Delta, q), 1\} = (1 - p)\lambda, & \text{if } q > 0, \\
\Pr\{\Delta+1, q-1)|(\Delta, q), 1\} = p(1 - \lambda), & \text{if } q > 0, \\
\Pr\{(1, q-1)|(\Delta, q), 1\} = (1 - p)(1 - \lambda), & \text{if } q > 0, \\
\Pr\{(\Delta+1, 1)|(\Delta, 0), 1\} = p\lambda, & \text{if } q = 0, \\
\Pr\{(1, 1)|(\Delta, 0), 1\} = (1 - p)\lambda, & \text{if } q = 0, \\
\Pr\{(\Delta+1, 0)|(\Delta, 0), 0\} = p(1 - \lambda), & \text{if } q = 0, \\
\Pr\{(1, 0)|(\Delta, 0), 0\} = (1 - p)(1 - \lambda), & \text{if } q = 0.
\end{cases}
\tag{8}
$$

  In both cases, the evolution of AoI still follows Equation (2) and the evolution of battery state follows Equation (3).
- **One-step cost**. For the current state $x = (\Delta, q)$, the one-step cost $C(x, a)$ of taking action a is expressed by

$$
C(x, a) = \Delta + \omega C_r a (1 - u(q)).
\tag{9}
$$

After the above modeling, the original problem (6) is transformed into obtaining the optimal policy for the MDP to minimize the average cost in an infinite horizon:

$$
\min_{\pi \in \Pi} \limsup_{T \to \infty} \frac{1}{T} \mathbb{E}_\pi \left\{ \sum_{t=0}^{T-1} C(x[t], a[t]) \right\}.
\tag{10}
$$

Denote Π_{SD} as the set of stationary deterministic policies. Given observation $(\Delta[t], q[t]) = (\Delta, q)$, the policy $\pi \in \Pi_{SD}$ selects action $a[t] = \pi(\Delta, q)$, where $\pi(\cdot) : (\Delta, q) \to \{0, 1\}$ is a deterministic function from state space $\mathcal{S}$ to action space $\mathcal{A}$. In the next section, we prove that there is an optimal stationary deterministic policy for the above unconstrained MDP with infinite countable state and action space.

3.2. The Existence of the Optimal Stationary Deterministic Policy

According to [15], we need to first address a discounted cost MDP, then relate it to the original average cost problem. Given an initial state $\mathbf{x}[0] = \hat{\mathbf{x}}$, the total expected discounted cost under a policy π is given by

$$V_\gamma^\pi(\hat{\mathbf{x}}) = \limsup_{T \to \infty} \mathbb{E}_\pi \left\{ \sum_{t=0}^{T-1} \gamma^t C(\mathbf{x}[t], a[t]) \,\middle|\, \mathbf{x}[0] = \hat{\mathbf{x}} \right\}, \tag{11}$$

where the discounted factor is $\gamma \in (0,1)$. Therefore, the problem of minimizing the expected discounted cost can be formulated as

$$V_\gamma(\hat{\mathbf{x}}) \triangleq \min_{\pi \in \Pi} V_\gamma^\pi(\hat{\mathbf{x}}), \tag{12}$$

where value function $V_\gamma(\hat{\mathbf{x}})$ denotes the minimum expected discounted cost. The policy is γ-optimal if it minimizes the above discounted cost. The optimality equation of $V_\gamma(\hat{\mathbf{x}})$ is introduced in Proposition 1.

Proposition 1.

(a) *The optimal expected discounted cost $V_\gamma(\hat{\mathbf{x}})$ satisfies the Bellman equation as follows:*

$$V_\gamma(\hat{\mathbf{x}}) = \min_{a \in \mathcal{A}} Q_\gamma(\hat{\mathbf{x}}, a), \tag{13}$$

where the state–action value function $Q_\gamma(\hat{\mathbf{x}}, a)$ is defined as

$$Q_\gamma(\hat{\mathbf{x}}, a) = C(\hat{\mathbf{x}}, a) + \gamma \sum_{\mathbf{x}' \in \mathcal{S}} \Pr(\mathbf{x}'|\hat{\mathbf{x}}, a) V_\gamma(\mathbf{x}'). \tag{14}$$

(b) *The policy π determined by the right hand side of* (13) *is γ-optimal, and $\pi \in \Pi_{SD}$.*
(c) *$V_\gamma(\hat{\mathbf{x}})$ can be solved by value iteration algorithm. Specifically, let $V_{\gamma,n}(\hat{\mathbf{x}})$ be the cost-to-go function and $V_{\gamma,0}(\hat{\mathbf{x}}) = 0$ for all state $\hat{\mathbf{x}} \in \mathcal{S}$. For all $n \geq 1$, we have:*

$$V_{\gamma,n}(\hat{\mathbf{x}}) = \min_{a \in \mathcal{A}} Q_{\gamma,n}(\hat{\mathbf{x}}, a), \tag{15}$$

where $Q_{\gamma,n}(\hat{\mathbf{x}}, a)$ is obtained as follows:

$$Q_{\gamma,n}(\hat{\mathbf{x}}, a) = C(\hat{\mathbf{x}}, a) + \gamma \sum_{\mathbf{x}' \in \mathcal{S}} \Pr(\mathbf{x}'|\hat{\mathbf{x}}, a) V_{\gamma,n-1}(\mathbf{x}'). \tag{16}$$

Then the equation $\lim_{n \to \infty} V_{\gamma,n}(\hat{\mathbf{x}}) = V_\gamma(\hat{\mathbf{x}})$ holds for every state $\hat{\mathbf{x}}$ and γ.

Proof. See Appendix A. $\square$

Now, we can show the monotonic properties of $V_\gamma(\hat{\mathbf{x}})$ in the following lemma by using (c) in Proposition 1.

Lemma 1. *Given fixed channel erasure probability p and EH probability λ, then*

(a) *value function $V_\gamma(\Delta, q)$ is **non-decreasing** in Δ, i.e., for any $1 \leq \Delta_1 \leq \Delta_2$ and any battery state $q \in \mathcal{B}$, we have*

$$V_\gamma(\Delta_1, q) \leq V_\gamma(\Delta_2, q), \tag{17}$$

and

$$V_\gamma(\Delta_2, q) - V_\gamma(\Delta_1, q) \geq \Delta_2 - \Delta_1. \tag{18}$$

(b)   value function $V_\gamma(\Delta, q)$ is **non-increasing** in q, i.e., for AoI $\Delta \geq 1$ and any battery state $q \in \{0, 1, \ldots, B-1\}$, we have

$$V_\gamma(\Delta, q) \geq V_\gamma(\Delta, q+1), \tag{19}$$

Proof. See Appendix B. $\square$

Based on the Proposition 1 and Lemma 1, we will verify the existence of the optimal stationary deterministic policy for the average cost problem (10) in the following theorem.

Theorem 1. *There exists an optimal policy $\pi^\star \in \Pi_{SD}$ for the average cost MDP in (10). Moreover, for every state x, there exists a value function $V(\cdot) : \mathcal{S} \rightarrow \mathbb{R}$ and a unique constant $g^\star \in \mathbb{R}$ such that:*

$$g^\star + V(x) = \min_{a \in \mathcal{A}} \left\{ C(x, a) + \sum_{x' \in \mathcal{S}} \Pr(x'|x, a) V(x') \right\}, \tag{20}$$

where $g^\star$ is the optimal average cost of problem (10) and satisfies $g^\star = \lim_{\gamma \to 1} (1 - \gamma) V_\gamma(x)$ for every state x, and the value function $V(x)$ satisfies

$$V(x) = \lim_{\gamma \to 1} \gamma V_\gamma(x) = \lim_{\gamma \to 1} V_\gamma(x) - g^\star = \limsup_{T \to \infty} \frac{1}{T} \mathbb{E}_\pi \left\{ \sum_{t=0}^{T-1} [C(x[t], a[t]) - g^\star] \right\}. \tag{21}$$

Proof. See Appendix C. $\square$

Based on Theorem 1, we have the following corollary:

Corollary 1. *The state–action value function $Q(x, a)$ for the average cost is given as follows:*

$$Q(x, a) = C(x, a) + \sum_{x' \in \mathcal{S}} \Pr(x'|x, a) V(x'), \tag{22}$$

which is similar to $Q_\gamma(x, a)$ in (14) by letting $\gamma \to 1$. Then the optimal policy $\pi^\star \in \Pi_{SD}$ for the average cost MDP in (10) can be expressed as follows:

$$\pi^\star(x) = \arg \min_{a \in \mathcal{A}} Q(x, a), \forall x \in \mathcal{S}. \tag{23}$$

3.3. Structure Analysis of Optimal Policy

Before analyzing the structure of the optimal policy $\pi^\star$, we first prove some monotonic properties of the value function $V(x)$ on different dimensions, which is summarized in the following lemma.

Lemma 2. *Given fixed channel erasure probability p and EH probability λ, then*

(a)   value function $V(\Delta, q)$ is **non-decreasing** in Δ, i.e., for any $1 \leq \Delta_1 \leq \Delta_2$ and any battery state $q \in \mathcal{B}$, we have

$$V(\Delta_1, q) \leq V(\Delta_2, q), \tag{24}$$

and

$$V(\Delta_2, q) - V(\Delta_1, q) \geq \Delta_2 - \Delta_1. \tag{25}$$

(b)   value function $V(\Delta, q)$ is **non-increasing** in q, i.e., for AoI $\Delta \geq 1$ and any battery state $q \in \{0, 1, \ldots, B-1\}$, we have

$$V(\Delta, q) \geq V(\Delta, q+1), \tag{26}$$

Proof. According to the (21), $V(x) = \lim_{\gamma \to 1} V_\gamma(\mathbf{x}) - g$. Therefore, the monotonic properties of $V_\gamma(\mathbf{x})$ in Lemma 1 are also valid for $V(x)$, which completes the proof. $\square$

Based on Lemma 2, we will derive the **proportional differential property** of the value function in Lemma 3.

Lemma 3. *Given fixed channel erasure probability p and EH probability λ, then value function $V(\Delta, q)$ has the **proportional differential property**, i.e., the inequality*

$$\frac{V(\Delta + 1, q + 1) - V(\Delta, q + 1)}{V(\Delta + 1, q) - V(\Delta, q)} \geq p \tag{27}$$

holds for AoI $\Delta \geq 1$ and any battery state $q \in \{0, 1, ..., B - 1\}$.

Proof. See Appendix D. $\square$

With Corollary 1, Lemmas 2 and 3, we directly provide our main result in the following theorem.

Theorem 2. *Assuming that the channel erasure probability p and EH probability λ are both fixed , there exists a threshold $\Delta_q \in \mathbb{Z}^+$ for given battery state q, such that when $\Delta < \Delta_q$, the optimal action $\pi^\star(\Delta, q) = 0$, i.e., the sensor keeps idle; when $\Delta \geq \Delta_q$, the optimal action $\pi^\star(\Delta, q) = 1$, i.e., the sensor chooses to generate and transmit a new update.*

Proof. See Appendix E. $\square$

Theorem 2 reveals the threshold structure of the optimal policy: if the optimal action in a certain state is to generate and transmit an update, then in the state with the same battery state and larger AoI, the optimal action must be the same. Note that the threshold Δ_q is actually determined by the channel erasure probability p, EH probability λ and pre-defined weighting factor ω. The closed-form expression of the threshold is difficult to be derived due to the complex transition probabilities. In the next section, we will show how to compute the optimal policy numerically.

3.4. Modified Relative Value Iteration Algorithm Design

In this section, we will propose a computationally efficient algorithm to find the optimal stationary deterministic policy based on the threshold structure.

Since the state space $\mathcal{S}$ is infinite, we will use a truncated space $\mathcal{S}^N$ for approximation in practice, where $\mathcal{S}^N = \{(\Delta, q) | \Delta \leq N, \Delta \in \mathbb{Z}^+, q \in \mathcal{B}\}$. It can be proved that when N is large enough, the optimal policy of the approximated MDP will be identical to that of the original problem [6].

However, the value iteration algorithm in Proposition 1 for the discounted cost problem cannot be applied to the average cost problem by letting $\gamma = 1$. It does not converge because the value function $V(\cdot)$ in (20) is not unique. One can check if $V(\cdot)$ satisfies (20), a new function $V'(\cdot) = V(\cdot) + c$ also satisfies (20), where $c \in \mathbb{R}$. Therefore, we introduce a *relative value iterative* (RVI) algorithm to obtain the optimal policy of the approximate average cost MDP [45]. We choose a reference state $\hat{x} \in \mathcal{S}^N$ and set $V_0(\mathbf{x}) = 0$ for all states $\mathbf{x} \in \mathcal{S}^N$ Then for all $n \geq 0$, we have

$$V_{n+1}(\mathbf{x}) = \min_{a \in \mathcal{A}} Q_{n+1}(\mathbf{x}, a), \tag{28}$$

and $Q_{n+1}(\mathbf{x}, a)$ is obtained as follows:

$$Q_{n+1}(\mathbf{x}, a) = C(\mathbf{x}, a) + \sum_{\mathbf{x}' \in \mathcal{S}^N} \Pr(\mathbf{x}' | \mathbf{x}, a) h_n(\mathbf{x}'), \tag{29}$$

where the differential value function is $h_n(\mathbf{x}) = V_n(\mathbf{x}) - V_n(\hat{\mathbf{x}})$. The equation $\lim_{n\to\infty} Q_n(\mathbf{x}, a) = Q(\mathbf{x}, a)$ holds for every state $\mathbf{x} \in \mathcal{S}^N$ and action $a \in \mathcal{A}$. Finally, we compute the optimal policy by

$$\pi^\star(\mathbf{x}) = \arg\min_{a \in \mathcal{A}} Q(\mathbf{x}, a). \tag{30}$$

Note that the optimal policy is still of a threshold structure. The corresponding proof is similar to that of Theorem 2.

Moreover, based on the RVI algorithm, we can exploit this threshold structure to reduce the computational complexity. When the optimal policy of a state $\mathbf{x}' = (\Delta', q')$ is 1, the optimal policy of state $\mathbf{x}' \in \{(\Delta, q) | \Delta > \Delta', \Delta \le N, q = q'\}$ will also be 1 without the need to calculate (30). Therefore, we propose a modified RVI algorithm, and the details are given in Algorithm 1.

Algorithm 1 Modified relative value iteration algorithm.

Input:
    Iteration number K,
    Iteration threshold ϵ,
    Maximum of AoI N,
    Maximum of battery state B,
    Reference state $\hat{\mathbf{x}}$.
Output:
    Optimal policy $\pi^\star(\mathbf{x})$ for all state $\mathbf{x}$.

1: **Initialization:** $h_0(\mathbf{x}) = 0$, for all $\mathbf{x} \in \mathcal{S}^N$
2: **for** episodes $n = 0, 1, 2, \ldots, K$ **do**
3:    **for** state $\mathbf{x} \in \mathcal{S}^N$ **do**
4:      **for** action $a \in \mathcal{A}$ **do**
5:        $Q_n(\mathbf{x}, a) \leftarrow C(\mathbf{x}, a) + \sum_{\mathbf{x}' \in \mathcal{S}^N} \Pr(\mathbf{x}'|\mathbf{x}, a) h_n(\mathbf{x}')$ `// Update the state-action value function.`
6:      **end for**
7:      $V_{n+1}(\mathbf{x}) \leftarrow \min_{a \in \mathcal{A}} Q_n(\mathbf{x}, a)$ `// Update the value function.`
8:      $h_{n+1}(\mathbf{x}) \leftarrow V_{n+1}(\mathbf{x}) - V_{n+1}(\hat{\mathbf{x}})$ `// Update the differential value function.`
9:    **end for**
10: **if** $\|h_{n+1}(\mathbf{x}) - h_n(\mathbf{x})\| \le \epsilon, \forall \mathbf{x} \in \mathcal{S}^N$ **then**
11:    **for** $\mathbf{x} = (\Delta, q) \in \mathcal{S}^N$ **do**
12:      **if** $\pi^\star(\Delta - 1, q) = 1$ **then**
13:        $\pi^\star(\mathbf{x}) \leftarrow 1,$ `// Leverage the threshold structure of the optimal policy.`
14:      **else**
15:        $\pi^\star(\mathbf{x}) \leftarrow \arg\min_{a \in \mathcal{A}} Q_n(\mathbf{x}, a)$
16:      **end if**
17:    **end for**
18:    **break**
19: **end if**
20: **end for**

---

4. Minimize Average Cost in an Unknown Environment

In the previous sections, we assumed that the channel erasure probability p and EH probability λ are known in advance. Thus, the *model-based* RVI method can be employed to obtain the optimal updating policy. However, statistics such as p and λ may be unknown and even time variant in many practical scenarios, which makes it impossible to apply modified RVI algorithm because the transition probabilities are not explicit and Equation (29) cannot be applied to estimate the state-action value function $Q(\mathbf{x}, a)$. In the field of reinforcement learning, alternatively, *model-free* methods can solve MDP problems with unknown transition probabilities. An example of a model-free algorithm is *Q-learning* [46]. Q-learning finds an optimal policy in the sense of maximizing the expected value of the total reward over any and all successive steps. However, it is only designed for discounted MDP. For the average cost problem in (10), we employ an average cost Q-learning algorithm. The basic idea of this algorithm comes from the *SMART* algorithm in [47], which is a model-free reinforcement learning algorithm proposed for semi-Markov decision problems (SMDP) under the average-reward criterion. We modify it to fit the average cost MDP problem.

The state–action value function $Q(\mathbf{x}, a)$ is essential for solving the optimal policy. When the model is unknown, as long as $Q(\mathbf{x}, a)$ can be estimated accurately, the optimal policy can also be obtained immediately by (30). So the key question is how to estimate the $Q(\mathbf{x}, a)$ function, or equivalently, the value of all state–action pairs. Similar to Q-learning, the average cost Q-learning algorithm uses the minimum value of the next state–action pairs to update the value of the current state–action pair. Moreover, it needs to estimate the shift value g by averaging all immediate cost.

Specifically, the average cost Q-learning algorithm learns $Q(\mathbf{x}, a)$ by episodes. Each episode contains several iterations, and each iteration corresponds to one time slot. Then in the nth time slot of an episode, the algorithm first observes the current state $\mathbf{x}[n] = (\Delta[n], q[n])$, selects an action $a[n]$ according to the ϵ-greedy policy:

$$
a[n] = \begin{cases} \arg\min\limits_{a \in \mathcal{A}} Q(\mathbf{x}[n], a), & \text{with probability } 1 - \epsilon, \\ \text{random action}, & \text{otherwise}. \end{cases} \tag{31}
$$

By (9), the immediate cost $C[n] = \Delta[n] + \omega C_r a[n](1 - u(q[n]))$ occurs, and the system will transit to the next state $\mathbf{x}[n + 1]$. The value of $Q(\mathbf{x}[n], a[n])$ is updated as follows:

$$
Q(\mathbf{x}[n], a[n]) = (1 - \alpha[n])Q(\mathbf{x}[n], a[n]) + \alpha[n](C[n] - g + \min\limits_{a \in \mathcal{A}} Q(\mathbf{x}[n + 1], a)), \tag{32}
$$

where $\alpha[n]$ is the learning rate. The shift value g is updated as follows:

$$
g = (1 - \beta[n])g + \beta[n]C[n] \tag{33}
$$

where $\beta[n] = \frac{1}{n}$. The details are given in Algorithm 2. We leverage the parameter ϵ to balance exploration and exploitation. As the number of epochs increases, the learned $Q(\mathbf{x}, a)$ value will approach its true value, so we can gradually decrease ϵ to 0 to reduce invalid exploration. At the same time, the shift value g will also be close to the optimal average cost $g^\star$ in (20). Note that in [47], the shift value g is updated only in a non-exploratory time slot. Here we update it by simply averaging all cost, similar to [48]. The performance comparison of the average cost Q learing algorithm and modified RVI algorithm is shown in the next section.

Algorithm 2 Average cost Q-learning algorithm.

Input:
    Maximum number of episodes K,
    Maximum iteration number of an episode N_e,
    Maximum of AoI N,
    Maximum of battery state B,
    Initial value of $Q^{N \times B \times 2} \leftarrow \mathbf{0}$,
    Initial value of $\epsilon \leftarrow 0$,
    Initial value of the shift value $g \leftarrow 0$.
Output:
    Learned policy $\pi(\mathbf{x})$ for all state $\mathbf{x}$,
    Average cost $g^\star$ by following the policy π.

1: **for** episodes $k = 0, 1, 2, \ldots, K$ **do**
2:    $g \leftarrow 0$ `// Initialize the shift value at the beginning of every episode.`
3:    **for** $n = 1, 2, \ldots, N_e$ **do**
4:       Observe the current state $\mathbf{x}[n]$
5:       Select an action $a[n]$ according to ϵ-greedy policy in (31)
6:       Calculate immediate cost $C[n] \leftarrow \Delta[n] + \omega C_r a[n](1 - u(q[n]))$
7:       Observe the next state $\mathbf{x}[n+1]$
8:       $\alpha[n] \leftarrow \frac{1}{\sqrt{n}}$
9:       $Q(\mathbf{x}[n], a[n]) \leftarrow (1 - \alpha[n])Q(\mathbf{x}[n], a[n]) + \alpha[n](C[n] - g + \min_{a \in \mathcal{A}} Q(\mathbf{x}[n+1], a))$ `//` `Update the state-action value function.`
10:      $\beta[n] \leftarrow \frac{1}{n}$
11:      $g \leftarrow (1 - \beta[n])g + \beta[n]C[n]$ `// Update the shift value.`
12:    **end for**
13:    Decrease ϵ
14: **end for**
15: **for** $\mathbf{x} = (\Delta, q) \in \mathcal{S}^N$ **do**
16:    $\pi(\mathbf{x}) \leftarrow \arg\min_{a \in \mathcal{A}} Q(\mathbf{x}, a)$ `// Calculate the learned policy` π.
17: **end for**

---

5. Numerical Results

In this section, we first show the threshold structure of the optimal policy by the simulation results. Then we compare the performance of the optimal policy with the following representative policies under different system parameters:

- Zero-wait policy [4]. The sensor generates and transmits an update in every time slot.
- Periodic policy. The sensor periodically generates and sends updates to the destination.
- Randomized policy. The sensor chooses to send an update or stay idle in each time slot with the same probability.
- Energy first policy. The sensor only uses the harvested energy, that is, as long as the battery state is not zero, it will choose to sense and send updates, otherwise it will remain idle.

Moreover, we will show the average cost Q-learning algorithm performs very close to the modified RVI with known statistics. We will also compare age and reliable energy cost trade-off curves of the optimal updating policies under EH supply, reliable energy supply and mixed energy supplies. Finally, we compare the performance of the optimal policy under the only EH supply and unit-sized battery setting with the prior results in [23,25].

5.1. Simulation Setup

In our simulations, we set the maximum of AoI $N = 500$, and the maximum of battery state $B = 20$. So the finite state space $\mathcal{S}^N = \{(\Delta, q) | \Delta \leq 500, \Delta \in \mathbb{Z}^+, q \in \mathcal{B}\}$. The cost of reliable energy C_r for one update is equal to 2. For the modified RVI algorithm, we set the iteration number $K = 1000$, iteration threshold $\epsilon = 10^{-5}$ and reference state $\hat{\mathbf{x}} = (1, B)$.

The optimal policy and other baseline policies are run for $T = 10{,}000$ time slots to compute the average cost. For the average cost Q-learning algorithm, we set the total number of episodes $K = 1000$, and the maximum iteration number in an episode $N_e = 100{,}000$.

5.2. Results

Figure 3 shows the optimal policy under different system parameters. All the subfigures in Figure 3 exhibit the threshold structure described in Theorem 2. Intuitively, when ω is very small, the optimal action for every state should be 1, and when ω is very large, the optimal action for battery state $q = 0$ should be 0. It can be observed from Figure 3a,b that when ω is small (i.e., $\omega = 0.1$), the optimal policy is to update for every state, which is exactly the zero-wait policy. Figure 3 also shows that when ω is relatively large (e.g., $\omega = 10$), and the AoI is small, even if the battery state is not zero, the optimal action in the corresponding state is to keep idle. When the AoI is large or the battery state is large, the optimal action is to measure and send updates. Moreover, in all the subfigures, the threshold Δ_q keeps monotonically non-increasing with the battery state q. However, this conclusion has not been rigorously proven.

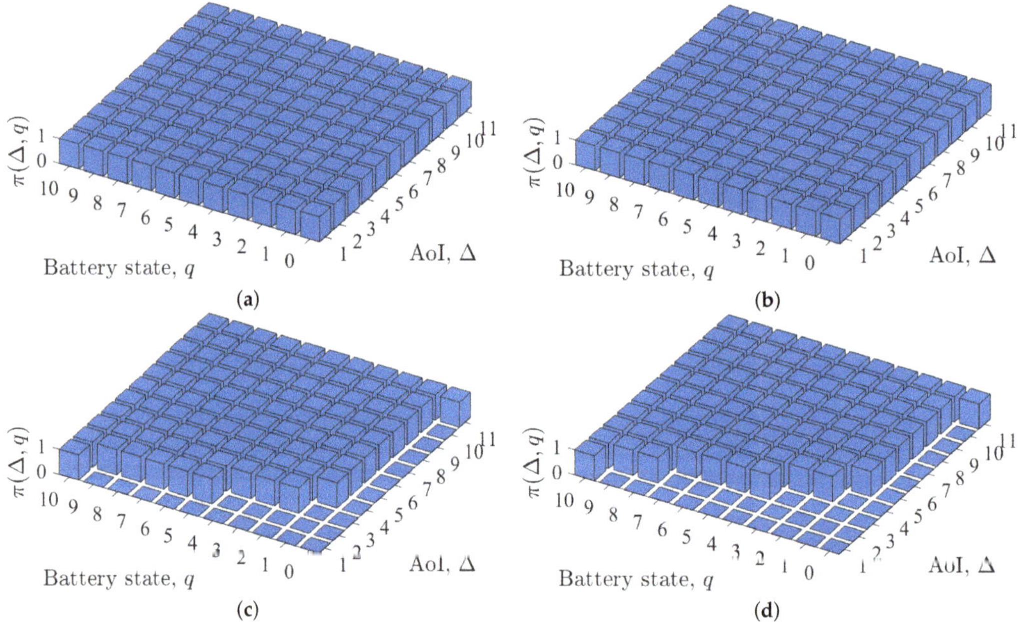

Figure 3. Optimal policy conditioned on different parameters: (**a**) $\omega = 0.1$, $p = 0.2$, $\lambda = 0.5$, (**b**) $\omega = 0.1$, $p = 0.4$, $\lambda = 0.5$, (**c**) $\omega = 10$, $p = 0.2$, $\lambda = 0.5$ and (**d**) $\omega = 10$, $p = 0.4$, $\lambda = 0.5$.

Figure 4 shows the time average cost with respect to ω under different policies. Here, we set the period of the periodic policy to 5 and 10 for comparison without loss of generality. It can be found that under different weighting factor ω, the optimal policy proposed in this paper can obtain the minimum long-term average cost compared with the other policies, which indicates the best trade-off between the average AoI and the cost of reliable energy. When ω tends to 0, the zero-wait policy tends to be optimal. Since there is no need to consider the update cost brought by paid reliable backup energy, the optimal policy should maximize the utilization of the updating opportunities.

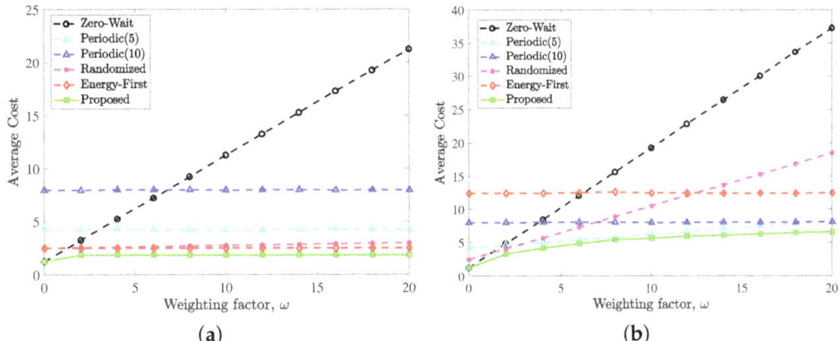

Figure 4. Performance comparison of the proposed optimal policy, zero-wait policy, periodic policy (period = 5), periodic policy (period = 10), randomized policy and energy first policy versus the weighting factor ω with simulation conditions: (**a**) $p = 0.2$, $\lambda = 0.5$ and (**b**) $p = 0.2$, $\lambda = 0.1$.

It can also be observed from Figure 4 that the growth of the optimal policy curve slows down as ω increases. This is because the optimal policy in the case of large ω does not tend to use the reliable energy when battery state $q = 0$, but prefers to wait for harvested energy, as shown in Figure 3. Since the EH probability is constant, the average AoI does not change much, resulting in no significant increase in the total average cost. Comparing Figure 4a,b, it is found that the larger the λ, the smaller the average cost variation with ω. This is because there is not much opportunity for the sensor to use reliable energy in the case of sufficient harvested energy.

Figure 5 reveals the impact of EH probabilities λ. In Figure 5a,b, we set $p = 0.2$, $\omega = 10$ and $p = 0.2, \omega = 1$, respectively.

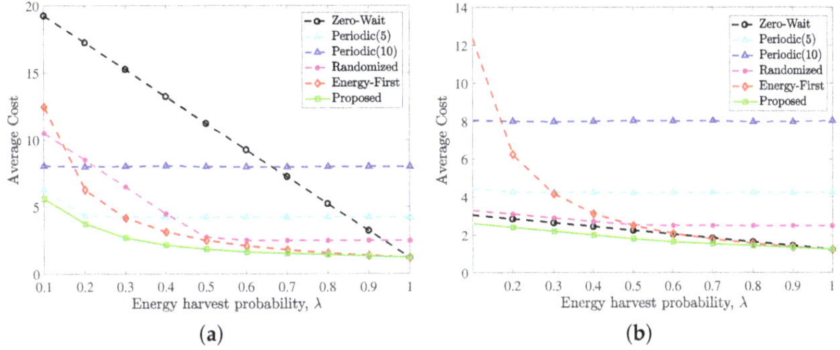

Figure 5. Performance comparison of the proposed optimal policy, zero-wait policy, periodic policy (period = 5), periodic policy (period = 10), randomized policy and energy first policy versus the EH probability λ with simulation conditions: (**a**) $p = 0.2$, $\omega = 10$ and (**b**) $p = 0.2$, $\omega = 1$.

It can also be found from both Figure 5a,b that the proposed optimal update policy outperforms all other policies under different EH probability. The interesting point is that when the EH probability tends to 1, i.e., energy arrives in each time slot, the performance of the zero-wait policy and the energy first policy is equal to the optimal policy, while there is still a performance gap between the optimal policy and the other two polices. This is intuitive because when the free harvested energy is sufficient, the optimal policy must be to generate and transmit updates in every time slot. However, the periodic policy and the randomized policy still keep idle in many time slots, which will lead to a higher average AoI and thus increase the average cost. Results show that the performance of zero-wait policy approaches the optimal policy for large λ, which is consistent with our findings in Figure 4.

In Figure 6, we compare the five policies under different channel erasure probability p.

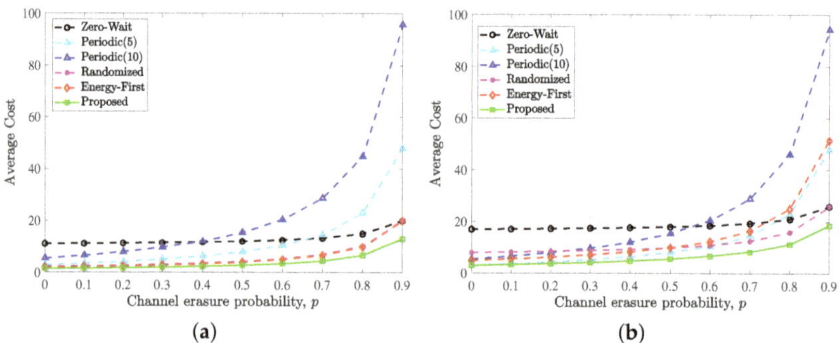

Figure 6. Performance comparison of the proposed optimal policy, zero-wait policy, periodic policy (period = 5), periodic policy (period = 10), randomized policy and energy first policy versus the erasure probability p with simulation conditions: (**a**) $\lambda = 0.5$, $\omega = 10$ and (**b**) $\lambda = 0.2$, $\omega = 10$.

It can be found that when the erasure probability increases from 0 to 0.9, the proposed optimal update policy always performs better than the other baseline policies. As p tends to 1, the average cost under all policies theoretically tends to infinity because all updates will be erased by the noisy channel and cannot be received by the destination. The simulation results confirmed this conjecture. Comparing Figure 6a,b, we can observe that when λ is large, the energy-first strategy will be close to the optimal strategy, which is also illustrated in Figure 5.

Figure 7 shows the performance of the average cost Q-learning algorithm. In every episode, the shift value g of the last inner iteration is recorded as the average cost. It can be found from Figure 7a that the average cost achieved by Algorithm 2 converges to that obtained by the modified RVI algorithm under different EH probability λ and channel erasure probability p. The age–energy trade-off is shown in Figure 7b. By fixing λ and p and changing ω from 0 to 1000, we run the modified RVI algorithm and average cost Q-learning algorithm to obtain the corresponding trade-off curve. It can be found that the curve obtained by the average cost Q-learning algorithm is very close to the optimal trade-off curve under the same condition, which further verifies the near-optimal performance of the average cost Q-learning algorithm in an unknown environment.

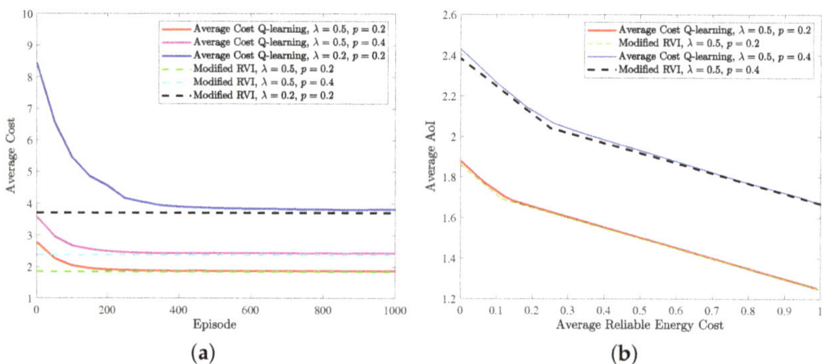

Figure 7. (**a**) Performance of the average cost Q-learning with respect to the modified RVI algorithm under different system parameters ($\omega = 10$); (**b**) age–energy trade-off curves computed by the average cost Q-learning and modified RVI algorithm.

Figure 8 shows the optimal age and reliable energy cost trade-off curves for different energy supplies. By fixing EH probability λ and channel erasure probability p and changing ω from 0 to 10,000, we run the modified RVI algorithm to get the optimal trade-off curve for mixed energy supplies. By letting EH probability $\lambda = 0$ and following the same steps, we can obtain the optimal trade-off curve for reliable energy supply. By letting weighting factor ω go to infinity, we can theoretically obtain the optimal trade-off "curve" corresponding to the EH supply. The "curve" contains only one point because the reliable energy consumption can only be 0 for the EH supply case. It should be noted that ω cannot be infinite in a simulation. Instead, we can set ω to a relatively large number (e.g., 10,000). To facilitate comparison, the channel erasure probability is set as $p = 0.2$, and the EH probability λ is set as 0.1, 0.3 and 0.7. It can be observed that the curves for the mixed energy supplies are always at the lower left of the curve for relying solely on reliable energy, which indicates that under the same average AoI, the reliable energy required by the system under the mixed energy supplies is smaller, and under the same reliable energy consumption, the AoI of the system under the mixed energy supplies is lower. The mixed energy design also achieves lower AoI than that with only EH, at the cost of paying for reliable energy. The optimal updating policy proposed in this paper makes full use of the harvested energy.

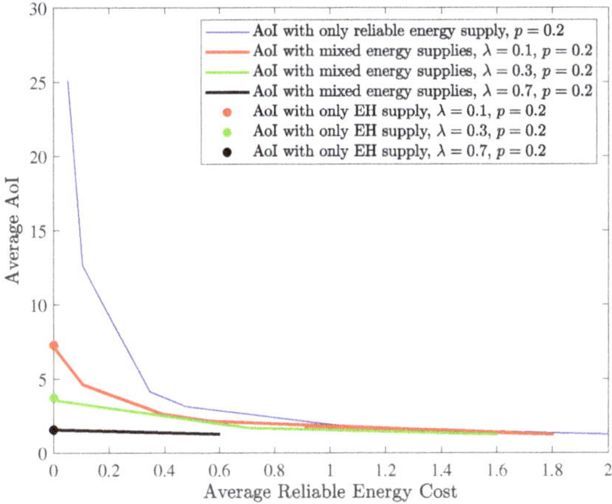

Figure 8. Age-reliable energy trade-off for different energy supplies: mixed energy supply, reliable energy supply and EH supply. The channel erasure probability $p = 0.2$, and the EH probability λ is set as 0.1, 0.3 and 0.7, respectively.

Figure 9 compares the performance of the optimal policy with the prior results in [23,25] for a special case where the sensor only uses the harvested energy and the battery capacity $B = 1$. Both [23,25] considered a continuous-time model, i.e., the energy arrival process is a Poisson process with an arrival rate of Λ energy units per *time unit* (TU), and proved that the optimal policies are threshold structure, in which a new update is generated and transmitted only if the time until the next energy arrival since the latest successful transmission exceeds a certain threshold. Specifically, [23] (Theorem 4, Equation (13)) provided the average AoI and threshold in closed-form under the optimal update policy for any energy arrival rate Λ in the error-free channel case. It is interesting that the optimal average AoI and the corresponding threshold are equal. Ref. [25] (Theorem 4, Equation (14)) extended the results of [23] to an error-prone channel case, while the energy arrival rate Λ is assumed to be 1. So we first show the results of [23,25] vs. different channel erasure probability p in Figure 9, where the energy arrival rate $\Lambda = 1$. It should be emphasized that the unit of the average AoI and threshold is TU. According to Theorems 1 and 2 in this paper, the optimal update policy exists and admits a threshold structure for any EH probability λ, channel erasure probability p, weighting factor ω and battery capacity B. This conclusion is based on the discrete-time model, i.e., the energy arrives as a Bernoulli process with parameter λ, which is different from the continuous-time model in [23,25], and the reliable backup energy is also considered. However, by the choice of some parameters (large ω, small λ), our results can be a good approximation of the results in [23,25]. First, by choosing a large ω, the reliable energy will almost never be used, and equivalently, only the EH supply exists. Secondly, by choosing a small λ, the Poisson process can be approximated as a Bernoulli process. This is because for a Poisson process with parameter Λ, we can discretize a TU into n small time slots of equal length, then according to probability theory, when n is large enough, the energy arrival process within a time slot can be approximated as a Bernoulli process with parameter $\lambda = \Lambda/n$, which is relatively small. In our simulation, we set the battery capacity $B = 1$, and take $\lambda = 0.1$ (i.e., $n = 10$) and $\omega = 10{,}000$. By changing the channel erasure probability p, we can run the modified RVI algorithm to compute the minimum average AoI and the optimal threshold. It needs to be mentioned that the unit of them is a time slot. For comparison, we need to divide their values by n to obtain the average AoI and threshold in TU. The final results are shown by the dashed lines in Figure 9. It can be observed that the results of this paper are extremely close to the explicit results in [23,25], which verifies the correctness of the analysis and also reflects the generality of our system model.

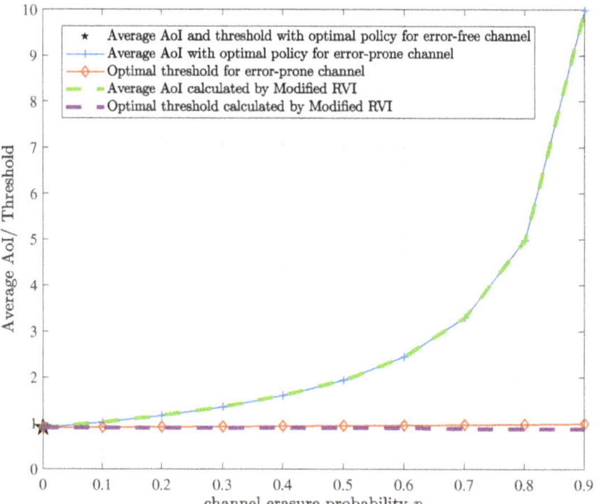

Figure 9. AoI and threshold with the proposed optimal policy for a special case where the sensor only uses the harvested energy and the battery capacity $B = 1$, and those with a unit-sized battery in [23,25] (error-free channel case and error-prone channel case, respectively), vs. the channel erasure probability p.

6. Conclusions

In this paper, we studied the optimal updating policy for an information update system, where a wireless sensor sends updates over an erasure channel using both harvested energy and reliable backup energy. Theoretical analysis indicates the threshold structure of the optimal policy and simulation results verify its performance. For the practical case where the statistics, such as the EH probability and channel erasure probability, are unknown in advance, a learning-based algorithm is proposed to compute the updating policy. Simulation results show its performance is close to that of the optimal policy. With the optimal policy, the design of mixed energy supplies can make full use of harvested energy and achieve the best age–energy trade-off. In future work, we will focus on the timeliness of the multi-sensor system under mixed energy supplies.

Author Contributions: Conceptualization, L.W., F.P., X.C. and S.Z.; methodology, L.W. and F.P.; software, L.W. and F.P.; validation, L.W. and F.P.; formal analysis, L.W. and F.P.; investigation, L.W.; writing—original draft preparation, L.W.; writing—review and editing, F.P., X.C. and S.Z.; visualization, L.W.; supervision, S.Z.; project administration, S.Z.; funding acquisition, S.Z. All authors have read and agreed to the published version of the manuscript.

Funding: This work is supported by Key Research and Development Program of China under Grant 2019YFE0113200&2019YFE0196600, Tsinghua University-China Mobile Communications Group Co.,Ltd. Joint Institute, Huawei Company Cooperation Project under Contract No. TC20210519013.

Institutional Review Board Statement: Not applicable.

Informed Consent Statement: Not applicable.

Data Availability Statement: Not applicable.

Conflicts of Interest: The authors declare no conflict of interest.

Abbreviations

The following abbreviations are used in this manuscript:

AoI	Age of Information
EH	Energy Harvesting
MDP	Markov Decision Process
RVI	Relative Value Iteration
SMART	Semi-Markov Average Reward Technique
SMDP	Semi-Markov Decision problems

Appendix A. Proof of Proposition 1

According to [15], the proof of Proposition 1 is equivalent to proving that there is a stationary deterministic policy π such that the expected discounted cost $V_\gamma^\pi(\mathbf{x})$ is finite for all $\mathbf{x}$, γ. So we can select a policy π which chooses to keep idle in each time slot. Then by (11), for any state $\mathbf{x} = (\Delta, q) \in \mathcal{S}$ and $\gamma \in (0,1)$, we have

$$
\begin{aligned}
V_\gamma^\pi(\mathbf{x}) &= \mathbb{E}_\pi\left\{ \sum_{t=0}^\infty \gamma^t C(\mathbf{x}[t], a[t]) | \mathbf{x}[0] = \mathbf{x} \right\} \\
&= \sum_{t=0}^\infty \gamma^t C(\mathbf{x}[t], a[t]) \\
&= \sum_{t=0}^\infty \gamma^t (\Delta + t) \\
&= \frac{1}{1-\gamma}\left(\Delta + \frac{\gamma}{1-\gamma}\right) < \infty,
\end{aligned}
\tag{A1}
$$

which completes the proof.

Appendix B. Proof of Lemma 1

The proof requires the use of value iteration algorithm(VIA) and mathematical induction. According to (c) in Proposition 1, The specific iteration process of VIA is as follows:

$$
\begin{cases}
V_{\gamma,0}(\mathbf{x}) = 0, \\
Q_{\gamma,k}(\mathbf{x}, a) = C(\mathbf{x}, a) + \gamma \sum_{\mathbf{x}' \in \mathcal{S}} \Pr(\mathbf{x}'|\mathbf{x}, a) V_{\gamma,k}(\mathbf{x}'), \\
V_{\gamma,k+1}(\mathbf{x}) = \min_{a \in \mathcal{A}} Q_{\gamma,k}(\mathbf{x}, a),
\end{cases}
\tag{A2}
$$

where $k \in \mathbb{Z}^+$. For any state $\mathbf{x} \in \mathcal{S}$, $V_{\gamma,k}(\mathbf{x})$ will converge when k goes into infinity:

$$
\lim_{k \to \infty} V_{\gamma,k}(\mathbf{x}) = V_\gamma(\mathbf{x}).
\tag{A3}
$$

Then we will use mathematical induction to prove the monotonicity of the value function in each component.

First let us tackle part (a) of Lemma 1.

For (17), we can verify that the inequality $V_{\gamma,1}(\Delta_1, q) \leq V_{\gamma,1}(\Delta_2, q)$ holds when $k = 1$. Then we assume that at the kth step of the induction method, the following formula holds:

$$V_{\gamma,k}(\Delta_1, q) \leq V_{\gamma,k}(\Delta_2, q), \forall \Delta_1 \leq \Delta_2. \tag{A4}$$

So the next formula that needs to be verified is

$$V_{\gamma,k+1}(\Delta_1, q) \leq V_{\gamma,k+1}(\Delta_2, q), \forall \Delta_1 \leq \Delta_2 \tag{A5}$$

Since $V_{\gamma,k+1}(\mathbf{x}) = \min_{a \in \mathcal{A}} Q_{\gamma,k}(\mathbf{x}, a)$, we need to bring out $Q_{\gamma,k}(\mathbf{x}, a)$ first. Due to the complexity of the transition probabilities and one-step cost function, we need to discuss the following three cases: $q = 0, 0 < q < B$ and $q = B$. For the sake of brevity, we only give the calculation details of the case $0 < q < B$, and the other two cases can be verified by following the exact same steps.

According to transition probability (7) and (8), we have the state-action value function $Q_{\gamma,k}(\Delta, q, 0)$ and $Q_{\gamma,k}(\Delta, q, 1)$ as follows:

$$Q_{\gamma,k}(\Delta, q, 0) = \Delta + \gamma\lambda V_{\gamma,k}(\Delta + 1, q + 1) + \gamma(1 - \lambda)V_{\gamma,k}(\Delta + 1, q), \tag{A6}$$

and

$$\begin{aligned}
Q_{\gamma,k}(\Delta, q, 1) = &\Delta + \gamma p\lambda V_{\gamma,k}(\Delta + 1, q) + \gamma p(1 - \lambda)V_{\gamma,k}(\Delta + 1, q - 1) \\
&+ \gamma(1 - p)\lambda V_{\gamma,k}(1, q) + \gamma(1 - p)(1 - \lambda)V_{\gamma,k}(1, q - 1).
\end{aligned} \tag{A7}$$

Because $V_{\gamma,k}(\Delta, q)$ is assumed to be non-decreasing function with respect to Δ for any fixed q, it is obvious that both $Q_{\gamma,k}(\Delta, q, 0)$ and $Q_{\gamma,k}(\Delta, q, 1)$ are non-decreasing with respect to Δ. Therefore, for any $\Delta_1 \leq \Delta_2$, we have

$$\begin{aligned}
V_{\gamma,k+1}(\Delta_1, q) &= \min_{a \in \mathcal{A}}\{Q_{\gamma,k}(\Delta_1, q, a)\} \\
&= \min\{Q_{\gamma,k}(\Delta_1, q, 0), Q_{\gamma,k}(\Delta_1, q, 1)\} \\
&\leq \min\{Q_{\gamma,k}(\Delta_2, q, 0), Q_{\gamma,k}(\Delta_2, q, 1)\} \\
&= V_{\gamma,k+1}(\Delta_2, q).
\end{aligned} \tag{A8}$$

As a result, with the induction we prove that $V_{\gamma,k}(\Delta, q)$ is a non-decreasing function with respect to Δ for any $q \in \{1, \ldots, B - 1\}$, i.e., the Equation (A4) holds. When k goes to infinity, combining (A3) and (A4), we prove that (17) holds in the case $0 < q < B$. In the other two cases, (17) still holds. So we have proved that (17) holds for any $q \in \mathcal{B}$.

For (18), it is easy to yield

$$\begin{aligned}
Q_\gamma(\Delta_2, q, 0) - Q_\gamma(\Delta_1, q, 0) = &\Delta_2 - \Delta_1 \\
&+ \gamma\lambda[V_\gamma(\Delta_2 + 1, q + 1) - V_\gamma(\Delta_1 + 1, q + 1)] \\
&+ \gamma(1 - \lambda)[V_\gamma(\Delta_2 + 1, q) - V_\gamma(\Delta_1 + 1, q)] \\
&\overset{(a)}{\geq} \Delta_2 - \Delta_1,
\end{aligned} \tag{A9}$$

and

$$
\begin{aligned}
Q_\gamma(\Delta_2, q, 1) - Q_\gamma(\Delta_1, q, 1) &= \Delta_2 - \Delta_1 \\
&+ \gamma p \lambda [V_\gamma(\Delta_2 + 1, q) - V_\gamma(\Delta_1 + 1, q)] \\
&+ \gamma p(1 - \lambda)[V_\gamma(\Delta_2 + 1, q - 1) - V_\gamma(\Delta_1 + 1, q - 1)] \\
&+ \gamma(1 - p)\lambda[V_\gamma(1, q) - V_\gamma(1, q)] \\
&+ \gamma(1 - p)(1 - \lambda)[V_\gamma(1, q - 1) - V_\gamma(1, q - 1)] \\
&\overset{(b)}{\geq} \Delta_2 - \Delta_1,
\end{aligned}
\tag{A10}
$$

where (a) and (b) are due to (17). Since $V_\gamma(\mathbf{x}) = \min_{a \in \mathcal{A}} Q_\gamma(\mathbf{x}, a)$, we prove that Equation (18) holds for all $q \in \{1, \ldots, B - 1\}$. Through the same proof process, it can also be verified that (18) is also valid when $q = 0$ and $q = B$. Therefore, we have completed the proof of part (a).

Second, we will tackle the part (b) of Lemma 1.

For (19), according to the exact same mathematical induction we have applied to (17), we can also verify that Equation (19) holds. Due to limited space, the details are omitted here.

Hence, we have completed the whole proof.

Appendix C. Proof of Theorem 1

By Proposition 4 in [15], it suffices to show that the following four conditions hold:

(1): For every state $\mathbf{x}$ and discount factor γ, the discount value function $V_\gamma(\mathbf{x})$ is finite.
(2): There exists a non-negative value L such that $L \leq h_\gamma(\mathbf{x})$ for all $\mathbf{x}$ and γ, where $h_\gamma(\mathbf{x}) = V_\gamma(\mathbf{x}) - V_\gamma(\hat{\mathbf{x}})$, and $\hat{\mathbf{x}}$ is a reference state.
(3): There exists a non-negative value $M(\mathbf{x})$, such that $h_\gamma(\mathbf{x}) \leq M(\mathbf{x})$ for every $\mathbf{x}$ and γ.
(4): The inequality $\sum_{\mathbf{x}' \in \mathcal{S}} Pr(\mathbf{x}'|\mathbf{x}, a)M(\mathbf{x}') < \infty$ holds for all $\mathbf{x}$ and a.

For condition (1), recall that we have verified that there exists a stationary deterministic policy π such that the expected discounted cost V_γ^π is finite in the proof of Proposition 1, and here we will extend this conclusion to any policy $\pi \in \Pi$. For any non-anticipative policy π and state $\mathbf{x} = (\Delta, q)$, we have

$$
C(\mathbf{x}[t], a[t]) = \Delta + \omega C_r a(1 - u(t)) \leq \Delta + \omega C_r.
\tag{A11}
$$

Since the AoI grows linearly at most, for any state $\mathbf{x} = (\Delta, q)$ and discounted factor γ, we have

$$
\begin{aligned}
V_\gamma(\mathbf{x}) = \min_{\pi \in \Pi} V_\gamma^\pi(\mathbf{x}) &= \min_{\pi \in \Pi} \mathbb{E}_\pi \left\{ \sum_{t=0}^\infty \gamma^t C(\mathbf{x}[t], a[t]) | \mathbf{x}[0] = (\Delta, q) \right\} \\
&\leq \sum_{t=0}^\infty \gamma^t (\Delta + t + \omega C_r) \\
&= \frac{1}{1 - \gamma}(\Delta + \omega C_r + \frac{\gamma}{1 - \gamma}) < \infty,
\end{aligned}
\tag{A12}
$$

which verifies condition (1).

Next let us focus on condition (2). By (17) and (19) in Lemma 1, $V_\gamma(\Delta, q)$ is non-decreasing with regard to age Δ and non-increasing with regard to battery state q. Hence, we can choose $L = 0$ and reference state $\hat{\mathbf{x}} = (1, B)$. Then we have $L = 0 \leq V_\gamma(\mathbf{x}) - V_\gamma(\hat{\mathbf{x}}) = h_\gamma(\mathbf{x})$, which verifies condition (2).

To prove that condition (3) holds, we need to introduce the following lemma:

Lemma A1. *Denote $\hat{x} = (1, B)$ as the reference state, and $T = \inf\{t : t \geq 0, x[t] = \hat{x}\}$ as the first hitting time from the initial state x to $\hat{x}$. Under the following lazy policy π':*

$$\pi'(\Delta, q) = \begin{cases} 1, & \text{if } q = B, \\ 0, & \text{otherwise,} \end{cases} \tag{A13}$$

the expected discounted cost from x to $\hat{x}$ is finite for all initial state $x \in \mathcal{S}$, i.e.,

$$C^{\pi'}(x) = \mathbb{E}_{\pi'}\left\{ \sum_{t=0}^{T-1} \gamma^t C(x[t], a[t]) | x[0] = x \right\} < \infty. \tag{A14}$$

Note that if $x = \hat{x}$, $C^{\pi'}(x) = 0$.

Proof. see Appendix F. $\square$

Considering a mixed non-anticipative policy π^m consisting of π' and optimal policy $\pi^\star$ for (12) from the initial state $\mathbf{x}$ as follows,

$$\pi^m(x[t]) = \begin{cases} \pi'(x[t]), & \text{if } t < T, \\ \pi^\star(x[t]), & \text{otherwise,} \end{cases} \tag{A15}$$

we have

$$V_\gamma(\mathbf{x}) \leq V_\gamma^{\pi^m}(\mathbf{x}) = \mathbb{E}_{\pi'}\left\{ \sum_{t=0}^{T-1} \gamma^t C(\mathbf{x}[t], a[t]) | \mathbf{x}[0] = \mathbf{x} \right\} + \mathbb{E}_{\pi^\star}\left\{ \sum_{t=T}^{\infty} \gamma^t C(\mathbf{x}[t], a[t]) | \mathbf{x}[T] = \hat{\mathbf{x}} \right\}$$

$$\overset{(a)}{=} C^{\pi'}(\mathbf{x}) + \mathbb{E}_{\pi^\star}\left\{ \gamma^T V_\gamma(\hat{\mathbf{x}}) \right\}$$

$$\overset{(b)}{\leq} C^{\pi'}(\mathbf{x}) + V_\gamma(\hat{\mathbf{x}}), \tag{A16}$$

where (a) is due to (A14) and (12), (b) is due to $\gamma \in (0, 1)$. Recall the definition of $h_\gamma(\mathbf{x})$, by setting $M(\mathbf{x}) = C^{\pi'}(\mathbf{x})$, the condition (3) holds.

Based on Lemma A1, $M(\mathbf{x}) = C^{\pi'}(\mathbf{x}) < \infty$ holds for any state $\mathbf{x}$. Since there will be finite possible states after transition from $\mathbf{x}$ under any action, the sum of finite $M(\cdot)$ is also finite. Hence, condition (4) holds.

Appendix D. Proof of Lemma 3

For (27), an equivalent transformation is made as follows.

$$V(\Delta + 1, q + 1) + pV(\Delta, q) \geq V(\Delta, q + 1) + pV(\Delta + 1, q). \tag{A17}$$

For every state $\mathbf{x}$, we have

$$V(\mathbf{x}) = \min_{a \in \mathcal{A}} Q(\mathbf{x}, a) = \min\{Q(\mathbf{x}, 0), Q(\mathbf{x}, 1)\}. \tag{A18}$$

So every value function in (A17) has two possible values. In order to prove Equation (A17), theoretically we need to discuss $2^4 = 16$ cases, which is obviously a bit too cumbersome. Here we use a little trick, that is, as long as we prove that for the $2^2 = 4$ possible combinations on the left hand side(LHS) of (A17), there exists a combination on the right hand side (RHS) of (A17) to make "$\geq$" hold, then we complete the proof. For convenience, we make a mapping by using four numbers to sequentially represent the action taken by the minimum state–action value function in Equation (A17). For example, "1010" represents the following:

$$Q(\Delta + 1, q + 1, 1) + pQ(\Delta, q, 0) \geq Q(\Delta, q + 1, 1) + pQ(\Delta + 1, q, 0), \tag{A19}$$

So according to the previous trick, we only need to verify "0000", "1010", "0101", and "1111" to prove Equation (A17). For brevity, we only show the verification process of "1010" in the following proof. The other three cases can also be proved by exactly the same steps.

Now we start to apply VIA and mathematical induction. Assuming that $V_0(\mathbf{x}) = 0$ for any states $\mathbf{x}$, it is easy to yield

$$V_1(\Delta + 1, q + 1) + pV_1(\Delta, q) \geq V_1(\Delta, q + 1) + pV_1(\Delta + 1, q), \tag{A20}$$

for any $q \in \{0, 1, ..., B - 1\}$ and $\Delta \in \mathbb{Z}^+$. By induction, assuming that for any $q \in \{0, 1, \ldots, B - 1\}$ and $\Delta \in \mathbb{Z}^+$, we have:

$$V_k(\Delta + 1, q + 1) + pV_k(\Delta, q) \geq V_k(\Delta, q + 1) + pV_k(\Delta + 1, q). \tag{A21}$$

What we need to do is to verify that Equation (A21) still holds in the next value iteration. Based on our previous analysis, we will focus on the "1010" case. For $\Delta \in \mathbb{Z}^+$ and $q \in \{0, 1, \ldots, B - 1\}$, we have

$$
\begin{aligned}
&Q_k(\Delta + 1, q + 1, 1) + pQ_k(\Delta, q, 0) \\
&\quad - [Q_k(\Delta, q + 1, 1) + pQ_k(\Delta + 1, q, 0)] \\
=&\Delta + 1 + p\lambda V_k(\Delta + 2, q + 1) + p(1 - \lambda)V_k(\Delta + 2, q) \\
&+ (1 - p)\lambda V_k(1, q + 1) + (1 - p)(1 - \lambda)V_k(1, q) \\
&+ p[\Delta + \omega C_r + \lambda V_k(\Delta + 1, q + 1) + (1 - \lambda)V_k(\Delta + 1, q)] \\
&- \Delta - p\lambda V_k(\Delta + 1, q + 1) - p(1 - \lambda)V_k(\Delta + 1, q) \\
&- (1 - p)\lambda V_k(1, q + 1) - (1 - p)(1 - \lambda)V_k(1, q) \\
&- p[\Delta + 1 + \omega C_r + \lambda V_k(\Delta + 2, q + 1) - (1 - \lambda)V_k(\Delta + 2, q)] \\
=&1 - p \geq 0. \tag{A22}
\end{aligned}
$$

Therefore, by the similar step, we can verify the other three cases and confirm that the following formula

$$V_{k+1}(\Delta + 1, q + 1) + pV_{k+1}(\Delta, q) \geq V_{k+1}(\Delta, q + 1) + pV_{k+1}(\Delta + 1, q) \tag{A23}$$

holds for any $\Delta \in \mathbb{Z}^+$ and $q \in \{0, 1, \ldots, B - 1\}$. Therefore, by induction, we prove that for any k, the Equation (A21) holds. Take the limits of k on both sides, then we are able to prove that (A17) holds, which indicates that (27) holds. Therefore, we have completed the proof.

Appendix E. Proof of Theorem 2

By Corollary 1, the optimal policy is of a threshold structure if $Q(\mathbf{x}, a)$ has a *sub-modular* structure, that is,

$$Q(\Delta, q, 0) - Q(\Delta, q, 1) \leq Q(\Delta + 1, q, 0) - Q(\Delta + 1, q, 1). \tag{A24}$$

We will divide the whole proof into the following three cases:
Case 1. When $q = 0$, for any $\Delta \in \mathbb{Z}^+$ we have:

$$Q(\Delta, q, 0) - Q(\Delta, q, 1)$$
$$=\Delta + \lambda V(\Delta+1, q+1) + (1-\lambda)V(\Delta+1, q)$$
$$- \Delta - \omega C_r - p\lambda V(\Delta+1, q+1) + p(1-\lambda)V(\Delta+1, q)$$
$$- (1-p)\lambda V(1, q+1) - (1-p)(1-\lambda)V(1, q)$$
$$=(1-p)\lambda(V(\Delta+1, q+1) - V(1, q+1))$$
$$+ (1-p)(1-\lambda)(V(\Delta+1, q) - V(1, q)) - \omega C_r. \tag{A25}$$

Therefore, we have

$$Q(\Delta+1, q, 0) - Q(\Delta+1, q, 1) - [Q(\Delta, q, 0) - Q(\Delta, q, 1)]$$
$$=(1-p)\lambda(V(\Delta+2, q+1) - V(\Delta+1, q+1))$$
$$+ (1-p)(1-\lambda)(V(\Delta+2, q) - V(\Delta, q))$$
$$\overset{(a)}{\geq} 0, \tag{A26}$$

where the last inequality (a) is due to (24) in Lemma 2.

Case 2. When $q \in \{1, \ldots, B-1\}$, for any $\Delta \in \mathbb{Z}^+$ we have

$$Q(\Delta+1, q, 0) - Q(\Delta+1, q, 1) - [Q(\Delta, q, 0) - Q(\Delta, q, 1)]$$
$$=Q(\Delta+1, q, 0) - Q(\Delta, q, 0) - [Q(\Delta+1, q, 1) - Q(\Delta, q, 1)]$$
$$=\lambda[V(\Delta+2, q+1) - V(\Delta+1, q+1)]$$
$$- p\lambda[V(\Delta+2, q) - V(\Delta+1, q)]$$
$$+ (1-\lambda)[V(\Delta+2, q) - V(\Delta+1, q)]$$
$$- p(1-\lambda)[V(\Delta+2, q-1) - V(\Delta+1, q-1)]$$
$$\overset{(a)}{\geq} 0, \tag{A27}$$

where the last inequality (a) is due to (27) in Lemma 3.

Case 3. When $q = B$, for any $\Delta \in \mathbb{Z}^+$ we have

$$Q(\Delta+1, q, 0) - Q(\Delta+1, q, 1) - [Q(\Delta, q, 0) - Q(\Delta, q, 1)]$$
$$=Q(\Delta+1, q, 0) - Q(\Delta, q, 0) - [Q(\Delta+1, q, 1) - Q(\Delta, q, 1)]$$
$$=(1-\lambda)[V(\Delta+2, q) - V(\Delta+1, q)]$$
$$- p(1-\lambda)[V(\Delta+2, q-1) - V(\Delta+1, q-1)]$$
$$\overset{(a)}{\geq} 0, \tag{A28}$$

where the last inequality (a) is also due to (27) in Lemma 3.

Therefore, we have completed the whole proof.

Appendix F. Proof of Lemma A1

Before dealing with the expected discounted cost $C^{\pi'}(\mathbf{x})$, we need to find the probability distribution of the first hitting time T, which is determined by the transition probabilities of system states. Under the lazy policy π', we can formulate a two-dimensional Markov chain to describe the dynamic changes of system states. The state transition probabilities of the formulated Markov chain is extremely complicated, and we can simplify it by combining some states, as depicted in Figure A1.

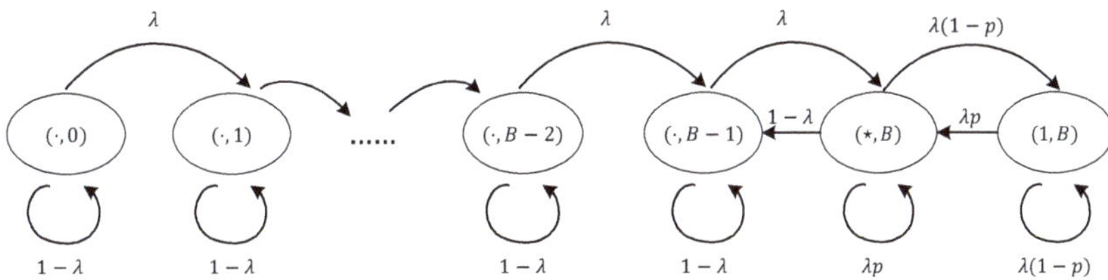

Figure A1. A simplified Markov chain of system states under the lazy policy. Note that $(1, B)$ is the reference state. $(\cdot, 1)$ means the state set $\{(\Delta, q) | \Delta \in \mathbb{Z}^+, q = 1\}$, $(\star, B)$ means the state set $\{(\Delta, q) | \Delta \in \mathbb{Z}^+, \Delta > 1, q = B\}$ and so on for the rest.

According to the simplified Markov chain, the initial state $\mathbf{x}$ can be divided into three cases: $(\star, B)$, $(\cdot, B - 1)$, and $(\cdot, q)$ where $q < B - 1$. Note that for the special case $\mathbf{x} = \hat{\mathbf{x}}$, $C^{\pi'}(x)$ is set to be 0. First, we focus on the case $\mathbf{x} = (\cdot, q)$, where $q < B - 1$. Suppose it takes $T = k$ time slots for state $\mathbf{x}$ to transit to state $\hat{\mathbf{x}}$ for the first time. Then state $\mathbf{x}' = (\cdot, B - 1)$ must be passed during these k time slots. Therefore, we can divide the entire transition process into two parts: state $\mathbf{x}$ first visits state $\mathbf{x}'$ after k_1 time slots, and then starts from state $\mathbf{x}'$ and enters state $\hat{\mathbf{x}}$ for the first time after $k_2 = k - k_1$ time slots. Denote $f^{(n)}_{\mathbf{x}_1, \mathbf{x}_2}$ as the first hitting probability from state $\mathbf{x}_1$ to state $\mathbf{x}_2$ after n time slots, then we have

$$f^{(k)}_{\mathbf{x}, \hat{\mathbf{x}}} = \sum_{k_1=0}^{k} f^{(k_1)}_{\mathbf{x}, \mathbf{x}'} f^{(k_2)}_{\mathbf{x}', \hat{\mathbf{x}}}. \tag{A29}$$

When the initial state first transits to state $\mathbf{x}'$, the total energy arrivals must be exactly $B - q - 1$. Hence, the first hitting probability $f^{(k_1)}_{\mathbf{x}, \mathbf{x}'}$ from state $\mathbf{x}$ to state $\mathbf{x}'$ can be expressed as follows:

$$
\begin{aligned}
f^{(k_1)}_{\mathbf{x}, \mathbf{x}'} &= \binom{k_1 - 1}{B - q - 2} \lambda^{B-q-2}(1 - \lambda)^{k_1 - 1 - (B - q - 2)}\lambda \\
&= \binom{k_1 - 1}{B - q - 2}\left(\frac{\lambda}{1 - \lambda}\right)^{B-q-1}(1 - \lambda)^{k_1} \\
&\overset{(a)}{\le} (k_1 - 1)^{B-q-2}\left(\frac{\lambda}{1 - \lambda}\right)^{B-q-1}(1 - \lambda)^{k_1},
\end{aligned}
\tag{A30}
$$

where $k_1 \ge B - q - 1$. The inequality (a) in (A30) is due to combination $\binom{N}{r} \le N^r, \forall N \ge r$. For any $k_1 < B - q - 1$, we have $f^{(k_1)}_{\mathbf{x}\mathbf{x}'} = 0$.

After entering state $\mathbf{x}'$, the system state will always change between states $\mathbf{x}' = (\cdot, B - 1)$ and $(\star, B)$ before entering state $\hat{\mathbf{x}}$ for the first time. By mathematical induction, $f^{(k_2)}_{\mathbf{x}', \hat{\mathbf{x}}}$ is given as follows:

$$
\begin{aligned}
f^{(k_2)}_{\mathbf{x}', \hat{\mathbf{x}}} &= \begin{bmatrix} 1 - \lambda & \lambda \end{bmatrix} \begin{bmatrix} 1 - \lambda & \lambda \\ 1 - \lambda & \lambda p \end{bmatrix}^{k_2 - 2} \begin{bmatrix} 0 \\ \lambda(1 - p) \end{bmatrix} \\
&= (1 - p)\lambda^2 \frac{\beta_1^{k_2 - 1} - \beta_2^{k_2 - 1}}{\beta_1 - \beta_2} \\
&= (1 - p)\lambda^2 \sum_{i=0}^{k_2 - 2} \beta_1^i \beta_2^{k_2 - 2 - i} \\
&\overset{(a)}{<} (1 - p)\lambda^2(k_2 - 1)\beta_1^{k_2 - 2},
\end{aligned}
\tag{A31}
$$

where $k_2 \geq 2$, β_1 and β_2 are the eigenvalues of the matrix $\begin{bmatrix} 1-\lambda & \lambda \\ 1-\lambda & \lambda p \end{bmatrix}$ and satisfy $-1 < \beta_2 < 0 < 1-\lambda < \beta_1 < 1$. The last inequality (a) of (A31) is due to $\beta_2 < 0 < \beta_1$ and $|\beta_2| < |\beta_1|$. For any $k_2 < 2$, we have $f_{\mathbf{x}',\hat{\mathbf{x}}}^{(k_2)} = 0$.

Therefore, we will verify the discounted cost from the initial state $\mathbf{x}$ to reference state $\hat{\mathbf{x}}$ is finite as follows:

$$
\begin{aligned}
C^{\pi'}(\mathbf{x}) &= \mathbb{E}_{\pi'}\left\{ \sum_{t=0}^{T-1} \gamma^t C(\mathbf{x}[t], a[t]) \,\middle|\, \mathbf{x}[0] = \mathbf{x} \right\} \\
&\overset{(a)}{\leq} \sum_{k=0}^{\infty} f_{\mathbf{x},\hat{\mathbf{x}}}^{(k)} \Big[\sum_{t=0}^{k} (\Delta + t + \omega C_r) \Big] \\
&\overset{(b)}{=} \sum_{k=B-q+1}^{\infty} \sum_{k_1=B-q-1}^{k} f_{\mathbf{x},\mathbf{x}'}^{(k_1)} f_{\mathbf{x}',\hat{\mathbf{x}}}^{(k_2)} \Big[\sum_{t=0}^{k} (\Delta + t + \omega C_r) \Big] \\
&\overset{(c)}{\leq} (1-p)\lambda^2 \frac{\left(\frac{1-\lambda}{\beta_1}\right)^{B-q-1}}{1 - \frac{1-\lambda}{\beta_1}} \sum_{k=2}^{\infty} \beta_1^{k-2} k^{B-q-1} \Big[\sum_{t=0}^{k} (\Delta + t + \omega C_r) \Big] \\
&\overset{(d)}{<} \infty.
\end{aligned}
\tag{A32}
$$

where inequality (a) is due to (A11), equality (b) is due to (A29), inequality (c) is due to (A30) and (A31), and inequality (d) is due to $0 < \beta_1 < 1$.

For the other two case where the initial state is $(\cdot, B-1)$ or $(\star, B)$, the discounted cost to the reference state for the first time can also be verified to be finite by similar steps. Therefore, we have completed the proof of Lemma A1.

References

1. Kaul, S.; Yates, R.; Gruteser, M. Real-time status: How often should one update? In Proceedings of the IEEE INFOCOM, Orlando, FL, USA, 25–30 March 2012; pp. 2731–2735.
2. Sun, Y.; Kadota, I.; Talak, R.; Modiano, E. Age of information: A new metric for information freshness. *Synth. Lect. Commun. Netw.* **2019**, *12*, 1–224. [CrossRef]
3. Yates, R.D.; Sun, Y.; Brown, D.R.; Kaul, S.K.; Modiano, E.; Ulukus, S. Age of information: An introduction and survey. *IEEE J. Sel. Areas Commun.* **2021**, *39*, 1183–1210. [CrossRef]
4. Sun, Y.; Uysal-Biyikoglu, E.; Yates, R.D.; Koksal, C.E.; Shroff, N.B. Update or wait: How to keep your data fresh. *IEEE Trans. Inf. Theory* **2017**, *63*, 7492–7508. [CrossRef]
5. Kadota, I.; Sinha, A.; Uysal-Biyikoglu, E.; Singh, R.; Modiano, E. Scheduling policies for minimizing age of information in broadcast wireless networks. *IEEE/ACM Trans. Netw.* **2018**, *26*, 2637–2650. [CrossRef]
6. Hou, Y.P.; Modiano, E.; Duan, L. Scheduling algorithms for minimizing age of information in wireless broadcast networks with random arrivals. *IEEE Trans. Mob. Comput.* **2019**, *19*, 2903–2915. [CrossRef]
7. Tang, H.; Wang, J.; Song, L.; Song, J. Minimizing age of information with power constraints: Multi-user opportunistic scheduling in multi-state time-varying channels. *IEEE J. Sel. Areas Commun.* **2020**, *38*, 854–868. [CrossRef]
8. Jackson, N.; Adkins, J.; Dutta, P. Capacity over capacitance for reliable energy harvesting sensors. In Proceedings of the 18th International Conference on Information Processing in Sensor Networks, Montreal, QC, Canada, 16–18 April 2019; pp. 193–204.
9. Ma, D.; Lan, G.; Hassan, M.; Hu, W.; Das, S.K. Sensing, computing, and communications for energy harvesting IoTs: A survey. *IEEE Commun. Surv. Tutorials* **2019**, *22*, 1222–1250. [CrossRef]
10. Sudevalayam, S.; Kulkarni, P. Energy harvesting sensor nodes: Survey and implications. *IEEE Commun. Surv. Tutorials* **2010**, *13*, 443–461. [CrossRef]
11. TEXAS Instruments. BQ25505 Ultra Low-Power Boost Charger with Battery Management and Autonomous Power Multiplexer for Primary Battery in Energy Harvester Applications. *BQ25505 Datasheet* **2019**, *3*. Available online: https://www.ti.com/lit/ds/symlink/bq25505.pdf (accessed on 10 March 2019).
12. Wu, X.; Tan, L.; Tang, S. Optimal Energy Supplementary and Data Transmission Schedule for Energy Harvesting Transmitter With Reliable Energy Backup. *IEEE Access* **2020**, *8*, 161838–161846. [CrossRef]
13. Wu, J.; Chen, W. Delay-Optimal Scheduling for Energy Harvesting Aided mmWave Communications with Random Blocking. In Proceedings of the ICC 2020—2020 IEEE International Conference on Communications (ICC), Dublin, Ireland, 7–11 June 2020; IEEE: Piscataway, NJ, USA, 2020; pp. 1–6.

14. Draskovic, S.; Thiele, L. Optimal Power Management for Energy Harvesting Systems with A Backup Power Source. In Proceedings of the 2021 10th Mediterranean Conference on Embedded Computing (MECO), Budva, Montenegro, 7–10 June 2021; IEEE: Piscataway, NJ, USA, 2021; pp. 1–9.
15. Sennott, L.I. Average cost optimal stationary policies in infinite state Markov decision processes with unbounded costs. *Oper. Res.* **1989**, *37*, 626–633. [CrossRef]
16. Yates, R.D. Lazy is timely: Status updates by an energy harvesting source. In Proceedings of the 2015 IEEE International Symposium on Information Theory (ISIT), Hong Kong, China, 14–19 June 2015; IEEE: Piscataway, NJ, USA, 2015; pp. 3008–3012.
17. Bacinoglu, B.T.; Ceran, E.T.; Uysal-Biyikoglu, E. Age of information under energy replenishment constraints. In Proceedings of the 2015 Information Theory and Applications Workshop (ITA), San Diego, CA, USA, 1–6 February 2015; IEEE: Piscataway, NJ, USA, 2015; pp. 25–31.
18. Arafa, A.; Ulukus, S. Age-minimal transmission in energy harvesting two-hop networks. In Proceedings of the GLOBECOM 2017—2017 IEEE Global Communications Conference, Singapore, 4–8 December 2017; IEEE: Piscataway, NJ, USA, 2017; pp. 1–6.
19. Arafa, A.; Ulukus, S. Timely updates in energy harvesting two-hop networks: Offline and online policies. *IEEE Trans. Wirel. Commun.* **2019**, *18*, 4017–4030. [CrossRef]
20. Arafa, A.; Ulukus, S. Age minimization in energy harvesting communications: Energy-controlled delays. In Proceedings of the 2017 51st Asilomar Conference on Signals, Systems, and Computers, Pacific Grove, CA, USA, 29 October–1 November 2017; IEEE: Piscataway, NJ, USA, 2017; pp. 1801–1805.
21. Wu, X.; Yang, J.; Wu, J. Optimal status update for age of information minimization with an energy harvesting source. *IEEE Trans. Green Commun. Netw.* **2017**, *2*, 193–204. [CrossRef]
22. Arafa, A.; Yang, J.; Ulukus, S.; Poor, H.V. Age-minimal transmission for energy harvesting sensors with finite batteries: Online policies. *IEEE Trans. Inf. Theory* **2019**, *66*, 534–556. [CrossRef]
23. Bacinoglu, B.T.; Sun, Y.; Uysal, E.; Mutlu, V. Optimal status updating with a finite-battery energy harvesting source. *J. Commun. Netw.* **2019**, *21*, 280–294. [CrossRef]
24. Feng, S.; Yang, J. Age of information minimization for an energy harvesting source with updating erasures: Without and with feedback. *IEEE Trans. Commun.* **2021**, *69*, 5091–5105. [CrossRef]
25. Arafa, A.; Yang, J.; Ulukus, S.; Poor, H.V. Timely Status Updating Over Erasure Channels Using an Energy Harvesting Sensor: Single and Multiple Sources. *IEEE Trans. Green Commun. Netw.* **2021**, *6*, 6–19 . [CrossRef]
26. Baknina, A.; Ulukus, S. Coded status updates in an energy harvesting erasure channel. In Proceedings of the 2018 52nd Annual Conference on Information Sciences and Systems (CISS), Princeton, NJ, USA, 21–23 March 2018; IEEE: Piscataway, NJ, USA, 2018; pp. 1–6.
27. Baknina, A.; Ozel, O.; Yang, J.; Ulukus, S.; Yener, A. Sending information through status updates. In Proceedings of the 2018 IEEE International Symposium on Information Theory (ISIT), Vail, CO, USA, 17–22 June 2018; IEEE: Piscataway, NJ, USA, 2018; pp. 2271–2275.
28. Ceran, E.T.; Gündüz, D.; György, A. Reinforcement learning to minimize age of information with an energy harvesting sensor with HARQ and sensing cost. In Proceedings of the IEEE INFOCOM 2019—IEEE Conference on Computer Communications Workshops (INFOCOM WKSHPS), Paris, France, 29 April–2 May 2019; IEEE: Piscataway, NJ, USA, 2019; pp. 656–661.
29. Hentati, A.; Frigon, J.F.; Ajib, W. Energy harvesting wireless sensor networks with channel estimation: Delay and packet loss performance analysis. *IEEE Trans. Veh. Technol.* **2019**, *69*, 1956–1969. [CrossRef]
30. Leng, S.; Yener, A. Age of Information Minimization for an Energy Harvesting Cognitive Radio. *IEEE Trans. Cogn. Commun. Netw.* **2019**, *5*, 427–439. [CrossRef]
31. Zheng, X.; Zhou, S.; Jiang, Z.; Niu, Z. Closed-form analysis of non-linear age of information in status updates with an energy harvesting transmitter. *IEEE Trans. Wirel. Commun.* **2019**, *18*, 4129–4142. [CrossRef]
32. Lu, Y.; Xiong, K.; Fan, P.; Zhong, Z.; Letaief, K.B. Online transmission policy in wireless powered networks with urgency-aware age of information. In Proceedings of the 2019 15th International Wireless Communications & Mobile Computing Conference (IWCMC), Tangier, Morocco, 24–28 June 2019; IEEE: Piscataway, NJ, USA, 2019; pp. 1096–1101.
33. Saurav, K.; Vaze, R. Online energy minimization under a peak age of information constraint. In Proceedings of the 2021 19th International Symposium on Modeling and Optimization in Mobile, Ad hoc, and Wireless Networks (WiOpt), Virtual, 18–21 October 2021; IEEE: Piscataway, NJ, USA, 2021; pp. 1–8.
34. Abd-Elmagid, M.A.; Dhillon, H.S. Closed-form characterization of the MGF of AoI in energy harvesting status update systems. *IEEE Trans. Inf. Theory* **2022**, *68*, 3896–3919. [CrossRef]
35. Gong, J.; Chen, X.; Ma, X. Energy-age tradeoff in status update communication systems with retransmission. In Proceedings of the 2018 IEEE Global Communications Conference (GLOBECOM), Abu Dhabi, United Arab Emirates, 9–13 December 2018; IEEE: Piscataway, NJ, USA, 2018; pp. 1–6.
36. Huang, H.; Qiao, D.; Gursoy, M.C. Age-energy tradeoff in fading channels with packet-based transmissions. In Proceedings of the IEEE INFOCOM 2020—IEEE Conference on Computer Communications Workshops (INFOCOM WKSHPS), Toronto, ON, Canada, 6–9 July 2020; IEEE: Piscataway, NJ, USA, 2020; pp. 323–328.
37. Gu, Y.; Chen, H.; Zhou, Y.; Li, Y.; Vucetic, B. Timely status update in Internet of Things monitoring systems: An age-energy tradeoff. *IEEE Internet Things J.* **2019**, *6*, 5324–5335. [CrossRef]

38. Nath, S.; Wu, J.; Yang, J. Optimum energy efficiency and age-of-information tradeoff in multicast scheduling. In Proceedings of the 2018 IEEE International Conference on Communications (ICC), Kansas City, MO, USA, 20–24 May 2018; IEEE: Piscataway, NJ, USA, 2018; pp. 1–6.
39. Gong, J.; Zhu, J.; Chen, X.; Ma, X. Sleep, Sense or Transmit: Energy-Age Tradeoff for Status Update with Two-Thresholds Optimal Policy. *IEEE Trans. Wirel. Commun.* **2021**, *21*, 1751–1765. [CrossRef]
40. Wang, L.; Peng, F.; Chen, X.; Zhou, S. Optimal Update for Energy Harvesting Sensor with Reliable Backup Energy. *arXiv* **2022**, arXiv:2201.01686.
41. Valentini, R.; Levorato, M. Optimal aging-aware channel access control for wireless networks with energy harvesting. In Proceedings of the 2016 IEEE International Symposium on Information Theory (ISIT), Barcelona, Spain, 10–15 July 2016; IEEE: Piscataway, NJ, USA, 2016; pp. 2754–2758.
42. Dong, Y.; Fan, P.; Letaief, K.B. Energy harvesting powered sensing in IoT: Timeliness versus distortion. *IEEE Internet Things J.* **2020**, *7*, 10897–10911. [CrossRef]
43. Gindullina, E.; Badia, L.; Gündüz, D. Age-of-information with information source diversity in an energy harvesting system. *IEEE Trans. Green Commun. Netw.* **2021**, *5*, 1529–1540. [CrossRef]
44. Sennott, L.I. Constrained average cost Markov decision chains. *Probab. Eng. Informational Sci.* **1993**, *7*, 69–83. [CrossRef]
45. Puterman, M.L. *Markov Decision Processes: Discrete Stochastic Dynamic Programming*; John Wiley & Sons: Hoboken, NJ, USA, 2014.
46. Sutton, R.S.; Barto, A.G. *Reinforcement Learning: An Introduction*; MIT Press: Cambridge, MA, USA, 2018.
47. Das, T.K.; Gosavi, A.; Mahadevan, S.; Marchalleck, N. Solving semi-Markov decision problems using average reward reinforcement learning. *Manag. Sci.* **1999**, *45*, 560–574. [CrossRef]
48. Ceran, E.T.; Gündüz, D.; György, A. Average age of information with hybrid ARQ under a resource constraint. *IEEE Trans. Wirel. Commun.* **2019**, *18*, 1900–1913. [CrossRef]

Article

Age Analysis of Status Updating System with Probabilistic Packet Preemption

Jixiang Zhang and Yinfei Xu *

School of Information Science and Engineering, Southeast University, Nanjing 210096, China; zhangjx@seu.edu.cn
* Correspondence: yinfeixu@seu.edu.cn

Abstract: The age of information (AoI) metric was proposed to measure the freshness of messages obtained at the terminal node of a status updating system. In this paper, the AoI of a discrete time status updating system with probabilistic packet preemption is investigated by analyzing the steady state of a three-dimensional discrete stochastic process. We assume that the queue used in the system is $Ber/Geo/1/2^*/\eta$, which represents that the system size is 2 and the packet in the buffer can be preempted by a fresher packet with probability η. Instead of considering the system's AoI separately, we use a three-dimensional state vector (n, m, l) to simultaneously track the real-time changes of the AoI, the age of a packet in the server, and the age of a packet waiting in the buffer. We give the explicit expression of the system's average AoI and show that the average AoI of the system without packet preemption is obtained by letting $\eta = 0$. When η is set to 1, the mean of the AoI of the system with a $Ber/Geo/1/2^*$ queue is obtained as well. Combining the results we have obtained and comparing them with corresponding average continuous AoIs, we propose a possible relationship between the average discrete AoI with the $Ber/Geo/1/c$ queue and the average continuous AoI with the $M/M/1/c$ queue. For each of two extreme cases where $\eta = 0$ and $\eta = 1$, we also determine the stationary distribution of AoI using the probability generation function (PGF) method. The relations between the average AoI and the packet preemption probability η, as well as the AoI's distribution curves in two extreme cases, are illustrated by numerical simulations. Notice that the probabilistic packet preemption may occur, for example, in an energy harvest (EH) node of a wireless sensor network, where the packet in the buffer can be replaced only when the node collects enough energy. In particular, to exhibit the usefulness of our idea and methods and highlight the merits of considering discrete time systems, in this paper, we provide detailed discussions showing how the results about continuous AoI are derived by analyzing the corresponding discrete time system and how the discrete age analysis is generalized to the system with multiple sources. In terms of packet service process, we also propose an idea to analyze the AoI of a system when the service time distribution is arbitrary.

Keywords: age of information; discrete time status updating system; probabilistic preemption; probability generation function; stationary distribution

Citation: Zhang, J.; Xu, Y. Age Analysis of Status Updating System with Probabilistic Packet Preemption. *Entropy* **2022**, *24*, 785. https://doi.org/10.3390/e24060785

Academic Editors: Anthony Ephremides and Yin Sun

Received: 1 May 2022
Accepted: 1 June 2022
Published: 2 June 2022

Publisher's Note: MDPI stays neutral with regard to jurisdictional claims in published maps and institutional affiliations.

1. Introduction

The freshness of transmitted messages has attracted increased attention in the design of practical communication systems. Messages obtained by a controller in a real-time monitor system may be used to perform traffic scheduling or resource allocation, and for such applications, the system's timeliness is crucial for the scheduler to make the right response and for precise control. The age of information (AoI) metric was proposed in [1] as the time elapsed since the generation time of the last received packet in the destination, which has been used widely in recent years to measure the packet's freshness and characterize the timeliness of various communication networks. A simple introduction to the AoI theory can be found in [2], and in [3], the authors made a detailed summary about the analytical results of age of information, along with employing the AoI optimization in many cyber-physical applications.

1.1. Related Work

For a status updating system with simple queue models, such as $M/M/1$, $M/D/1$, and $D/M/1$ queues, the expression of average AoI was obtained in [4–7]. In particular, in [7], the authors considered a queue using Last-Come-First-Served (LCFS) discipline, and the newer packet from the source could preempt the packet currently in service. The influence of different packet management strategies on a system's average AoI was investigated in [8,9], where only one or two packets can be stored in the system. Specifically, the average AoI of a system with three queues—that is, $M/M/1/1$, $M/M/1/2$, and $M/M/1/2^*$—was determined. The difference between last two queues lies in whether the packet waiting in the buffer can be substituted by following packets from the source. For two cases with a system size equal to 2, it was shown that updating the waiting packet with a fresher one can always result in a lower average AoI, which is apparent because transmitting the packet with a smaller age is helpful for improving the timeliness of the information transmission systems. Apart from these, the benefit of introducing a proper packet deadline, both deterministic and random, to reduce the average age of information was proved in [10–12]. Controlling packet preemptions to improve the freshness of a transmitted message was discussed in [13–15]. The authors of [16] showed that the average AoI can be significantly improved when adding a period of waiting time before the service of a new packet begins. Assuming there are two parallel servers in the status updating system, the expressions of the average AoI were determined in [17]. A freshness-based cache updating in a parallel relay network was considered in [18]. Notice that when more than one server was present, the updating packet could reach the destination through different paths. In these situations, since a packet generated behind may be transmitted to destination via a short-delay path, it is possible that this packet arrives at the receiver before the packets generated before it. Recently, many papers have been launched considering the AoI of status updating networks with simple structures, such as the status updating system with multiple sources [19–25], the system with more than one hop transmission [26–30], and the system in which the packet transmission is assisted by a relay [31–35]. Recently, using the SHS method, the AoI of an arbitrarily connected network named the gossip network was discussed in [36,37]. For each of the above systems, the average performance of the AoI was characterized, and even some properties of the AoI's distribution were obtained in certain papers. For example, for the age on a line network of preemptive memoryless servers, in [38], the author proved that the age at a node is identical in distribution to the sum of independent exponential service times by calculating the Moment Generation Function (MGF) of the defined age vector. In [39,40], the distribution of AoI was studied in a wireless networked control system with two-hop packet transmission. The authors devised the problem of minimizing the tail of the AoI distribution with respect to the sampling rate under a First-Come First-Serve (FCFS) queuing discipline. In [41], for the phase-type (PH-type) interarrival time or packet service time, the authors numerically obtained the exact distribution of the (peak) age of information for the system with $PH/PH/1/1$ and $M/PH/1/2$ queues. Within the paper, they used the sample path arguments and the theory of Markov Fluid Queues (MFQ). Except for the works we mentioned above which focus on obtaining analytical results of the AoI for status updating systems with various queue models, even more papers have been published in which the authors considered designing optimal systems under different timeliness requirements, such as in [42–51]. In such problems, usually the age of information is used as a freshness metric and is studied as the optimization objective.

1.2. Discussion of Existing Methods

As far as we know, at least three methods have been proposed to analyze the AoI of a continuous time status updating system. The first one is the method based on the graph of the AoI stochastic process, which was given in [2]. The time average AoI is obtained by calculating the area below the sample path of the AoI process. Using the common assumption that the age process is ergodic, this time average AoI converges to the AoI's mean as the observation time tends to infinity. It shows that the average AoI of a status updating system is determined by

$$\mathbb{E}[\Delta] = \frac{\mathbb{E}[YT] + \mathbb{E}[Y^2]/2}{\mathbb{E}[Y]} \tag{1}$$

when the packet arrival process and the distribution of service time are specified, in which the notation Y denotes the interarrival time between two successive updating packets, and T represents the packet system time. Secondly, in [6], the authors illustrated the usage of the Stochastic Hybrid System (SHS) approach to the analysis of system's stationary AoI. They employed a continuous state vector to track the real-time age of the updating packets from the source and described all the possible state vector transfers under the system's random dynamics—for example, if a new packet arrives, whether the packet service is completed. Then, the steady state of the multiple-dimensional continuous time Markov process was characterized by a group of differential equations, and the first few of the AoI's moments could be obtained using the theory of SHS [52]. This method was used later to determine the average AoI of more general systems, including the system with multiple sources, packet preemption, and even stochastic energy harvesting at certain system nodes. The last method was introduced in [5], where the authors proposed a novel description of the AoI process and characterized its sample paths using a new point process. They proved that the stationary distribution of the AoI can be represented in terms of the distributions of the system's delay and the peak AoI. From this point of view, large numbers of analytical formulas about the AoI's stationary distribution were obtained (in the form of its Laplace Stieljes Transform (LST)) for single-server systems. In addition, we found that the same method has been used to consider the distribution of discrete time (peak) AoI in [53,54], where the z-transform of the (peak) AoI's distribution was derived for the system with some discrete queues.

Although plenty of results have been obtained using the methods mentioned above, interested readers may find that most of the results are heavily dependent on the assumptions that the packet arrivals form a Poisson process and the service time distribution are exponential, especially for the SHS method. The memoryless property of both interarrival time distribution and the distribution of packet service time dramatically simplifies the age analysis of the considered status updating system. So far, the first method based on the graphical argument of the AoI process is used only to calculate the AoI's mean, but it seems that the theory of Level Crossing in [55] may be useful when considering the AoI's distribution from the sample paths themselves. The level crossing method has been used to derive the steady state probability density function of queue waiting in several variants of the $M/G/1$ queue. It is worth trying to determine whether this theory can be used to find the stationary distribution of continuous AoI. Using the SHS method, similarly, only the first few of the AoI's moments can be calculated. In order to obtain the distribution property of system's AoI, one has to solve the system of differential equations, which is extremely hard in general. At last, in [5], the authors pointed out that the general formula they proposed holds sample-path-wise, regardless of the service discipline or the distributions of interarrival and packet service times; however, the results they obtained are not straight-forward, as they only derived the LST of the AoI's stationary distribution, while computing the explicit expression of this distribution is also a hard problem due to the difficulty of computing the inverse of the LST. On the other hand, it is unknown if the method and the obtained formula can be generalized to more general status updating systems, not just for the system with a single server.

In the following part, we introduce the idea and methods to analyze the AoI of discrete time status updating systems and talk about their merits compared with those ways dealing with continuous time age of information. By an explicit example, we show how the results of continuous AoI can be obtained by considering the corresponding discrete time systems.

1.3. Analysis of Discrete Time AoI: Idea and Methods

We propose the idea and methods to characterize the steady state AoI of a discrete time status updating system, in which the packet arrivals, the packet service, and AoI declines are considered in discrete time slots. Although there are not many, there are still some works analyzing the AoI of a discrete system with different queue models. To our best knowledge, the analysis of discrete AoI was proposed for the first time in [56]. Using the proof techniques and tools developed to analyze continuous AoI, the authors obtained the average (peak) AoI of a $Ber/G/1$ and $G/G/\infty$ queue modeled discrete time status updating system. The notation "Ber" represents that the packet arrival or the service of the packet forms a Bernoulli stochastic process; equivalently, in each time slot, a packet arrives (or the packet service is completed), which is independent and occurs with an identical probability. Later, using the similar description of the age process's sample path as in [5], in [53,54] the expression of the discrete AoI's distribution was obtained for the system with a First-Come First-Served (FCFS) queue, the preemptive Last-Come First-Served (LCFS) queue, and the bufferless status updating system. Discrete time systems with multiple sources are considered in [57]. Under the assumption of Bernoulli packet arrivals and a common general discrete phase-type service time distribution across all the sources, the authors obtained the exact per-source distributions of AoI and peak AoI in matrix-geometric form for three different queueing disciplines, i.e., nonpreemptive bufferless, preemptive bufferless, and nonpreemptive single buffer with replacement.

In our work [58], we obtain the explicit formula of average discrete AoI, $\overline{\Delta}_{Ber/Geo/1/1}$ for a bufferless status updating system (actually, the service time distribution in [58] is arbitrary) by defining a two-dimensional age process which characterizes the AoI at the destination and the age of packet in service as a whole. The idea we proposed in [58] can be regarded as the discretization of the SHS method, which is shown to be equally powerful and more flexible when applied to more general systems. We describe all the possible state transfers for every initial state vector and then establish the stationary equations of the defined two-dimensional discrete age process. These equations are solved completely in [58]; thus, the distribution of AoI can be determined explicitly as one of the marginal distributions of the two-dimensional age process's stationary distribution. Given the AoI's distribution, the mean, the variance, and the tail probabilities of the AoI can be easily calculated. The idea of constituting multiple-dimensional age processes is then used in [59] to obtain the mean and the distribution of the infinite size state updating system. In [60], the distributions of the AoI of a system with $Ber/Geo/1/1$, $Ber/Geo/1/2$, and $Ber/Geo/1/2^*$ queues are derived explicitly using the method of solving equations. In this paper, the AoIs of a system with $Ber/Geo/1/2$ and $Ber/Geo/1/2^*$ queues are considered simultaneously, which are connected together by the probabilistic packet preemption in the system's buffer. In addition, in order to avoid the tedious calculation required to solve the stationary equations and calculate the marginal distribution, we define the Probability Generation Function (PGF) of the multiple-dimensional stationary distribution, from which both the AoI's mean and its stationary distribution can be obtained effectively. For the system's average AoI, in Table 1, we list the results we have obtained about the discrete AoI and the corresponding expressions of the continuous system's average AoI. The average AoI $\overline{\Delta}_{Ber/Geo/1/1}$ was obtained in [58], and the other two expressions will be derived in the current paper. Apart from the AoI's mean, we also determine the distribution of the discrete AoI $\Delta_{Ber/Geo/1/2}$ and $\Delta_{Ber/Geo/1/2^*}$ by writing the PGF as the power series.

As mentioned above, one can see the similarity between our idea and the SHS method, and one may mistakenly think that we simply change the continuous time into discrete time slots. The power of combining multiple-dimensional state vector descriptions with the PGF method may be underestimated due to the simple assumptions used in the current paper—that is, the packet arrivals form a Bernoulli process and the packet service time is geometrically distributed. It is known that in order to obtain the complete statistical information, not just the mean of the stationary AoI by the method of SHS, one has to solve a group of differential equations, which may be possible for some systems with simple queues but generally is impossible. In addition, the usage of SHS analysis is heavily restricted because it requires that both the packet arrival process and the packet service process are memoryless, i.e., the interarrival time and the packet service time have to be i.i.d. exponential random variables. In the following, we explain the merits of considering a discrete time system in two aspects.

Table 1. Some formulas of the average continuous and average discrete age of information.

Average Continuous and Average Discrete AoIs
$\overline{\Delta}_{M/M/1/1} = \frac{1}{\mu}\left(1 + \frac{1}{\rho} + \frac{\rho}{1+\rho}\right)$
$\overline{\Delta}_{Ber/Geo/1/1} = \frac{1}{\gamma}\left((1-\gamma) + \frac{1}{\rho_d} + \frac{\rho_d}{1/(1-\gamma)+\rho_d}\right)$
$\overline{\Delta}_{M/M/1/2} = \frac{1}{\mu}\left(1 + \frac{1}{\rho} + \frac{2\rho^2}{1+\rho+\rho^2}\right)$
$\overline{\Delta}_{Ber/Geo/1/2} = \frac{1}{\gamma}\left((1-\gamma) + \frac{1}{\rho_d} + \frac{2\rho_d^2(1-\gamma)(1-\gamma/2)}{1+\rho_d(1-2\gamma)+\rho_d^2(1-\gamma)^2}\right)$
$\overline{\Delta}_{M/M/1/2^*} = \frac{1}{\mu}\left(1 + \frac{1}{\rho} + \frac{\rho^2(1+3\rho+\rho^2)}{(1+\rho+\rho^2)(1+\rho)^2}\right)$
$\overline{\Delta}_{Ber/Geo/1/2^*} = \frac{1}{\gamma}\left((1-\gamma) + \frac{1}{\rho_d} + \frac{\rho_d^2(1-\gamma)[1+3\rho_d(1-\gamma)+\rho_d^2(1-\gamma)(1-2\gamma)]}{[1+\rho_d(1-2\gamma)+\rho_d^2(1-\gamma)^2][1+\rho_d(1-\gamma)]^2}\right)$

(1)   Calculation: reducing the complexity

Observing that when all the state transitions are described in discrete time slots, the stationary equations characterizing the steady state of the defined age process become a set of linear equations, which is more likely to be solved compared with those differential equations, we show in this paper that these linear equations can be dealt with using the PGF method even more easily and more effectively. In our another work, we have determined the explicit expression of average AoI and the corresponding AoI's distribution assuming the $Ber/Geo/1/c$ queue is used in the status updating system, where the system's size c can be arbitrary. For the cases $c = 3$ and 4, we obtain that

$$\overline{\Delta}_{Ber/Geo/1/3} = \frac{1}{\gamma}\left((1-\gamma) + \frac{1}{\rho_d} + \frac{\rho_d^2(1-\gamma^2) + 3\rho_d^3(1-5\gamma/3+\gamma^2/3)}{1 + \rho_d(1-3\gamma) + \rho_d^2(1-3\gamma+3\gamma^2) + \rho_d^3(1-\gamma)^3}\right) \tag{2}$$

and

$$\overline{\Delta}_{Ber/Geo/1/4} = \frac{1}{\gamma}\left((1-\gamma) + \frac{1}{\rho_d}\right.$$

$$\left. + \frac{\rho_d^2(1-\gamma) + 2\rho_d^3(1-\gamma)(1-2\gamma) + 4\rho_d^4(1-\gamma)(1-11\gamma/4+9\gamma^2/4-\gamma^3/4)}{1 + \rho_d(1-4\gamma) + \rho_d^2(1-4\gamma+6\gamma^2) + \rho_d^3(1-4\gamma+6\gamma^2-4\gamma^3) + \rho_d^4(1-\gamma)^4}\right) \tag{3}$$

Although we have not mentioned this yet, the readers should find that those expressions of average continuous and average discrete AoI given in Table 1 are quite similar. We propose the following possible relationship:

$$\mu \cdot \overline{\Delta}_{M/M/1/c} = \gamma \cdot \overline{\Delta}_{Ber/Geo/1/c}\big|_{\gamma=0}, \text{ then replacing } \rho_d \text{ with } \rho \tag{4}$$

The relation (4) holds at least for $c = 1$, $c = 2$, and $c = 2^*$. Note that the relation (4) is given only by observation, and it is not easy to prove that (4) is applicable in general situations, because the average continuous AoI $\overline{\Delta}_{M/M/1/c}$ is temporarily unknown. If Equation (4) is fortunately applicable in general, which we hope, then from expressions (2) and (3), immediately we have

$$\overline{\Delta}_{M/M/1/3} = \frac{1}{\mu}\left(1 + \frac{1}{\rho} + \frac{\rho^2 + 3\rho^3}{1 + \rho + \rho^2 + \rho^3}\right) \tag{5}$$

and

$$\overline{\Delta}_{M/M/1/4} = \frac{1}{\mu}\left(1 + \frac{1}{\rho} + \frac{\rho^2 + 2\rho^3 + 4\rho^4}{1 + \rho + \rho^2 + \rho^3 + \rho^4}\right) \tag{6}$$

Notice that the average continuous AoIs (5) and (6) are not derived using any of the three methods we discussed earlier—that is, the method based on the sample path of the AoI process, the SHS, and the method proposed in [5,54]. On the contrary, we first characterize the stationary AoI of the corresponding discrete time system and then obtain the expression of the continuous AoI's mean through relationship (4). There is no doubt that the formulas of $\overline{\Delta}_{M/M/1/3}$ and $\overline{\Delta}_{M/M/1/4}$ can be obtained using AoI's SHS analysis; however, the general formula of AoI's mean, i.e., $\overline{\Delta}_{M/M/1/c}$ for arbitrary size c is temporarily unknown. Furthermore, the stationary distribution of discrete AoI can also be determined explicitly from the PGF defined for the considered system, while the distribution properties of the continuous AoI cannot be revealed easily through either the graphical method or the AoI's SHS analysis. Although it is not possible to accurately reprint the continuous AoI's distribution in every position using the discrete approximation, the difference between them can be reasonably small when the length of the time slot is short enough. In the current paper, we determine the distribution expressions of discrete AoI for the system with $Ber/Geo/1/2$ and $Ber/Geo/1/2^*$ queues. Unlike in [5,54], these expressions are straight-forward and not expressed in the form of other transformations.

According to vabove discussions, from the perspective of deriving the average AoI or obtaining the AoI's distribution, considering the status updating system in the discrete time model is of great significance. To a certain extent, we can even conclude that our method is stronger since more specific results about AoI have been obtained.

(2)   Generalization: In terms of system structure and service time distribution

Recently, using the SHS method, the age analysis has been generalized to the status updating networks with a simple structure, especially the system with multiple sources. In this part, we briefly explain how the discrete age of information is characterized in the multiple-source bufferless system and the two-source system equipped with a size 1 buffer. The system models are depicted in Figure 1.

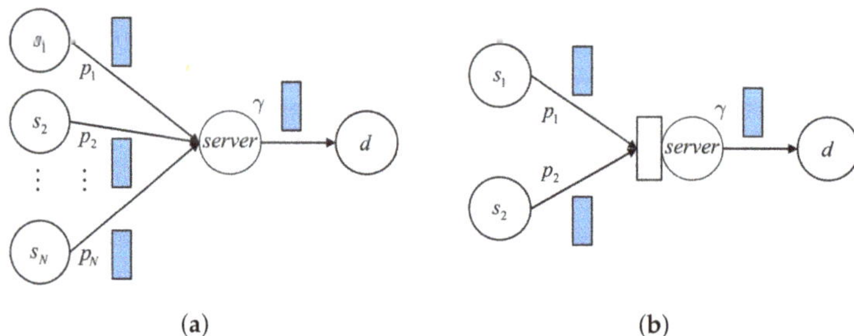

Figure 1. (**a**) Status updating system with multiple sources and bufferless server. (**b**) Status updating system with two sources and a size 1 buffer.

Specifically, we assume the packets arrive at the beginning of one time slot; whether the packet service is completed is determined at the end of the time slot. Since the system's random dynamics are considered in time slots, it is possible that more than one packet arrives to the server (buffer) from different sources in a time slot. The server has to choose one of these and discard the other packets if the system does not have a buffer. This packet collision problem can be solved by assigning priorities to the packets from different sources; then, the packet with the highest priority is selected and put into the server.

In the bufferless system given in the first picture of Figure 1, let r_i be the priority of source s_i, $1 \leq i \leq N$, and assume $r_1 > r_2 > \cdots > r_N$—that is, the priority of source s_i is over that of s_j if $i < j$. In each time slot, source s_i generates a new packet with probability p_i, and the packet generation process is independent of all other sources. Actually, this situation is exactly the generalization of our work in [58] when the status updating system has multiple independent sources. For the given $i \in [1, N]$, it shows that the AoI process corresponding to source s_i can be analyzed separately and is thus similar to work [58], showing that a two-dimensional state vector (n_i, m_i) is sufficient to track the real-time changes of AoI_i and the age of the packet in the server from source s_i. In this system, we observe that it does not matter whether the service of the packet from s_i can be preempted by other packets with higher priorities. The state vector transfers from every (n_i, m_i) can be described as in [58], but the transition probabilities need to be modified. For example, for $n_i > m_i \geq 1$, we have

$$\text{State vector at next time slot} = \begin{cases} (n_i + 1, m_i + 1) & \text{the packet service is not completed,} \\ (m_i + 1, 0) & \text{the service of the packet is over.} \end{cases} \tag{7}$$

if the service process cannot be preempted. In contrast, it can be decided that

$$\text{State vector at next time slot} = \begin{cases} (n_i + 1, m_i + 1) & \text{no packets of higher priorities arrive, the service is not over,} \\ (n_i + 1, 0) & \text{one packet with higher priority comes,} \\ (m_i + 1, 0) & \text{no packets with higher priorities arrive, the service is over.} \end{cases} \tag{8}$$

when packet service preemption is allowable. After all the state transfers are described and their transition probabilities are determined, we can obtain the stationary equations, which can be solved completely as in [58] or by using the PGF method as in this paper. Like [58], the service time distribution in this case can be arbitrary.

Although there are multiple sources, it can be seen that the age analysis of each source is easy when the status updating system has no buffer. Notice that in this case, no queue is formed before the server; thus, there is no chance that the packets from different sources are combined. As a result, the packets from every source are totally divided, and the AoI of each source can be analyzed separately. The situation is much more difficult if the system has a buffer. As an example, we consider the AoI of each source of a two-source system, which is depicted in the second picture of Figure 1.

We can define a six-dimensional state vector $(n_1, n_2, m_1, m_2, l_1, l_2)$ to describe the AoI of two sources simultaneously, where the state components represent the values of two AoIs at the destination, the age of a packet in the server, and the age of a packet in the system's buffer. In every position of the system, apart from the "age", it is necessary to indicate which source the packet comes from. Therefore, a three-dimensional state vector (n, m, l) that does not include this information is not sufficient. Notice that at any time, at most one of m_1 and m_2 are non-zero. This is the same for the parameters l_1 and l_2. When there is a buffer in front of the server, apparently a queue is formed if a packet arrives and finds that the server is currently busy. Each one of two packets in the system (one is in the server and the other is in the buffer) may come from source s_1 or s_2. Of course, these two packets may belong to different sources. Although the problem becomes complex, theoretically, all the state transfers of every initial six-dimensional state vector can be determined explicitly, since the randomness that causes the state vector transfers are limited to random packet arrival, the service of the packet, and the additional packet preemption. Then, according to the balance of probabilities in the steady state, the stationary equations are established; this solves the first half of the AoI analysis. Details of the latter half—that is, deriving the average AoI from the group of stationary equations—can be found in the procedures in this article.

We find that in [61], the authors obtained the average continuous AoI for the same two-source status updating system in Figure 1b using the SHS method. They added another assumption that the packet in the server and the packet in the buffer must belong to different sources in their second and the third considered situation and named the policies "source-aware packet management". As we have mentioned above, although the packets from two sources are still combined, after adding this restriction, the complexity of the problem has been greatly reduced.

In fact, the state vector defined for a discrete time system has a very clear physical meaning. For the status updating system with the FCFS queue, the first parameter denotes the AoI and the other state components represent the ages of packets in the server and in the buffer of the system. Thus, a $(c+1)$-dimensional state vector is needed if the size of the system is equal to c. Compared with analyzing the AoI of discrete systems, in the SHS method, the defined state vector is sometimes easier, such as in the system with multi-sources. In [61], in order to characterize the AoI of one source in a two-source system, the authors used a four-dimensional state vector $[x_0(t), x_1(t), x_2(t), x_3(t)]$ that describes the evolutions of AoI when different random events occur. As mentioned before, we use the six-dimensional state vector $(n_1, n_2, m_1, m_2, l_1, l_2)$ describing the random changes of both source 1 and source 2. The parameter n_1 or n_2 can also be deleted if only one of two sources are analyzed. In our proposed method, we show the correspondence between the dimension of the state vector and the size of the discrete system; this may not be a unique way to define the discrete state vector. Although considering the AoI of the discrete time system has higher computational complexity, the biggest advantage of discrete AoI analysis is that it can obtain the stationary distribution of the AoI.

Except the simple status updating networks given in Figure 1, we have also obtained the average discrete AoI for a status updating system with two-stage service, where for simplicity, in front of each server, no buffer is equipped. For the system with two parallel servers, the age analysis is more difficult, since some packets may become "ineffective" if one packet is generated later but arrives to the destination earlier. Some policies need to be identified to deal with these packets—for instance, deleting the packet directly once it becomes ineffective. If nothing is done, when an ineffective packet is obtained at the receiver, the value of AoI will not be reduced.

Another direction of generalization we shall talk about is the distribution of packet service time (while the packet arrival process is still Bernoulli). Now, taking the size 2 status updating system as an example, we explain how the service time distribution is relaxed to be an arbitrary distribution. Using a three-dimensional state vector (n, m, l), we can fully describe the random dynamics including the AoI at the receiver and the age of two packets in the system if both the packet interarrival time and the service time have memoryless properties. In each time slot, the changes of the AoI's value and the packet ages depend on random packet arrival, which is memoryless and independent, and whether the packet service is over. When the service time distribution is arbitrary, the probability that the service is completed in one time slot is related to the time this packet has experienced in the server. Let S be the random variable of service time, and we represent the general distribution as

$$\Pr\{S = i\} = q_i \qquad (i \geq 1) \tag{9}$$

We assume that, before the current time slot, the packet has stayed in server for j time slots; then, the probabilities that determine the state vector transfers should be the following two conditional probabilities:

$$\Pr\{S = j+1 | S > j\} \text{ and } \Pr\{S > j+1 | S > j\} \tag{10}$$

Therefore, if we have knowledge about this passed service time j, as before, all the state transfers of state vector (n, m, l) can be completely described and the age analysis becomes feasible. Since no one of the three parameters n, m, and l can provide this information, it is natural to introduce an extra component, say k, to denote the service time that the packet has consumed and constitute the four-dimensional state vector (n, m, l, k). In this way, the possible state transfers of this four-dimensional state vector can be described and the transition probabilities can also be determined. For example, let the initial state vector be (n, m, l, k)—we have the state transfers and transition probabilities as

$$
\textit{Next state vector} = \begin{cases} (n+1, m+1, l+1, k+1) & \text{the service is not over with prob. } 1 - q_{k+1} / \sum_{i=k+1}^{\infty} q_i \\ (m+1, l+1, 0, 0) & \text{the service completes with prob. } q_{k+1} / \sum_{i=k+1}^{\infty} q_i \end{cases} \tag{11}
$$

where we assume the queue discipline is FCFS and there is no packet preemption. We show that the four parameters n, m, l, and k satisfy the relationships $n > m > l \geq 0$ and $n > m \geq k \geq 0$. The first one holds because n, m, and l are three ages of packets generated in chronological order, and $n > m \geq k$ is satisfied since the packet system time m must be larger than or equal to the service time of the packet, which is denoted by k. These relations determine which vectors are qualified state vectors. Although we show that the state transfers can be analyzed and the group of stationary equations can be determined by considering the balance of those stationary probabilities; however, it can be expected that solving these equations is not easy. Since the service time probabilities q_is are arbitrary, the expression of the average AoI, as we can determine in later work, will not be closed-formed. It is also important to note that the PGF method cannot be used when the service time distribution is not geometric, because the transition probabilities is no longer the same for different state vectors and thus cannot be the common factor.

Summarizing the above discussions, we have proved that on the basis of original memoryless status updating system, by introducing an extra component to denote the time the packet has consumed in the server, the age analysis becomes feasible for the situation where the packet service time is arbitrarily distributed. Although it may be difficult to obtain the expressions of the system's average AoI, the idea is still applicable when we generalize the size 2 system to a status updating system with arbitrary size c. In one of our works, we have shown that for a size c discrete time status updating system with Bernoulli packet arrivals and geometrically distributed service time, in order to fully characterize the real time transfers of the system's AoI and all the packet ages, a $(c+1)$ dimensional state vector $(n, m_1, \cdots, m_c)$ should be defined. By adding an extra state component k that records the service time the packet has experienced in the server, according to previous discussions, the age analysis can be generalized to a size c status updating system whose service time distribution is arbitrary (at least we can establish all the stationary equations).

We have to attribute the above idea to [62], in which the authors considered the timely transmission of the updates over an erasure channel. They assume that each update consists of k symbols and the symbol erasure in each time slot is an i.i.d. Bernoulli process. The aim of [62] is to design an optimal online transmission scheme to minimize the time average AoI, and the problem is formulated as a Markov Decision Process (MDP). Although the optimization of AoI is not our interest, the state tuple (δ_t, d_t, l_t) defined in Section 2. A is very enlightening, based on which the transmission policy at the next time slot is determined. At the t-th time slot, the notation δ_t denotes the value of AoI, d_t is the age of the next update, i.e., the packet at the head of the queue, and l_t records the number of symbols that has been obtained successfully up to this time slot—these symbols belong to the update that is transmitted currently. A similar timely source coding problem was also discussed in [63], in which the authors also pointed out that the length of the encoded update is equivalent to the service time of the update, and their considered system behaves as a discrete time $Geo/G/1$ queue (we use the notation $Ber/G/1$). Therefore, the role of l_t in [62] can be regarded (or redefined) as the service time that the current update has consumed. By adding this knowledge, the distribution of the source in these papers and the service time distribution in the discrete time status updating system which we study in this part can be arbitrary.

In previous paragraphs, we explain the idea and methods used to study the AoI of discrete time status updating systems. We have shown how the discrete AoI is characterized for the basic system, the system with multiple sources, and the system whose service time distribution is arbitrary. As part of AoI theory, we believe that discrete AoI deserves more attention, and it is meaningful to establish analytical results including the AoI's mean and its distribution for more general systems. In particular, the proposed possible relationship in (4) shows that discussing discrete AoI not only has independent theoretical significance but also helps to determine certain results about continuous AoI. If one problem is difficult in the continuous time model, it is a choice to consider it in discrete time settings.

1.4. The Work in the Current Paper

We have discussed numerous topics of discrete AoI in the previous subsection, and it is inappropriate to consider all the issues in one article. In this paper, we focus on the stationary AoI of a discrete time system with a $Ber/Geo/1/2$ and $Ber/Geo/1/2^*$ queue and discuss both in a single model. We assume the packet in the buffer can be probabilistically preempted by the fresher packets from the source and define the queue model in this scenario as $Ber/Geo/1/2^*/\eta$, where η is the preemption probability. In the literature of AoI, the probabilistic packet preemption (replacement) has been studied in [64]. In [65], the probabilistic preemption was considered in the scenario where a CPU is used frequently to deal with the unpredictable tasks. Then, for the case of $\eta = 0$, the queue model of the system reduces to $Ber/Geo/1/2$, while when η is equal to 1, the status updating system with $Ber/Geo/1/2^*$ queue is obtained. For the general case, we derive the explicit expression of the system's average AoI. By writing the defined PGF as the power series, for two extreme cases of $\eta = 0$ and $\eta = 1$, the distribution expressions of two discrete AoIs are determined as well.

The rest of the paper is organized as follows. In Section 2, we describe the model of a discrete time status updating system with probabilistic packet preemption. The stationary distribution and the mean of the system's AoI are also defined. By analyzing the steady state of a three-dimensional stochastic age process, in Section 3, we obtain the explicit formula of the average AoI under general preemption probability using the probability generation function (PGF) method. In Section 4, let $\eta = 0$ and $\eta = 1$, and we determine the average AoIs $\overline{\Delta}_{Ber/Geo/1/2}$ and $\overline{\Delta}_{Ber/Geo/1/2^*}$ from the general expression derived previously in Section 3. Furthermore, in order to obtain the stationary distribution of two discrete AoIs, we write the PGF as power series. Then, the coefficient before x^n gives the probability that the AoI takes value n for each $n \geq 1$. Numerical results are placed in Section 5. For the general case, we illustrate the relationships between the average AoI and η and the traffic intensity ρ_d, respectively. In addition, the mean and the cumulative probabilities of three discrete AoIs including $\Delta_{Ber/Geo/1/1}$, $\Delta_{Ber/Geo/1/2}$, and $\Delta_{Ber/Geo/1/2^*}$ are depicted. These average discrete AoIs and their corresponding average continuous AoIs are also numerically compared in Section 5. Finally, we conclude the paper in Section 6.

2. System Model and Problem Formulation

We depict the model of the status updating system which uses the $Ber/Geo/1/2^*/\eta$ queue in Figure 2, in which the packet in the system's buffer can be preempted by a fresher packet from the source s with probability η. The packet arrivals to the transmitter are assumed to form a Bernoulli stochastic process—that is, in each time slot, a new packet comes with an identical probability, which we denote by p. Packet service time follows the geometric distribution with intensity γ. The updated packet generated at s is transmitted to the destination d through the transmitter, in which a random period of time is consumed. The age of information (AoI) at d is defined as the time elapsed since the generation time of the last obtained packet. Within the time when no packet is received, the value of AoI increases by 1 after each time slot ends. Every time a packet passes the transmitter and arrives to d, the AoI will be reduced to the system time of the obtained packet, which is actually equal to the instantaneous age of this packet.

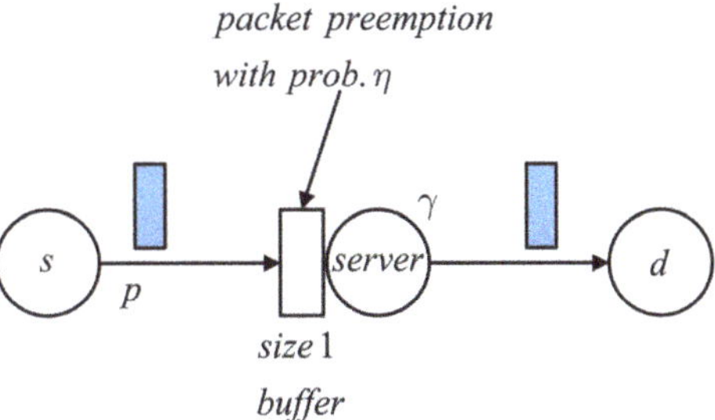

Figure 2. The model of the discrete time status updating system with probabilistic packet preemption in the system's buffer.

Let $a(k)$ be the value of AoI in the kth time slot. The AoI at the next time slot, $a(k+1)$, is determined by

$$a(k+1) = \begin{cases} a(k)+1 & \text{if no packet is obtained,} \\ a(k)+1-Y_j & \text{when } j\text{th generated packet arrives to } d. \end{cases} \tag{12}$$

where Y_j is the interarrival time between the $(j-1)$th and jth arriving packet.

Notice that these $(j-1)$th and jth packets may be generated discontinuously, since between them, some updating packets may be discarded when they arrive and find the system full. Actually, this is exactly the difference between the finite and infinite status updating systems. Based on this observation, in [59], we have determined the average AoI and its stationary distribution for an infinite size status updating system with Bernoulli packet arrivals and geometric service time.

Denote the stationary AoI for the system with probabilistic packet preemption as $\Delta_{Ber/Geo/1/2^*/\eta}$. We define the time average AoI as follows, which is equal to the mean of the AoI because the age process is assumed to be ergodic. We have

$$\overline{\Delta}_{Ber/Geo/1/2^*/\eta} = \lim_{T\to\infty} \frac{1}{T} \sum_{k=1}^{T} a(k) \tag{13}$$

$$= \lim_{T\to\infty} \frac{1}{T} \sum_{i=1}^{M_T} i \cdot |\{1 \le k \le T : a(k) = i\}| \tag{14}$$

$$= \sum_{i=1}^{\infty} i \cdot \pi_i \tag{15}$$

where $|\{1 \le k \le T : a(k) = i\}|$ is the times that the AoI takes value i, and $M_T = \max_{1 \le k \le T} a(k)$ is the maximal discrete AoI in T time slots. For each $i \ge 1$,

$$\pi_i = \lim_{T\to\infty} \frac{|\{1 \le k \le T : a(k) = i\}|}{T} \tag{16}$$

is the probability that the stationary AoI takes value i. In fact, the probability distribution $\{\pi_i, i \ge 1\}$ forms the stationary distribution of the AoI $\Delta_{Ber/Geo/1/2^*/\eta}$.

The randomness of both packet arrivals and the service time in the server, along with the probabilistic packet preemption in the system's buffer, together make the AoI at the destination change randomly. After one time slot, the value of AoI may increase by 1 if no packet is obtained or drop to the age of the obtained packet at that time if one such packet is successfully received. In order to fully describe these random dynamics of AoI, we propose to use a three-dimensional state vector to simultaneously record the changes of the AoI, the age of a packet in the server, and the age of the packet waiting in the buffer, and then constitute the three-dimensional stochastic process. Next, the steady-state of this multiple-dimensional age process is analyzed. To obtain the mean and the distribution of AoI, we define the PGF corresponding to the stationary distribution of the three-dimensional age process, from which both the AoI's mean and its distribution can be obtained. The detailed analysis of the system's AoI is given in Section 3.

3. AoI Analysis for Status Updating System with Probabilistic Packet Preemption

Define the three-dimensional state vector (n, m, l), where we use n to denote the AoI at destination d, and the other two parameters m and l are the ages of the packets in the system's server and the buffer. In the kth time slot, if the server is busy while the buffer is empty, then n_k and m_k are greater than 0 but $l_k = 0$. When both the server and the buffer are empty, we have $m_k = l_k = 0$. In this case, the entire system is empty.

Consider the following three-dimensional age process

$$AoI_{PP} = \{(n_k, m_k, l_k) : n_k > m_k \geq l_k \geq 0, k \in \mathbb{N}\} \tag{17}$$

where the subscript "PP" in expression (17) is the abbreviation of probabilistic preemption. Notice that when the system is empty, the last two parameters m_k and l_k are both equal to 0. When there are two packets in the system, i.e., one is in the server and the other is in the buffer, we show that the state components satisfy $n_k > m_k > l_k \geq 1$, since in a path from the source to the receiver, the packet ahead always has a greater age. It is clearly shown later that this relationship facilitates the derivation of probability generation function $H_{PP}(x)$, which is defined in Equation (20).

Define three random variables A, B, and F to represent whether a packet is generated in a time slot, if the service of the packet is completed, and if the arriving packet replaces the original one in the buffer. For each possible initial state vector, according to different realizations of r.v.s (A, B, F), the state transfers of the three-dimensional state vector (n, m, l) can be described specifically. We list all of them using Table 2. For example, the third row of the table shows that a packet of age l is in the buffer and a new packet arrives, since the r.v. A takes value 1. However, $F = 0$ means that this new packet will not substitute the original one; meanwhile, $B = 0$ implies that the packet service is not over at this time slot. Summarizing all these events, the beginning state vector (n, m, l) will transfer to $(n + 1, m + 1, l + 1)$ at the next time slot, and the transition probability is determined as $p(1 - \gamma)(1 - \eta)$. The other cases in the third column of Table 2 are obtained through similar discussions.

From the state transfers given in Table 2 and the corresponding transition probabilities, we can establish all the stationary equations that characterize the steady-state of age process AoI_{PP}. Let $\pi_{(n,m,l)}$, $n > m \geq l \geq 0$ be the probability that the process stays at the state vector (n, m, l) when it reaches the steady-state; we show that these stationary probabilities $\pi_{(n,m,l)}$ satisfy the following equations.

$$\begin{cases} \pi_{(n,m,l)} = \pi_{(n-1,m-1,l-1)}[(1-p)(1-\gamma) + p(1-\gamma)(1-\eta)] & (n > m > l \geq 2) \\ \pi_{(n,m,1)} = \pi_{(n-1,m-1,0)}p(1-\gamma) + \sum_{j=1}^{m-2} \pi_{(n-1,m-1,j)}p(1-\gamma)\eta & (n > m \geq 3) \\ \pi_{(n,2,1)} = \pi_{(n-1,1,0)}p(1-\gamma) & (n \geq 3) \\ \pi_{(n,m,0)} = \pi_{(n-1,m-1,0)}(1-p)(1-\gamma) \\ \qquad + \sum_{k=n}^{\infty} \pi_{(k,n-1,m-1)}[(1-p)\gamma + p\gamma(1-\eta)] & (n > m \geq 2) \\ \pi_{(n,1,0)} = \pi_{(n-1,0,0)}p(1-\gamma) \\ \qquad + \sum_{k=n}^{\infty} \pi_{(k,n-1,0)}p\gamma + \sum_{k=n}^{\infty}\sum_{j=1}^{n-2} \pi_{(k,n-1,j)}p\gamma\eta & (n \geq 3) \\ \pi_{(2,1,0)} = \pi_{(1,0,0)}p(1-\gamma) + \sum_{k=2}^{\infty} \pi_{(k,1,0)}p\gamma \\ \pi_{(n,0,0)} = \pi_{(n-1,0,0)}(1-p) + \sum_{k=n}^{\infty} \pi_{(k,n-1,0)}(1-p)\gamma & (n \geq 2) \\ \pi_{(1,0,0)} = \sum_{n=1}^{\infty} \pi_{(n,0,0)}p\gamma \end{cases} \tag{18}$$

Table 2. The state vector transfers of age process AoI_{PP}.

Initial State Vector	Considered r.v.s	Realizations and Next State Vector
$(n, m, l),$ $n > m > l \geq 1$	$(A = 0, B)$	$(0,0){:}(n+1, m+1, l+1)$
		$(0,1){:}(m+1, l+1, 0)$
	$(A = 1, B, F)$	$(1,0,0){:}(n+1, m+1, l+1)$
		$(1,0,1){:}(n+1, m+1, 1)$
		$(1,1,0){:}(m+1, l+1, 0)$
		$(1,1,1){:}(m+1, 1, 0)$
$(n, m, 0),$ $n > m \geq 1$	(A, B)	$(0,0){:}(n+1, m+1, 0)$
		$(0,1){:}(m+1, 0, 0)$
		$(1,0){:}(n+1, m+1, 1)$
		$(1,1){:}(m+1, 1, 0)$
$(n, 0, 0),$ $n \geq 1$	$A = 0$	$(n+1, 0, 0)$
	$(A = 1, B)$	$(1,0){:}(n+1, 1, 0)$
		$(1,1){:}(1, 0, 0)$

We explain the stationary equations only for a part of the state vectors and show that the other equations in (18) can be determined in a similar manner. Firstly, for the fifth row of (18), the state vector $(n, 1, 0)$ can be obtained from $(n-1, 0, 0)$ assuming that a new packet arrives and enters the server directly, but the service does not end in a single time slot. Next, from the current state vector $(k, n-1, 0)$, $k \geq n$, if the service of the age $(n-1)$ packet is completed and a new packet arrives at the same time slot, it is observed that the packet of age $(n-1)$ will be sent to the receiver, which makes the AoI change to n at next time slot. The new packet enters the server; thus, the middle parameter of the state vector changes to 1. This gives the expected state $(n, 1, 0)$. Since in this case, the buffer is empty, when a new packet comes, it occupies the buffer directly, and no packet preemption occurs. At last, we consider the situation where the age process begins with an arbitrary state $(k, n-1, j)$ where $k > n-1 > j \geq 1$. As long as the packet service is completed and at the same time a new packet arrives preempting the original one in the buffer, again, we will obtain the state vector $(n, 1, 0)$ after one time slot. Combining all of the above cases, the stationary equation corresponding to $(n, 1, 0)$ is finally determined. In addition to the fifth row, we also explain the last equation in (18). Observing that in order to obtain the state vector $(1, 0, 0)$, the receiver needs a packet of age 1, the system has then to be emptied. This state can be transferred to only from $(n, 0, 0)$, and the service time of the newly arrived packet is restricted to be only one time slot.

To derive the expression of the average AoI $\overline{\Delta}_{Ber/Geo/1/2^*/\eta}$, we do not solve Equation (18) although this approach is feasible for the AoI analysis of tje current system. In our work [60], we analyzed the AoI of a status updating system with $Ber/Geo/1/1$, $Ber/Geo/1/2$, and $Ber/Geo/1/2^*$ queues, and the expression of the AoI's stationary distribution was determined for each case. There, we completely solved the stationary equations for each system and obtained the explicit expression for every stationary probability. Notice that this work can be regarded as a discrete correspondence of the packet management of continuous AoI in [8,9]. Assuming all the probabilities $\pi_{(n,m,l)}$ have been determined by solving Equation (18), we have

$$\Pr\left\{\Delta_{Ber/Geo/1/2^*/\eta} = n\right\} = \begin{cases} \pi_{(1,0,0)} & (n = 1) \\ \pi_{(n,0,0)} + \sum_{l=0}^{n-2} \sum_{m=l+1}^{n-1} \pi_{(n,m,l)} & (n \geq 2) \end{cases} \tag{19}$$

since the probability that the AoI takes each n is equal to the sum of all the stationary probabilities with the identical first component. Equation (19) gives the stationary distribution of the AoI, from which we can calculate the average value of AoI as

$$\overline{\Delta}_{Ber/Geo/1/2^*/\eta} = \sum_{m=1}^{\infty} n \cdot \Pr\left\{\Delta_{Ber/Geo/1/2^*/\eta} = n\right\}$$

However, the number of calculations to solve Equation (18) may be large, and apart from this, extra computations are required to determine the AoI's distribution according to Formula (19). Since the AoI is denoted by the first component, to obtain the distribution of AoI, we need to sum all the other state components. Notice that when the dimension of defined state vector is bigger, more calculations are required to determine the AoI's distribution. Therefore, we must determine the mean of AoI and its distribution in another way, i.e., the probability generation function (PGF) method.

For $0 < x \le 1$, define the probability generation function

$$H_{PP}(x) = \sum_{n=1}^{\infty} x^n \Pr\left\{\Delta_{Ber/Geo/1/2^*/\eta} = n\right\} \tag{20}$$

and we write $H_{PP}(x)$ further as

$$H_{PP}(x) = x\Pr\left\{\Delta_{Ber/Geo/1/2^*/\eta} = 1\right\} + \sum_{n=2}^{\infty} x^n \Pr\left\{\Delta_{Ber/Geo/1/2^*/\eta} = n\right\}$$

$$= x\pi_{(1,0,0)} + \sum_{n=2}^{\infty} x^n \left\{\pi_{(n,0,0)} + \sum_{l=0}^{n-2} \sum_{m=l+1}^{n-1} \pi_{(n,m,l)}\right\} \tag{21}$$

$$= \sum_{n=1}^{\infty} x^n \pi_{(n,0,0)} + \sum_{n=2}^{\infty} x^n \sum_{l=0}^{n-2} \sum_{m=l+1}^{n-1} \pi_{(n,m,l)} \tag{22}$$

$$= \sum_{n=1}^{\infty} x^n \pi_{(n,0,0)} + \sum_{l=0}^{\infty} \sum_{m=l+1}^{\infty} \sum_{n=m+1}^{\infty} x^n \pi_{(n,m,l)} \tag{23}$$

$$= \sum_{n=1}^{\infty} x^n \pi_{(n,0,0)} + \sum_{m=1}^{\infty} \sum_{n=m+1}^{\infty} x^n \pi_{(n,m,0)} + \sum_{l=1}^{\infty} \sum_{m=l+1}^{\infty} \sum_{n=m+1}^{\infty} x^n \pi_{(n,m,l)} \tag{24}$$

where in (21) we have used the probability expressions (19). Equation (23) is obtained by exchanging the summation order in (22). In Equation (24), we divide the PGF $H_{PP}(x)$ into three parts. It can be seen in the following paragraphs that the entire function (20) is obtained by determining these parts separately.

According to expression (20), immediately, we have

$$H_{PP}(1) = 1, \quad \left.\frac{\mathrm{d}H_{PP}(x)}{\mathrm{d}x}\right|_{x=1} = \overline{\Delta}_{Ber/Geo/1/2^*/\eta} \tag{25}$$

That is, the average AoI can be obtained from the PGF's derivative at point $x = 1$, and the probability that the steady state AoI equals n is determined by the coefficient before the term x^n for every $n \ge 1$.

Now, we determine the PGF $H_{PP}(x)$. For $0 < x \le 1$, define the functions

$$h_1(x) = \sum_{n=1}^{\infty} x^n \pi_{(n,0,0)}$$
$$h_2(x) = \sum_{m=1}^{\infty} \sum_{n=m+1}^{\infty} x^n \pi_{(n,m,0)}$$
$$h_3(x) = \sum_{l=1}^{\infty} \sum_{m=l+1}^{\infty} \sum_{n=m+1}^{\infty} x^n \pi_{(n,m,l)}$$

and

$$h_2^{(m)}(x) = \sum_{m=1}^{\infty} \sum_{n=m+1}^{\infty} x^m \pi_{(n,m,0)}$$

We first give the following lemma, from which the PGF $H_{PP}(x)$ can be determined completely.

Lemma 1. *For the functions $h_i(x)$, $1 \leq i \leq 3$ and $h_2^{(m)}(x)$, we have*

$$h_1(x) = \frac{p\gamma M_1 x}{1 - (1-p)x} + \frac{(1-p)\gamma x}{1 - (1-p)x} h_2^{(m)}(x) \tag{26}$$

$$h_2(x) = \frac{p(1-\gamma)x}{1 - (1-p)(1-\gamma)x} h_1(x) + \frac{p\gamma x}{[1 - (1-p)(1-\gamma)x][1 - (1-\gamma)x]} h_2^{(m)}(x) \tag{27}$$

$$h_3(x) = \frac{p(1-\gamma)x}{1 - (1-\gamma)x} h_2(x) \tag{28}$$

and it is determined that

$$h_2^{(m)}(x) = \frac{[\gamma + p^2(1-\gamma)\eta - (1-p)(1-\gamma)(1-p\eta)\gamma x] M_2 x}{[1 - (1-p)(1-\gamma)x][1 - (1-\gamma)(1-p\eta)x]} \tag{29}$$

in which the numbers M_1 and M_2 are given as

$$M_1 = \frac{(1-p)\gamma^2}{(p + \gamma - 2p\gamma)\gamma + p^2(1-\gamma)^2} \tag{30}$$

$$M_2 = \frac{p\gamma(1-\gamma)}{(p + \gamma - 2p\gamma)\gamma + p^2(1-\gamma)^2} \tag{31}$$

Proof. Lemma 1 is proved in Appendix A. $\square$

Using Lemma 1, we calculate the PGF $H_{PP}(x)$ as follows. Equation (24) shows

$$\begin{aligned}
H_{PP}(x) &= h_1(x) + h_2(x) + h_3(x) \\
&= h_1(x) + h_2(x) + \frac{p(1-\gamma)x}{1 - (1-\gamma)x} h_2(x) \\
&= h_1(x) + \frac{1 - (1-p)(1-\gamma)x}{1 - (1-\gamma)x} \left(\frac{p(1-\gamma)x}{1 - (1-p)(1-\gamma)x} h_1(x) \right. \\
&\quad \left. + \frac{p\gamma x}{[1 - (1-p)(1-\gamma)x][1 - (1-\gamma)x]} h_2^{(m)}(x) \right) \\
&= \frac{1 - (1-p)(1-\gamma)x}{1 - (1-\gamma)x} h_1(x) + \frac{p\gamma x}{[1 - (1-\gamma)x]^2} h_2^{(m)}(x)
\end{aligned} \tag{32}$$

where in (32) we have substituted Equation (27).

Using Equation (26) and merging the same terms, eventually, we obtain

$$\begin{aligned}
H_{PP}(x) = {} & \frac{p\gamma M_1 x[1 - (1-p)(1-\gamma)x]}{[1 - (1-p)x][1 - (1-\gamma)x]} \\
& + \frac{\gamma x\{1 - (1-p)[2(1-\gamma) + p\gamma]x + (1-p)^2(1-\gamma)^2 x^2\}}{[1 - (1-p)x][1 - (1-\gamma)x]^2} h_2^{(m)}(x)
\end{aligned} \tag{33}$$

in which the function $h_2^{(m)}(x)$ is given in Equation (29).

According to Formula (25), the average AoI of the system with probabilistic packet preemption is calculated in Theorem 1.

Theorem 1. *For the discrete time state updating system with a Ber/Geo/1/2*/η queue, assuming the packet waiting in the buffer can be preempted by following fresher packets with probability η, then the average age of information of this system is determined as*

$$\overline{\Delta}_{Ber/Geo/1/2^*/\eta} = \frac{(p+\gamma-p\gamma)(p+\gamma)-p\gamma}{p\gamma}M_1 + \frac{(p+\gamma-p\gamma)^2-p\gamma(1-p)}{p\gamma}\cdot\left.\frac{dh_2^{(m)}(x)}{dx}\right|_{x=1}$$

$$+\frac{\left\{(p+\gamma-p\gamma)[1-3(1-p)(1-\gamma)]-2p\gamma(1-p)\right\}p\gamma+\text{Poly}_1}{p^2\gamma^2}M_2 \tag{34}$$

in which we define

$$\text{Poly}_1 = [(p+\gamma-p\gamma)^2-p\gamma(1-p)][2p(1-\gamma)+(1-p)\gamma] \tag{35}$$

and the derivative of $h_2^{(m)}(x)$ at point 1 is calculated as

$$\left.\frac{dh_2^{(m)}(x)}{dx}\right|_{x=1} = \left(\frac{1}{p+\gamma-p\gamma}+\frac{p(1-\gamma)(1-p\eta)}{(p+\gamma-p\gamma)[\gamma+p(1-\gamma)\eta]}\right)M_2 \tag{36}$$

Let $p = \rho_d\cdot\gamma$ and substitute numbers M_1, M_2; the average AoI is also written as

$$\overline{\Delta}_{Ber/Geo/1/2^*/\eta} = \frac{1+\rho_d(1-2\gamma)+\rho_d^2(1-\gamma)^2+2\rho_d^3(1-\gamma)^2}{\rho_d\gamma[1+\rho_d(1-2\gamma)+\rho_d^2(1-\gamma)^2]}$$

$$+\frac{(1-\gamma)+\rho_d(1-\gamma)(1-2\gamma)+\rho_d^2(1-\gamma)(1-\gamma+\gamma^2)}{1+\rho_d(1-2\gamma)+\rho_d^2(1-\gamma)^2}\left\{\frac{(2-\gamma)-\rho_d\gamma(1-\gamma)}{\gamma[1+\rho_d(1-\gamma)]}\right.$$

$$\left.-\frac{(1-\gamma)-\rho_d(1-\gamma)[1+\gamma-(1-\gamma)\eta]+\rho_d^2\gamma^2(1-\gamma)\eta}{\gamma[1+\rho_d(1-\gamma)(1+\eta)+\rho_d^2(1-\gamma)^2\eta]}\right\} \tag{37}$$

where $\rho_d = p/\gamma$ is defined as the discrete traffic load.

Proof. The average AoI is determined by first computing the derivative of $H_{PP}(x)$ in (33) and then letting $x = 1$. Replacing parameter p with $\rho_d\cdot\gamma$, expression (37) is obtained eventually. Although a certain amount of calculation is required, all the computations are straight-forward.

Here, we only provide the details from obtaining Equation (36). From (29), we have

$$\left.\frac{dh_2^{(m)}(x)}{dx}\right|_{x=1} = \left.\frac{d}{dx}\frac{[\gamma+p^2(1-\gamma)\eta-(1-p)(1-\gamma)(1-p\eta)\gamma x]M_2 x}{[1-(1-p)(1-\gamma)x][1-(1-\gamma)(1-p\eta)x]}\right|_{x=1}$$

$$= \left.\frac{d}{dx}\left(M_2\frac{[\gamma+p^2(1-\gamma)\eta]x-(1-p)(1-\gamma)(1-p\eta)\gamma x^2}{1-[(1-p)(1-\gamma)+(1-\gamma)(1-p\eta)]x+(1-p)(1-\gamma)^2(1-p\eta)x^2}\right)\right|_{x=1}$$

$$= M_2\frac{\text{Poly}_2\cdot(p+\gamma-p\gamma)[\gamma+p(1-\gamma)\eta]+[\gamma+p^2(1-\gamma)\eta-(1-p)(1-\gamma)(1-p\eta)\gamma]\cdot\text{Poly}_3}{(p+\gamma-p\gamma)^2[\gamma+p(1-\gamma)\eta]^2} \tag{38}$$

where

$$\text{Poly}_2 = \gamma+p^2(1-\gamma)\eta-2(1-p)(1-\gamma)(1-p\eta)\gamma$$

and

$$\text{Poly}_3 = (1-p)(1-\gamma)+(1-\gamma)(1-p\eta)-2(1-p)(1-\gamma)^2(1-p\eta)$$

$$= (1-p)(1-\gamma)[\gamma+p(1-\gamma)\eta]+(p+\gamma-p\gamma)(1-\gamma)(1-p\eta)$$

Notice that

$$\gamma + p^2(1-\gamma)\eta - (1-p)(1-\gamma)(1-p\eta)\gamma$$
$$= \gamma + p(1-\gamma)\eta - p(1-p)(1-\gamma)\eta - (1-p)(1-\gamma)(1-p\eta)\gamma$$
$$= \gamma + p(1-\gamma)\eta - (1-p)(1-\gamma)[\gamma + p(1-\gamma)\eta]$$
$$= (p + \gamma - p\gamma)[\gamma + p(1-\gamma)\eta] \tag{39}$$

Substituting (39) into (38) results in

$$\left. \frac{\mathrm{d}h_2^{(m)}(x)}{\mathrm{d}x} \right|_{x=1} = M_2 \left(1 - \frac{(1-p)(1-\gamma)(1-p\eta)\gamma}{(p+\gamma-p\gamma)[\gamma+p(1-\gamma)\eta]} + \frac{(1-p)(1-\gamma)}{p+\gamma-p\gamma} + \frac{(1-\gamma)(1-p\eta)}{\gamma+p(1-\gamma)\eta} \right)$$
$$= M_2 \left(\frac{1}{p+\gamma-p\gamma} + \frac{p(1-\gamma)(1-p\eta)}{(p+\gamma-p\gamma)[\gamma+p(1-\gamma)\eta]} \right) \tag{40}$$

which is exactly Equation (36). $\square$

Notice that in definition (20), for each $n \geq 1$, the coefficient of x^n is the probability that the AoI equals n. In order to obtain these coefficients, we decompose the PGF $H_{PP}(x)$ into power series. This shows that

$$H_{PP}(x) = \frac{p\gamma M_1 x}{\gamma - p} \left(\frac{(1-p)\gamma}{1-(1-p)x} - \frac{p(1-\gamma)}{1-(1-\gamma)x} \right)$$
$$- \left(\frac{(1-p)^2\gamma^2 M_2 x^2}{(\gamma-p)[1-(1-p)x]} - \frac{p\gamma(1-p)(1-\gamma)M_2 x^2}{(\gamma-p)[1-(1-\gamma)x]} + \frac{p\gamma M_2 x^2}{[1-(1-\gamma)x]^2} \right)$$
$$\times \left(\frac{\eta(1-p)(p+\gamma-p\gamma)}{(1-\eta)[1-(1-p)(1-\gamma)x]} - \frac{(1-p\eta)[\gamma+p(1-\gamma)\eta]}{(1-\eta)[1-(1-\gamma)(1-p\eta)x]} \right) \tag{41}$$

when the preemption probability $\eta \neq 1$, while for the case $\eta = 1$, we have

$$H_{PP}(x) = \frac{p\gamma M_1 x}{\gamma - p} \left(\frac{(1-p)\gamma}{1-(1-p)x} - \frac{p(1-\gamma)}{1-(1-\gamma)x} \right)$$
$$+ \left(\frac{(1-p)^2\gamma^2 M_2 x^2}{(\gamma-p)[1-(1-p)x]} - \frac{p\gamma(1-p)(1-\gamma)M_2 x^2}{(\gamma-p)[1-(1-\gamma)x]} + \frac{p\gamma M_2 x^2}{[1-(1-\gamma)x]^2} \right)$$
$$\times \frac{\gamma + p^2(1-\gamma) - (1-p)^2(1-\gamma)\gamma x}{[1-(1-p)(1-\gamma)x]^2} \tag{42}$$

The details of obtaining Equations (41) and (42) are given in Appendix B. In Section 4, along with the average value of the AoI, we determine the AoI's stationary distribution for two extreme cases: $\eta = 0$ and $\eta = 1$.

4. Stationary Age of Information under Two Extreme Cases

In this section, we determine the average AoI of the status updating system without packet preemption by setting $\eta = 0$, and when the preemption probability η is equal to 1, the mean of the AoI for the $Ber/Geo/1/2^*$ queue modeled system is also derived. In addition, using Equations (41) and (42), the stationary distributions of the discrete AoI for two cases are also obtained.

Theorem 2. *Assuming the packet arrivals form a Bernoulli process and the service time is geometrically distributed, the average AoIs of the discrete time status updating system with Ber/Geo/1/2 and Ber/Geo/1/2* queues are calculated as*

$$\overline{\Delta}_{Ber/Geo/1/2} = \frac{1}{\gamma}\left((1-\gamma) + \frac{1}{\rho_d} + \frac{2\rho_d^2(1-\gamma)(1-\gamma/2)}{1+\rho_d(1-2\gamma)+\rho_d^2(1-\gamma)^2} \right) \tag{43}$$

and

$$\overline{\Delta}_{Ber/Geo/1/2^*} = \frac{1}{\gamma}\left((1-\gamma) + \frac{1}{\rho_d} + \frac{\rho_d^2(1-\gamma)\left[1+3\rho_d(1-\gamma)+\rho_d^2(1-\gamma)(1-2\gamma)\right]}{\left[1+\rho_d(1-2\gamma)+\rho_d^2(1-\gamma)^2\right]\left[1+\rho_d(1-\gamma)\right]^2} \right) \tag{44}$$

For each $n \geq 1$, the distribution of the AoI $\Delta_{Ber/Geo/1/2}$ is given by

$$\Pr\{\Delta_{Ber/Geo/1/2} = n\} = \frac{p\gamma^2(1-p)M_1}{(\gamma-p)^2}(1-p)^n - \frac{(\gamma-p^2)(1-p)\gamma^2 M_2}{(\gamma-p)^2}(1-\gamma)^{n-1}$$
$$- \frac{p\gamma^2(1-p)M_2}{\gamma-p}(n-1)(1-\gamma)^{n-1} + \frac{p\gamma^2 M_2}{2}n(n-1)(1-\gamma)^{n-2} \tag{45}$$

while when the system has full packet preemption, we show that

$$\Pr\{\Delta_{Ber/Geo/1/2^*} = n\}$$
$$= \frac{p\gamma M_1}{\gamma-p}\left(\gamma(1-p)^n - p(1-\gamma)^n\right)$$
$$+ \frac{(1-p)[\gamma^2 + p(1-\gamma)(p+\gamma)]M_2}{\gamma-p}\left((1-p)^{n-1} - [(1-p)(1-\gamma)]^{n-1}\right)$$
$$- \frac{(1-p)\gamma[p+2(1-p)\gamma]M_2}{\gamma-p}\left((1-\gamma)^{n-1} - [(1-p)(1-\gamma)]^{n-1}\right) + p\gamma M_2\Big(A(1-\gamma)^{n-2}$$
$$+ B(n-1)(1-\gamma)^{n-2} + C[(1-p)(1-\gamma)]^{n-2} + D(n-1)[(1-p)(1-\gamma)]^{n-2}\Big) \tag{46}$$

in which the coefficients A, B, C, and D are determined by

$$A = \frac{2-p}{p^3}\left((1-p)^2\gamma - \frac{2(1-p)[\gamma+p^2(1-\gamma)]}{2-p} \right) \tag{47}$$

$$C = -\frac{(1-p)(2-p)}{p^3}\left((1-p)^2\gamma - \frac{2(1-p)[\gamma+p^2(1-\gamma)]}{2-p} \right) \tag{48}$$

$$D = -\frac{p^2(1-p)\cdot A + (1-p)^2[\gamma+p^2(1-\gamma)]}{p(2-p)} \tag{49}$$

and

$$B = [\gamma + p^2(1-\gamma)] - A - C - D \tag{50}$$

Proof. We first derive two average AoIs in Equations (43) and (44) from the general expression (37). Let η be 0; then, no packet preemption will occur in the system's buffer. The system's queue model reduces to $Ber/Geo/1/2$, and from (37), we can obtain the average AoI $\overline{\Delta}_{Ber/Geo/1/2}$.

In this case, it is easy to show the last two terms within the brace of (37) can be calculated to be $1/\gamma$. Thus, we have

$$\overline{\Delta}_{Ber/Geo/1/2} = \overline{\Delta}_{Ber/Geo/1/2^*/\eta}\Big|_{\eta=0}$$

$$= \frac{1 + \rho_d(1-2\gamma) + \rho_d^2(1-\gamma)^2 + 2\rho_d^3(1-\gamma)^2}{\rho_d\gamma[1 + \rho_d(1-2\gamma) + \rho_d^2(1-\gamma)^2]}$$

$$+ \frac{(1-\gamma) + \rho_d(1-\gamma)(1-2\gamma) + \rho_d^2(1-\gamma)(1-\gamma+\gamma^2)}{\gamma[1 + \rho_d(1-2\gamma) + \rho_d^2(1-\gamma)^2]}$$

$$= \frac{1 + \rho_d(2-3\gamma) + \rho_d^2(1-\gamma)(2-3\gamma) + \rho_d^3(1-\gamma)(3-3\gamma+\gamma^2)}{\rho_d\gamma[1 + \rho_d(1-2\gamma) + \rho_d^2(1-\gamma)^2]}$$

$$= \frac{1}{\rho_d\gamma}\left(1 + \rho_d(1-\gamma) + \frac{2\rho_d^3(1-\gamma)(1-\gamma/2)}{1 + \rho_d(1-2\gamma) + \rho_d^2(1-\gamma)^2}\right) \tag{51}$$

$$= \frac{1}{\gamma}\left((1-\gamma) + \frac{1}{\rho_d} + \frac{2\rho_d^2(1-\gamma)(1-\gamma/2)}{1 + \rho_d(1-2\gamma) + \rho_d^2(1-\gamma)^2}\right) \tag{52}$$

where in Equation (51) we use the method of long division.

For the other extreme case of $\eta = 1$, obviously the general expression (37) gives the average AoI $\overline{\Delta}_{Ber/Geo/1/2^*}$. Similarly, we first determine the value of the last two terms within the brace. We show that the difference of the last two terms equals

$$\frac{1}{\gamma} - \frac{\rho_d^2(1-\gamma)}{\gamma[1 + \rho_d(1-\gamma)^2]} \tag{53}$$

thus, the average AoI $\overline{\Delta}_{Ber/Geo/1/2^*}$ is calculated as

$$\overline{\Delta}_{Ber/Geo/1/2^*} = \overline{\Delta}_{Ber/Geo/1/2^*/\eta}\Big|_{\eta=1}$$

$$= \overline{\Delta}_{Ber/Geo/1/2} - \frac{(1-\gamma) + \rho_d(1-\gamma)(1-2\gamma) + \rho_d^2(1-\gamma)(1-\gamma+\gamma^2)}{1 + \rho_d(1-2\gamma) + \rho_d^2(1-\gamma)^2} \cdot \frac{\rho_d^2(1-\gamma)}{\gamma[1 + \rho_d(1-\gamma)]^2} \tag{54}$$

$$= \frac{1}{\gamma}\left((1-\gamma) + \frac{1}{\rho_d} + \frac{\rho_d^2(1-\gamma)(2-\gamma)}{1 + \rho_d(1-2\gamma) + \rho_d^2(1-\gamma)^2}\right.$$

$$\left. - \frac{\rho_d^2(1-\gamma)^2 + \rho_d^3(1-\gamma)^2(1-2\gamma) + \rho_d^4(1-\gamma)^2(1-\gamma+\gamma^2)}{[1 + \rho_d(1-2\gamma) + \rho_d^2(1-\gamma)^2][1 + \rho_d(1-\gamma)]^2}\right) \tag{55}$$

$$= \frac{1}{\gamma}\left((1-\gamma) + \frac{1}{\rho_d} + \frac{\rho_d^2(1-\gamma)[1 + 3\rho_d(1-\gamma) + \rho_d^2(1-\gamma)(1-2\gamma)]}{1 + \rho_d(1-2\gamma) + \rho_d^2(1-\gamma)^2}\right) \tag{56}$$

In Equation (54), since in the case of $\eta = 0$, the difference of the last two terms is $1/\gamma$, the average AoI $\overline{\Delta}_{Ber/Geo/1/2}$ is obtained. This equation also gives the exact gap between two average AoIs of the system with and without packet preemption. Notice that the latter term in (54) is always positive; then, the average AoI must become lower when the packet preemption strategy is applied. Equation (52) is substituted in (55), and in Equation (56), the expression of the average AoI $\overline{\Delta}_{Ber/Geo/1/2^*}$ is finally determined.

Next, the distribution of the discrete AoI is calculated. Before the expressions (45) and (46) are derived, we first verify that both (45) and (46) are proper probability distributions by providing a specific numerical example.

Numerical results of two AoI distributions. Let $p = 1/4$ and $\gamma = 1/2$.

Firstly, from Equations (30) and (31), two numbers M_1 and M_2 are determined to be

$$M_1 = \frac{12}{17}, \quad M_2 = \frac{4}{17}.$$

after some simple calculations, for each $n \geq 1$, the expression (45) gives

$$\Pr\{\Delta_{Ber/Geo/1/2} = n\} = \frac{9}{17}\left(\frac{3}{4}\right)^n - \frac{21}{68}\left(\frac{1}{2}\right)^n - \frac{3}{68}(n-1)\left(\frac{1}{2}\right)^{n-1} + \frac{1}{136}n(n-1)\left(\frac{1}{2}\right)^{n-2} \tag{57}$$

To obtain the numerical result of Equation (46), it is necessary to determine the four coefficients A, B, C, and D according to expressions (47)–(50). We directly find that

$$A = -\frac{39}{2}, \quad C = \frac{117}{8}, \quad D = \frac{45}{32}, \quad B = 4.$$

After some extra computations, it is shown that

$$\Pr\{\Delta_{Ber/Geo/1/2^*} = n\} = \frac{1}{2}\left(\frac{3}{4}\right)^n - \frac{105}{34}\left(\frac{1}{2}\right)^n$$
$$+ \frac{57}{17}\left(\frac{3}{8}\right)^n + \frac{2}{17}(n-1)\left(\frac{1}{2}\right)^{n-2} + \frac{45}{1088}(n-1)\left(\frac{3}{8}\right)^{n-2} \tag{58}$$

It can be checked directly that the sum of both (57) and (58) from $n = 1$ to ∞ are equal to 1. Therefore, expressions (45) and (46) indeed form the proper probability distributions.

In the following, by decomposing (41) and (42) further into several simplest rational fractions, we derive the explicit expressions of AoI distributions for the system with and without packet preemption.

First of all, for $\eta = 0$, it is easy to prove that the last part of (41) is equal to

$$\frac{-\gamma}{1 - (1 - \gamma)x}$$

thus, we have following equations. This shows that

$$H_{PP}(x)\big|_{\eta=0}$$

$$= \frac{p\gamma M_1 x}{\gamma - p}\left(\frac{(1-p)\gamma}{1-(1-p)x} - \frac{p(1-\gamma)}{1-(1-\gamma)x}\right)$$

$$- \left(\frac{(1-p)^2\gamma^2 M_2 x^2}{(\gamma-p)[1-(1-p)x]} - \frac{p\gamma(1-p)(1-\gamma)M_2 x^2}{(\gamma-p)[1-(1-\gamma)x]} + \frac{p\gamma M_2 x^2}{[1-(1-\gamma)x]^2}\right) \cdot \frac{-\gamma}{1-(1-\gamma)x}$$

$$= \frac{p\gamma^2(1-p)M_1 x}{(\gamma-p)[1-(1-p)x]} - \frac{p^2\gamma(1-\gamma)M_1 x}{(\gamma-p)[1-(1-\gamma)x]}$$

$$+ \frac{(1-p)^2\gamma^3 M_2 x^2}{(\gamma-p)[1-(1-p)x][1-(1-\gamma)x]} - \frac{p\gamma^2(1-p)(1-\gamma)M_2 x^2}{(\gamma-p)[1-(1-\gamma)x]^2} + \frac{p\gamma^2 M_2 x^2}{[1-(1-\gamma)x]^3}$$

$$= \frac{p\gamma^2(1-p)M_1 x}{(\gamma-p)[1-(1-p)x]} - \frac{p^2\gamma(1-\gamma)M_1 x}{(\gamma-p)[1-(1-\gamma)x]}$$

$$+ \frac{(1-p)^2\gamma^3 M_2 x^2}{(\gamma-p)}\left(\frac{1-p}{(\gamma-p)[1-(1-p)x]} - \frac{1-\gamma}{(\gamma-p)[1-(1-\gamma)x]}\right)$$

$$- \frac{p\gamma^2(1-p)(1-\gamma)M_2 x^2}{(\gamma-p)[1-(1-\gamma)x]^2} + \frac{p\gamma^2 M_2 x^2}{[1-(1-\gamma)x]^3}$$

$$= \left(\frac{p\gamma^2(1-p)M_1 x}{\gamma-p} + \frac{(1-p)^3\gamma^3 M_2 x^2}{(\gamma-p)^2}\right)\sum_{n=0}^{\infty}[(1-p)x]^n$$

$$- \left(\frac{p^2\gamma(1-\gamma)M_1 x}{\gamma-p} + \frac{(1-p)^2(1-\gamma)\gamma^3 M_2 x^2}{(\gamma-p)^2}\right)\sum_{n=0}^{\infty}[(1-\gamma)x]^n$$

$$- \frac{p\gamma^2(1-p)(1-\gamma)M_2 x^2}{\gamma-p}\sum_{n=1}^{\infty}n[(1-\gamma)x]^{n-1} + \frac{p\gamma^2 M_2 x^2}{2}\sum_{n=2}^{\infty}n(n-1)[(1-\gamma)x]^{n-2} \tag{59}$$

Taking the coefficient before x^n, we find that

$$\Pr\{\Delta_{Ber/Geo/1/2} = n\}$$

$$= \left(\frac{p\gamma^2(1-p)M_1}{\gamma-p}(1-p)^{n-1} + \frac{(1-p)^3\gamma^3 M_2}{(\gamma-p)^2}(1-p)^{n-2}\right)$$

$$- \left(\frac{p^2\gamma(1-\gamma)M_1}{\gamma-p}(1-\gamma)^{n-1} + \frac{(1-p)^2(1-\gamma)\gamma^3 M_2}{(\gamma-p)^2}(1-\gamma)^{n-2}\right)$$

$$- \frac{p\gamma^2(1-p)(1-\gamma)M_2}{\gamma-p}(n-1)(1-\gamma)^{n-2} + \frac{p\gamma^2 M_2}{2}n(n-1)(1-\gamma)^{n-2}$$

$$= \frac{p\gamma^2(1-p)M_1}{(\gamma-p)^2}(1-p)^n - \frac{(\gamma-p^2)(1-p)\gamma^2 M_2}{(\gamma-p)^2}(1-\gamma)^{n-1}$$

$$- \frac{p\gamma^2(1-p)M_2}{\gamma-p}(n-1)(1-\gamma)^{n-1} + \frac{p\gamma^2 M_2}{2}n(n-1)(1-\gamma)^{n-2} \tag{60}$$

This gives the stationary distribution (45) for the system without packet preemption.

On the other hand, when η is equal to 1, the system has full packet preemption. Factoring the PGF in Equation (42), we can also determine the stationary distribution of the AoI $\Delta_{Ber/Geo/1/2^*}$ by taking the coefficients of terms x^n for each $n \geq 1$. We give the explicit decomposition below, from which the distribution of AoI for the system with packet preemption is obtained explicitly.

From Equation (42), we show that

$$H_{PP}(x)|_{\eta=1}$$
$$= \frac{p\gamma M_1 x}{\gamma - p}\left(\frac{(1-p)\gamma}{1-(1-p)x} - \frac{p(1-\gamma)}{1-(1-\gamma)x}\right) + \frac{(1-p)^2\gamma^2 M_2 x^2}{\gamma - p}\left(\frac{\gamma^2 + p(1-\gamma)(p+\gamma)}{\gamma^2[1-(1-p)x]}\right.$$
$$\left. - \frac{(1-\gamma)[\gamma^2 + p(1-\gamma)(p+\gamma)]}{\gamma^2[1-(1-p)(1-\gamma)x]} - \frac{p(1-\gamma)(p+\gamma-p\gamma)}{\gamma[1-(1-p)(1-\gamma)x]^2}\right)$$
$$- \frac{p\gamma(1-p)(1-\gamma)M_2 x^2}{\gamma - p}\left(\frac{p+2(1-p)\gamma}{p[1-(1-\gamma)x]} - \frac{(1-p)[p+2(1-p)\gamma]}{p[1-(1-p)(1-\gamma)x]}\right.$$
$$\left. - \frac{(1-p)(p+\gamma-p\gamma)}{[1-(1-p)(1-\gamma)x]^2}\right) + p\gamma M_2 x^2\left(\frac{A}{1-(1-\gamma)x} + \frac{B}{[1-(1-\gamma)x]^2}\right.$$
$$\left. + \frac{C}{1-(1-p)(1-\gamma)x} + \frac{D}{[1-(1-p)(1-\gamma)x]^2}\right) \tag{61}$$

in which we determine

$$A = \frac{2-p}{p^3}\left((1-p)^2\gamma - \frac{2(1-p)[\gamma + p^2(1-\gamma)]}{2-p}\right) \tag{62}$$

$$C = -\frac{(1-p)(2-p)}{p^3}\left((1-p)^2\gamma - \frac{2(1-p)[\gamma + p^2(1-\gamma)]}{2-p}\right) \tag{63}$$

$$D = -\frac{p^2(1-p)\cdot A + (1-p)^2[\gamma + p^2(1-\gamma)]}{p(2-p)} \tag{64}$$

and

$$B = [\gamma + p^2(1-\gamma)] - A - C - D \tag{65}$$

Obtaining the second and the third row of (61) is not hard, while for the last row, we give some derivation details in Appendix C. Following the same procedures as those used to obtain (60), according to Equation (61), the probability that a stationary AoI equals each n is determined by the coefficient of the term x^n.

$$H_{PP}(x)|_{\eta=1}$$
$$= \frac{p\gamma M_1}{\gamma - p}\left((1-p)\gamma(1-p)^{n-1} - p(1-\gamma)(1-\gamma)^{n-1}\right) + \frac{(1-p)^2\gamma^2 M_2}{\gamma - p}$$
$$\times \left(\frac{\gamma^2 + p(1-\gamma)(p+\gamma)}{\gamma^2}(1-p)^{n-2} - \frac{(1-\gamma)[\gamma^2 + p(1-\gamma)(p+\gamma)]}{\gamma^2}[(1-p)(1-\gamma)]^{n-2}\right.$$
$$\left. - \frac{p(1-\gamma)(p+\gamma-p\gamma)}{\gamma}(n-1)[(1-p)(1-\gamma)]^{n-2}\right) - \frac{p\gamma(1-p)(1-\gamma)M_2}{\gamma - p}$$
$$\times \left(\frac{p+2(1-p)\gamma}{p}(1-\gamma)^{n-2} - \frac{(1-p)[p+2(1-p)\gamma]}{p}[(1-p)(1-\gamma)]^{n-2}\right.$$
$$\left. - (1-p)(p+\gamma-p\gamma)(n-1)[(1-p)(1-\gamma)]^{n-2}\right) + p\gamma M_2\left(A(1-\gamma)^{n-2}\right.$$
$$\left. + B(n-1)(1-\gamma)^{n-2} + C[(1-p)(1-\gamma)]^{n-2} + D(n-1)[(1-p)(1-\gamma)]^{n-2}\right)$$
$$= \frac{p\gamma M_1}{\gamma - p}\left(\gamma(1-p)^n - p(1-\gamma)^n\right)$$
$$+ \frac{(1-p)[\gamma^2 + p(1-\gamma)(p+\gamma)]M_2}{\gamma - p}\left((1-p)^{n-1} - [(1-p)(1-\gamma)]^{n-1}\right)$$
$$- \frac{(1-p)\gamma[p+2(1-p)\gamma]M_2}{\gamma - p}\left((1-\gamma)^{n-1} - [(1-p)(1-\gamma)]^{n-1}\right) + p\gamma M_2\left(A(1-\gamma)^{n-2}\right.$$
$$\left. + B(n-1)(1-\gamma)^{n-2} + C[(1-p)(1-\gamma)]^{n-2} + D(n-1)[(1-p)(1-\gamma)]^{n-2}\right) \tag{66}$$

which determines the distribution of AoI $\Delta_{Ber/Geo/1/2^*}$.

So far, in Equations (52), (56), (60) and (66), we have obtained all the results in Theorem 2; thus, the proof is completed. $\square$

Actually, we have obtained the explicit expression of AoI's distribution for the system with packet preemption in our early work [60]. Earlier in this paper, we explain that solving the stationary equations is feasible for the easy situations but cannot be generalized when the system structure or queue models become complex. In [60], we focused on the discrete time system with three queues, i.e., the $Ber/Geo/1/1$, $Ber/Geo/1/2$, and $Ber/Geo/1/2^*$, and named them *"discrete packet management strategies"*. There, we determined the AoI's stationary distribution for each system, and all the cases are dealt with by solving the stationary equations directly. Although the calculations are long—even tedious—these methods still have great significance, especially when the general status updating system is considered where the packet arrival process or the packet service process is arbitrary. It is with these methods that the analysis of discrete AoI can break through the limitation of the memoryless property that is imposed on the packet arrival and packet service processes in the SHS approach.

In [9], based on graphical arguments of the age process, the authors determined the average continuous AoI for the system with $M/M/1/2$ and $M/M/1/2^*$ queues as

$$\Delta_{M/M/1/2} = \frac{1}{\mu}\left(1 + \frac{1}{\rho} + \frac{2\rho^2}{1+\rho+\rho^2}\right) \tag{67}$$

and

$$\Delta_{M/M/1/2^*} = \frac{1}{\mu}\left(1 + \frac{1}{\rho} + \frac{\rho^2(1+3\rho+\rho^2)}{(1+\rho+\rho^2)(1+\rho)^2}\right) \tag{68}$$

In addition, in previous work [58], we have proved that the mean of the AoI for a bufferless discrete time status updating system is equal to

$$\overline{\Delta}_{Ber/Geo/1/1} = \frac{1}{\gamma}\left((1-\gamma) + \frac{1}{\rho_d} + \frac{\rho_d}{1/(1-\gamma)+\rho_d}\right) \tag{69}$$

while the corresponding continuous system with an $M/M/1/1$ queue has the average AoI

$$\overline{\Delta}_{M/M/1/1} = \frac{1}{\mu}\left(1 + \frac{1}{\rho} + \frac{\rho}{1+\rho}\right) \tag{70}$$

which was also given in [9].

We list Equations (43), (44) and (67)–(70) in Table 3—notice that this table has been given previously in Table 1 except for the last row, which gives the average continuous and discrete AoI for an infinite size status updating system. The mean of the discrete AoI $\Delta_{Ber/Geo/1/\infty}$ was obtained recently in our work [59]. It is observed that apart from some additional product factors, the expressions of discrete AoI means for the system with Bernoulli packet arrivals and geometric service times are identical to those of the continuous system's average AoI, which uses the Poisson-exponential assumptions. So far, we have obtained enough evidence to propose the following relationship between the mean of discrete and continuous AoIs:

$$\mu \cdot \overline{\Delta}_{M/M/1/c} = \gamma \cdot \overline{\Delta}_{Ber/Geo/1/c}\big|_{\gamma=0} \text{ then replacing } \rho_d \text{ by } \rho \tag{71}$$

It is interesting and meaningful to verify the correspondence (71) by calculating the average AoI for more continuous time and discrete time systems. For example, determining the mean of AoI assuming the $M/M/1/c$ queue is used in the continuous time system and for the discrete time systems with general $Ber/Geo/1/c$ queues, where the system's size c is larger than 2.

Table 3. Some formulas of the average continuous and average discrete age of information.

Average Continuous and Average Discrete AoIs
$\overline{\Delta}_{M/M/1/1} = \frac{1}{\mu}\left(1 + \frac{1}{\rho} + \frac{\rho}{1+\rho}\right)$ $\overline{\Delta}_{Ber/Geo/1/1} = \frac{1}{\gamma}\left((1-\gamma) + \frac{1}{\rho_d} + \frac{\rho_d}{1/(1-\gamma)+\rho_d}\right)$
$\overline{\Delta}_{M/M/1/2} = \frac{1}{\mu}\left(1 + \frac{1}{\rho} + \frac{2\rho^2}{1+\rho+\rho^2}\right)$ $\overline{\Delta}_{Ber/Geo/1/2} = \frac{1}{\gamma}\left((1-\gamma) + \frac{1}{\rho_d} + \frac{2\rho_d^2(1-\gamma)(1-\gamma/2)}{1+\rho_d(1-2\gamma)+\rho_d^2(1-\gamma)^2}\right)$
$\overline{\Delta}_{M/M/1/2^*} = \frac{1}{\mu}\left(1 + \frac{1}{\rho} + \frac{\rho^2(1+3\rho+\rho^2)}{(1+\rho+\rho^2)(1+\rho)^2}\right)$ $\overline{\Delta}_{Ber/Geo/1/2^*} = \frac{1}{\gamma}\left((1-\gamma) + \frac{1}{\rho_d} + \frac{\rho_d^2(1-\gamma)[1+3\rho_d(1-\gamma)+\rho_d^2(1-\gamma)(1-2\gamma)]}{[1+\rho_d(1-2\gamma)+\rho_d^2(1-\gamma)^2][1+\rho_d(1-\gamma)]^2}\right)$
$\overline{\Delta}_{M/M/1/\infty} = \frac{1}{\mu}\left(1 + \frac{1}{\rho} + \frac{\rho^2}{1-\rho}\right)$ $\overline{\Delta}_{Ber/Geo/1/\infty} = \frac{1}{\gamma}\left((1-\gamma) + \frac{1}{\rho_d} + \frac{\rho_d^2(1-\gamma)}{1-\rho_d}\right)$

5. Numerical Simulation

We provide the numerical results in this section. For general preemption probability, in the first two plots of Figure 3, we illustrate the relationships between the average AoI $\overline{\Delta}_{Ber/Geo/1/2^*/\eta}$ and the packet preemption probability η, and the traffic load ρ_d, respectively. The means of three discrete AoIs including $\overline{\Delta}_{Ber/Geo/1/1}$, $\overline{\Delta}_{Ber/Geo/1/2}$, and $\overline{\Delta}_{Ber/Geo/1/2^*}$ are plotted in Figure 3c. For comparison, we also provide the numerical simulations of corresponding average continuous AoIs. At last, for three discrete AoIs, we depict their distribution curves and the cumulative probabilities in Figure 4. Notice that in our work [58], the distribution of the AoI for the bufferless system was obtained as

$$\Pr\{\Delta_{Ber/Geo/1/1} = n\} = \frac{p(1-p)\gamma^3[(1-p)^n - (1-\gamma)^n]}{(p+\gamma-p\gamma)(\gamma-p)^2} - \frac{(p\gamma)^2 n(1-\gamma)^n}{(p+\gamma-p\gamma)(\gamma-p)} \quad (72)$$

For three different traffic loads ρ_d, we first draw the graphs between the average AoI $\overline{\Delta}_{Ber/Geo/1/2^*/\eta}$ and the preemption probability η. It is understandable that replacing the packet in a buffer with a fresher one can decrease the average AoI at the destination, and the numerical results in Figure 3a show that this trend is consistent as the preemption probability becomes large; that is, the mean of the AoI is decreasing monotonically when η increases. We mark the values at two extreme points where $\eta = 0$ and $\eta = 1$, which gives the average AoI $\overline{\Delta}_{Ber/Geo/1/2}$ and $\overline{\Delta}_{Ber/Geo/1/2^*}$. Notice that the closer to $\eta = 0$, the more similar the behavior of the system becomes to that of a system using $Ber/Geo/1/2$ queues, and when η gradually gets to 1, a status updating system with $Ber/Geo/1/2^*$ queue is finally obtained. The three curves in Figure 3a also show that as the traffic load ρ_d increases from 0.4 to 0.45, the average AoI of the system with probabilistic packet preemption is reduced; thus, the timeliness performance is improved.

In Figure 3b, for three settings of preemption probabilities, i.e., $\eta = 0$, $\eta = 0.5$ and $\eta = 1$, the relationships of the average AoI versus traffic intensity ρ_d are illustrated. The topmost curve gives the average AoI of the system without packet preemption because for the case of $\eta = 0$, the average AoI reduces to $\overline{\Delta}_{Ber/Geo/1/2}$. On the other hand, the curve at the bottom corresponds to the AoI's mean of the system that has full packet preemption. In order to make the differences among these graphs more significant, we draw the results in the range $\rho_d \geq 0.45$. Three curves in Figure 3b clearly show that the timeliness of the system with complete packet preemption is the best, since when η is set to 1, the system's average AoI is the lowest. Since the results in Figure 3a show that the average AoI is monotonically decreasing when η increases, the graphs of the AoI's mean for a system with probabilistic packet preemption is located between the blue and the black lines in Figure 3b, such as the red line, which denotes the average AoI $\overline{\Delta}_{Ber/Geo/1/2^*/0.5}$. In addition, the gaps between these curves are not significant for small ρ_ds but become large as ρ_d increases.

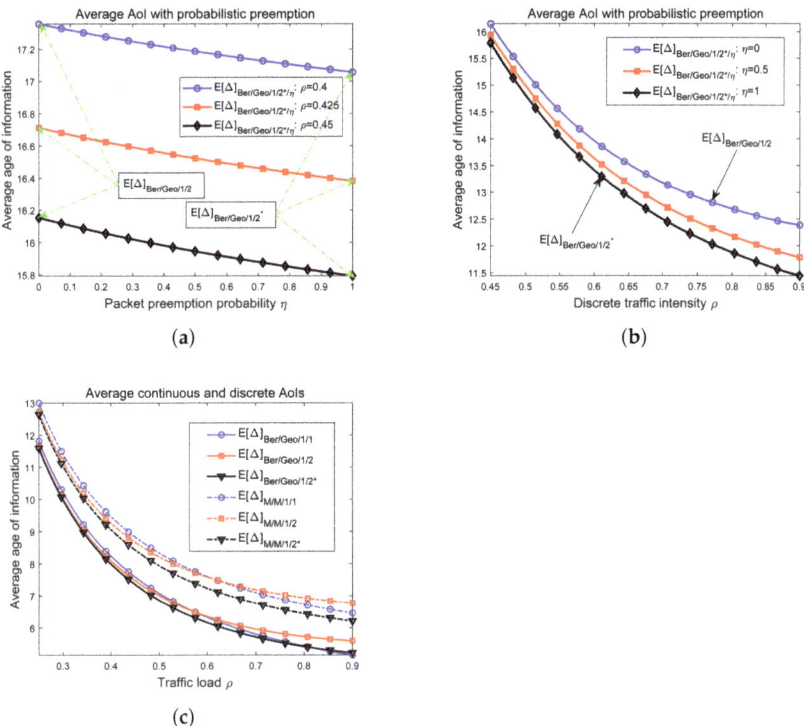

Figure 3. (**a**) Average AoI versus preemption probability η (different traffic load). (**b**) Average AoI versus traffic load ρ_d (different preemption intensity). (**c**) Comparisons of average discrete AoI and average continuous AoI.

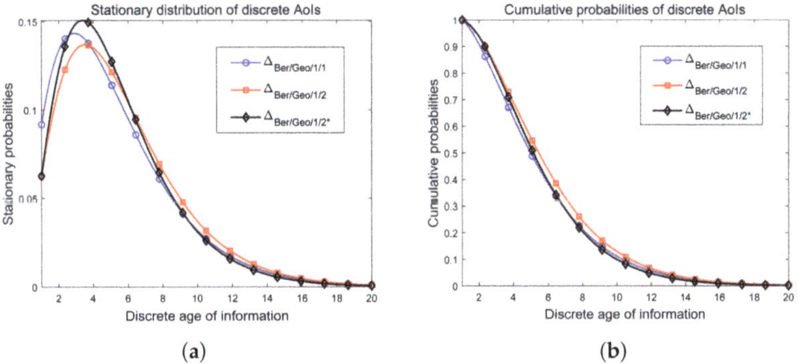

Figure 4. (**a**) Stationary distributions of discrete AoI for bufferless system and the system with and without packet preemption. (**b**) The cumulative probabilities of three discrete AoIs.

From $\rho_d = 0.15$ to 0.9, we depict both the average discrete AoIs and the corresponding continuous average AoIs in Figure 3c for a bufferless system and size 2 status updating system with and without preemption. Continuous AoIs are denoted by dashed lines, and we use solid lines to represent the discrete AoIs. First of all, all the curves are decreasing as ρ_d becomes large, and the gaps between them are gradually apparent. For three continuous AoIs, it is observed that the average AoI $\overline{\Delta}_{M/M/1/2^*}$ is the lowest in all the range of traffic load ρ. For the other two status updating systems, it is found that the system with an $M/M/1/2$ queue has a lower average AoI when ρ takes small values, while for high ρ, the average AoI of the bufferless system is smaller, and thus the timeliness is better. These results are the same for the graphs of discrete AoIs. Notice that when the discrete traffic intensity ρ_d is extremely large (near 0.9), the numerical results show that the average AoI $\overline{\Delta}_{Ber/Geo/1/1}$ can be even smaller than $\overline{\Delta}_{Ber/Geo/1/2^*}$.

In Figure 3c, both continuous and discrete average AoIs are monotonically decreasing in the whole range of ρ_d; however, the monotonicity of the curve between the average AoI and ρ_d can only be maintained for small-size status updating systems. It is known that the average AoI of an infinite size system, i.e., $\overline{\Delta}_{M/M/1/\infty}$, is not monotonic when the traffic load varies from 0 to 1. Thus, for a size c status updating system with Bernoulli packet arrivals and geometrically distributed service time, there must be a *critical size* c^* such that when $c < c^*$, the mean of the system's AoI $\overline{\Delta}_{Ber/Geo/1/c}$ is monotonically decreasing as ρ_d tends to 1. In contrast, for those cases where the system size $c \geq c^*$, the curve has a valley, and an optimal ρ_d exists at which the average AoI is minimized. Similarly, for the continuous average AoI $\overline{\Delta}_{M/M/1/c}$ of the system with general size c, a c^* also exists so that $\overline{\Delta}_{M/M/1/c}$ is always decreasing when $c < c^*$ and the graph of $\overline{\Delta}_{M/M/1/c}$ first falls and then rises for those cs where $c \geq c^*$. In addition, from the alternation of $\overline{\Delta}_{M/M/1/1}$ and $\overline{\Delta}_{M/M/1/2}$, and also of $\overline{\Delta}_{Ber/Geo/1/1}$ and $\overline{\Delta}_{Ber/Geo/1/2}$, we can infer that, as a function of c, the graphs of $\overline{\Delta}_{M/M/1/c}$ and $\overline{\Delta}_{Ber/Geo/1/c}$ are not monotonic.

At last, the distribution curves and cumulative probabilities of three discrete AoIs are depicted in Figure 4, in which we set a relatively large ρ_d to make the difference between them clear. On the whole, these curves are similar. In Figure 4a, from the distributions of AoI $\Delta_{Ber/Geo/1/1}$ to that of $\Delta_{Ber/Geo/1/2}$, the peak of the curve decreases and the point at which the peak stationary probability is achieved moves slightly to the right. As the AoI becomes large, the distribution curve of the system with the $Ber/Geo/1/1$ queue drops more sharply. The distribution corresponding to $\Delta_{Ber/Geo/1/2^*}$ has the largest peak value pf all of three discrete AoIs, and the descent speed is the fastest when the value of AoI is large. In addition, it seems that this maximal probability is taken at the same discrete AoI as that of the distribution of $\Delta_{Ber/Geo/1/2}$. We also provide the cumulative probabilities of three discrete AoIs in Figure 4b.

6. Conclusions

In this paper, we consider the stationary AoI of a size 2 status updating system where the packet waiting in the buffer can be preempted by fresher packets with the given probability η. We show that this phenomenon may occur in the energy-harvest (EH) nodes of wireless sensor networks where the charging process is stochastic. We constitute a three-dimensional age process and derive the general expression of the system's average AoI using the PGF method. Let $\eta = 0$ and $\eta = 1$; the mean of two discrete AoIs $\overline{\Delta}_{Ber/Geo/1/2}$ and $\overline{\Delta}_{Ber/Geo/1/2}$ are determined, and the exact distribution expressions of both AoIs are also obtained by writing the PGF as the power series.

We propose the idea and methods for the analysis of discrete AoIs—that is, constituting multiple-dimensional age processes and applying the PGF method. A detailed introduction is given to exhibit the usage of the idea and methods to more discrete time status updating systems. With this paper, we have shown how the AoI of basic discrete system is characterized, while in further work, we will focus on the age analysis of systems with a more general structure, such as systems with multi-sources and systems with multi-hop packet transmission. As one part of the AoI theory, we believe that the research into discrete AoI deserves more attention.

Author Contributions: Writing—original draft preparation, J.Z.; writing—review and editing, Y.X. All authors have read and agreed to the published version of the manuscript.

Funding: This research was funded in part by the National Natural Science Foundation of China under Grant 61901105 and Grant 62171119, in part by the Natural Science Foundation of Jiangsu Province under Grant BK20190343, in part by the Zhi Shan Young Scholar Program of Southeast University.

Institutional Review Board Statement: Not applicable.

Informed Consent Statement: Not applicable.

Data Availability Statement: Not applicable.

Conflicts of Interest: The authors declare no conflict of interest.

Appendix A. Proof of Theorem 1

In this appendix, we derive all the results in Lemma 1 from the stationary Equations (18).

Define that

$$M_1 = \sum_{n=1}^{\infty} \pi_{(n,0,0)}, \quad M_2 = \sum_{m=1}^{\infty} \sum_{n=m+1}^{\infty} \pi_{(n,m,0)}, \quad \text{and } M_3 = \sum_{l=1}^{\infty} \sum_{m=l+1}^{\infty} \sum_{n=m+1}^{\infty} \pi_{(n,m,l)},$$

These three numbers are determined at the first place.

According to the last two rows of (18), we have

$$\begin{aligned}
M_1 &= \pi_{(1,0,0)} + \sum_{n=2}^{\infty} \pi_{(n,0,0)} \\
&= \left(\sum_{n=1}^{\infty} \pi_{(n,0,0)} \right) p\gamma + \sum_{n=2}^{\infty} \left\{ \pi_{(n-1,0,0)}(1-p) + \sum_{k=n}^{\infty} \pi_{(k,n-1,0)}(1-p)\gamma \right\} \\
&= p\gamma M_1 + (1-p)M_1 + \sum_{\tilde{m}=1}^{\infty} \sum_{\tilde{n}=\tilde{m}+1}^{\infty} \pi_{(\tilde{n},\tilde{m},0)}(1-p)\gamma \qquad\qquad\text{(A1)} \\
&= p\gamma M_1 + (1-p)M_1 + (1-p)\gamma M_2 \qquad\qquad\qquad\qquad\text{(A2)}
\end{aligned}$$

where in (A1), we have used the substitutions $k = \tilde{n}$ and $n - 1 = \tilde{m}$. From Equation (A2), we obtain the first relation

$$p(1-\gamma)M_1 = (1-p)\gamma M_2 \qquad\qquad\text{(A3)}$$

Next, we deal with the number M_3 as follows.

$$\begin{aligned}
M_3 &= \sum_{l=1}^{\infty} \sum_{m=l+1}^{\infty} \sum_{n=m+1}^{\infty} \pi_{(n,m,l)} \\
&= \sum_{m=2}^{\infty} \sum_{n=m+1}^{\infty} \pi_{(n,m,1)} + \sum_{l=2}^{\infty} \sum_{m=l+1}^{\infty} \sum_{n=m+1}^{\infty} \pi_{(n,m,l)} \qquad\text{(A4)}
\end{aligned}$$

Using the first row of (18), the latter sum in Equation (A4) equals

$$\begin{aligned}
&\sum_{l=2}^{\infty} \sum_{m=l+1}^{\infty} \sum_{n=m+1}^{\infty} \pi_{(n-1,m-1,l-1)}(1-\gamma)(1-p\eta) \\
&= \sum_{l=1}^{\infty} \sum_{m=l+1}^{\infty} \sum_{n=m+1}^{\infty} \pi_{(n,m,l)}(1-\gamma)(1-p\eta) \\
&= (1-\gamma)(1-p\eta)M_3 \qquad\qquad\text{(A5)}
\end{aligned}$$

and from the second and the third row of (18), the first part of (A4) is calculated as

$$\begin{aligned}
&\sum_{n=3}^{\infty} \pi_{(n,2,1)} + \sum_{m=3}^{\infty} \sum_{n=m+1}^{\infty} \pi_{(n,m,1)} \\
&= \sum_{n=3}^{\infty} \pi_{(n-1,1,0)} p(1-\gamma) \\
&\quad + \sum_{m=3}^{\infty} \sum_{n=m+1}^{\infty} \left\{ \pi_{(n-1,m-1,0)} p(1-\gamma) + \sum_{j=1}^{m-2} \pi_{(n-1,m-1,j)} p(1-\gamma)\eta \right\} \\
&= \sum_{n=2}^{\infty} \pi_{(n,1,0)} p(1-\gamma) \\
&\quad + \sum_{m=2}^{\infty} \sum_{n=m+1}^{\infty} \pi_{(n,m,0)} p(1-\gamma) + \sum_{m=3}^{\infty} \sum_{n=m+1}^{\infty} \sum_{j=1}^{m-2} \pi_{(n-1,m-1,j)} p(1-\gamma)\eta \\
&= \sum_{m=1}^{\infty} \sum_{n=m+1}^{\infty} \pi_{(n,m,0)} p(1-\gamma) + \sum_{\tilde{m}=2}^{\infty} \sum_{\tilde{n}=\tilde{m}+1}^{\infty} \sum_{\tilde{l}=1}^{\tilde{m}-1} \pi_{(\tilde{n},\tilde{m},\tilde{l})} p(1-\gamma)\eta \qquad\text{(A6)} \\
&= p(1-\gamma)M_2 + p(1-\gamma)\eta M_3 \qquad\qquad\qquad\qquad\qquad\qquad\text{(A7)}
\end{aligned}$$

where in (A6), we let $n - 1 = \tilde{n}$, $m - 1 = \tilde{m}$, and $j = \tilde{l}$. Equations (A4), (A5) and (A7) together give

$$p(1 - \gamma)M_2 = \gamma M_3 \tag{A8}$$

Since the sum of all the stationary probabilities equals 1, thus we have

$$M_1 + M_2 + M_3 = 1 \tag{A9}$$

Combining Equations (A3), (A8) and (A9), we can solve that

$$M_1 = \frac{(1 - p)\gamma^2}{(p + \gamma - 2p\gamma)\gamma + p^2(1 - \gamma)^2} \tag{A10}$$

$$M_2 = \frac{p\gamma(1 - \gamma)}{(p + \gamma - 2p\gamma)\gamma + p^2(1 - \gamma)^2} \tag{A11}$$

$$M_3 = \frac{p^2(1 - \gamma)^2}{(p + \gamma - 2p\gamma)\gamma + p^2(1 - \gamma)^2} \tag{A12}$$

We mention that from the fourth, the fifth, and the sixth equation in (18), another relation can also be obtained for the second number M_2, which is given directly as

$$M_2 = p(1 - \gamma)M_1 + p\gamma M_2 + p\gamma\eta M_3 + (1 - p)(1 - \gamma)M_2 + (1 - p\eta)\gamma M_3 \tag{A13}$$

which is reduced to (A8) when eliminating M_1 using the relation (A3).

Then, the relationships between functions $h_i(x)$, $1 \leq i \leq 3$ and $h_2^{(m)}(x)$ are determined through similar procedures. First of all, we see that

$$\begin{aligned}
h_1(x) &= \sum_{m=1}^{\infty} x^n \pi_{(n,0,0)} \\
&= x\pi_{(1,0,0)} + \sum_{m=2}^{\infty} x^n \left\{ \pi_{(n-1,0,0)}(1 - p) + \sum_{k=n}^{\infty} \pi_{(k,n-1,0)}(1 - p)\gamma \right\} \\
&= p\gamma M_1 x + \sum_{m=1}^{\infty} x^{n+1}\pi_{(n,0,0)}(1 - p) + \sum_{n=2}^{\infty} x^n \sum_{k=n}^{\infty} \pi_{(k,n-1,0)}(1 - p)\gamma \\
&= p\gamma M_1 x + (1 - p)xh_1(x) + \sum_{\tilde{m}=1}^{\infty} x^{\tilde{m}+1} \sum_{\tilde{n}=\tilde{m}+1}^{\infty} \pi_{(\tilde{n},\tilde{m},0)}(1 - p)\gamma \tag{A14} \\
&= p\gamma M_1 x + (1 - p)xh_1(x) + (1 - p)\gamma x h_2^{(m)}(x) \tag{A15}
\end{aligned}$$

in which we denote $k = \tilde{n}$ and $n - 1 = \tilde{m}$.

From (A15), we obtain

$$h_1(x) = \frac{p\gamma M_1 x}{1 - (1 - p)x} + \frac{(1 - p)\gamma x}{1 - (1 - p)x} h_2^{(m)}(x) \tag{A16}$$

Using stationary Equation (18), we determine function $h_2(x)$ in the following.

$$\begin{aligned}
h_2(x) &= \sum_{m=1}^{\infty} \sum_{n=m+1}^{\infty} x^n \pi_{(n,m,0)} \\
&= \sum_{n=2}^{\infty} x^n \pi_{(n,1,0)} + \sum_{m=2}^{\infty} \sum_{n=m+1}^{\infty} x^n \\
&\quad \times \left\{ \pi_{(n-1,m-1,0)}(1 - p)(1 - \gamma) + \sum_{k=n}^{\infty} \pi_{(k,n-1,m-1)}(1 - p\eta)\gamma \right\} \\
&= x^2 \pi_{(2,1,0)} + \sum_{n=3}^{\infty} x^n \left\{ \pi_{(n-1,0,0)}p(1 - \gamma) + \sum_{k=n}^{\infty} \pi_{(k,n-1,0)}p\gamma \right. \\
&\quad \left. + \sum_{k=n}^{\infty} \sum_{j=1}^{n-2} \pi_{(k,n-1,j)}p\gamma\eta \right\} + \sum_{m=1}^{\infty} \sum_{n=m+1}^{\infty} x^{n+1} \pi_{(n,m,0)}(1 - p)(1 - \gamma) \\
&\quad + \sum_{m=2}^{\infty} \sum_{n=m+1}^{\infty} x^n \sum_{k=n}^{\infty} \pi_{(k,n-1,m-1)}(1 - p\eta)\gamma \\
&= x^2 \left\{ \pi_{(1,0,0)}p(1 - \gamma) + \sum_{k=2}^{\infty} \pi_{(k,1,0)}p\gamma \right\} + \sum_{n=2}^{\infty} x^{n+1} \pi_{(n,0,0)}p(1 - \gamma) \\
&\quad + \sum_{n=3}^{\infty} x^n \sum_{k=n}^{\infty} \pi_{(k,n-1,0)}p\gamma + \sum_{n=3}^{\infty} x^n \sum_{k=n}^{\infty} \sum_{j=1}^{n-2} \pi_{(k,n-1,j)}p\gamma\eta \\
&\quad + (1 - p)(1 - \gamma)xh_2(x) + (1 - p\eta)\gamma x h_3^{(m)}(x) \tag{A17}
\end{aligned}$$

Let $k = \tilde{n}$, $n - 1 = \tilde{m}$, and $m - 1 = \tilde{l}$, we have

$$\sum_{m=2}^{\infty} \sum_{n=m+1}^{\infty} x^n \sum_{k=n}^{\infty} \pi_{(k,n-1,m-1)}(1 - p\eta)\gamma$$
$$= \sum_{\tilde{l}=1}^{\infty} \sum_{\tilde{m}=\tilde{l}+1}^{\infty} x^{\tilde{m}+1} \sum_{\tilde{n}=\tilde{m}+1}^{\infty} \pi_{(\tilde{n},\tilde{m},\tilde{l})}(1 - p\eta)\gamma$$
$$= (1 - p\eta)\gamma x h_3^{(m)}(x) \tag{A18}$$

where we define

$$h_3^{(m)}(x) = \sum_{l=1}^{\infty} \sum_{m=l+1}^{\infty} \sum_{n=m+1}^{\infty} x^m \pi_{(n,m,l)} \tag{A19}$$

and the last term in Equation (A17) is obtained.

Continuing the calculation of (A17), we obtain that

$$h_2(x) = p(1 - \gamma)x h_1(x) + p\gamma x h_2^{(m)}(x) + \sum_{\tilde{m}=2}^{\infty} x^{\tilde{m}+1} \sum_{\tilde{n}=\tilde{m}+1}^{\infty} \sum_{\tilde{l}=1}^{\tilde{m}-1} \pi_{(\tilde{n},\tilde{m},\tilde{l})} p\gamma\eta$$
$$+ (1 - p)(1 - \gamma)x h_2(x) + (1 - p\eta)\gamma x h_3^{(m)}(x) \tag{A20}$$

In (A20), we have used the substitutions $k = \tilde{n}$, $n - 1 = \tilde{m}$, and $j = \tilde{l}$.

Except for the factor $p\gamma\eta x$, the sum in (A20) is equal to

$$x^2 \sum_{n=3}^{\infty} \pi_{(n,2,1)} + x^3 \sum_{n=4}^{\infty} \left[\pi_{(n,3,1)} + \pi_{(n,3,2)}\right] + x^4 \sum_{n=5}^{\infty} \left[\pi_{(n,4,1)} + \pi_{(n,4,2)} + \pi_{(n,4,3)}\right] + \cdots$$
$$= \sum_{m=2}^{\infty} x^m \sum_{n=m+1}^{\infty} \pi_{(n,m,1)} + \sum_{m=3}^{\infty} x^m \sum_{n=m+1}^{\infty} \pi_{(n,m,2)} + \sum_{m=4}^{\infty} x^m \sum_{n=m+1}^{\infty} \pi_{(n,m,3)} + \cdots$$
$$= \sum_{l=1}^{\infty} \sum_{m=l+1}^{\infty} x^m \sum_{n=m+1}^{\infty} \pi_{(n,m,l)}$$
$$= h_3^{(m)}(x) \tag{A21}$$

Substituting the result (A21) into Equation (A20) and merging the same terms gives

$$h_2(x)[1 - (1 - p)(1 - \gamma)x] = p(1 - \gamma)x h_1(x) + p\gamma x h_2^{(m)}(x) + \gamma x h_3^{(m)}(x) \tag{A22}$$

We compute $h_3^{(m)}(x)$ while determining the other function $h_2^{(m)}(x)$ in the end. As before, from the equations in (18), we find that

$$h_3^{(m)}(x) = \sum_{l=1}^{\infty} \sum_{m=l+1}^{\infty} \sum_{n=m+1}^{\infty} x^m \pi_{(n,m,l)}$$
$$= \sum_{m=2}^{\infty} \sum_{n=m+1}^{\infty} x^m \pi_{(n,m,1)} + \sum_{l=2}^{\infty} \sum_{m=l+1}^{\infty} \sum_{n=m+1}^{\infty} x^m \pi_{(n-1,m-1,l-1)}(1 - \gamma)(1 - p\eta)$$
$$= \sum_{n=3}^{\infty} x^2 \pi_{(n,2,1)} + \sum_{m=3}^{\infty} \sum_{n=m+1}^{\infty} x^m \left\{ \pi_{(n-1,m-1,0)} p(1 - \gamma) \right.$$
$$\left. + \sum_{j=1}^{m-2} \pi_{(n-1,m-1,j)} p(1 - \gamma)\eta \right\} + (1 - \gamma)(1 - p\eta)x h_3^{(m)}(x)$$
$$= \sum_{n=3}^{\infty} x^2 \pi_{(n-1,1,0)} p(1 - \gamma) + \sum_{m=2}^{\infty} \sum_{n=m+1}^{\infty} x^{m+1} \pi_{(n,m,0)} p(1 - \gamma)$$
$$+ \sum_{m=3}^{\infty} \sum_{n=m+1}^{\infty} x^m \sum_{j=1}^{m-2} \pi_{(n-1,m-1,j)} p(1 - \gamma)\eta + (1 - \gamma)(1 - p\eta)x h_3^{(m)}(x)$$
$$= p(1 - \gamma)x h_2^{(m)}(x) + p(1 - \gamma)\eta x h_3^{(m)}(x) + (1 - \gamma)(1 - p\eta)x h_3^{(m)}(x)$$
$$= p(1 - \gamma)x h_2^{(m)}(x) + (1 - \gamma)x h_3^{(m)}(x) \tag{A23}$$

which shows that

$$h_3^{(m)}(x) = \frac{p(1 - \gamma)x}{1 - (1 - \gamma)x} h_2^{(m)}(x) \tag{A24}$$

Combining Equations (A22) and (A24) yields the relation

$$h_2(x) = \frac{p(1 - \gamma)x}{1 - (1 - p)(1 - \gamma)x} h_1(x) + \frac{p\gamma x}{[1 - (1 - p)(1 - \gamma)x][1 - (1 - \gamma)x]} h_2^{(m)}(x) \tag{A25}$$

Now, we deal with function $h_3(x)$. Similarly to the process of obtaining (A23), we have

$$
\begin{aligned}
h_3(x) &= \sum_{m=2}^{\infty}\sum_{n=m+1}^{\infty} x^n \pi_{(n,m,1)} + \sum_{l=2}^{\infty}\sum_{m=l+1}^{\infty}\sum_{n=m+1}^{\infty} x^n \pi_{(n-1,m-1,l-1)}(1-\gamma)(1-p\eta) \\
&= \sum_{n=3}^{\infty} x^n \pi_{(n,2,1)} + \sum_{m=3}^{\infty}\sum_{n=m+1}^{\infty} x^n \Big\{ \pi_{(n-1,m-1,0)}p(1-\gamma) \\
&\quad + \sum_{j=1}^{m-2} \pi_{(n-1,m-1,j)}p(1-\gamma)\eta \Big\} + (1-\gamma)(1-p\eta)xh_3(x) \\
&= \sum_{n=3}^{\infty} x^n \pi_{(n-1,1,0)}p(1-\gamma) + \sum_{m=2}^{\infty}\sum_{n=m+1}^{\infty} x^{n+1} \pi_{(n,m,0)}p(1-\gamma) \\
&\quad + \sum_{m=3}^{\infty}\sum_{n=m+1}^{\infty} x^n \sum_{j=1}^{m-2} \pi_{(n-1,m-1,j)}p(1-\gamma)\eta + (1-\gamma)(1-p\eta)xh_3(x) \\
&= p(1-\gamma)xh_2(x) + p(1-\gamma)\eta xh_3(x) + (1-\gamma)(1-p\eta)xh_3(x) \\
&= p(1-\gamma)xh_2(x) + (1-\gamma)xh_3(x)
\end{aligned}
\tag{A26}
$$

from which we derive

$$
h_3(x) = \frac{p(1-\gamma)x}{1-(1-\gamma)x}h_2(x)
\tag{A27}
$$

So far, we have obtained the relations (26)–(28) in Equations (A16), (A25) and (A27). To complete the proof of Lemma 1, the remaining part is the determination of the last function $h_2^{(m)}(x)$. It is shown that

$$
\begin{aligned}
h_2^{(m)}(x) &= \sum_{m=1}^{\infty}\sum_{n=m+1}^{\infty} x^m \pi_{(n,m,0)} \\
&= \sum_{n=2}^{\infty} x\pi_{(n,1,0)} + \sum_{m=2}^{\infty}\sum_{n=m+1}^{\infty} x^m \\
&\quad \times \Big\{ \pi_{(n-1,m-1,0)}(1-p)(1-\gamma) + \sum_{k=n}^{\infty} \pi_{(k,n-1,m-1)}(1-p\eta)\gamma \Big\} \\
&= x\pi_{(2,1,0)} + \sum_{n=3}^{\infty} x\Big\{ \pi_{(n-1,0,0)}p(1-\gamma) + \sum_{k=n}^{\infty} \pi_{(k,n-1,0)}p\gamma \\
&\quad + \sum_{k=n}^{\infty}\sum_{j=1}^{n-2} \pi_{(k,n-1,j)}p\gamma\eta \Big\} + \sum_{m=1}^{\infty}\sum_{n=m+1}^{\infty} x^{m+1} \pi_{(n,m,0)}(1-p)(1-\gamma) \\
&\quad + \sum_{m=2}^{\infty}\sum_{n=m+1}^{\infty} x^m \sum_{k=n}^{\infty} \pi_{(k,n-1,m-1)}(1-p\eta)\gamma \\
&= x\Big\{ \pi_{(1,0,0)}p(1-\gamma) + \sum_{k=2}^{\infty} \pi_{(k,1,0)}p\gamma \Big\} + \sum_{n=2}^{\infty} x\pi_{(n,0,0)}p(1-\gamma) \\
&\quad + \sum_{n=3}^{\infty} x\sum_{k=n}^{\infty} \pi_{(k,n-1,0)}p\gamma + \sum_{n=3}^{\infty} x\sum_{k=n}^{\infty}\sum_{j=1}^{n-2} \pi_{(k,n-1,j)}p\gamma\eta \\
&\quad + (1-p)(1-\gamma)xh_2^{(m)}(x) + (1-p\eta)\gamma xh_3^{(l)}(x)
\end{aligned}
\tag{A28}
$$

Similarly, we use the substitutions $k = \tilde{n}$, $n-1 = \tilde{m}$, and $m-1 = \tilde{l}$, and we write

$$
\begin{aligned}
&\sum_{m=2}^{\infty}\sum_{n=m+1}^{\infty} x^m \sum_{k=n}^{\infty} \pi_{(k,n-1,m-1)}(1-p\eta)\gamma \\
&- \sum_{\tilde{l}=1}^{\infty}\sum_{\tilde{m}=\tilde{l}+1}^{\infty} x^{\tilde{l}+1} \sum_{\tilde{n}=\tilde{m}+1}^{\infty} \pi_{(\tilde{n},\tilde{m},\tilde{l})}(1-p\eta)\gamma \\
&= (1-p\eta)\gamma xh_3^{(l)}(x)
\end{aligned}
\tag{A29}
$$

Thus, the last term in Equation (A28) is obtained, where $h_3^{(l)}(x)$ is defined and determined as

$$
\begin{aligned}
h_3^{(l)}(x) &= \sum_{l=1}^{\infty}\sum_{m=l+1}^{\infty} x^l \sum_{n=m+1}^{\infty} \pi_{(n,m,l)} \\
&= \sum_{m=2}^{\infty} x\sum_{n=m+1}^{\infty} \pi_{(n,m,1)} + \sum_{l=2}^{\infty}\sum_{m=l+1}^{\infty} x^l \sum_{n=m+1}^{\infty} \pi_{(n-1,m-1,l-1)}(1-\gamma)(1-p\eta) \\
&= x\sum_{n=3}^{\infty} \pi_{(n,2,1)} + \sum_{m=3}^{\infty} x\sum_{n=m+1}^{\infty} \Big\{ \pi_{(n-1,m-1,0)}p(1-\gamma) \\
&\quad + \sum_{j=1}^{m-2} \pi_{(n-1,m-1,j)}p(1-\gamma)\eta \Big\} + (1-\gamma)(1-p\eta)xh_3^{(l)}(x) \\
&= x\sum_{n=3}^{\infty} \pi_{(n-1,1,0)}p(1-\gamma) + \sum_{m=2}^{\infty} x\sum_{n=m+1}^{\infty} \pi_{(n,m,0)}p(1-\gamma) \\
&\quad + \sum_{m=3}^{\infty} x\sum_{n=m+1}^{\infty} \sum_{j=1}^{m-2} \pi_{(n-1,m-1,j)}p(1-\gamma)\eta + (1-\gamma)(1-p\eta)xh_3^{(l)}(x) \\
&= p(1-\gamma)M_2x + p(1-\gamma)\eta M_3x + (1-\gamma)(1-p\eta)xh_3^{(l)}(x) \\
&= [\gamma + p(1-\gamma)\eta]M_3x + (1-\gamma)(1-p\eta)xh_3^{(l)}(x)
\end{aligned}
\tag{A30}
$$

In Equation (A30), we have used the relation (A8), which says $p(1-\gamma)M_2 = \gamma M_3$. From (A30), we obtain the result

$$h_3^{(l)}(x) = \frac{[\gamma + p(1-\gamma)\eta]M_3 x}{1 - (1-\gamma)(1-p\eta)x} \tag{A31}$$

Coming back to the calculation of function $h_2^{(m)}(x)$, Equation (A28) shows

$$h_2^{(m)}(x)[1-(1-p)(1-\gamma)x] = p(1-\gamma)M_1 x + p\gamma M_2 x + p\gamma\eta M_3 x + (1-p\eta)\gamma x h_3^{(l)}(x)$$

$$= \left[\gamma + p^2(1-\gamma)\eta\right]M_2 x + (1-p\eta)\gamma x h_3^{(l)}(x), \tag{A32}$$

in which we have used Equations (A3) and (A8) to replace numbers M_1 and M_3. Substituting expression (A31), after some extra operations, we determine that

$$h_2^{(m)}(x) = \frac{\left[\gamma + p^2(1-\gamma)\eta - (1-p)(1-\gamma)(1-p\eta)\gamma x\right]M_2 x}{[1-(1-p)(1-\gamma)x][1-(1-\gamma)(1-p\eta)x]} \tag{A33}$$

This eventually completes the proof of Lemma 1.

Appendix B. Proof of Equations (41) and (42)

In Equation (33), we show that

$$H_{PP}(x) = \frac{p\gamma M_1 x[1-(1-p)(1-\gamma)x]}{[1-(1-p)x][1-(1-\gamma)x]}$$

$$+ \frac{\gamma x\left\{1-(1-p)[2(1-\gamma)+p\gamma]x+(1-p)^2(1-\gamma)^2 x^2\right\}}{[1-(1-p)x][1-(1-\gamma)x]^2}h_2^{(m)}(x) \tag{A34}$$

Equations (41) and (42) are obtained by decomposing each part of (A34). For the first part, we assume

$$\frac{1-(1-p)(1-\gamma)x}{[1-(1-p)x][1-(1-\gamma)x]} = \frac{A}{1-(1-p)x} + \frac{B}{1-(1-\gamma)x}$$

$$= \frac{(A+B)-[A(1-\gamma)+B(1-p)]x}{[1-(1-p)x][1-(1-\gamma)x]} \tag{A35}$$

and according to the coefficients of corresponding terms, we obtain

$$A+B = 1, \quad A(1-\gamma)+B(1-p) = (1-p)(1-\gamma) \tag{A36}$$

which determine A and B as

$$A = \frac{(1-p)\gamma}{\gamma-p}, \text{ and } B = \frac{p(1-\gamma)}{\gamma-p} \tag{A37}$$

Therefore,

$$\frac{p\gamma M_1 x[1-(1-p)(1-\gamma)x]}{[1-(1-p)x][1-(1-\gamma)x]} = \frac{p\gamma M_1 x}{\gamma-p}\left(\frac{(1-p)\gamma}{1-(1-p)x} - \frac{p(1-\gamma)}{1-(1-\gamma)x}\right) \tag{A38}$$

For the second part of expression (A34), let

$$\frac{1 - (1-p)[2(1-\gamma) + p\gamma]x + (1-p)^2(1-\gamma)^2 x^2}{[1 - (1-p)x][1 - (1-\gamma)x]^2}$$

$$= \frac{A}{1 - (1-p)x} + \frac{B}{1 - (1-\gamma)x} + \frac{C}{[1 - (1-\gamma)x]^2}$$

$$= \frac{(A+B+C) - [(A+B)(1-\gamma) + A(1-\gamma) + B(1-p) + C(1-p)]x + c_2 x^2}{[1 - (1-p)x][1 - (1-\gamma)x]^2}$$

in which

$$c_2 = [A(1-\gamma) + B(1-p)](1-\gamma)$$

Thus, we have

$$\begin{cases} 1 = A + B + C \\ (1-p)[2(1-\gamma) + p\gamma] = (A+B)(1-\gamma) + A(1-\gamma) + B(1-p) + C(1-p) \\ (1-p)^2(1-\gamma) = A(1-\gamma) + B(1-p) \end{cases} \tag{A39}$$

Substituting the third relation and $A + B = 1 - C$ into the second equation, we obtain

$$(1-p)[2(1-\gamma) + p\gamma] = (1-C)(1-\gamma) + (1-p)^2(1-\gamma) + C(1-p) \tag{A40}$$

from which we can solve that $C = p$.

Then, according to the equations

$$A + B = 1 - p \quad \text{and} \quad A(1-\gamma) + B(1-p) = (1-p)^2(1-\gamma)$$

the other two numbers are obtained to be

$$A = \frac{(1-p)^2\gamma}{\gamma - p}, \quad B = -\frac{p(1-p)(1-\gamma)}{\gamma - p} \tag{A41}$$

Thus, the factorization of the second part is obtained.

The last part—that is, the function $h_2^{(m)}(x)$—is dealt with similarly. Omitting the straight-forward calculations, we directly find that

$$h_2^{(m)}(x) = -\frac{\eta(1-p)(p+\gamma-p\gamma)}{(1-\eta)[1-(1-p)(1-\gamma)x]} + \frac{(1-p\eta)[\gamma + p(1-\gamma)\eta]}{(1-\eta)[1-(1-\gamma)(1-p\eta)x]} \tag{A42}$$

Notice that $(1-\eta)$ is contained in the denominator of fractions in Equation (A42), thus, $\eta \neq 1$. When $\eta = 1$, Equation (29) shows that

$$h_2^{(m)}(x)\Big|_{\eta=1} = \frac{\gamma + p^2(1-\gamma) - (1-p)^2(1-\gamma)\gamma x}{[1-(1-p)(1-\gamma)x]^2} \tag{A43}$$

Summarizing the above results, both the Equations (41) and (42) are determined.

Appendix C. Factorization of Last Part of Equation (42)

We write

$$\frac{p\gamma M_2 x^2}{[1-(1-\gamma)x]^2} \cdot \frac{\gamma + p^2(1-\gamma) - (1-p)^2(1-\gamma)\gamma x}{[1-(1-p)(1-\gamma)x]^2}$$

$$= p\gamma M_2 x^2 \left(\frac{A}{1-(1-\gamma)x} + \frac{B}{[1-(1-\gamma)x]^2} + \frac{C}{1-(1-p)(1-\gamma)x} + \frac{D}{[1-(1-p)(1-\gamma)x]^2} \right)$$

$$= p\gamma M_2 x^2 \left(\frac{(A+B) - A(1-\gamma)x}{[1-(1-\gamma)x]^2} + \frac{(C+D) - C(1-p)(1-\gamma)x}{[1-(1-p)(1-\gamma)x]^2} \right) \tag{A44}$$

Merging the terms in the bracket of (A44), according to corresponding coefficients, it is shown that

$$
\begin{cases}
A + B + C + D = \gamma + p^2(1 - \gamma) \\
2(A + B)(1 - p) + 2(C + D) + A + C(1 - p) = (1 - p)^2\gamma \\
(A + B)(1 - p)^2 + 2A(1 - p) + (C + D) + 2C(1 - p) = 0 \\
A(1 - p) + C = 0
\end{cases}
\tag{A45}
$$

The last row of (A45) shows that $C = -A(1 - p)$, and using the first relationship, the second and the third row of Equation (A45) are equivalent to

$$
\begin{cases}
2(1 - p)[\gamma + p^2(1 - \gamma) - (C + D)] + 2(C + D) + A - A(1 - p)^2 = (1 - p)^2\gamma \\
(1 - p)^2[\gamma + p^2(1 - \gamma) - (C + D)] + 2A(1 - p) + (C + D) - 2A(1 - p)^2 = 0
\end{cases}
\tag{A46}
$$

which gives

$$
p(2 - p)(C + D) = -2p(1 - p)A - (1 - p)^2[\gamma + p^2(1 - \gamma)]
\tag{A47}
$$

and

$$
2p(C + D) = -p(2 - p)A + (1 - p)^2\gamma - 2(1 - p)[\gamma + p^2(1 - \gamma)]
\tag{A48}
$$

Combining Equations (A47) and (A48), the coefficient A is solved as

$$
A = \frac{2 - p}{p^3}\left((1 - p)^2\gamma - \frac{2(1 - p)[\gamma + p^2(1 - \gamma)]}{2 - p}\right)
\tag{A49}
$$

and immediately

$$
C = -A(1 - p) = -\frac{(1 - p)(2 - p)}{p^3}\left((1 - p)^2\gamma - \frac{2(1 - p)[\gamma + p^2(1 - \gamma)]}{2 - p}\right)
\tag{A50}
$$

Using Equation (A47), we have

$$
\begin{aligned}
D &= \frac{-2p(1 - p)A - (1 - p)^2[\gamma + p^2(1 - \gamma)]}{p(2 - p)} - C \\
&= \frac{-2p(1 - p)A - (1 - p)^2[\gamma + p^2(1 - \gamma)]}{p(2 - p)} + A(1 - p) \\
&= \frac{p^2(1 - p)A - (1 - p)^2[\gamma + p^2(1 - \gamma)]}{p(2 - p)}
\end{aligned}
\tag{A51}
$$

and in the end, the last number B is determined by

$$
B = [\gamma + p^2(1 - \gamma)] - A - C - D
\tag{A52}
$$

So far, all the coefficients are obtained and the decomposition is totally determined.

References

1. Kaul, S.; Gruteser, M.; Rai, V.; Kenney, J. Minimizing age of information in vehicular networks. In Proceedings of the 2011 8th Annual IEEE Communications Society Conference on Sensor, Mesh and Ad Hoc Communications and Networks, Salt Lake City, UT, USA, 27–30 June 2011; pp. 350–358. [CrossRef]
2. Kosta, A.; Pappas, N.; Angelakis, V. Age of Information: A New Concept, Metric, and Tool. *Found. Trends Netw.* **2017**, *12*, 162–259. [CrossRef]
3. Yates, R.D.; Sun, Y.; Brown, D.R.; Kaul, S.K.; Modiano, E.; Ulukus, S. Age of Information: An Introduction and Survey. *IEEE J. Sel. Areas Commun.* **2021**, *39*, 1183–1210. [CrossRef]
4. Kaul, S.; Yates, R.; Gruteser, M. Real-time status: How often should one update? In Proceedings of the 2012 Proceedings IEEE INFOCOM, Orlando, FL, USA, 25–30 March 2012; pp. 2731–2735. [CrossRef]

5. Inoue, Y.; Masuyama, H.; Takine, T.; Tanaka, T. A General Formula for the Stationary Distribution of the Age of Information and Its Application to Single-Server Queues. *IEEE Trans. Inf. Theory* **2019**, *65*, 8305–8324. [CrossRef]
6. Yates, R.D.; Kaul, S.K. The Age of Information: Real-Time Status Updating by Multiple Sources. *IEEE Trans. Inf. Theory* **2019**, *65*, 1807–1827. [CrossRef]
7. Kaul, S.K.; Yates, R.D.; Gruteser, M. Status updates through queues. In Proceedings of the 2012 46th Annual Conference on Information Sciences and Systems (CISS), Princeton, NJ, USA, 21–23 March 2012; pp. 1–6. [CrossRef]
8. Costa, M.; Codreanu, M.; Ephremides, A. Age of information with packet management. In Proceedings of the 2014 IEEE International Symposium on Information Theory, Honolulu, HI, USA, 29 June–4 July 2014; pp. 1583–1587. [CrossRef]
9. Costa, M.; Codreanu, M.; Ephremides, A. On the Age of Information in Status Update Systems With Packet Management. *IEEE Trans. Inf. Theory* **2016**, *62*, 1897–1910. [CrossRef]
10. Kam, C.; Kompella, S.; Nguyen, G.D.; Wieselthier, J.E.; Ephremides, A. Age of information with a packet deadline. In Proceedings of the 2016 IEEE International Symposium on Information Theory (ISIT), Barcelona, Spain, 10–15 July 2016; pp. 2564–2568. [CrossRef]
11. Kam, C.; Kompella, S.; Nguyen, G.D.; Wieselthier, J.E.; Ephremides, A. On the Age of Information With Packet Deadlines. *IEEE Trans. Inf. Theory* **2018**, *64*, 6419–6428. [CrossRef]
12. Inoue, Y. Analysis of the Age of Information with Packet Deadline and Infinite Buffer Capacity. In Proceedings of the 2018 IEEE International Symposium on Information Theory (ISIT), Vail, CO, USA, 17–22 June 2018; pp. 2639–2643. [CrossRef]
13. Kavitha, V.; Altman, E.; Saha, I. Controlling Packet Drops to Improve Freshness of information. *arXiv* **2018**, arXiv:1807.09325v1.
14. Wang, B.; Feng, S.; Yang, J. When to preempt? Age of information minimization under link capacity constraint. *J. Commun. Netw.* **2019**, *21*, 220–232. [CrossRef]
15. Arafa, A.; Yates, R.D.; Poor, H.V. Timely Cloud Computing: Preemption and Waiting. In Proceedings of the 2019 57th Annual Allerton Conference on Communication, Control, and Computing (Allerton), Monticello, IL, USA, 24–27 September 2019; pp. 528–535. [CrossRef]
16. Zou, P.; Ozel, O.; Subramaniam, S. Waiting Before Serving: A Companion to Packet Management in Status Update Systems. *IEEE Trans. Inf. Theory* **2020**, *66*, 3864–3877. [CrossRef]
17. Kam, C.; Kompella, S.; Nguyen, G.D.; Ephremides, A. Effect of Message Transmission Path Diversity on Status Age. *IEEE Trans. Inf. Theory* **2016**, *62*, 1360–1374. [CrossRef]
18. Kaswan, P.; Bastopcu, M.; Ulukus, S. Freshness Based Cache Updating in Parallel Relay Networks. In Proceedings of the 2021 IEEE International Symposium on Information Theory (ISIT), Melbourne, Australia, 12–20 July 2021; pp. 3355–3360. [CrossRef]
19. Moltafet, M.; Leinonen, M.; Codreanu, M. On the Age of Information in Multi-Source Queueing Models. *IEEE Trans. Commun.* **2020**, *68*, 5003–5017. [CrossRef]
20. Abd-Elmagid, M.A.; Dhillon, H.S. Age of Information in Multi-source Updating Systems Powered by Energy Harvesting. *IEEE J. Sel. Areas Inf. Theory* **2022**, *3*, 98–112. [CrossRef]
21. Chen, Z.; Pappas, N.; Björnson, E.; Larsson, E.G. Optimizing Information Freshness in a Multiple Access Channel With Heterogeneous Devices. *IEEE Open J. Commun. Soc.* **2021**, *2*, 456–470. [CrossRef]
22. Jiang, Y.; Miyoshi, N. Joint Performance Analysis of Ages of Information in a Multi-Source Pushout Server. *IEEE Trans. Inf. Theory* **2022**, *68*, 965–975. [CrossRef]
23. Tang, Z.; Sun, Z.; Yang, N.; Zhou, X. Age of Information of Multi-Source Systems with Packet Management. In Proceedings of the 2020 IEEE International Conference on Communications Workshops (ICC Workshops), Dublin, Ireland, 7–11 June 2020; pp. 1–6. [CrossRef]
24. Farazi, S.; Klein, A.G.; Brown, D.R. Average Age of Information in Multi-Source Self-Preemptive Status Update Systems with Packet Delivery Errors. In Proceedings of the 2019 53rd Asilomar Conference on Signals, Systems, and Computers, Pacific Grove, CA, USA, 3–6 November 2019; pp. 396–400. [CrossRef]
25. Abd-Elmagid, M.A.; Dhillon, H.S. A Stochastic Hybrid Systems Approach to the Joint Distribution of Ages of Information in Networks. *arXiv* **2022**, arXiv:2205.07448v1.
26. He, T.; Chin, K.W.; Zhang, Z.; Liu, T.; Wen, J. Optimizing Information Freshness in RF-Powered Multi-Hop Wireless Networks. *IEEE Trans. Wirel. Commun.* **2022**. [CrossRef]
27. Gu, Y.; Wang, Q.; Chen, H.; Li, Y.; Vucetic, B. Optimizing Information Freshness in Two-Hop Status Update Systems under a Resource Constraint. *IEEE J. Sel. Areas Commun.* **2021**, *39*, 1380–1392. [CrossRef]
28. Ayan, O.; Gürsu, H.M.; Papa, A.; Kellerer, W. Probability Analysis of Age of Information in Multi-Hop Networks. *IEEE Netw. Lett.* **2020**, *2*, 76–80. [CrossRef]
29. Farazi, S.; Klein, A.G.; Brown, D.R. Fundamental bounds on the age of information in multi-hop global status update networks. *J. Commun. Netw.* **2019**, *21*, 268–279. [CrossRef]
30. Talak, R.; Karaman, S.; Modiano, E. Minimizing age-of-information in multi-hop wireless networks. In Proceedings of the 55th Annual Allerton Conference on Communication, Control, and Computing (Allerton), Monticello, IL, USA, 3–6 October 2017; pp. 486–493. [CrossRef]
31. Moradian, M.; Dadlani, A. Average Age of Information in Two-Way Relay Networks with Service Preemptions. In Proceedings of the IEEE GLOBECOM, Madrid, Spain, 7–11 December 2021; pp. 01–06. [CrossRef]

32. Zakeri, A.; Moltafet, M.; Leinonen, M.; Codreanu, M. Minimizing AoI in Resource-Constrained Multi-Source Relaying Systems with Stochastic Arrivals. In Proceedings of the IEEE GLOBECOM, Madrid, Spain, 7–11 December 2021. [CrossRef]
33. Yuan, X.; Zhu, Y.; Jiang, H.; Hu, Y.; Schmeink, A. Data Freshness Optimization in Relaying Network Operating with Finite Blocklength Codes. In Proceedings of the IEEE GLOBECOM, Madrid, Spain, 7–11 December 2021.
34. Li, B.; Wang, Q.; Chen, H.; Zhou, Y.; Li, Y. Optimizing Information Freshness for Cooperative IoT Systems With Stochastic Arrivals. *IEEE Internet Things J.* **2021**, *8*, 14485–14500. [CrossRef]
35. Zheng, Y.; Hu, J.; Yang, K. Average Age of Information in Wireless Powered Relay Aided Communication Network. *IEEE Internet Things J.* **2021**. [CrossRef]
36. Buyukates, B.; Bastopcu, M.; Ulukus, S. Version Age of Information in Clustered Gossip Networks. *IEEE J. Sel. Areas Inf. Theory* **2022**, *3*, 85–97. [CrossRef]
37. Yates, R.D. The Age of Gossip in Networks. In Proceedings of the 2021 IEEE International Symposium on Information Theory (ISIT), Melbourne, Australia, 12–20 July 2021; pp. 2984–2989. [CrossRef]
38. Yates, R.D. The Age of Information in Networks: Moments, Distributions, and Sampling. *IEEE Trans. Inf. Theory* **2020**, *66*, 5712–5728. [CrossRef]
39. Champati, J.P.; Al-Zubaidy, H.; Gross, J. Statistical guarantee optimization for age of information for the D/G/1 queue. In Proceedings of the IEEE Conference on Computer Communications Workshops (INFOCOM WKSHPS), Honolulu, HI, USA, 15–19 April 2018; pp. 130–135. [CrossRef]
40. Champati, J.P.; Al-Zubaidy, H.; Gross, J. Statistical Guarantee Optimization for AoI in Single-Hop and Two-Hop FCFS Systems With Periodic Arrivals. *IEEE Trans. Commun.* **2021**, *69*, 365–381. [CrossRef]
41. Akar, N.; Doğan, O.; Atay, E.U. Finding the Exact Distribution of (Peak) Age of Information for Queues of PH/PH/1/1 and M/PH/1/2 Type. *IEEE Trans. Commun.* **2020**, *68*, 5661–5672. [CrossRef]
42. Talak, R.; Karaman, S.; Modiano, E. Optimizing Information Freshness in Wireless Networks under General Interference Constraints. *IEEE-ACM Trans. Netw.* **2020**, *28*, 15–28. [CrossRef]
43. Abd-Elmagid, M.A.; Dhillon, H.S.; Pappas, N. A Reinforcement Learning Framework for Optimizing Age of Information in RF-Powered Communication Systems. *IEEE Trans. Commun.* **2020**, *68*, 4747–4760. [CrossRef]
44. Kadota, I.; Sinha, A.; Modiano, E. Scheduling Algorithms for Optimizing Age of Information in Wireless Networks with Throughput Constraints. *IEEE-ACM Trans. Netw.* **2019**, *27*, 1359–1372. [CrossRef]
45. Yang, H.H.; Arafa, A.; Quek, T.Q.S.; Poor, H.V. Optimizing Information Freshness in Wireless Networks: A Stochastic Geometry Approach. *IEEE. Trans. Mob. Comput.* **2021**, *20*, 2269–2280. [CrossRef]
46. He, Q.; Dán, G.; Fodor, V. Joint Assignment and Scheduling for Minimizing Age of Correlated Information. *IEEE-ACM Trans. Netw.* **2019**, *27*, 1887–1900. [CrossRef]
47. Hsu, Y.P.; Modiano, E.; Duan, L. Scheduling Algorithms for Minimizing Age of Information in Wireless Broadcast Networks with Random Arrivals. *IEEE. Trans. Mob. Comput.* **2020**, *19*, 2903–2915. [CrossRef]
48. Kadota, I.; Modiano, E. Minimizing the Age of Information in Wireless Networks with Stochastic Arrivals. *IEEE. Trans. Mob. Comput.* **2021**, *20*, 1173–1185. [CrossRef]
49. Kadota, I.; Sinha, A.; Uysal-Biyikoglu, E.; Singh, R.; Modiano, E. Scheduling Policies for Minimizing Age of Information in Broadcast Wireless Networks. *IEEE-ACM Trans. Netw.* **2018**, *26*, 2637–2650. [CrossRef]
50. Maatouk, A.; Kriouile, S.; Assad, M.; Ephremides, A. On the Optimality of the Whittle's Index Policy for Minimizing the Age of Information. *IEEE Trans. Wirel. Commun.* **2021**, *20*, 1263–1277. [CrossRef]
51. Zhou, B.; Saad, W. Joint Status Sampling and Updating for Minimizing Age of Information in the Internet of Things. *IEEE Trans. Commun.* **2019**, *67*, 7468–7482. [CrossRef]
52. Hespanha, J.P. Modelling and analysis of stochastic hybrid systems. *IEE Proc.-Control Theory Appl.* **2006**, *153*, 520–535. [CrossRef]
53. Kosta, A.; Pappas, N.; Ephremides, A.; Angelakis, V. Non-linear Age of Information in a Discrete Time Queue: Stationary Distribution and Average Performance Analysis. In Proceedings of the 2020 IEEE International Conference on Communications (ICC), Dublin, Ireland, 7–11 June 2020; pp. 1–6. [CrossRef]
54. Kosta, A.; Pappas, N.; Ephremides, A.; Angelakis, V. The Age of Information in a Discrete Time Queue: Stationary Distribution and Non-Linear Age Mean Analysis. *IEEE J. Sel. Areas Commun.* **2021**, *39*, 1352–1364. [CrossRef]
55. Brill, P. *Level Crossing Methods in Stochastic Models*; Springer: New York, NY, USA, 2017.
56. Tripathi, V.; Talak, R.; Modiano, E. Age of Information for Discrete Time Queues. *arXiv* **2019**, arXiv:1901.10463v1.
57. Akar, N.; Doğan, O. Discrete-Time Queueing Model of Age of Information With Multiple Information Sources. *IEEE Internet Things J.* **2021**, *8*, 14531–14542. [CrossRef]
58. Zhang, J.; Xu, Y. On Age of Information for Discrete Time Status Updating System With Ber/G/1/1 Queues. In Proceedings of the IEEE Information Theory Workshop (ITW), Kanazawa, Japan, 17–21 October 2021; pp. 1–6.
59. Zhang, J.; Xu, Y. On Discrete Age of Information of Infinite Size Status Updating System. *arXiv* **2022**, arXiv:2204.11532v1.
60. Zhang, J. Discrete Packet Management: Analysis of Age of Information of Discrete Time Status Updating Systems. *arXiv* **2022**, arXiv:2204.13333v1.
61. Moltafet, M.; Leinonen, M.; Codreanu, M. Average AoI in Multi-Source Systems With Source-Aware Packet Management. *IEEE Trans. Commun.* **2021**, *69*, 1121–1133. [CrossRef]

62. Feng, S.; Yang, J. Age-Optimal Transmission of Rateless Codes in an Erasure Channel. In Proceedings of the 2019 IEEE International Conference on Communications (ICC), Shanghai, China, 20–24 May 2019.
63. Zhong, J.; Yates, R.D.; Soljanin, E. Timely Lossless Source Coding for Randomly Arriving Symbols. In Proceedings of the 2018 IEEE Information Theory Workshop (ITW), Guangzhou, China, 25–29 November 2018.
64. Doğan, O.; Akar, N. The Multi-Source Probabilistically Preemptive M/PH/1/1 Queue with Packet Errors. *IEEE Trans. Commun.* **2021**, *69*, 7297–7308. [CrossRef]
65. Thekkilakattil, A.; Dobrin, R.; Punnekkat, S. Probabilistic preemption control using frequency scaling for sporadic real-time tasks. In Proceedings of the 7th IEEE International Symposium on Industrial Embedded Systems (SIES'12), Karlsruhe, Germany, 20–22 June 2012; pp. 158–165. [CrossRef]

Article

Using Timeliness in Tracking Infections [†]

Melih Bastopcu [1] and Sennur Ulukus [2,*]

[1] Coordinated Science Laboratory, University of Illinois Urbana-Champaign, Urbana, IL 61801, USA; bastopcu@illinois.edu

[2] Department of Electrical and Computer Engineering, University of Maryland, College Park, MD 20742, USA

* Correspondence: ulukus@umd.edu

[†] This paper is an extended version of our paper published in 2021 IEEE INFOCOM 2021—IEEE International Conference on Computer Communications (online), 10–13 May 2021.

Abstract: We consider real-time timely tracking of infection status (e.g., COVID-19) of individuals in a population. In this work, a health care provider wants to detect both infected people and people who have recovered from the disease as quickly as possible. In order to measure the timeliness of the tracking process, we use the long-term average difference between the actual infection status of the people and their real-time estimate by the health care provider based on the most recent test results. We first find an analytical expression for this average difference for given test rates, infection rates and recovery rates of people. Next, we propose an alternating minimization-based algorithm to find the test rates that minimize the average difference. We observe that if the total test rate is limited, instead of testing all members of the population equally, only a portion of the population may be tested in unequal rates calculated based on their infection and recovery rates. Next, we characterize the average difference when the test measurements are erroneous (i.e., noisy). Further, we consider the case where the infection status of individuals may be dependent, which occurs when an infected person spreads the disease to another person if they are not detected and isolated by the health care provider. In addition, we consider an age of incorrect information-based error metric where the staleness metric increases linearly over time as long as the health care provider does not detect the changes in the infection status of the people. Through extensive numerical results, we observe that increasing the total test rate helps track the infection status better. In addition, an increased population size increases diversity of people with different infection and recovery rates, which may be exploited to spend testing capacity more efficiently, thereby improving the system performance. Depending on the health care provider's preferences, test rate allocation can be adjusted to detect either the infected people or the recovered people more quickly. In order to combat any errors in the test, it may be more advantageous for the health care provider to not test everyone, and instead, apply additional tests to a selected portion of the population. In the case of people with dependent infection status, as we increase the total test rate, the health care provider detects the infected people more quickly, and thus, the average time that a person stays infected decreases. Finally, the error metric needs to be chosen carefully to meet the priorities of the health care provider, as the error metric used greatly influences who will be tested and at what test rate.

Keywords: timely infection tracking; age of information; timely tracking of multiple processes; Markovian infection spread model

Citation: Bastopcu, M.; Ulukus, S. Using Timeliness in Tracking Infections. *Entropy* **2022**, *24*, 779. https://doi.org/10.3390/e24060779

Academic Editor: Luca Faes

Received: 8 March 2022
Accepted: 27 May 2022
Published: 31 May 2022

Publisher's Note: MDPI stays neutral with regard to jurisdictional claims in published maps and institutional affiliations.

1. Introduction

We consider the problem of timely tracking of an infectious disease, e.g., COVID-19, in a population of n people. In this problem, a health care provider wants to detect infected people as quickly as possible in order to take precautions such as isolating them from the rest of the population. The health care provider also wants to detect people who have recovered from the disease as soon as possible since these people need to return to work which is especially critical in sectors such as education, food retail, public transportation, etc. Ideally, the health care provider should test all people all the time. However, as the total test rate is limited, the question is how frequently the health care provider should apply tests on these people when their infection and recovery rates are known. In a broader sense, this problem is related to timely tracking of multiple processes in a resource-constrained setting where each process takes binary values of 0 and 1 with different change rates.

Recent studies have shown that people who have recovered from infectious diseases such as COVID-19 can be reinfected. Furthermore, the recovery times of individuals may vary significantly. For these reasons, in this problem, the ith person becomes infected with rate λ_i which is independent of the others. Similarly, the ith person recovers from the disease with rate μ_i. We note that the index i may represent a specific individual or a group of individuals that share common features such as age, gender, and profession. Depending on the demographics, coefficients λ_i and μ_i may be statistically known by the health care provider. We denote the infection status of the ith person as $x_i(t)$ (shown with the black curves on the left in Figure 1) which takes the value 1 when the person is infected and the value 0 when the person is healthy. The health care provider applies tests to people marked as healthy with rate s_i and to people marked as infected with rate c_i. Based on the test results, the health care provider forms an estimate for the infection status of the ith person denoted by $\hat{x}_i(t)$ (shown with the blue curves on the right in Figure 1) which takes the value 1 when the most recent test result is positive and the value 0 when it is negative.

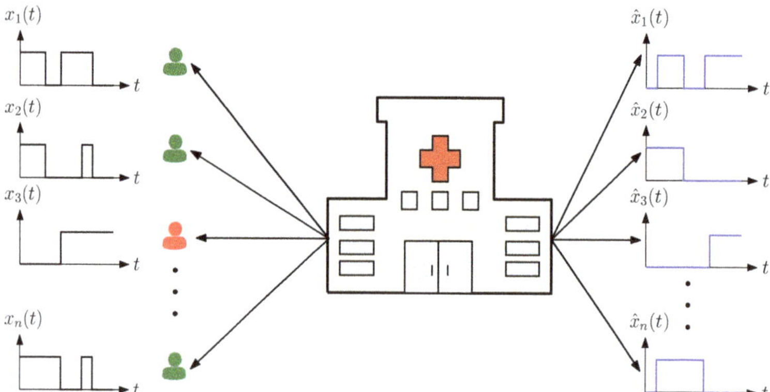

Figure 1. System model. There are n people whose infection status are given by $x_i(t)$. The health care provider applies tests on these people. Based on the test results, estimations for the infection status $\hat{x}_i(t)$ are generated. Infected people are shown in red and healthy people are shown in green.

We measure the timeliness of the tracking process by the difference between the actual infection status of people and the real-time estimate of the health care provider which is based on the most recent test results. The difference can occur in two different cases: (i) when the person is sick ($x_i(t) = 1$) and the health care provider maps this person as healthy ($\hat{x}_i(t) = 0$), and (ii) when the person recovers from the disease ($x_i(t) = 0$) but the health care provider still considers this person as infected ($\hat{x}_i(t) = 1$). The former case represents the error due to late detection of infected people, while the latter case represents the error due to late detection of healed people. Depending on the health care provider's preferences, detecting infected people may be more important than detecting recovered people (controlling infection), or the other way around (returning people to workforce).

The age of information was proposed to measure timeliness of information in communication systems, and has been studied in the context of queueing systems [1–8], multi-hop and multi-cast networks [9–17], social networks [18], timely remote estimation of random processes [19–25], energy harvesting systems [26–40], wireless fading channels [41,42], scheduling in networks [43–55], lossless and lossy source and channel coding [56–66], vehicular, IoT and UAV systems [67–70], caching systems [71–82], computation-intensive systems [83–90], learning systems [91–93], gossip networks [94–97] and so forth. A more detailed review of the age of information literature can be found in references [98–100]. Most relevant to our work, the real-time timely estimation of single and multiple counting processes [19,25], a Wiener process [20], a random walk process [101], and binary and multiple states Markov sources [23,51,102] have been studied. The study that is closest to our work is reference [23], where the remote estimation of a symmetric binary Markov source is studied in a time-slotted system by finding the optimal sampling policies via formulating a Markov Decision Process (MDP) for real-time error, AoI and AoII metrics. Different from [23], in our work, we consider real-time timely estimation of multiple non-symmetric binary sources for a continuous time system. In our work, the sampler (health care provider) does not know the states of the sources (infection status of people), and thus, takes the samples (applies medical tests) randomly (exponential random variables) with fixed rates. Thus, we optimize the test rates of people to minimize the real-time estimation error.

In this paper, we consider the real-time timely tracking of infection status of n people. We first find an analytical expression for the long-term average difference between the actual infection status of people and the estimate of the health care provider based on test results. Then, we propose an alternating minimization-based algorithm to identify the test rates s_i and c_i for all people. We observe that if the total test rate is limited, we may not apply tests on all people equally. Next, we provide an alternative method to characterize the average difference, by finding the steady state of a Markov chain defined by $(x_i(t), \hat{x}_i(t))$. By using this alternative method, we determine the average estimation error when there are errors in the test measurements expressed by a false positive rate p and a false negative rate q. Next, we consider the infection status of two people where an infected person may spread the disease to another person if the infection has not been detected by the health care provider to consequently isolate the infected person. Finally, we consider an age of incorrect information-based error metric where the estimation error increases linearly over time when the health care provider has not detected the changes in the infection status of the people.

Through extensive numerical results, we observe that increasing the total test rate helps track the infection status of people better, and increasing the size of the population increases diversity which may be exploited to improve the performance. Depending on the health care provider's priorities, we can allocate additional tests to people marked as healthy to detect the infections faster or to people marked as infected to detect the recoveries more quickly. In order to combat the test errors, the health care provider may prefer to apply tests to only a selected portion of the population with higher test rates. When the infection status of a person depends on that of another person, the average time that a person remains infected can be reduced by increasing the total test rate as it helps to detect the infected people more quickly. Finally, we observe that depending on the error metric used, the test rate distribution among the population differs greatly, and thus, we should choose an error metric that aligns with the priorities of the health care provider.

2. System Model

We consider a population of n people. We denote the infection status of the ith person at time t as $x_i(t)$ (black curve in Figure 2a) which takes binary values 0 or 1 as follows,

$$
x_i(t) = \begin{cases} 1, & \text{if the } i\text{th person is infected at time } t, \\ 0, & \text{otherwise.} \end{cases} \tag{1}
$$

In this paper, we consider a model where each person can be infected multiple times after recovering from the disease. We denote the time interval that the ith person stays healthy for the jth time as $W_i(j)$ which is exponentially distributed with rate λ_i. We denote the recovery time for the ith person after being infected with the virus for the jth time as $R_i(j)$ which is exponentially distributed with rate μ_i.

A health care provider wants to track the infection status of each person. Based on the test results at times $t_{i,\ell}$, the health care provider generates an estimate for the status of the ith person denoted as $\hat{x}_i(t)$ (blue curve in Figure 2a) by

$$\hat{x}_i(t) = x_i(t_{i,\ell}), \quad t_{i,\ell} \le t < t_{i,\ell+1}. \tag{2}$$

When $\hat{x}_i(t)$ is 1, the health care provider applies the next test to the ith person after an exponentially distributed time with rate c_i. When $\hat{x}_i(t)$ is 0, the next test is applied to the ith person after an exponentially distributed time with rate s_i.

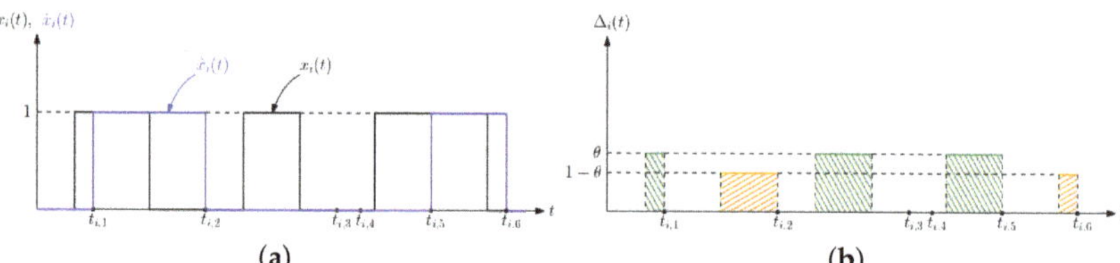

Figure 2. (**a**) A sample evolution of $x_i(t)$ and $\hat{x}_i(t)$, and (**b**) the corresponding $\Delta_i(t)$ in (5). Green areas correspond to the error caused by $\Delta_{i1}(t)$ in (3). Orange areas correspond to the error caused by $\Delta_{i2}(t)$ in (4).

An estimation error happens when the actual infection status of the ith person, $x_i(t)$, is different than the estimate of the health care provider, $\hat{x}_i(t)$, at time t. This could happen in two ways: when $x_i(t) = 1$ and $\hat{x}_i(t) = 0$, i.e., when the ith person is sick, but remains undetected by the health care provider, and when $x_i(t) = 0$ and $\hat{x}_i(t) = 1$, i.e., when the ith person has recovered, but the health care provider is unaware that the ith person has recovered.

We denote the error caused by the former case, i.e., when $x_i(t) = 1$ and $\hat{x}_i(t) = 0$, by $\Delta_{i1}(t)$ (green areas in Figure 2b),

$$\Delta_{i1}(t) = \max\{x_i(t) - \hat{x}_i(t), 0\}, \tag{3}$$

and we denote the error caused by the latter case, i.e., when $x_i(t) = 0$ and $\hat{x}_i(t) = 1$, by $\Delta_{i2}(t)$ (orange areas in Figure 2b),

$$\Delta_{i2}(t) = \max\{\hat{x}_i(t) - x_i(t), 0\}. \tag{4}$$

Then, the total estimation error for the ith person $\Delta_i(t)$ is

$$\Delta_i(t) = \theta \Delta_{i1}(t) + (1 - \theta)\Delta_{i2}(t), \tag{5}$$

where θ is the importance factor in $[0, 1]$. A large θ gives more importance to the detection of infected people, and a small θ gives more importance to the detection of recovered people.

We define the long-term weighted average difference between $x_i(t)$ and $\hat{x}_i(t)$ as

$$\Delta_i = \lim_{T \to \infty} \frac{1}{T} \int_0^T \Delta_i(t)dt. \tag{6}$$

Then, the overall average difference of all people Δ is

$$\Delta = \frac{1}{n}\sum_{i=1}^{n} \Delta_i. \tag{7}$$

Our aim is to track the infection status of all people. Due to limited resources, there is a total test rate constraint $\sum_{i=1}^{n} s_i + \sum_{i=1}^{n} c_i \le C$. Thus, our aim is to find the optimal test rates s_i and c_i to minimize Δ in (7) while satisfying this total test rate constraint. We formulate the following problem,

$$\min_{\{s_i,c_i\}} \quad \Delta$$
$$\text{s.t.} \quad \sum_{i=1}^{n} s_i + \sum_{i=1}^{n} c_i \leq C$$
$$s_i \geq 0, \quad c_i \geq 0, \quad i = 1,\ldots,n. \tag{8}$$

We provide a summary of the list of the variables used in this work in Table 1. In the next section, we find the total average difference Δ.

Table 1. List of variables used in this work.

Variables	Definition of the Variables
Sections 2–4	
n	number of people in the population
$x_i(t)$	infection status of the ith person at time t
$\hat{x}_i(t)$	estimation of $x_i(t)$ at the health care provider
λ_i, μ_i	infection and recovery rates for the ith person
c_i, s_i	test rates applied to the ith person when $\hat{x}_i(t) = 1$, and $\hat{x}_i(t) = 0$
$\Delta_i(t)$	total estimation error for the ith person at time t
θ	importance factor in $[0,1]$
Δ_i	the long-time weighted average for the ith person
C	total test rate constraint
Section 5	
Δ_i^e	the long-time average difference for the ith person with erroneous test measurements
q	false-negative testing probability with $0 \leq q < \frac{1}{2}$
p	false-positive testing probability with $0 \leq p < \frac{1}{2}$
v_i	test rate applied to the ith person with erroneous test measurements
Section 6	
λ, μ	individual infection and recovery rate of a person
λ_{12}	the rate of spreading the virus from an undetected infected person to a healthy person
c, s	test rates applied to people when $\hat{x}_i(t) = 1$, and $\hat{x}_i(t) = 0$
Δ_i^d	the long-time average difference for the ith person with dependent infection rates
Section 7	
w_i	test rate applied to the ith person for AoII-based error metric
Δ_i^s	the long-time average difference for the ith person with AoII-based error metric

3. Average Difference Analysis

In this section, we provide a probabilistic analysis to characterize the average difference Δ. In Section 5.1, we give an alternative method to find Δ by analyzing the steady-state distribution of the Markov chain induced by the states $(x_i(t), \hat{x}_i(t))$. Here, we first find analytical expressions for $\Delta_{i1}(t)$ in (3) and $\Delta_{i2}(t)$ in (4) when $s_i > 0$ and $c_i > 0$. We note that $\Delta_{i1}(t)$ can be equal to 1 when $\hat{x}_i(t) = 0$ and is always equal to 0 when $\hat{x}_i(t) = 1$. Assume that at time 0, both $x_i(0)$ and $\hat{x}_i(0)$ are 0. After an exponentially distributed time with rate λ_i, which is denoted by W_i, the ith person is infected, and thus $x_i(t)$ becomes 1. At that time, since $\hat{x}_i(t) = 0$, $\Delta_{i1}(t)$ becomes 1. Further, $\Delta_{i1}(t)$ will be equal to 0 again either when the ith person recovers from the disease which happens after R_i which is exponentially distributed with rate μ_i or when the health care provider performs a test on the ith person after D_i, which is exponentially distributed with rate s_i. We define $T_m(i)$ as the earliest time at which one of these two cases happens, i.e., $T_m(i) = \min\{R_i, D_i\}$ (which is shown by the green areas in Figure 3a). We note that $T_m(i)$ is also exponentially distributed with rate $\mu_i + s_i$, and we have $\mathbb{P}(T_m(i) = R_i) = \frac{\mu_i}{\mu_i + s_i}$ and $\mathbb{P}(T_m(i) = D_i) = \frac{s_i}{\mu_i + s_i}$. If the ith person recovers from the disease before testing, we return to the initial case where both $x_i(t)$ and $\hat{x}_i(t)$ are equal to 0 again. In this case, the cycle repeats itself, i.e., the ith person becomes sick again after W_i and $\Delta_{i1}(t)$ remains as 1 until either the person recovers or the health care provider performs a test which takes another $T_m(i)$ duration. If the health care provider performs a test before the person recovers, then $\hat{x}_i(t)$ becomes 1. We denote the time interval for which $\hat{x}_i(t)$ stays at 0 as I_{i1} which is given by

$$I_{i1} = \sum_{\ell=1}^{K_1} T_m(i, \ell) + W_i(\ell), \tag{9}$$

where K_1 is geometric with rate $\mathbb{P}(T_m(i) = D_i) = \frac{s_i}{\mu_i + s_i}$. Due to [103] (Prob. 9.4.1), $\sum_{\ell=1}^{K_1} T_m(i, \ell)$ and $\sum_{\ell=1}^{K_1} W_i(\ell)$ are exponentially distributed with rates s_i and $\frac{\lambda_i s_i}{\mu_i + s_i}$, respectively. As $\mathbb{E}[I_{i1}] = \mathbb{E}[\sum_{\ell=1}^{K_1} T_m(i, \ell)] + \mathbb{E}[\sum_{\ell=1}^{K_1} W_i(\ell)]$, we have

$$\mathbb{E}[I_{i1}] = \frac{1}{s_i} + \frac{s_i + \mu_i}{s_i \lambda_i}. \tag{10}$$

When $\hat{x}_i(t) = 1$, the health care provider marks the ith person as infected. The ith person recovers from the virus after R_i. After the ith person recovers, either the health care provider performs a test after Z_i which is exponentially distributed with rate c_i or the ith person is reinfected with the virus which takes W_i time. We define $T_u(i)$ as the earliest time at which one of these two cases happens, i.e., $T_u(i) = \min\{W_i, Z_i\}$ (which is shown by the orange areas in Figure 3b). Similarly, we note that $T_u(i)$ is exponentially distributed with rate $\lambda_i + c_i$, and we have $\mathbb{P}(T_u(i) = W_i) = \frac{\lambda_i}{\lambda_i + c_i}$ and $\mathbb{P}(T_u(i) = Z_i) = \frac{c_i}{\lambda_i + c_i}$. If the person is reinfected with the virus before a test is applied, this cycle repeats itself, i.e., the ith person recovers after another R_i, and then either a test is applied to the ith person, or the person is infected again which takes another $T_u(i)$. If the health care provider performs a test to the ith person before the person is reinfected, the health care provider marks the ith person as healthy again, i.e., $\hat{x}_i(t)$ becomes 0. We denote the time interval that $\hat{x}_i(t)$ is equal to 1 as I_{i2} which is given by

$$I_{i2} = \sum_{\ell=1}^{K_2} T_u(i, \ell) + R_i(\ell), \tag{11}$$

where K_2 is geometric with rate $\mathbb{P}(T_u(i) = Z_i) = \frac{c_i}{\lambda_i + c_i}$. Similarly, $\sum_{\ell=1}^{K_2} T_u(i, \ell)$ and $\sum_{\ell=1}^{K_2} R_i(\ell)$ are exponentially distributed with rates c_i and $\frac{c_i \mu_i}{\lambda_i + c_i}$, respectively. As $\mathbb{E}[I_{i2}] = \mathbb{E}[\sum_{\ell=1}^{K_2} T_u(i, \ell)] + \mathbb{E}[\sum_{\ell=1}^{K_2} R_i(\ell)]$, we have

$$\mathbb{E}[I_{i2}] = \frac{1}{c_i} + \frac{c_i + \lambda_i}{c_i \mu_i}. \tag{12}$$

We denote the time interval between the jth and $(j+1)$th times that $\hat{x}_i(t)$ changes from 1 to 0 as the jth cycle $I_i(j)$ where $I_i(j) = I_{i1}(j) + I_{i2}(j)$. We note that $\Delta_{i1}(t)$ is always equal to 0 during $I_{i2}(j)$, i.e., $\hat{x}_i(t) = 1$, and $\Delta_{i1}(t)$ is equal to 1 when $x_i(t) = 1$ in $I_{i1}(j)$. We denote the total time duration when $\Delta_{i1}(t)$ is equal to 1 as $T_{e,1}(i,j)$ during the jth cycle where $T_{e,1}(i,j) = \sum_{\ell=1}^{K_1} T_m(i,\ell)$. Thus, we have $\mathbb{E}[T_{e,1}(i)] = \frac{1}{s_i}$. Then, using ergodicity, similar to [80], Δ_{i1} is equal to

$$\Delta_{i1} = \frac{\mathbb{E}[T_{e,1}(i)]}{\mathbb{E}[I_i]} = \frac{\mathbb{E}[T_{e,1}(i)]}{\mathbb{E}[I_{i1}] + \mathbb{E}[I_{i2}]}. \tag{13}$$

Thus, we have

$$\Delta_{i1} = \frac{\mu_i \lambda_i}{\mu_i + \lambda_i} \frac{c_i}{\mu_i c_i + \lambda_i s_i + c_i s_i}. \tag{14}$$

Next, we find Δ_{i2}. We note that $\Delta_{i2}(t)$ is equal to 1 when $x_i(t) = 0$ in $I_{i2}(j)$ and is always equal to 0 during $I_{i1}(j)$. Similarly, we denote the total time duration where $\Delta_{i2}(t)$ is equal to 1 in the jth cycle $I_i(j)$ as $T_{e,2}(i,j)$ which is equal to $T_{e,2}(i,j) = \sum_{\ell=1}^{K_2} T_u(i,\ell)$. Thus, we have $\mathbb{E}[T_{e,2}(i)] = \frac{1}{c_i}$. Then, similar to Δ_{i1} in (13), Δ_{i2} is equal to

$$\Delta_{i2} = \frac{\mu_i \lambda_i}{\mu_i + \lambda_i} \frac{s_i}{\mu_i c_i + \lambda_i s_i + c_i s_i}. \tag{15}$$

By using (5), (14), and (15), we obtain Δ_i as

$$\Delta_i = \frac{\mu_i \lambda_i}{\mu_i + \lambda_i} \frac{\theta c_i + (1-\theta) s_i}{\mu_i c_i + \lambda_i s_i + c_i s_i}. \tag{16}$$

Then, by inserting (16) in (7), we obtain Δ. In the next section, we solve the optimization problem in (8).

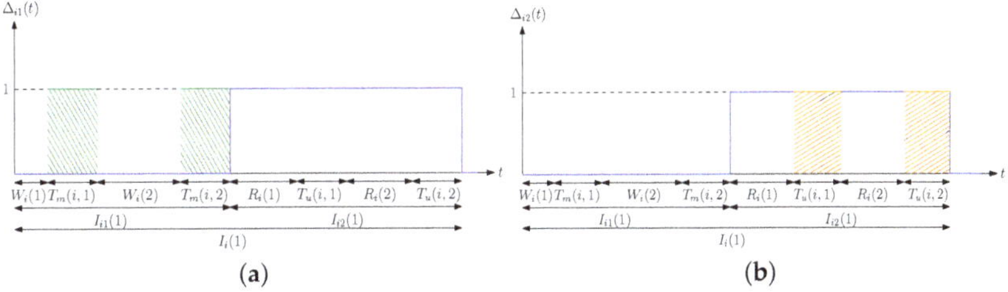

Figure 3. A sample evolution of (**a**) $\Delta_{i1}(t)$, and (**b**) $\Delta_{i2}(t)$ in a typical cycle.

4. Optimization of Average Difference

In this section, we solve the optimization problem in (8). Using Δ_i in (16) in (7), we rewrite (8) as

$$\min_{\{s_i,c_i\}} \quad \sum_{i=1}^{n} \frac{\mu_i \lambda_i}{\mu_i + \lambda_i} \frac{\theta c_i + (1-\theta) s_i}{\mu_i c_i + \lambda_i s_i + c_i s_i}$$

$$\text{s.t.} \quad \sum_{i=1}^{n} s_i + \sum_{i=1}^{n} c_i \leq C$$

$$s_i \geq 0, \quad c_i \geq 0, \quad i = 1, \dots, n. \tag{17}$$

We define the Lagrangian function [104] for (17) as

$$\mathcal{L} = \sum_{i=1}^{n} \frac{\mu_i \lambda_i}{\mu_i + \lambda_i} \frac{\theta c_i + (1-\theta)s_i}{\mu_i c_i + \lambda_i s_i + c_i s_i} + \beta \left(\sum_{i=1}^{n} s_i + c_i - C \right) - \sum_{i=1}^{n} \nu_i s_i - \sum_{i=1}^{n} \eta_i c_i, \tag{18}$$

where $\beta \geq 0$, $\nu_i \geq 0$, and $\eta_i \geq 0$. The KKT conditions are

$$\frac{\partial \mathcal{L}}{\partial s_i} = \frac{\mu_i \lambda_i c_i}{\mu_i + \lambda_i} \frac{(1-\theta)\mu_i - \theta(c_i + \lambda_i)}{(\mu_i c_i + \lambda_i s_i + s_i c_i)^2} + \beta - \nu_i = 0, \tag{19}$$

$$\frac{\partial \mathcal{L}}{\partial c_i} = \frac{\mu_i \lambda_i s_i}{\mu_i + \lambda_i} \frac{\theta \lambda_i - (1-\theta)(\mu_i + s_i)}{(\mu_i c_i + \lambda_i s_i + s_i c_i)^2} + \beta - \eta_i = 0, \tag{20}$$

for all i. The complementary slackness conditions are

$$\beta \left(\sum_{i=1}^{n} s_i + c_i - C \right) = 0, \quad \nu_i s_i = 0, \quad \eta_i c_i = 0. \tag{21}$$

First, we find s_i. From (19), we have

$$(\mu_i c_i + \lambda_i s_i + s_i c_i)^2 = \frac{\mu_i \lambda_i c_i}{\mu_i + \lambda_i} \frac{\theta(c_i + \lambda_i) - (1-\theta)\mu_i}{\beta - \nu_i}. \tag{22}$$

When $\theta(c_i + \lambda_i) \geq (1-\theta)\mu_i$, we solve (22) for s_i as

$$s_i = \frac{\mu_i c_i}{\lambda_i + c_i} \left(\sqrt{\frac{1}{\mu_i c_i} \frac{\lambda_i}{\mu_i + \lambda_i} \frac{\theta(c_i + \lambda_i) - (1-\theta)\mu_i}{\beta}} - 1 \right)^{+}, \tag{23}$$

where we used the fact that we either have $s_i > 0$ and $\nu_i = 0$, or $s_i = 0$ and $\nu_i \geq 0$, due to (21). Here, $(\cdot)^{+} = \max(\cdot, 0)$. On the other hand, when $\theta(c_i + \lambda_i) < (1-\theta)\mu_i$, we have $\frac{\partial \Delta_i}{\partial s_i} > 0$, and thus it is optimal to choose $s_i = 0$ as our aim is to minimize Δ in (7). In this case, when $s_i = 0$, we have $\Delta_i = \frac{\theta \lambda_i}{\mu_i + \lambda_i}$ which is independent of the value of c_i. As we obtain the same Δ_i for all values of c_i, and the total update rate is limited, i.e., $\sum_{i=1}^{n} s_i + c_i \leq C$, in this case, it is optimal to choose $c_i = 0$ as well (i.e., when $s_i = 0$).

Next, we find c_i. From (20), we have

$$(\mu_i c_i + \lambda_i s_i + s_i c_i)^2 = \frac{\mu_i \lambda_i s_i}{\mu_i + \lambda_i} \frac{(1-\theta)(\mu_i + s_i) - \theta \lambda_i}{\beta - \eta_i}. \tag{24}$$

When $(1-\theta)(\mu_i + s_i) \geq \theta \lambda_i$, we solve (24) for c_i as

$$c_i = \frac{\lambda_i s_i}{\mu_i + s_i} \left(\sqrt{\frac{1}{\lambda_i s_i} \frac{\mu_i}{\mu_i + \lambda_i} \frac{(1-\theta)(s_i + \mu_i) - \theta \lambda_i}{\beta}} - 1 \right)^{+}, \tag{25}$$

where we used the fact that we either have $c_i > 0$ and $\eta_i = 0$, or $c_i = 0$ and $\eta_i \geq 0$, due to (21). Similarly, when $(1-\theta)(s_i + \mu_i) < \theta \lambda_i$, we have $\frac{\partial \Delta_i}{\partial c_i} > 0$. Thus, in this case, it is optimal to choose $c_i = 0$. When $c_i = 0$, we have $\Delta_i = \frac{(1-\theta)\mu_i}{\mu_i + \lambda_i}$ which is independent of the value of s_i. Thus, it is optimal to choose $s_i = 0$ when $c_i = 0$.

From (23), if $\frac{1}{\mu_i c_i} \frac{\lambda_i}{\mu_i + \lambda_i} (\theta(c_i + \lambda_i) - (1-\theta)\mu_i) \leq \beta$, we must have $s_i = 0$. Thus, for a given c_i, the optimal test rate allocation policy for s_i is a *threshold policy* where s_i's with small $\frac{1}{\mu_i c_i} \frac{\lambda_i}{\mu_i + \lambda_i} (\theta(c_i + \lambda_i) - (1-\theta)\mu_i)$ are equal to zero. Similarly, from (25), if $\frac{1}{\lambda_i s_i} \frac{\mu_i}{\mu_i + \lambda_i} ((1-\theta)(s_i + \mu_i) - \theta \lambda_i) \leq \beta$, we must have $c_i = 0$. Thus, for a given s_i, the optimal policy to determine c_i is a *threshold policy* where c_i's with small $\frac{1}{\lambda_i s_i} \frac{\mu_i}{\mu_i + \lambda_i} ((1-\theta)(s_i + \mu_i) - \theta \lambda_i)$ are equal to zero.

Next, we show that in the optimal policy, if $s_i > 0$ and $c_i > 0$ for some i, then the total test rate constraint must be satisfied with equality, i.e., $\sum_{i=1}^n s_i + c_i = C$.

Lemma 1. *In the optimal policy, if $s_i > 0$ and $c_i > 0$ for some i, then we have $\sum_{i=1}^n s_i + c_i = C$.*

Proof of Lemma 1. The derivatives of Δ_i with respect to s_i and c_i are

$$\frac{\partial \Delta_i}{\partial s_i} = \frac{\mu_i \lambda_i c_i}{\mu_i + \lambda_i} \frac{(1-\theta)\mu_i - \theta(c_i + \lambda_i)}{(c_i \mu_i + s_i c_i + \lambda_i s_i)^2}, \tag{26}$$

$$\frac{\partial \Delta_i}{\partial c_i} = \frac{\mu_i \lambda_i s_i}{\mu_i + \lambda_i} \frac{\theta \lambda_i - (1-\theta)(s_i + \mu_i)}{(c_i \mu_i + s_i c_i + \lambda_i s_i)^2}. \tag{27}$$

We note that $s_i > 0$ in (23) implies that $\theta(c_i + \lambda_i) > (1-\theta)\mu_i$. In this case, we have $\frac{\partial \Delta_i}{\partial s_i} < 0$. Similarly, $c_i > 0$ in (25) implies that $(1-\theta)(s_i + \mu_i) > \theta \lambda_i$. Thus, we have $\frac{\partial \Delta_i}{\partial c_i} < 0$. Therefore, in the optimal policy, if we have $s_i > 0$ and $c_i > 0$ for some i, then we must have $\sum_{i=1}^n s_i + c_i = C$. Otherwise, we can further decrease Δ in (7) by increasing c_i or s_i. $\square$

Next, we propose an alternating minimization-based algorithm for finding s_i and c_i. For this purpose, for given initial (s_i, c_i) pairs, we define ϕ_i as

$$\phi_i = \begin{cases} \frac{1}{\mu_i c_i} \frac{\lambda_i}{\mu_i + \lambda_i} (\theta(c_i + \lambda_i) - (1-\theta)\mu_i), & i=1,\ldots,n, \\ \frac{1}{\lambda_i s_i} \frac{\mu_i}{\mu_i + \lambda_i} ((1-\theta)(s_i + \mu_i) - \theta \lambda_i), & i=n+1,\ldots,2n. \end{cases} \tag{28}$$

Then, we define u_i as

$$u_i = \begin{cases} \frac{\mu_i c_i}{\lambda_i + c_i} \left(\sqrt{\frac{\phi_i}{\beta}} - 1 \right)^+, & i = 1,\ldots,n, \\ \frac{\lambda_i s_i}{\mu_i + s_i} \left(\sqrt{\frac{\phi_i}{\beta}} - 1 \right)^+, & i = n+1,\ldots,2n. \end{cases} \tag{29}$$

From (23) and (25), $s_i = u_i$ and $c_i = u_{n+i}$, for $i = 1,\ldots,n$.

Next, we find s_i and c_i by determining β in (29). First, assume that, in the optimal policy, there is an i such that $s_i > 0$ and $c_i > 0$. Thus, by Lemma 1, we must have $\sum_{i=1}^n s_i + c_i = C$. We initially take random (s_i, c_i) pairs such that $\sum_{i=1}^n s_i + c_i = C$. Then, given the initial (s_i, c_i) pairs, we immediately choose $u_i = 0$ for $\phi_i < 0$. For the remaining u_i with $\phi_i \geq 0$, we apply a solution method similar to that in [80]. By assuming $\phi_i \geq \beta$, i.e., by disregarding $(\cdot)^+$ in (29), we solve $\sum_{i=1}^{2n} u_i = C$ for β. Then, we compare the smallest ϕ_i which is larger than zero in (28) with β. If we have $\phi_i \geq \beta$, then it implies that $u_i \geq 0$ for all remaining i. Thus, we have obtained u_i values for given initial (s_i, c_i) pairs. If the smallest ϕ_i which is larger than zero is smaller than β, then the corresponding u_i is negative and we should choose $u_i = 0$ for the smallest non-negative ϕ_i. Then, we repeat this procedure until the smallest non-negative ϕ_i is larger than β. After determining all u_i, we obtain $s_i = u_i$ and $c_i = u_{n+i}$ for $i = 1,\ldots,n$. Then, with the updated values of (s_i, c_i) pairs, we keep finding u_i's until the KKT conditions in (19) and (20) are satisfied.

We note that for indices (persons) i for which (s_i, c_i) are zero, the health care provider does not perform any tests, and maps these people as either always infected, i.e., $\hat{x}_i(t) = 1$ for all t, or always healthy, i.e., $\hat{x}_i(t) = 0$. If $\hat{x}_i(t) = 0$ for all t, $\Delta_i = \frac{\theta \lambda_i}{\mu_i + \lambda_i}$, and if $\hat{x}_i(t) = 1$ for all t, $\Delta_i = \frac{(1-\theta)\mu_i}{\mu_i + \lambda_i}$. Thus, for such i, the health care provider should choose $\hat{x}_i(t) = 0$ for all t, if $\frac{\theta \lambda_i}{\mu_i + \lambda_i} < \frac{(1-\theta)\mu_i}{\mu_i + \lambda_i}$, and should choose $\hat{x}_i(t) = 1$ for all t, otherwise, without performing any tests.

Finally, we note that the problem in (17) is not a convex optimization problem as the objective function is not jointly convex in s_i and c_i. Therefore, the solutions obtained via the proposed method may not be globally optimal. For this reason, we select different initial starting points and apply the proposed alternating minimization-based algorithm and choose the solution that achieves the smallest Δ in (7).

In the next section, we first provide an alternative method to find the average difference Δ in (6) and then characterize the average difference for the erroneous test measurements.

5. Average Difference for the Case with Erroneous Test Measurements

We note that the infection status of the ith person and its estimate at the health care provider form a continuous time Markov chain (Section 7.5 of [105]) with the states $(x_i(t), \hat{x}_i(t)) \in \{(0,0), (0,1), (1,0), (1,1)\}$. In this section, by finding the steady-state distribution for $(x_i(t), \hat{x}_i(t))$, we provide an alternative method to find Δ in (6). Then, we consider the case with erroneous test measurements. For this case, we characterize the long-term average difference for the ith person denoted by Δ_i^e.

5.1. An Alternative Method to Characterize Average Difference

When there is no error in the tests, the state transition graph is shown in Figure 4a. Assuming that $s_i > 0$, $c_i > 0$, every state is accessible from any other state, and thus, the Markov chain induced by the system is irreducible. Note that in Section 4, we see that the testing rates for some people can be equal to 0, i.e., $s_i = 0$ and $c_i = 0$. For these people, we choose $\hat{x}_i(t)$ to be either always 0 or 1, i.e., consider them as always healthy or sick all the time. Depending on the choice of $\hat{x}_i(t)$, when $s_i = 0$ and $c_i = 0$, either the states $(0,0)$ and $(1,0)$, or the states $(0,1)$ and $(1,1)$ will be transient, and thus, have 0 probability in the steady state. By using small time-step approximation to a discrete time Markov chain, one can show that the self transition probabilities are non-zero, and thus, the Markov chain induced by the system is also aperiodic (Section 7.5 of [105]). Therefore, the Markov chain shown in Figure 4a admits a unique stationary distribution given by $\pi = \{\pi_{00}, \pi_{01}, \pi_{10}, \pi_{11}\}$. We find the stationary distribution by writing the local-balance equations which are given as

$$\pi_{00}\lambda_i = \pi_{10}\mu_i + \pi_{01}c_i, \tag{30}$$

$$\pi_{10}(\mu_i + s_i) = \pi_{00}\lambda_i, \tag{31}$$

$$\pi_{01}(c_i + \lambda_i) = \pi_{11}\mu_i, \tag{32}$$

$$\pi_{11}\mu_i = \pi_{10}s_i + \pi_{01}\lambda_i. \tag{33}$$

By using (30)–(33) and $\sum_{k=1}^{2}\sum_{\ell=1}^{2}\pi_{k\ell} = 1$, we find the steady-state distribution π as

$$\pi_{01} = \frac{\mu_i\lambda_i}{\mu_i + \lambda_i} \frac{s_i}{\mu_i c_i + \lambda_i s_i + c_i s_i}, \tag{34}$$

$$\pi_{10} = \frac{\mu_i\lambda_i}{\mu_i + \lambda_i} \frac{c_i}{\mu_i c_i + \lambda_i s_i + c_i s_i}, \tag{35}$$

and $\pi_{00} = \frac{\mu_i + s_i}{\lambda_i}\pi_{10}$, and $\pi_{11} = \frac{c_i + \lambda_i}{\mu_i}\pi_{01}$. We note that Δ_{i1} in (14) is also equal to π_{10} in (35), i.e., we have $\Delta_{i1} = \pi_{10}$. Similarly, Δ_{i2} in (15) is equal to π_{01} in (34). Thus, by observing that the states $(x_i(t), \hat{x}_i(t))$ form a continuous time Markov chain, we can find the average difference Δ in (6) by finding the steady-state distribution for π. This method will be particularly useful in the following section where we consider the case with erroneous test measurements.

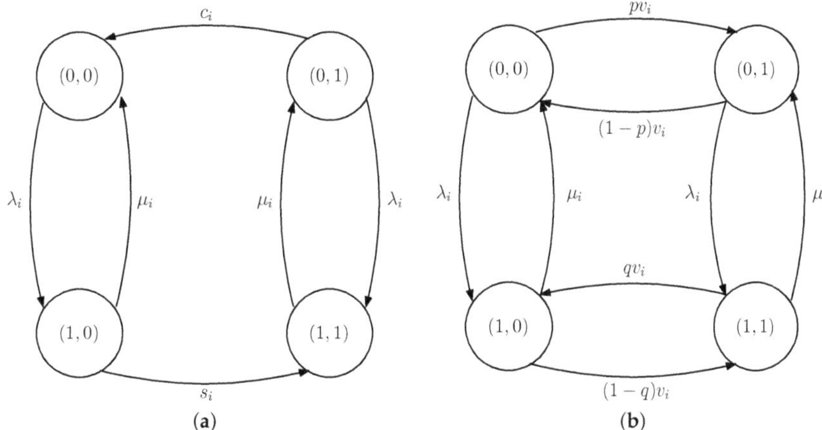

Figure 4. Transition graphs of the states $(x_i(t), \hat{x}_i(t))$ (**a**) when there is no error in the tests, and (**b**) when there are errors in the tests.

5.2. Average Difference with Erroneous Test Measurements

In this section, we consider the case where the test measurements can be erroneous. When a test in applied to an infected person, i.e., when $x_i(t) = 1$, the test result will be 0 with probability q and 1 with probability $1 - q$, where $0 \leq q < \frac{1}{2}$. In other words, the false-negative probability is equal to q. Similarly, when a test is applied to a healthy person, i.e., when $x_i(t) = 0$, the test result will be 1 with probability p and 0 with probability $1 - p$, where $0 \leq p < \frac{1}{2}$. Thus, the false-positive probability is equal to p. The probability distribution for the test measurements is provided in Table 2.

Table 2. The probability distribution for successful and false test measurements.

$x_i(t) \setminus \hat{x}_i(t)$	0	1
0	$1 - p$	p
1	q	$1 - q$

In this section, we consider the case where the health care provider applies only one test rate v_i to the ith person, whether the person is currently marked as healthy or infected. That is, we do not consider separate testing rates of s_i and c_i for healthy and infected people as we did before, instead, here both s_i and c_i are equal o v_i. Since the health care provider applies the same test rate for the ith person, here we do not consider the importance factor θ either. Then, we define the long-term average difference for the ith person with the error on the test measurements as follows, where the superscript e stands for "erroneous".

$$\Delta_i^e = \Delta_{i1}^e + \Delta_{i2}^e, \tag{36}$$

and the definitions of Δ_{i1}^e and Δ_{i2}^e follow similarly from (13). We note that with the test rates v_i and errors on the test measurements, the states $(x_i(t), \hat{x}_i(t))$ form a continuous time Markov chain, and the corresponding state transition graph is shown in Figure 4b. Assuming that $v_i > 0$, one can show that there is a unique steady-state distribution $\boldsymbol{\pi}^e = \{\pi_{00}^e, \pi_{01}^e, \pi_{10}^e, \pi_{11}^e\}$ which can be found by solving the local balance equations which are given as follows

$$\pi_{00}^e(v_i p + \lambda_i) = \pi_{01}^e v_i(1 - p) + \pi_{10}^e \mu_i, \tag{37}$$

$$\pi_{10}^e(v_i(1 - q) + \mu_i) = \pi_{00}^e \lambda_i + \pi_{11}^e v_i q, \tag{38}$$

$$\pi_{01}^e(v_i(1 - p) + \lambda_i) = \pi_{00}^e v_i p + \pi_{11}^e \mu_i, \tag{39}$$

$$\pi_{11}^e(v_i q + \mu_i) = \pi_{10}^e v_i(1 - q) + \pi_{01}^e \lambda_i. \tag{40}$$

Then, by using (37)–(40) and $\sum_{k=1}^2 \sum_{\ell=1}^2 \pi_{k\ell}^e = 1$, we find the steady-state distribution π^e as

$$\pi_{00}^e = \frac{\mu_i \lambda_i q + (1 - p)\mu_i(v_i + \mu_i)}{(\lambda_i + \mu_i)(\lambda_i + \mu_i + v_i)}, \tag{41}$$

$$\pi_{01}^e = \frac{\mu_i \lambda_i(1 - q) + p\mu_i(v_i + \mu_i)}{(\lambda_i + \mu_i)(\lambda_i + \mu_i + v_i)}, \tag{42}$$

$$\pi_{10}^e = \frac{\mu_i \lambda_i(1 - p) + q\lambda_i(v_i + \lambda_i)}{(\lambda_i + \mu_i)(\lambda_i + \mu_i + v_i)}, \tag{43}$$

$$\pi_{11}^e = \frac{\mu_i \lambda_i p + (1 - q)\lambda_i(v_i + \lambda_i)}{(\lambda_i + \mu_i)(\lambda_i + \mu_i + v_i)}. \tag{44}$$

We note that Δ_{i1}^e, and Δ_{i2}^e are equal to π_{10}^e in (43), and π_{01}^e in (42), respectively. Thus, if $v_i > 0$, then Δ_i^e in (36) becomes

$$\Delta_i^e = \frac{p\mu_i^2 + q\lambda_i^2 + (2 - p - q)\mu_i \lambda_i + v_i(p\mu_i + q\lambda_i)}{(\lambda_i + \mu_i)(\lambda_i + \mu_i + v_i)}. \tag{45}$$

We immediately note that if false-positive test probability p and false-negative test probability q are equal to 0, Δ_i^e becomes $\frac{2\mu_i \lambda_i}{(\lambda_i + \mu_i)(\lambda_i + \mu_i + v_i)}$ which is equal to $\Delta_{i1} + \Delta_{i2}$ provided in (14) and (15), respectively, when $v_i = s_i = c_i$. Then, $\frac{\partial \Delta_i^e}{\partial p} \geq 0$ is equivalent to $v_i + \mu_i - \lambda_i \geq 0$ and $\frac{\partial \Delta_i^e}{\partial q} \geq 0$ is equivalent to $v_i + \lambda_i - \mu_i \geq 0$ which means that depending on the values of v_i, μ_i, and λ_i, the long-term average difference Δ_i^e can be an increasing function of only p or only q, or both p and q, but Δ_i^e cannot be a decreasing function of both p and q. This is expected as false-negative and false-positive tests negatively affect the estimation process. One can also show that $\frac{\partial \Delta_i^e}{\partial v_i} < 0$ and $\frac{\partial^2 \Delta_i^e}{\partial v_i^2} > 0$ which means that Δ_i^e decreases with v_i and is a convex function of the test rate v_i.

Next, we consider the case when $v_i = 0$. Note that when $v_i = 0$, the health care provider either maps these people as always sick or always healthy depending on their infection and recovery rates. Thus, when $v_i = 0$ and depending on the estimate $\hat{x}_i(t)$, two of the states in Figure 4b will never be visited and thus, these states will have 0 steady-state probabilities. For this case, the steady states are given by $\bar{\pi}_{1,\hat{x}_i}^e$ and $\bar{\pi}_{0,\hat{x}_i}^e$. The local balance equation is $\lambda_i \bar{\pi}_{0,\hat{x}_i}^e = \mu_i \bar{\pi}_{1,\hat{x}_i}^e$. By using $\bar{\pi}_{0,\hat{x}_i}^e + \bar{\pi}_{1,\hat{x}_i}^e = 1$, we find the steady-state distribution as $\bar{\pi}_{0,\hat{x}_i}^e = \frac{\mu_i}{\mu_i + \lambda_i}$, and $\bar{\pi}_{1,\hat{x}_i}^e = \frac{\lambda_i}{\mu_i + \lambda_i}$. Thus, if $\mu_i < \lambda_i$, i.e., if people are infected more frequently, then the health care provider chooses its estimate as $\hat{x}_i(t) = 1$ and, $\Delta_i^e = \frac{\mu_i}{\mu_i + \lambda_i}$. If $\mu_i \geq \lambda_i$, i.e., if people stay healthy more often, then we have $\hat{x}_i(t) = 0$, and $\Delta_i^e = \frac{\lambda_i}{\mu_i + \lambda_i}$. Therefore, when $v_i = 0$, we have

$$\Delta_i^e = \min\left\{ \frac{\mu_i}{\mu_i + \lambda_i}, \frac{\lambda_i}{\mu_i + \lambda_i} \right\}. \tag{46}$$

In order to find the optimal test rates v_i in the case of errors on the test measurements, we formulate the following optimization problem

$$\min_{\{v_i\}} \quad \sum_{i=1}^{n} \mathbb{1}\{v_i > 0\} \frac{p\mu_i^2 + q\lambda_i^2 + (2 - p - q)\mu_i\lambda_i + v_i(p\mu_i + q\lambda_i)}{(\lambda_i + \mu_i)(\lambda_i + \mu_i + v_i)}$$

$$+ \mathbb{1}\{v_i = 0\} \min\left\{\frac{\mu_i}{\mu_i + \lambda_i}, \frac{\lambda_i}{\mu_i + \lambda_i}\right\}$$

$$\text{s.t.} \quad \sum_{i=1}^{n} v_i \leq C$$

$$v_i \geq 0, \quad i = 1, \ldots, n, \tag{47}$$

where the objective function is given by the summation of Δ_i^e in (45) when $v_i > 0$ and Δ_i^e in (46) when $v_i = 0$ over all people and $\mathbb{1}\{.\}$ is the indicator function taking value 1 when $\{\cdot\}$ is true and 0, otherwise. In (47), we have a constraint on the total test rate, i.e., $\sum_{i=1}^{n} v_i \leq C$. We note that the optimization problem in (47) is in general not convex due to the indicator function in the objective function. However, for a given set of $\mathbb{1}\{v_i = 0\}$, the optimization problem in (47) is convex and can be solved optimally. Thus, by solving the problem in (47) for all possible set of $\mathbb{1}\{v_i = 0\}$, we can determine the global optimal solution which requires to solve 2^n different optimization problems which can be impractical for large n. Because of this reason, next, we provide a greedy algorithm to solve the optimization problem in (47).

In the greedy solution, initially, assuming that $\mathbb{1}\{v_i > 0\} = 1$ for all i, we consider the following the optimization problem

$$\min_{\{v_i\}} \quad \sum_{i=1}^{n} \frac{p\mu_i^2 + q\lambda_i^2 + (2 - p - q)\mu_i\lambda_i + v_i(p\mu_i + q\lambda_i)}{(\lambda_i + \mu_i)(\lambda_i + \mu_i + v_i)}$$

$$\text{s.t.} \quad \sum_{i=1}^{n} v_i \leq C$$

$$v_i \geq 0, \quad i = 1, \ldots, n, \tag{48}$$

where the objective function in (48) is equal to Δ_i^e in (45). For this optimization problem, we define the Lagrangian function for (48) as

$$\mathcal{L} = \sum_{i=1}^{n} \frac{p\mu_i^2 + q\lambda_i^2 + (2 - p - q)\mu_i\lambda_i + v_i(p\mu_i + q\lambda_i)}{(\lambda_i + \mu_i)(\lambda_i + \mu_i + v_i)} + \bar{\beta}\left(\sum_{i=1}^{n} v_i - C\right) - \sum_{i=1}^{n} \bar{v}_i v_i, \tag{49}$$

where $\bar{\beta} \geq 0$, $\bar{v}_i \geq 0$. We note that the problem defined in (48) is a convex optimization problem, and thus we can find the optimal test rates v_i by analyzing the KKT and the complementary slackness conditions. The KKT conditions are given by

$$\frac{\partial \mathcal{L}}{\partial v_i} = \frac{-2(1 - p - q)\mu_i\lambda_i}{(\mu_i + \lambda_i)(\mu_i + \lambda_i + v_i)^2} + \bar{\beta} - \bar{v}_i = 0, \tag{50}$$

for all i. The complementary slackness conditions are

$$\bar{\beta}\left(\sum_{i=1}^{n} v_i - C\right) = 0, \quad \bar{v}_i v_i = 0. \tag{51}$$

By using (50) and (51), we find the optimal v_i values for the problem in (48) as

$$v_i = (\mu_i + \lambda_i)\left(\sqrt{\frac{\mu_i\lambda_i}{(\mu_i + \lambda_i)^3} \frac{2(1 - p - q)}{\bar{\beta}}} - 1\right)^+. \tag{52}$$

With the test rates v_i in (52) we find the average differences Δ_i^e in (45) and then compare them with Δ_i^e in (46) when $v_i = 0$. Due to the errors in the tests, Δ_i^e in (46) with $v_i = 0$

can be smaller than Δ_i^e in (45) with the test rates v_i found in (52). For these people, we choose index i where the difference between Δ_i^e in (45) with the v_i in (52) and Δ_i^e in (46) is the highest. Then, we take $v_i = 0$ as applying no test to this person can further decrease Δ_i^e. For the remaining people, we solve the optimization problem in (48). After obtaining the test rates for the remaining people, we again compare average differences Δ_i^e with the test rates in (52) and with no test and we choose $v_i = 0$ for the person where Δ_i^e can be further decreased. We repeat these steps until all Δ_i^es with $v_i > 0$ cannot be further decreased by choosing $v_i = 0$.

We note that the solution obtained in (52) has a *threshold* structure. As false-positive and -negative test rates increase, the term $\frac{2(1-p-q)}{\beta}$ in (52) becomes smaller. As a result, some people with higher $\sqrt{\frac{(\mu_i+\lambda_i)^3}{\mu_i\lambda_i}}$ may not be tested by the health care provider. Thus, when p and q are high, a smaller portion of the population is tested with higher test rates in order to combat the test errors.

6. Average Estimation Error with Dependent Infection Rates

In this section, we consider the case where we have two people whose infection rates depend on each other. When these two people are healthy, they can be individually infected with the virus after an exponential time with rate λ. When one of these two people is infected and this has not been detected by the health care provider, this person can infect the other healthy person after an exponential time with rate λ_{12} which has been illustrated in Figure 5. Thus, when both of the people are healthy, their individual infection rate is λ. However, when one of them is sick and this has not been detected by the health care provider, the healthy person's total infection rate is equal to $\lambda + \lambda_{12}$. On the other hand, if only one person is infected, i.e., $x_i(t) = 1$, which has also been detected by the health care provider, $\hat{x}_i(t) = 1$, then we assume that we isolate the infected person from the healthy one, and thus, the healthy person's infection rate remains as λ instead of $\lambda + \lambda_{12}$. When the people are infected, they recover from the disease after an exponential time with rate μ.

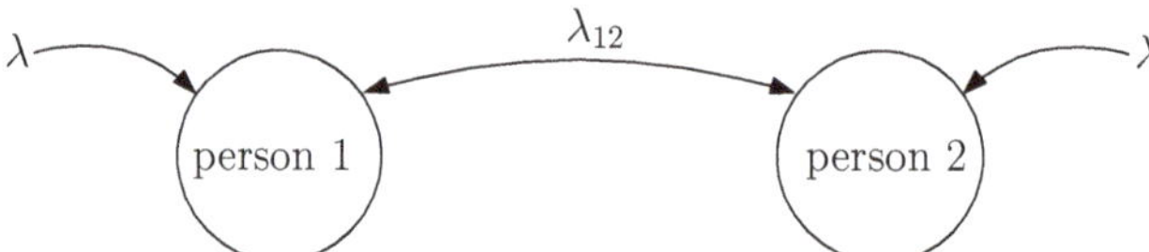

Figure 5. The infection rates of two people where the individual infection rate is equal to λ. When the infection has not been detected, these two people can infect each other with rate λ_{12}.

When the health care provider believes that a person is healthy, i.e., $\hat{x}_i(t) = 0$, the next test is applied to this person after an exponential time with rate s. When the health care provider believes that a person is sick, i.e., $\hat{x}_i(t) = 1$, the next test applied to this person after an exponential time with rate c. Here, we note that since the people are identical in terms of their infection and recovery rates, the health care provider applies the same test rates.

Similar to Section 5, we note that the states $\{x_1(t), \hat{x}_1(t), x_2(t), \hat{x}_2(t)\}$ form a continuous time Markov chain where the unique stationary distribution is given by $\boldsymbol{\pi}^d = \{\pi_{0000}^d, \pi_{0001}^d, \ldots, \pi_{1111}^d\}$. In order to find the stationary distribution, we write the local balance equations as follows

$$2\lambda \pi_{0000}^d = \mu \pi_{1000}^d + c\pi_{0100}^d + \mu \pi_{0010}^d + c\pi_{0001}^d, \tag{53}$$

$$(2\lambda + c)\pi_{0001}^d = \mu \pi_{0011}^d + c\pi_{0101}^d + \mu \pi_{1001}^d, \tag{54}$$

$$(\lambda + \lambda_{12} + \mu + s)\pi_{0010}^d = c\pi_{0110}^d + \mu \pi_{1010}^d + \lambda \pi_{0000}^d, \tag{55}$$

$$(\lambda + \mu)\pi_{0011}^d = c\pi_{0111}^d + \mu \pi_{1011}^d + s\pi_{0010}^d + \lambda \pi_{0001}^d, \tag{56}$$

$$(2\lambda + c)\pi^d_{0100} = c\pi^d_{0101} + \mu\pi^d_{0110} + \mu\pi^d_{1100}, \tag{57}$$

$$(2\lambda + 2c)\pi^d_{0101} = \mu\pi^d_{0111} + \mu\pi^d_{1101}, \tag{58}$$

$$(\lambda + \mu + s + c)\pi^d_{0110} = \lambda\pi^d_{0100} + \mu\pi^d_{1110}, \tag{59}$$

$$(\lambda + \mu + c)\pi^d_{0111} = s\pi^d_{0110} + \lambda\pi^d_{0101} + \mu\pi^d_{1111}, \tag{60}$$

$$(\lambda + \lambda_{12} + \mu + s)\pi^d_{1000} = \lambda\pi^d_{0000} + c\pi^d_{1001} + \mu\pi^d_{1010}, \tag{61}$$

$$(\lambda + \mu + s + c)\pi^d_{1001} = \mu\pi^d_{1011} + \lambda\pi^d_{0001}, \tag{62}$$

$$(2\mu + 2s)\pi^d_{1010} = (\lambda + \lambda_{12})\pi^d_{1000} + (\lambda + \lambda_{12})\pi^d_{0010}, \tag{63}$$

$$(2\mu + s)\pi^d_{1011} = s\pi^d_{1010} + \lambda\pi^d_{1001} + \lambda\pi^d_{0011}, \tag{64}$$

$$(\lambda + \mu)\pi^d_{1100} = s\pi^d_{1000} + \lambda\pi^d_{0100} + c\pi^d_{1101} + \mu\pi^d_{1110}, \tag{65}$$

$$(\lambda + \mu + c)\pi^d_{1101} = s\pi^d_{1001} + \lambda\pi^d_{0101} + \mu\pi^d_{1111}, \tag{66}$$

$$(2\mu + s)\pi^d_{1110} = \lambda\pi^d_{1100} + s\pi^d_{1010} + \lambda\pi^d_{0110}, \tag{67}$$

$$2\mu\pi^d_{1111} = s\pi^d_{1110} + \lambda\pi^d_{1101} + s\pi^d_{1011} + \lambda\pi^d_{0111}. \tag{68}$$

By using (53)–(68) and $\sum_{j=1}^{2}\sum_{\ell=1}^{2}\sum_{m=1}^{2}\sum_{h=1}^{2}\pi^d_{j\ell mh} = 1$, we find the stationary distribution π^d. We denote the long-term average estimation error for person i as Δ^d_i for $i = 1,2$, where the superscript d stands for "dependent", which is given by

$$\Delta^d_i = \Delta^d_{i1} + \Delta^d_{i2}, \tag{69}$$

where Δ^d_{i1} and Δ^d_{i2} follow from (13). Then, we have

$$\Delta^d_{11} = \pi^d_{1000} + \pi^d_{1001} + \pi^d_{1010} + \pi^d_{1011}, \tag{70}$$

$$\Delta^d_{12} = \pi^d_{0100} + \pi^d_{0101} + \pi^d_{0110} + \pi^d_{0111}, \tag{71}$$

$$\Delta^d_{21} = \pi^d_{0010} + \pi^d_{0110} + \pi^d_{1010} + \pi^d_{1110}, \tag{72}$$

$$\Delta^d_{22} = \pi^d_{0001} + \pi^d_{0101} + \pi^d_{1001} + \pi^d_{1101}. \tag{73}$$

In Section 8, for given infection, recovery and test rates, we numerically evaluate the stationary distribution and find the average difference Δ^d_i.

7. Age of Incorrect Information Based Error Metric

To date, we have considered an estimation error metric that takes the value 1 if the actual infection status of a person is different than the real-time estimation at the health care provider. Thus, the error metric takes values based on the information content. On the other hand, the traditional age metric introduced in [1] considers only the time passed since the most recently received status update packet is generated at the source. As a result, the traditional age metric does not consider the information content and age alone may not be a suitable performance metric for the problem considered in our work.

In the context of infection tracking, it is important to know how long the estimations at the health care provider have been different from the actual infection status of the people. However, the error metric that we have considered thus far does not have the time component, i.e., it only takes value 1 independent of the time duration that it has been off from the actual health status. Motivated by the AoII introduced in [51,102] which accounts for both the time and the information content, in this section, we consider the following error metric, where the superscript s stands for "synchronization" implied in AoII,

$$\Delta^s_i = (t - V_i(t))\mathbb{1}\{\hat{x}_i(t) \neq x_i(t)\}, \tag{74}$$

where $V_i(t)$ is the last time instant where the health care provider makes an accurate estimation of the health status for the ith person, i.e., the last time instant when $\Delta_i^s = 0$. Similarly, we define

$$\Delta_{i1}^s = (t - V_{i1}(t)) \max\{x_i(t) - \hat{x}_i(t), 0\}, \tag{75}$$

$$\Delta_{i2}^s = (t - V_{i2}(t)) \max\{\hat{x}_i(t) - x_i(t), 0\}, \tag{76}$$

where $V_{i1}(t)$ and $V_{i2}(t)$ are equal to the last time instants when Δ_{i1}^s and Δ_{i2}^s are equal to 0, respectively. A sample evolution of Δ_{i1}^s and Δ_{i2}^s is shown in Figure 6 and we note that $\Delta_i^s(t) = \Delta_{i1}^s(t) + \Delta_{i2}^s(t)$.

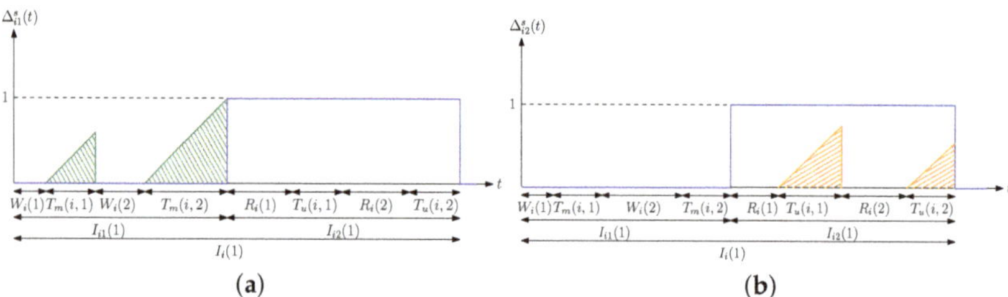

Figure 6. A sample evolution of (**a**) $\Delta_{i1}^s(t)$, and (**b**) $\Delta_{i2}^s(t)$ in a typical update cycle.

Similar to Section 3, the infection and the recovery rates of the ith person are λ_i and μ_i, respectively. In this section, the health care provider applies only one test rate for each person denoted by w_i. That is, we do not consider separate testing rates of s_i and c_i for healthy and infected people as we did previously, instead, here both s_i and c_i are equal o w_i. We first consider the case where $w_i > 0$. By following the steps in Section 3, one can show that $\mathbb{E}[I_{i1}] = \frac{1}{w_i} + \frac{w_i + \mu_i}{w_i \lambda_i}$ and $\mathbb{E}[I_{i2}] = \frac{1}{w_i} + \frac{w_i + \lambda_i}{w_i \mu_i}$ which can be obtained by substituting w_i instead of s_i and c_i in (10) and (12), respectively. Next, we denote the total area when $\Delta_{i1}^s(t) > 0$ as $A_{e,1}(i, j)$ during the jth cycle where $A_{e,1}(i, j) = \sum_{\ell=1}^{K_1} \frac{T_m(i,\ell)^2}{2}$ and K_1 has a geometric distribution with success rate $\frac{w_i}{\mu_i + w_i}$. Then, we have $\mathbb{E}[A_{e,1}(i)] = \frac{1}{w_i(w_i + \mu_i)}$. Similarly, we denote the total area when $\Delta_{i2}^s(t) > 0$ as $A_{e,2}(i, j)$ during the jth cycle where $A_{e,2}(i, j) = \sum_{\ell=1}^{K_2} \frac{T_u(i,\ell)^2}{2}$ and K_2 has a geometric distribution with success rate $\frac{w_i}{\lambda_i + w_i}$. Then, we have $\mathbb{E}[A_{e,2}(i)] = \frac{1}{w_i(w_i + \lambda_i)}$. By using ergodicity, the long-term average differences become $\Delta_{i1}^s = \frac{\mathbb{E}[A_{e,1}(i)]}{\mathbb{E}[I_{i1}] + \mathbb{E}[I_{i2}]}$ and $\Delta_{i2}^s = \frac{\mathbb{E}[A_{e,2}(i)]}{\mathbb{E}[I_{i1}] + \mathbb{E}[I_{i2}]}$ which gives

$$\Delta_i^s = \Delta_{i1}^s + \Delta_{i2}^s = \frac{\mu_i \lambda_i}{\mu_i + \lambda_i} \frac{2w_i + \mu_i + \lambda_i}{(w_i + \mu_i + \lambda_i)(w_i + \mu_i)(w_i + \lambda_i)}, \tag{77}$$

when $w_i > 0$. One can show that Δ_i^s is a decreasing function of w_i, i.e., $\frac{\partial \Delta_i^s}{\partial w_i} < 0$, and Δ_i^s is a convex function of w_i, i.e., $\frac{\partial^2 \Delta_i^s}{\partial w_i^2} > 0$.

When $w_i = 0$, we have $\mathbb{E}[I_i] = \frac{\mu_i \lambda_i}{\mu_i + \lambda_i}$, i.e., $\mathbb{E}[I_i]$ is equal to the expected time of a person's healthy and sick states. Since the health care provider applies no tests to test a person, it either estimates this person to be always sick ($\hat{x}_i(t) = 1$) or always healthy ($\hat{x}_i(t) = 0$). When $w_i = 0$ and $\hat{x}_i(t) = 1$, then $\Delta_i^s = \frac{1}{\mu_i} \frac{\lambda_i}{\mu_i + \lambda_i}$. When $w_i = 0$ and $\hat{x}_i(t) = 1$, we have $\Delta_i^s = \frac{1}{\lambda_i} \frac{\mu_i}{\mu_i + \lambda_i}$. If $\mu_i < \lambda_i$, then the health care provider $\hat{x}_i(t) = 1$, and $\hat{x}_i(t) = 0$, otherwise. Thus, when $w_i = 0$, we have $\Delta_i^s = \min\left\{ \frac{1}{\mu_i} \frac{\lambda_i}{\mu_i + \lambda_i}, \frac{1}{\lambda_i} \frac{\mu_i}{\mu_i + \lambda_i} \right\}$.

In order to find the optimal test rates, we formulate the following optimization problem

$$\min_{\{w_i\}} \quad \sum_{i=1}^{n} \mathbb{1}\{w_i > 0\} \frac{\mu_i \lambda_i}{\mu_i + \lambda_i} \frac{2w_i + \mu_i + \lambda_i}{(w_i + \mu_i + \lambda_i)(w_i + \mu_i)(w_i + \lambda_i)}$$
$$+ \mathbb{1}\{w_i = 0\} \min\left\{ \frac{1}{\mu_i} \frac{\lambda_i}{\mu_i + \lambda_i}, \frac{1}{\lambda_i} \frac{\mu_i}{\mu_i + \lambda_i} \right\}$$
$$\text{s.t.} \quad \sum_{i=1}^{n} w_i \leq C$$
$$w_i \geq 0, \quad i = 1, \ldots, n, \tag{78}$$

where the objective function in (78) is equal to the summation of Δ_i^s in (77) when $w_i > 0$ and Δ_i^s when $w_i = 0$ over all people. In order to solve the problem in (78), we follow the same greedy solution approach in Section 5. First, by assuming that $w_i > 0$, and thus, the average difference Δ_i^s is given in (77), we solve the following optimization problem

$$\min_{\{w_i\}} \quad \sum_{i=1}^{n} \frac{\mu_i \lambda_i}{\mu_i + \lambda_i} \frac{2w_i + \mu_i + \lambda_i}{(w_i + \mu_i + \lambda_i)(w_i + \mu_i)(w_i + \lambda_i)}$$
$$\text{s.t.} \quad \sum_{i=1}^{n} w_i \leq C$$
$$w_i \geq 0, \quad i = 1, \ldots, n. \tag{79}$$

Since the problem in (79) is a convex optimization problem, by defining Lagrangian function and analyzing the KKT and the complementary slackness conditions, we can find the optimal w_i values. In order to avoid being repetitive, we skip these optimization steps. Then, we compare Δ_i^s in (77) with w_i values found in (79) with $\min\{\frac{1}{\mu_i}\frac{\lambda_i}{\mu_i+\lambda_i}, \frac{1}{\lambda_i}\frac{\mu_i}{\mu_i+\lambda_i}\}$. If we can reduce Δ_i^s further, we choose $w_i = 0$ for the person with the highest improvement. Then, we solve the optimization problem in (79) for the remaining people. We repeat these steps until there is no improvement in Δ_i^s by choosing $w_i = 0$.

In the next section, we provide extensive numerical results to evaluate optimal test rates in various settings considered in this paper.

8. Numerical Results

In this section, we provide seven numerical results. For these examples, we take λ_i as

$$\lambda_i = ar^i, \quad i = 1, \ldots, n, \tag{80}$$

where $r = 0.9$ and a is such that $\sum_{i=1}^{n} \lambda_i = 6$. Furthermore, we take μ_i as

$$\mu_i = bq^i, \quad i = 1, \ldots, n, \tag{81}$$

where $q = 1.1$ and b is such that $\sum_{i=1}^{n} \mu_i = 4$. Since λ_i in (80) decreases with i, people with lower indices become infected more quickly compared to people with higher indices. Since μ_i in (81) increases with i, people with higher indices recover more quickly compared to people with lower indices. Thus, a person with a low index becomes infected quickly and recovers slowly.

In the first example, we take the total number of people as $n = 10$, the total test rate as $C = 16$, and $\theta = 0.5$. We start with randomly chosen s_i and c_i such that $\sum_{i=1}^{n} s_i + c_i = 16$, and apply the alternating minimization-based method proposed in Section 4. We repeat this process for 30 different initial (s_i, c_i) pairs and choose the solution that gives the smallest Δ. In Figure 7a, we observe that the first three people are never tested by the health care provider. We note that s_i, which is the test rate when $\hat{x}_i(t) = 0$, initially increases with i but then decreases with i. This means that people who become infected rarely are tested less frequently when they are marked as healthy. Similarly, we observe in Figure 7a that

c_i, which is the test rate when $\hat{x}_i(t) = 1$, monotonically increases with i. In other words, people who recover from the virus quickly are tested more frequently when they are marked as infected.

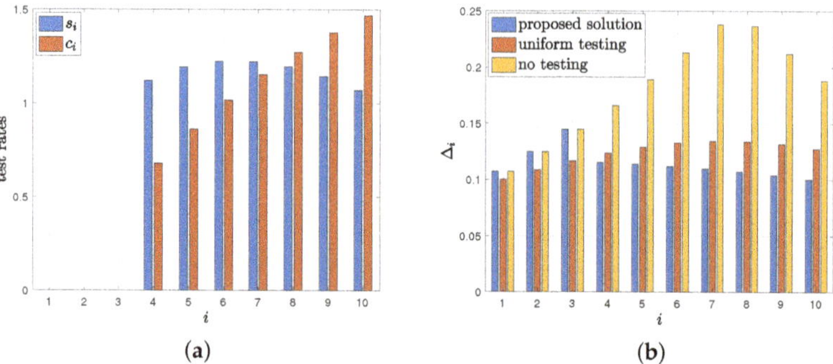

Figure 7. (**a**) Test rates s_i and c_i, (**b**) corresponding average difference Δ_i.

In Figure 7b, we plot Δ_i resulting from the solution found from the proposed algorithm, Δ_i when the health care provider applies tests to everyone in the population uniformly, i.e., $s_i = c_i = \frac{C}{2n}$ for all i, and Δ_i when the health care provider applies no tests, i.e., $s_i = c_i = 0$ for all i. In the case of no tests, we have $\Delta_i = \min\{\frac{\theta\lambda_i}{\mu_i+\lambda_i}, \frac{(1-\theta)\mu_i}{\mu_i+\lambda_i}\}$. We observe in Figure 7b that the health care provider applies tests on people whose Δ_i can be reduced the most as opposed to uniform testing where everyone is tested equally. Thus, the first three people who have the smallest Δ_i are not tested by the health care provider. With the proposed solution, by not testing the first three people, Δ_i are further reduced for the remaining people compared to uniform testing. For the people who are not tested, the health care provider chooses $\hat{x}_i(t) = 1$ all the time, i.e., marks these people always sick as $\frac{\theta\lambda_i}{\mu_i+\lambda_i} > \frac{(1-\theta)\mu_i}{\mu_i+\lambda_i}$. This is expected as these people have high λ_i and low μ_i, i.e., they are infected easily and they stay sick for a long time.

In the second example, we use the same set of variables except for the total test rate C. We vary the total test rate C in between 5 and 20. We plot Δ with respect to C in Figure 8. We observe that Δ decreases with C. Thus, with higher total test rates, the health care provider can track the infection status of the population better as expected.

In the third example, we use the same set of variables except for the total number of people n. In addition, we also use uniform infection and healing rates, i.e., $\lambda_i = \frac{6}{n}$ and $\mu_i = \frac{4}{n}$ for all i, for comparison with λ_i in (80) and μ_i in (81), while keeping the total infection and healing rates the same, i.e., $\sum_{i=1}^{n} \lambda_i = 6$ and $\sum_{i=1}^{n} \mu_i = 4$, for both cases. We vary the number of people n from 2 to 30. We observe in Figure 9 that when the infection and healing rates are uniform in the population, the health care provider can track the infection status with the same efficiency, even though the population size increases (while keeping the total infection and healing rates fixed). For the case of λ_i in (80) and μ_i in (81), when we increase the population size, we increase the number of people who rarely become sick, i.e., people with high i indices, and also people who rarely heal from the disease, i.e., people with small i indices. Thus, it becomes easier for the health care provider to track the infection status of the people. This is why when we use λ_i in (80) and μ_i in (81), we observe in Figure 9 that the health care provider can track the infection status of the people better, even though the population size increases.

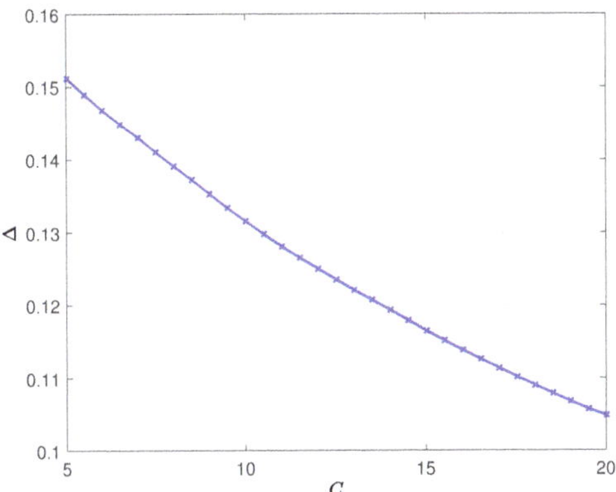

Figure 8. The average difference Δ with respect to total test rate C.

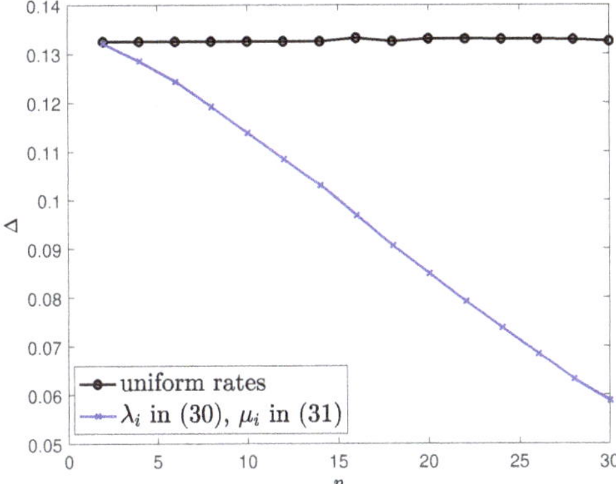

Figure 9. The average difference Δ with respect to number of people n. We use uniform infection and healing rates, i.e., $\lambda_i = \frac{6}{n}$ and $\mu_i = \frac{4}{n}$ for all i, and also λ_i in (80) and μ_i in (81) with $\sum_{i=1}^{n} \lambda_i = 6$ and $\sum_{i=1}^{n} \mu_i = 4$.

In the fourth example, we employ the same set of variables as the first example except for the importance factor θ. Here, we vary θ in between 0.2 and 0.7. We plot Δ in (7), $\bar{\Delta}_1$ which is $\bar{\Delta}_1 = \frac{1}{n} \sum_{i=1}^{n} \Delta_{i1}$, and $\bar{\Delta}_2$ which is $\bar{\Delta}_2 = \frac{1}{n} \sum_{i=1}^{n} \Delta_{i2}$ in Figure 10a. Note that $\bar{\Delta}_1$ represents the average difference when people are infected, but have not been detected by the health care provider, and $\bar{\Delta}_2$ represents the average difference when people have recovered, but the health care provider still marks them as infected. Note that when θ is high, we assign importance to minimization of $\bar{\Delta}_1$, i.e., the early detection of people with infection, and when θ is low, we give importance to minimization of $\bar{\Delta}_2$, i.e., the early detection of people who recovered from the disease. This is why we observe in Figure 10a that $\bar{\Delta}_1$ decreases with θ while $\bar{\Delta}_2$ increases with θ.

We plot the total test rates $\sum_{i=1}^{n} s_i$ and $\sum_{i=1}^{n} c_i$ in Figure 10b. We observe in Figure 10b that if it is more important to detect the infected people, i.e., if θ is high, then the health care provider should apply higher test rates to people who are marked as healthy. In other words, $\sum_{i=1}^{n} s_i$ increases with θ. Similarly, if it is more important to detect people who recovered from the disease, then the health care provider should apply high test rates to people who are marked as infected. That is, $\sum_{i=1}^{n} c_i$ is high when θ is low. Therefore, depending on the priorities of the health care provider, a suitable θ needs to be chosen.

In the fifth numerical result, we consider the case where there are errors in the test measurements, i.e., the model in Section 5. We take the total test rate as $C = 20$, and vary error rates in the test $p = q = \{0.1, 0.2, 0.4\}$. In Figure 11a, we provide the test rates v_i that we found as a result of our greedy policy in Section 5. When the error rates p and q are low, i.e., when $p = q = 0.1$, we see that the health care provider applies tests to everyone in the population and the corresponding Δ_i^e is lower than applying no test as shown in Figure 11b. As we increase the error rates, we observe that some people in the population start to be not tested by the health care provider, see Figure 11a when $p = q = \{0.2, 0.4\}$. In this case, the health care provider applies more tests to the remaining people to combat the test errors. However, although it applies more tests to the remaining people, we observe in Figure 11b that the achieved average difference Δ_i^e becomes higher as error rates increase.

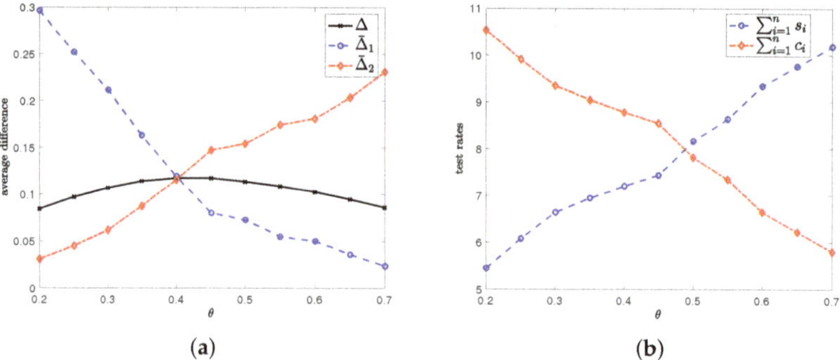

Figure 10. (a) Δ in (7), $\bar{\Delta}_1$ which is $\frac{1}{n} \sum_{i=1}^{n} \Delta_{i1}$, and $\bar{\Delta}_2$ which is $\frac{1}{n} \sum_{i=1}^{n} \Delta_{i2}$, (b) corresponding total test rates $\sum_{i=1}^{n} s_i$ and $\sum_{i=1}^{n} c_i$.

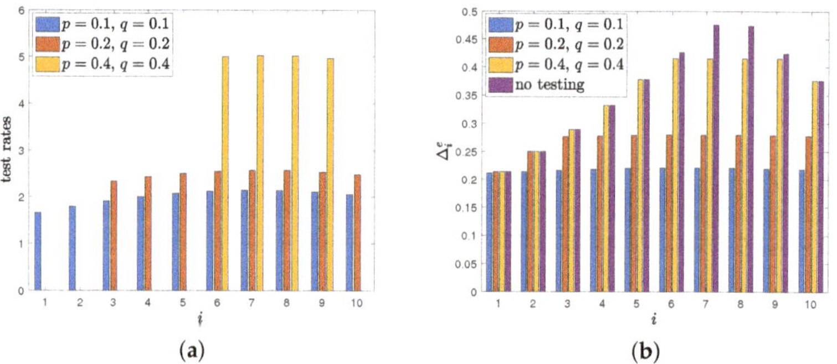

Figure 11. (a) Test rates v_i, (b) corresponding average difference Δ_i^e when there is error in the tests.

In the sixth numerical result, we consider the case where the infection status of the people depend on each other. In other words, when one person is infected, they can infect the other person with rate λ_{12} when they are not detected by the health care provider, i.e., the infection model in Section 6. For this example, first, we take $\mu = 5$, $\lambda = 2.5$, $s = c = \frac{C}{4}$ and vary $\lambda = \{2, \ldots, 200\}$ and $C = \{20, 40, 60\}$. If $\lambda_{12} = 0$, i.e., if the infection status of people are independent from each other, then the average time that person 1 or 2 is sick is equal to $\frac{\lambda}{\lambda + \mu} = \frac{1}{3}$. As we increase infection rate λ_{12} among the person 1 and 2, we see in Figure 12a that the average time that person 1 is sick increases. However, we note that as we increase the total test rate, the health care provider can detect a sick person more frequently, and this explains why the average infected time is low in Figure 12a when the test rate is high. Then, we consider $\lambda_{12} = \{5, 10, 15\}$ and vary the total test rates $\lambda = \{2, \ldots, 200\}$. We plot the average time that both person 1 and 2 stay as sick in Figure 12b. As we increase the total test rate, the health care provider detects the infected person more quickly, and thus, prohibits the infection from spreading. As a result, we observe that the average time that both people are infected decreases in C in Figure 12b. Since both people can be infected with the virus independent from each other with rate λ, the plots in Figure 12b do not drop to 0.

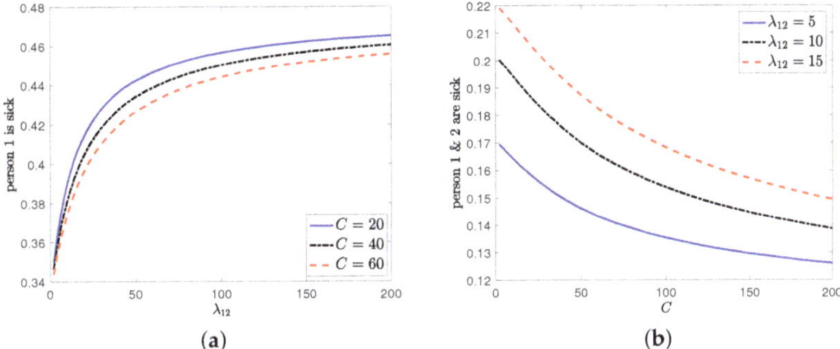

Figure 12. (**a**) The percentage of the time that person 1 stays as infected while we increase λ_{12}, (**b**) the percentage of the time that both person 1 and 2 stay as infected while we increase the total test rate C.

In the last numerical result, we consider the age of incorrect information-based error metric in Section 7. Here, the estimation error increases with the time that the health care provider does not detect the changes in the infection status of the people. As a result, the average difference expression given by Δ_i^s in (77) is different than Δ_i^e in (45) when $p = q = 0$. For this example, we consider the total test rate $C = 4$ and compare the normalized average differences given by $\frac{\Delta_i^s}{\sum_{i=1}^n \Delta_i^s}$, and $\frac{\Delta_i^e}{\sum_{i=1}^n \Delta_i^e}$ and the corresponding test rates w_i and v_i. In Figure 13b, depending on the error metric model, people who are tested by the health care provider show considerable variation in their test rates. For example, with the error metric Δ_i^s in (77), we apply tests to every third person while the same person is not tested with the error metric Δ_i^e in (45). In Figure 13a, we provide the normalized average difference values. Here, the average normalized error for the tested people exhibit similar values whereas the normalized difference may vary for the untested people. Thus, we should choose a suitable error metric that satisfies the priorities of the health care provider as it greatly affects who is tested and with which test rates.

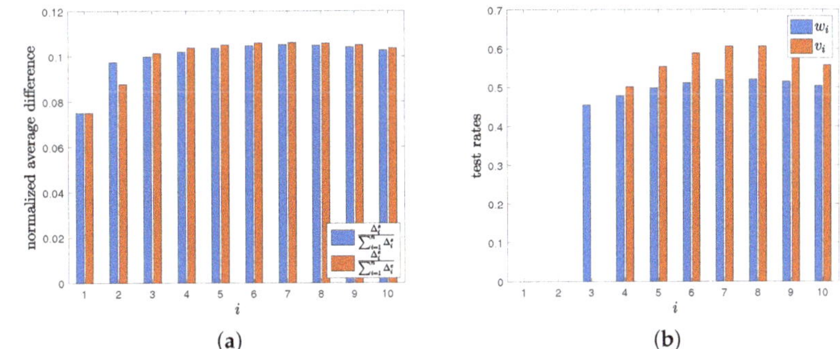

Figure 13. (**a**) The normalized average differences $\frac{\Delta_i^s}{\sum_{i=1}^n \Delta_i^s}$, and $\frac{\Delta_i^e}{\sum_{i=1}^n \Delta_i^e}$, and (**b**) the corresponding test rates w_i and v_i.

9. Conclusions and Discussion

We considered the timely tracking of infection status of individuals in a population. For exponential infection and healing processes with given rates, we determined the rates of exponential testing processes. We considered errors on the test measurements and observed that in order to combat the test errors, a limited portion of the population may be tested with higher test rates. Then, we studied a dependent infection spread model for two people, where one infected person can spread the virus to the other if it has not been detected by the health care provider. Finally, we studied an AoII-based error metric where the error function linearly increases over time as the changes in the infection status have not been detected by the health care provider. We observed in numerical results that the test rates depend on the individuals' infection and recovery rates, the individuals' last known state of being healthy or infected, as well as the health care provider's priorities of detecting infected people versus detecting recovered people more quickly.

In the literature, in order to model epidemics, population is partitioned into groups called *compartments*. One such example is the SIR model used in [106] with the compartments susceptible (S), infected (I), and recovered (R) which has been further developed by adding the states hospitalized (H), and death (D) in [107]. In these epidemic models, the transitions between the compartments are assumed to be Markovian. In [107], with epidemiological data, the delay distributions for the infected (I) to hospitalized (H), and infected (I) to death (D) are well approximated by exponential and gamma distributions, respectively. However, due to the lack of data availability the delay distribution for infected (I) to recovered (R) is modeled with gamma distribution with higher tolerance. In our work, we modeled infection and recovery times, i.e., the delays between recovered (R) to infected (I) and infected (I) to recovered (R) with exponential distributions. Therefore, more realistic infection tracking models can be developed by considering gamma distributions as observed in [107]. This more realistic model corresponds to the problem of real-time timely tracking of a binary Markov source in a serially connected network. The serially connected network model was studied in [8] with the traditional age of information metric. We note that considering the same networking model with the AoII-based error metric to track information dissemination of a binary Markov source represents a promising research direction and has direct applications to the real-time tracking of epidemic spread models. One can also study the extension of dependent infection spread model in Section 6 to $n > 2$ people as a future research direction.

Another interesting research direction could be to consider different kinds of tests with different false-positive and false-negative test rates. Regarding this problem, instead of having a total test rate capacity C, we may consider a total test budget K. Assuming that each test bears a different cost, the goal might be to identify how many tests the health care provider should obtain from each type. Here, one can study a trade-off between applying fewer tests with a small probability of error versus applying more tests to individuals with a high probability of error. Moreover, one can consider a scenario where the health care provider may prefer to apply different test types to individuals depending on their infection and recovery rates.

Author Contributions: Conceptualization, M.B. and S.U.; methodology, M.B. and S.U.; software, M.B.; validation, M.B. and S.U.; formal analysis, M.B. and S.U.; investigation, M.B. and S.U.; resources, M.B. and S.U.; data curation, M.B. and S.U.; writing—original draft preparation, M.B. and S.U.; writing—review and editing, M.B. and S.U.; visualization, M.B. and S.U.; supervision, S.U.; project administration, S.U.; funding acquisition, S.U. All authors have read and agreed to the published version of the manuscript.

Funding: This work was supported by NSF Grants CCF 17-13977 and ECCS 18-07348.

Data Availability Statement: Not applicable.

Conflicts of Interest: The authors declare no conflict of interest.

References

1. Kaul, S.K.; Yates, R.D.; Gruteser, M. Real-time status: How often should one update? In Proceedings of the 2012 Proceedings IEEE INFOCOM, Orlando, FL, USA, 25–30 March 2012.
2. Kadota, I.; Sinha, A.; Uysal-Biyikoglu, E.; Singh, R.; Modiano, E. Scheduling policies for minimizing age of information in broadcast wireless networks. *IEEE/ACM Trans. Netw.* **2018**, *26*, 2637–2650. [CrossRef]
3. Kam, C.; Kompella, S.; Nguyen, G.D.; Wieselthier, J.E.; Ephremides, A. Age of information with a packet deadline. In Proceedings of the 2016 IEEE International Symposium on Information Theory (ISIT), Barcelona, Spain, 10–15 July 2016.
4. Sun, Y.; Uysal-Biyikoglu, E.; Yates, R.D.; Koksal, C.E.; Shroff, N.B. Update or wait: How to keep your data fresh. *IEEE Trans. Inf. Theory* **2017**, *63*, 7492–7508. [CrossRef]
5. Najm, E.; Telatar, E. Status updates in a multi-stream M/G/1/1 preemptive queue. In Proceedings of the IEEE Infocom 2018-IEEE Conference On Computer Communications Workshops (Infocom Wkshps), Honolulu, HI, USA, 15–19 April 2018.
6. Soysal, A.; Ulukus, S. Age of information in G/G/1/1 systems: Age expressions, bounds, special cases, and optimization. *IEEE Trans. Inf. Theory* **2021**, *67*, 7477–7489. [CrossRef]
7. Buyukates, B.; Ulukus, S. Age of information with Gilbert-Elliot servers and samplers. *arXiv* **2020**, arXiv:2002.05711.
8. Yates, R.D. The age of information in networks: Moments, distributions, and sampling. *IEEE Trans. Inf. Theory* **2020**, *66*, 5712–5728. [CrossRef]
9. Talak, R.; Karaman, S.; Modiano, E. Minimizing age-of-information in multi-hop wireless networks. In Proceedings of the 2017 55th Annual Allerton Conference on Communication, Control, and Computing (Allerton), Monticello, IL, USA, 3–6 October 2017.
10. Tripathi, V.; Moharir, S. Age of information in multi-source systems. In Proceedings of the GLOBECOM 2017-2017 IEEE Global Communications Conference, Centre, Singapore, 4–8 December 2017.
11. Bedewy, A.M.; Sun, Y.; Shroff, N.B. The age of information in multihop networks. *IEEE/ACM Trans. Netw.* **2019**, *27*, 1248–1257. [CrossRef]
12. Zhong, J.; Yates, R.D.; Soljanin, E. Multicast with prioritized delivery: How fresh is your data? In Proceedings of the 2018 IEEE 19th International Workshop on Signal Processing Advances in Wireless Communications (SPAWC), Kalamata, Greece, 25–28 June 2018.
13. Buyukates, B.; Soysal, A.; Ulukus, S. Age of information in two-hop multicast networks. In Proceedings of the 2018 52nd Asilomar Conference on Signals, Systems, and Computers, Pacific Grove, CA, USA, 28–31 October 2018.
14. Buyukates, B.; Soysal, A.; Ulukus, S. Age of information in multihop multicast networks. *J. Commun. Netw.* **2019**, *21*, 256–267. [CrossRef]
15. Buyukates, B.; Soysal, A.; Ulukus, S. Age of information in multicast networks with multiple update streams. In Proceedings of the 2019 53rd Asilomar Conference on Signals, Systems, and Computers, Pacific Grove, CA, USA, 3–6 November 2019.
16. Krishnan, K.S.A.; Sharma, V. Minimizing age of information in a multihop wireless network. In Proceedings of the ICC 2020-2020 IEEE International Conference on Communications, Dublin, Ireland, 7–11 June 2020.
17. Farazi, S.; Klein, A.G.; Brown, D.R., III. Fundamental bounds on the age of information in multi-hop global status update networks. *J. Commun. Netw.* **2019**, *21*, 268–279. [CrossRef]
18. Ioannidis, S.; Chaintreau, A.; Massoulie, L. Optimal and scalable distribution of content updates over a mobile social network. In Proceedings of the IEEE Infocom, Rio De Janeiro, Brazil, 24 April 2009.
19. Wang, M.; Chen, W.; Ephremides, A. Reconstruction of counting process in real-time: The freshness of information through queues. In Proceedings of the ICC 2019-2019 IEEE International Conference on Communications (ICC), Shanghai, China, 20–24 May 2019.
20. Sun, Y.; Polyanskiy, Y.; Uysal-Biyikoglu, E. Remote estimation of the Wiener process over a channel with random delay. In Proceedings of the 2017 IEEE International Symposium on Information Theory (ISIT), Aachen, Germany, 25–30 June 2017.
21. Sun, Y.; Cyr, B. Information aging through queues: A mutual information perspective. In Proceedings of the 2018 IEEE 19th International Workshop on Signal Processing Advances in Wireless Communications (SPAWC), Kalamata, Greece, 25–28 June 2018.
22. Chakravorty, J.; Mahajan, A. Remote estimation over a packet-drop channel with Markovian state. *IEEE Trans. Autom. Control* **2020**, *65*, 2016–2031. [CrossRef]
23. Kam, C.; Kompella, S.; Ephremides, A. Age of incorrect information for remote estimation of a binary Markov source. In Proceedings of the IEEE INFOCOM 2020-IEEE Conference on Computer Communications Workshops (INFOCOM WKSHPS), Toronto, ON, Canada, 6–9 July 2020.
24. Arafa, A.; Banawan, K.; Seddik, K.G.; Poor, H.V. Sample, quantize and encode: Timely estimation over noisy channels. *IEEE Trans. Commun.* **2021**, *69*, 6485-6499. [CrossRef]
25. Bastopcu, M.; Ulukus, S. Who should Google Scholar update more often? In Proceedings of the IEEE INFOCOM 2020-IEEE Conference on Computer Communications Workshops (INFOCOM WKSHPS), Toronto, ON, Canada, 6–9 July 2020.
26. Bacinoglu, B.T.; Sun, Y.; Uysal-Biyikoglu, E.; Mutlu, V. Achieving the age-energy trade-off with a finite-battery energy harvesting source. In Proceedings of the 2018 IEEE International Symposium on Information Theory (ISIT), Vail, CO, USA, 17–22 June 2018.
27. Baknina, A.; Ozel, O.; Yang, J.; Ulukus, S.; Yener, A. Sending information through status updates. In Proceedings of the 2018 IEEE International Symposium on Information Theory (ISIT), Vail, CO, USA, 17–22 June 2018.
28. Baknina, A.; Ulukus, S. Coded status updates in an energy harvesting erasure channel. In Proceedings of the 2018 52nd Annual Conference on Information Sciences and Systems (CISS), Princeton, NJ, USA, 21–23 March 2018.

29. Wu, X.; Yang, J.; Wu, J. Optimal status update for age of information minimization with an energy harvesting source. *IEEE Trans. Green Commun. Netw.* **2018**, *2*, 193–204. [CrossRef]

30. Feng, S.; Yang, J. Optimal status updating for an energy harvesting sensor with a noisy channel. In Proceedings of the IEEE INFOCOM 2018-IEEE Conference on Computer Communications Workshops (INFOCOM WKSHPS), Honolulu, HI, USA, 15–19 April 2018.

31. Feng, S.; Yang, J. Minimizing age of information for an energy harvesting source with updating failures. In Proceedings of the 2018 IEEE International Symposium on Information Theory (ISIT), Vail, CO, USA, 17–22 June 2018.

32. Arafa, A.; Yang, J.; Ulukus, S.; Poor, H.V. Age-minimal online policies for energy harvesting sensors with incremental battery recharges. In Proceedings of the 2018 Information Theory and Applications Workshop (ITA), San Diego, CA, USA, 11–16 February 2018.

33. Arafa, A.; Yang, J.; Ulukus, S. Age-minimal online policies for energy harvesting sensors with random battery recharges. In Proceedings of the 2018 IEEE international conference on communications (ICC), Kansas City, MO, USA, 20–24 May 2018.

34. Arafa, A.; Yang, J.; Ulukus, S.; Poor, H.V. Age-minimal transmission for energy harvesting sensors with finite batteries: Online policies. *IEEE Trans. Inf. Theory* **2020**, *66*, 534–556. [CrossRef]

35. Arafa, A.; Yang, J.; Ulukus, S.; Poor, H.V. Online timely status updates with erasures for energy harvesting sensors. In Proceedings of the 2018 56th Annual Allerton Conference on Communication, Control, and Computing (Allerton), Monticello, IL, USA, 2–5 October 2018.

36. Arafa, A.; Yang, J.; Ulukus, S.; Poor, H.V. Using erasure feedback for online timely updating with an energy harvesting sensor. In Proceedings of the 2019 IEEE International Symposium on Information Theory (ISIT), Paris, France, 7–12 July 2019.

37. Arafa, A.; Yang, J.; Ulukus, S.; Poor, H.V. Timely status updating over erasure channels using an energy harvesting sensor: Single and multiple sources. *IEEE Trans. Green Commun. Netw.* **2022**, *6*, 6–19. [CrossRef]

38. Farazi, S.; Klein, A.G.; Brown, D.R., III. Average age of information for status update systems with an energy harvesting server. In Proceedings of the IEEE INFOCOM 2018-IEEE Conference on Computer Communications Workshops (INFOCOM WKSHPS), Honolulu, HI, USA 15–19 April 2018.

39. Leng, S.; Yener, A. Age of information minimization for an energy harvesting cognitive radio. *IEEE Trans. Cogn. Commun. Netw.* **2019**, *5*, 427–439. [CrossRef]

40. Chen, Z.; Pappas, N.; Bjornson, E.; Larsson, E.G. Age of information in a multiple access channel with heterogeneous traffic and an energy harvesting node. In Proceedings of the IEEE INFOCOM 2019-IEEE Conference on Computer Communications Workshops (INFOCOM WKSHPS), Paris, France, 29 April–2 May 2019.

41. Bhat, R.V.; Vaze, R.; Motani, M. Throughput maximization with an average age of information constraint in fading channels. *IEEE Trans. Wirel. Commun.* **2021**, *20*, 481–494. [CrossRef]

42. Ostman, J.; Devassy, R.; Durisi, G.; Uysal, E. Peak-age violation guarantees for the transmission of short packets over fading channels. In Proceedings of the IEEE INFOCOM 2019-IEEE Conference on Computer Communications Workshops (INFOCOM WKSHPS), Paris, France, 29 April–2 May 2019.

43. Bastopcu, M.; Ulukus, S. Age of information with soft updates. In Proceedings of the 2018 56th Annual Allerton Conference on Communication, Control, and Computing (Allerton), Monticello, IL, USA, 2–5 October 2018.

44. Bastopcu, M.; Ulukus, S. Minimizing age of information with soft updates. *J. Commun. Netw.* **2019**, *21*, 233–243. [CrossRef]

45. Buyukates, B.; Soysal, A.; Ulukus, S. Age of information scaling in large networks. In Proceedings of the 2019 IEEE Global Communications Conference (GLOBECOM), Waikoloa, HI, USA, 9–13 December 2019.

46. Buyukates, B.; Soysal, A.; Ulukus, S. Age of information scaling in large networks with hierarchical cooperation. In Proceedings of the 2019 IEEE Global Communications Conference (GLOBECOM), Waikoloa, HI, USA, 9–13 December 2019.

47. Buyukates, B.; Soysal, A.; Ulukus, S. Scaling laws for age of information in wireless networks. *IEEE Trans. Wirel. Commun.* **2021**, *20*, 2413–2427. [CrossRef]

48. Zhong, J.; Yates, R.D.; Soljanin, E. Minimizing content staleness in dynamo-style replicated storage systems. In Proceedings of the IEEE INFOCOM 2018-IEEE Conference on Computer Communications Workshops (INFOCOM WKSHPS), Honolulu, HI, USA, 15–19 April 2018.

49. Rajaraman, N.; Vaze, R.; Reddy, G. Not just age but age and quality of information. *IEEE J. Sel. Areas Commun.* **2021**, *39*, 1325–1338. [CrossRef]

50. Liu, Z.; Ji, B. Towards the tradeoff between service performance and information freshness. In Proceedings of the ICC 2019-2019 IEEE International Conference on Communications (ICC), Shanghai, China, 20–24 May 2019.

51. Maatouk, A.; Assaad, M.; Ephremides, A. The age of incorrect information: An enabler of semantics-empowered communication. *arXiv* **2020**, arXiv:2012.13214.

52. Uysal, E.; Kaya, O.; Ephremides, A.; Gross, J.; Codreanu, M.; Popovski, P.; Johansson, K.H. Semantic communications in networked systems. *arXiv* **2021**, arXiv:2103.05391.

53. Ayan, O.; Vilgelm, M.; Klügel, M.; Hirche, S.; Kellerer, W. Age-of-information vs. value-of-information scheduling for cellular networked control systems. In Proceedings of the 10th ACM/IEEE International Conference on Cyber-Physical Systems, Montreal, QC, Canada, 16–18 April 2019.

54. Bastopcu, M.; Ulukus, S. Timely group updating. In Proceedings of the 2021 55th Annual Conference on Information Sciences and Systems (CISS), Baltimore, MD, USA, 24–26 March 2021.

55. Banerjee, S.; Bhattacharjee, R.; Sinha, A. Fundamental limits of age-of-information in stationary and non-stationary environments. In Proceedings of the 2020 IEEE International Symposium on Information Theory (ISIT), Los Angeles, CA, USA, 21–26 June 2020.

56. Zhong, J.; Yates, R.D. Timeliness in lossless block coding. In Proceedings of the 2016 Data Compression Conference (DCC), Snowbird, UT, USA, 29 March–1 April 2016.

57. Zhong, J.; Yates, R.D.; Soljanin, E. Timely lossless source coding for randomly arriving symbols. In Proceedings of the 2018 IEEE Information Theory Workshop (ITW), Guangzhou, China, 25–29 November 2018.

58. Mayekar, P.; Parag, P.; Tyagi, H. Optimal lossless source codes for timely updates. In Proceedings of the 2018 IEEE International Symposium on Information Theory (ISIT), Vail, CO, USA, 17–22 June 2018.

59. Mayekar, P.; Parag, P.; Tyagi, H. Optimal source codes for timely updates. *IEEE Trans. Inf. Theory* **2020**, *66*, 3714–3731. [CrossRef]

60. Bastopcu, M.; Buyukates, B.; Ulukus, S. Optimal selective encoding for timely updates. In Proceedings of the 2020 54th Annual Conference on Information Sciences and Systems (CISS), Princeton, NJ, USA, 18–20 March 2020.

61. Buyukates, B.; Bastopcu, M.; Ulukus, S. Optimal selective encoding for timely updates with empty symbol. In Proceedings of the 2020 IEEE International Symposium on Information Theory (ISIT), Los Angeles, CA, USA, 21–26 June 2020.

62. Bastopcu, M.; Buyukates, B.; Ulukus, S. Selective encoding policies for maximizing information freshness. *IEEE Trans. Commun.* **2021**, *69*, 5714–5726. [CrossRef]

63. Ramirez, D.; Erkip, E.; Poor, H.V. Age of information with finite horizon and partial updates. In Proceedings of the ICASSP 2020-2020 IEEE International Conference on Acoustics, Speech and Signal Processing (ICASSP), Barcelona, Spain, 4–8 May 2020.

64. Arafa, A.; Banawan, K.; Seddik, K.G.; Poor, H.V. On timely channel coding with hybrid ARQ. In Proceedings of the 2019 IEEE Global Communications Conference (GLOBECOM), Big Island, HI, USA, 9–13 December 2019.

65. Arafa, A.; Wesel, R.D. Timely transmissions using optimized variable length coding. In Proceedings of the IEEE INFOCOM 2021-IEEE Conference on Computer Communications Workshops (INFOCOM WKSHPS), Vancouver, BC, Canada, 10–13 May 2021.

66. Bastopcu, M.; Ulukus, S. Partial updates: Losing information for freshness. In Proceedings of the 2020 IEEE International Symposium on Information Theory (ISIT), Los Angeles, CA, USA, 21–26 June 2020.

67. Abd-Elmagid, M.A.; Dhillon, H.S. Average peak age-of-information minimization in UAV-assisted IoT networks. *IEEE Trans. Veh. Technol.* **2019**, *68*, 2003–2008. [CrossRef]

68. Liu, J.; Wang, X.; Dai, H. Age-optimal trajectory planning for UAV-assisted data collection. In Proceedings of the IEEE INFOCOM 2018-IEEE Conference on Computer Communications Workshops (INFOCOM WKSHPS), Honolulu, HI, USA, 15–19 April 2018.

69. Abd-Elmagid, M.A.; Pappas, N.; Dhillon, H.S. On the role of age of information in the internet of things. *IEEE Commun. Mag.* **2019**, *57*, 72–77. [CrossRef]

70. Alabbasi, A.; Aggarwal, V. Joint information freshness and completion time optimization for vehicular networks. *IEEE Trans. Serv. Comput.* **2020**, *15*, 1–14. [CrossRef]

71. Gao, W.; Cao, G.; Srivatsa, M.; Iyengar, A. Distributed maintenance of cache freshness in opportunistic mobile networks. In Proceedings of the 2012 IEEE 32nd International Conference on Distributed Computing Systems, Macau, China, 18–21 June 2012.

72. Yates, R.D.; Ciblat, P.; Yener, A.; Wigger, M. Age-optimal constrained cache updating. In Proceedings of the 2017 IEEE International Symposium on Information Theory (ISIT), Aachen, Germany, 25–30 June 2017.

73. Kam, C.; Kompella, S.; Nguyen, G.D.; Wieselthier, J.E.; Ephremides, A. Information freshness and popularity in mobile caching. In Proceedings of the 2017 IEEE International Symposium on Information Theory (ISIT), Aachen, Germany, 25–30 June 2017.

74. Zhang, S.; Li, J.; Luo, H.; Gao, J.; Zhao, L.; Shen, X.S Towards fresh and low-latency content delivery in vehicular networks: An edge caching aspect. In Proceedings of the 2018 10th International Conference on Wireless Communications and Signal Processing (WCSP), Hangzhou, China, 18–20 October 2018.

75. Tang, H.; Ciblat, P.; Wang, J.; Wigger, M.; Yates, R. Age of information aware cache updating with file- and age-dependent update durations. In Proceedings of the 2020 18th International Symposium on Modeling and Optimization in Mobile, Ad Hoc, and Wireless Networks (WiOPT), Volos, Greece, 15–19 June 2020.

76. Zhong, J.; Yates, R.D.; Soljanin, E. Two freshness metrics for local cache refresh. In Proceedings of the 2018 IEEE International Symposium on Information Theory (ISIT), Vail, CO, USA, 17–22 June 2018.

77. Yang, L.; Zhong, Y.; Zheng, F.; Jin, S. Edge caching with real-time guarantees. *arXiv* **2019**, arXiv:1912.11847.

78. Bastopcu, M.; Ulukus, S. Maximizing information freshness in caching systems with limited cache storage capacity. In Proceedings of the 2020 54th Asilomar Conference on Signals, Systems, and Computers, Pacific Grove, CA, USA, 1–5 November 2020.

79. Bastopcu, M.; Ulukus, S. Cache freshness in information updating systems. In Proceedings of the 2021 55th Annual Conference on Information Sciences and Systems (CISS), Baltimore, MD, USA, 24–26 March 2021.

80. Bastopcu, M.; Ulukus, S. Information freshness in cache updating systems. *IEEE Trans. Wirel. Commun.* **2021**, *20*, 1861–1874. [CrossRef]

81. Kaswan, P.; Bastopcu, M.; Ulukus, S. Freshness based cache updating in parallel relay networks. In Proceedings of the 2021 IEEE International Symposium on Information Theory (ISIT), Melbourne, Australia, 12–20 July 2021.

82. Gu, Y.; Wang, Q.; Chen, H.; Li, Y.; Vucetic, B. Optimizing information freshness in two-hop status update systems under a resource constraint. *IEEE J. Sel. Areas Commun.* **2021**, *39*, 1380–1392. [CrossRef]

83. Kuang, Q.; Gong, J.; Chen, X.; Ma, X. Age-of-information for computation-intensive messages in mobile edge computing. *arXiv* **2019**, arXiv:1901.01854.

84. Gong, J.; Kuang, Q.; Chen, X.; Ma, X. Reducing age-of-information for computation-intensive messages via packet replacement. In Proceedings of the 2019 11th International Conference on Wireless Communications and Signal Processing (WCSP), Xi'an, China, 23–25 October 2019.

85. Zou, P.; Ozel, O.; Subramaniam, S. Trading off computation with transmission in status update systems. In Proceedings of the 2019 IEEE 30th Annual International Symposium on Personal, Indoor and Mobile Radio Communications (PIMRC), Istanbul, Turkey, 8–11 September 2019.

86. Bastopcu, M.; Ulukus, S. Age of information for updates with distortion. In Proceedings of the 2019 IEEE Information Theory Workshop (ITW), Gotland, Sweden, 25–28 August 2019.

87. Bastopcu, M.; Ulukus, S. Age of information for updates with distortion: Constant and age-dependent distortion constraints. *IEEE/ACM Trans. Netw.* **2021**, *29*, 2425–2438. [CrossRef]

88. Buyukates, B.; Ulukus, S. Timely updates in distributed computation systems with stragglers. In Proceedings of the 2020 54th Asilomar Conference on Signals, Systems, and Computers, Pacific Grove, CA, USA, 1–5 November 2020.

89. Buyukates, B.; Ulukus, S. Timely distributed computation with stragglers. *IEEE Trans. Commun.* **2020**, *68*, 5273–5282. [CrossRef]

90. Zou, P.; Ozel, O.; Subramaniam, S. Optimizing information freshness through computation–transmission tradeoff and queue management in edge computing. *IEEE/ACM Trans. Netw.* **2021**, *29*, 949–963. [CrossRef]

91. Buyukates, B.; Ulukus, S. Timely communication in federated learning. In Proceedings of the IEEE INFOCOM 2021-IEEE Conference on Computer Communications Workshops (INFOCOM WKSHPS), Vancouver, BC, Canada, 10–13 May 2021.

92. Ozfatura, E.; Buyukates, B.; Gündüz, D.; Ulukus, S. Age-based coded computation for bias reduction in distributed learning. In Proceedings of the GLOBECOM 2020-2020 IEEE Global Communications Conference, Taipei, Taiwan, 7–11 December 2020.

93. Ceran, E.T.; Gündüz, D.; György, A. A reinforcement learning approach to age of information in multi-user networks with HARQ. *IEEE J. Sel. Areas Commun.* **2021**, *39*, 1412–1426. [CrossRef]

94. Yates, R.D. The age of gossip in networks. In Proceedings of the 2021 IEEE International Symposium on Information Theory (ISIT), Melbourne, Australia, 12–20 July 2021.

95. Buyukates, B.; Bastopcu, M.; Ulukus, S. Age of gossip in networks with community structure. In Proceedings of the 2021 IEEE 22nd International Workshop on Signal Processing Advances in Wireless Communications (SPAWC), Lucca, Italy, 27–30 September 2021.

96. Bastopcu, M.; Buyukates, B.; Ulukus, S. Gossiping with binary freshness metric. In Proceedings of the 2021 IEEE Globecom Workshops (GC Wkshps), Madrid, Spain, 7–11 December 2021.

97. Kaswan, P.; Ulukus, S. Timely gossiping with file slicing and network coding. *arXiv* **2022**, arXiv:2202.00649.

98. Kosta, A.; Pappas, N.; Angelakis, V. Age of information: A new concept, metric, and tool. *Found. Trends Netw.* **2017**, *12*, 162–259. [CrossRef]

99. Sun, Y.; Kadota, I.; Talak, R.; Modiano, E. Age of information: A new metric for information freshness. *Synth. Lect. Commun. Netw.* **2019**, *12*, 1–224. [CrossRef]

100. Yates, R.D.; Sun, Y.; Brown, D.R., III; Kaul, S.K.; Modiano, E.; Ulukus, S. Age of information: An introduction and survey. *IEEE J. Sel. Areas Commun.* **2021**, *39*, 1183–1210. [CrossRef]

101. Yun, J.; Joo, C.; Eryilmaz, A. Optimal real-time monitoring of an information source under communication costs. In Proceedings of the 2018 IEEE Conference on Decision and Control (CDC), Miami Beach, FL, USA, 17–19 December 2018.

102. Maatouk, A.; Kriouile, S.; Assaad, M.; Ephremides, A. The age of incorrect information: A new performance metric for status updates. *IEEE/ACM Trans. Netw.* **2020**, *28*, 2215–2228. [CrossRef]

103. Yates, R.D.; Goodman, D.J. *Probability and Stochastic Processes*; Wiley: Hoboken, NJ, USA, 2014.

104. Boyd, S.P.; Vandenberghe, L. *Convex Optimization*; Cambridge University Press: Cambridge, UK, 2004.

105. Bertsekas, D.P.; Tsitsiklis, J.N. *Introduction to Probability*; Athena Scientific: Belmont, MA, USA, 2008.

106. Chen, Y.C.; Lu, P.E.; Chang, C.S.; Liu, T.H. A time-dependent SIR model for COVID-19 with undetectable infected persons. *IEEE Trans. Netw. Sci. Eng.* **2020**, *7*, 3279–3294. [CrossRef]

107. Olmez, S.Y.; Mori, J.; Miehling, E.; Başar, T.; Smith, R.L.; West, M.; Mehta, P.G. A data-informed approach for analysis, validation, and identification of COVID-19 models. In Proceedings of the 2021 American Control Conference (ACC), New Orleans, LA, USA, 25–28 May 2021.

Article

Implementation and Evaluation of Age-Aware Downlink Scheduling Policies in Push-Based and Pull-Based Communication

Tahir Kerem Oğuz [1,*], Elif Tuğçe Ceran [1], Elif Uysal [1] and Tolga Girici [2]

[1] Department of Electrical and Electronics Engineering, Middle East Technical University, Ankara 06800, Turkey; elifce@metu.edu.tr (E.T.C.); uelif@metu.edu.tr (E.U.)
[2] Department of Electrical and Electronics Engineering, TOBB University of Economics and Technology, Ankara 06560, Turkey; tgirici@etu.edu.tr
[*] Correspondence: kerem.oguz@alumni.metu.edu.tr

Abstract: As communication systems evolve to better cater to the needs of machine-type applications such as remote monitoring and networked control, advanced perspectives are required for the design of link layer protocols. The age of information (AoI) metric has firmly taken its place in the literature as a metric and tool to measure and control the data freshness demands of various applications. AoI measures the timeliness of transferred information from the point of view of the destination. In this study, we experimentally investigate AoI of multiple packet flows on a wireless multi-user link consisting of a transmitter (base station) and several receivers, implemented using software-defined radios (SDRs). We examine the performance of various scheduling policies under push-based and pull-based communication scenarios. For the push-based communication scenario, we implement age-aware scheduling policies from the literature and compare their performance with those of conventional scheduling methods. Then, we investigate the query age of information (QAoI) metric, an adaptation of the AoI concept for pull-based scenarios. We modify the former age-aware policies to propose variants that have a QAoI minimization objective. We share experimental results obtained in a simulation environment as well as on the SDR testbed.

Keywords: age of information; query age of information; wireless networks; software-defined radio; scheduling

Citation: Oğuz, T.K.; Ceran, E.T.; Uysal, E.; Girici, T. Implementation and Evaluation of Age-Aware Downlink Scheduling Policies in Push-Based and Pull-Based Communication. *Entropy* **2022**, *24*, 673. https://doi.org/10.3390/e24050673

Academic Editor: Syed A. Jafar

Received: 2 March 2022
Accepted: 23 March 2022
Published: 11 May 2022

Publisher's Note: MDPI stays neutral with regard to jurisdictional claims in published maps and institutional affiliations.

1. Introduction

The advent and the fast growth of the Internet of things (IoT) has further complicated the design of communication networks, in the presence of an increase in demand in networked services catered by the fifth generation (5G) evolution of communication networks. On the one hand, machine-type communications are typically less bandwidth hungry than typical multimedia services. On the other hand, IoT flows tend to be composed of many small packets generated by large numbers of end nodes, and they may have end-to-end freshness requirements that may be challenging to satisfy with conventional link or transport layer approaches based on optimizing throughput and delay. Increasing the sampling rate of IoT nodes to respond to freshness requirements or adopting first-come-first-served service policies can cause bottlenecks on the network, resulting in a reduction in quality of service. It has been argued in recent literature that optimizing data generation, transmission, and transport with respect to higher-level metrics such as Age of Information can prevent unnecessary network load, while improving the freshness of flows. In a broader perspective, there are proposals to encapsulate the significance or the value of the transferred information to the communication problem in certain "semantic metrics" and use these in the design of algorithms and protocols in all network layers, referred to as "semantic communication" [1].

Within the set of semantic metrics, the age of information (AoI) from the receiver's point of view is defined as the time elapsed since the generation of the newest status update that has been received by the destination [2]. AoI is gaining momentum as a key performance indicator (KPI) for machine-type communications (MTC). The primary reason for the interest in AoI is the growing demand for timely and fresh information in many emerging real-time and remote monitoring-based applications such as the Internet of things, vehicular networks, and cyber-physical systems.

AoI monitors the freshness of the entire information stream from the receiver's point of view. Hence, it reveals further aspects of the network compared to traditional metrics such as delay or throughput. For instance, the delay metric measures the timeliness from the transmitted packet's perspective. A low average delay does not mean a low average age in every case [3]. Continuous packet transmission policy (known as zero-wait policy in the literature) can optimize delay, but it may not provide age-optimality in the presence of FCFS (first-come-first-serve) queues [4]. Moreover, if the transmitter has an energy constraint, the inefficiency of the zero-wait policy becomes more apparent [3]. Improving the throughput alone can maximize the amount of data flow to the receiver node but may cause an overload of the queues within the network. Packets waiting in the queue result in outdated information reaching the receiver node. In this case, to reduce backlogs within queues, the packet generation rate should be decreased. However, an over-reduction of the packet generation rate would cause the receiver to be updated sporadically, which also leads to reduced AoI performance. This dilemma shows that AoI is a composite measure of both throughput and delay. For achieving optimal AoI, frequent packets must arrive regularly [5]. Consequently, solving the scheduling problem with an AoI minimization objective requires a novel formulation.

A significant portion of the AoI literature consists of studies involving push-based communication scenarios. In the push-based model, the generation of a new packet triggers the communication process. Then, the transmitter module sends the generated packet to the receiver module. The sequence of operations of the communication process proceeds from the information source to the destination. However, one of the network models often encountered in real-life scenarios is the pull-based model, where the query source requests (or queries) information from the receiver module. In this scenario, the initiator of the communication process is the query source that aims to pull information from the receiver module. The source of these queries could be users or applications that want to monitor the information source. In the pull-based network, the sequence of operations of the communication process proceeds from the destination to the source.

In this paper, we consider both push-based and pull-based status update systems and experimentally investigate the performance of several age-aware downlink scheduling policies in wireless multi-user networks. The main contribution of this study is to report one of the pioneering experimental studies of age-aware MAC layer scheduling policies. We have implemented a multi-user downlink network with a single base station and multiple receivers using software-defined radios (SDRs). This testbed implementation allowed us to examine push-based and pull-based scenarios and state-of-the-art scheduling policies. Along with the other well-known policies, we have proposed max-weight policies for different pull-based scenarios and provided extensive simulation and experimental results.

The rest of the paper is organized as follows. In Section 2, we present the related work. In Section 3, the system model is presented and the problems of minimizing the average AoI, QAoI, gmagmaild and EAoI in the network are formulated. Age-aware downlink scheduling policies are exhibited in Section 4, and the experimental setup is explained in detail in Section 5. Simulation and experimental results are presented in Section 6, and the paper is concluded in Section 7.

2. Related Work

There are numerous studies examining the AoI metric in the literature. The major works that stand out in the literature are those investigating the effects of different queuing types and developing scheduling policies to minimize average AoI in the network. An important concern when proposing a scheduling policy is the required computational load [5]. The work in [6] shows that a scheduling problem with an age minimization goal is an NP-hard problem in a multi-user network. In [7], age-aware scheduling policies are derived for the lossy channel case in a multi-user network. The greedy policy is inspected, and results indicate that the policy is optimal in the symmetric channel state for mean AoI minimization. In [8], the network is analyzed based on the peak-age and mean-age metrics, and a virtual-queue-based policy and an age-based policy are developed. The virtual-queue-based policy is shown to be peak-age optimal. The age-based policy is proved to be within a factor of four of optimal values for peak age as well as average age. In [5], Whittle's index (WI) policy and max-weight (MW) policy are proposed. The lower bound for AoI that can be calculated by using the statistical information of the network is derived. Lower and upper limits of AoI performances for WI and MW policies are calculated and proven to be within a factor of four of the optimal (upper limit is at most four times higher than the lower limit). There are also learning-based approaches in the literature to find an optimal age-aware policy for multi-user networks [9,10].

In the multi-user scheduling problem, the generation procedure of the packets has a significant impact on the AoI. In the literature, sources that generate a fresh packet at every time frame are referred to as "active sources" [8]. For a system model with active sources, whenever there is a transmission, the age of the corresponding flow will be reset to its minimum possible value (one frame duration in our setup). However, many realistic scenarios may be better modeled with a packet generation that is a stochastic process. For example, [11] studied a case where the packet generation procedure is a Bernoulli stochastic process and proposed scheduling policies suitable for that system model.

The queue service policy (e.g., LCFS (last-come-first-serve), FCFS (first-come-first-serve)) also has a significant effect on AoI [12,13]. For the active source case, queuing policies become even more important since sources load the network with the highest rate available. Queue management policy determines the behavior of the queue when the new packets arrive. If the queue is managed with an LCFS policy, the freshest packet will be at the top of queue, and the first packet that leaves the queue will be the one with the most up-to-date information. In FCFS queues, a new packet is added to the bottom end of the queue. To transmit the most timely packet, all packets in front of the last inserted packet must be sent for transmission. As a result, the most up-to-date packet loses time and becomes stale waiting the transmission of other packets in the queue.

The overwhelming majority of the AoI literature to date has emphasized theoretical studies. However, there are also studies on implementation in the literature [14–22]. For a survey of this implementation-oriented literature, see [23]. In [14–17], the experimental setup mostly lies between the transport layer and application layer. The effects of different wireless access technologies on end-to-end TCP/IP connections were measured by [14–16]. Studies in [18–20] cover a broader range of interconnection layers and capture the performances of novel age-based MAC layer algorithms. In [18], Wi-Fi protocol is implemented on SDRs. The uplink of a wireless network is taken into consideration, and the effect of utilizing the MW scheduling policy is investigated. The work presented in [19] experimentally investigates the effects of packet management policies on the performance of networked control systems. A test environment was developed by [20] to evaluate various ALOHA-like random access protocols. In our previous work in [21], we implemented a multi-user wireless network using SDRs. We compared the AoI performances of MW and WI policies with round-robin and greedy policies.

The time-average age metric weighs information freshness of all time frames equally. However, there are many types of real-world applications where the demand for timely information varies in time. For these, minimization of time-average age may not be the most relevant objective. In the literature, various semantic metrics alter this model, placing higher emphasis on selected time frames. For example, the age of incorrect information (AoII) metric focuses on the usefulness of the information and aims to maximize the freshness of non-repetitious information. In the AoII concept, obtaining redundant information is pointless for the receiver and does not reduce AoII. The objective is to minimize the age of differing information [24–26].

Query age of information (QAoI) is a recent semantic metric proposed to investigate pull-based scenarios from the AoI perspective [27–30]. QAoI considers a model where the freshness of information is valuable only at query moments. These queries are sent to the receiver modules in the network. Then, the receiver modules respond to these queries with the most up-to-date information. The source of the queries can be a user or an application that needs to obtain the most up-to-date information. In [29], the pull-based scenario is discussed, and the effective age of information (EAoI) metric is presented for the multi-user system model. Query generation is modeled as an independent Bernoulli process for each receiver, and the immediate EAoI is assumed to be zero for frames without queries. For the queried frames, immediate EAoI is related to the immediate AoI of the receiver under the proactive serving assumption as a query response procedure. According to the proactive serving method, the receiver module can wait for the query response for a frame if it is expecting a packet arrival within the frame. If a packet arrives at the end of the frame, the receiver sends the information in the newly received packet as a query response. In the study, WI-based scheduling policy is proposed for the multi-user system model, and the performance of the policy is demonstrated in the simulation environment.

The work in [28] presents the query-AoI metric for a single receiver in pull-based communication. The calculation of the QAoI metric presented in this study is similar to the EAoI. However, an instantaneous serving scenario is adopted instead of the proactive serving in [29]. In addition, the transmitter module is assumed to have an energy constraint, and the presence of the energy constraint turns the problem in another direction while increasing the value of the QAoI reduction per transmission. Within the scope of the study, the permanent query (PQ) model, which is a query generation procedure that approximates the studied problem with the standard AoI problem, and the query arrival process-aware (QAPA) model, which generates queries based on periodic or stochastic processes, are examined. The optimal scheduling policy in the PQ process is the same as the optimal scheduling policy for AoI. In the case of QAPA, the scheduling policy has information on the query process (either stochastic or deterministic) and can schedule accordingly.

A continuous-time status update model is investigated in [30], where a source node submits update packets to a channel with random transmission delay, and the query source tries to pull information from the receiver module according to a stochastic arrival process. The average QAoI is defined as the average AoI measured at query instants, and the system model is examined from both AoI and QAoI perspectives. Age-aware scheduling policies do not use the information about the query process and freshness equally for all frames. On the other hand, QAoI-aware scheduling policies use additional information about the query process in the scheduling decisions. This extra information allows the scheduling policy to distribute transmission attempts more efficiently and reduce the time spent in the FCFS queue. Eventually, from the query source's perspective, QAoI-aware policies can provide better AoI performance than AoI-aware policies.

To the best of our knowledge, this is the first work in the literature that considers practical implementation and evaluation of QAoI-aware scheduling policies. In addition, we propose and implement novel max-weight policies for the effective AoI and query-AoI system models and evaluate their performance in terms of AoI, EAoI, and QAoI, in both simulation and SDR environments. We have observed that the resulting EAoI-aware max-weight (EAoI-MW) policy has a similar EAoI performance to the WI but yields a higher network throughput. We have also observed that QAoI-aware max-weight (QAoI-MW) provides superior QAoI performance than AoI-aware policies.

3. System Model

We consider a wireless multi-user network where a common access point or a base station (BS) needs to send status update packets containing time-sensitive information to multiple receiver modules. Let M denote the total number of receivers. We also assume a discrete time system where time is divided into fixed-length frames denoted by $t \in \{1, \ldots, T\}$. In each frame, the base station (transmitter) is allowed to activate the connection for a single receiver $i \in \{1, \ldots, M\}$, and it cannot send packets to more than one receiver within a frame. A transmission attempt of a status update to a single user takes a constant time, which is assumed to be equal to the duration of one frame. Wireless channels between the receivers and the base station are unreliable. The state of each channel changes randomly from one time slot to the next and is modeled by a Bernoulli random variable. Channel states for each receiver are also independent of the others.

The packet generation scheme in the system follows the "active source" model. At the beginning of each frame, information sources generate new packets for each receiver, and these packets reach the BS immediately. The base station selects one of these packets for transmission and discards the others. There are no queuing-related delays between the information sources and the base station. If a receiver successfully receives a packet, the AoI of this receiver successfully drops to one since the newly formed packet at the information source reaches the receiver within a frame, without observing any delay.

In the system model, there are also query sources linked to each receiver. Each query source is independent of the other, and used to model the behavior of a real-life user or application interested in a particular time-sensitive piece of information at query instants. Query arrival frames to receivers can follow either a deterministic or stochastic pattern. When a query source requests information from a receiver, it sends a query. Then, the receiver responds to it with the latest information that the receiver holds. Query and response messages are transmitted without any errors.

The BS judiciously selects a receiver for transmission according to a stationary scheduling policy $\pi \in \Pi$ represented by $a_i(t)$, for all $i \in \{1, \ldots, M\}$ and $t \in \{1, \ldots, T\}$. If the receiver i is selected for transmission in frame t, then $a_i(t)$ will be equal to one. Otherwise, $a_i(t)$ will be equal to zero. Evaluation of $a_i(t)$ is given in (1).

$$a_i(t) = \begin{cases} 1 & \text{if the receiver } i \text{ is selected,} \\ 0 & \text{otherwise .} \end{cases} \tag{1}$$

If a successful transmission occurs, the base station is informed over an error-free channel in the same frame. By utilizing this knowledge, scheduling policy can keep track of the AoI of the receivers. Similarly, $c_i(t)$ is a binary variable indicating the random channel state of receiver i at frame t. If the channel status of the receiver i is ON, then the successful transmission can be made at frame t, and $c_i(t)$ will be equal to one. Otherwise, if the channel is not available for transmission, $c_i(t)$ will be equal to zero. We assume $c_i(t)$ is an independent and Bernoulli-distributed random variable and the probability of successful transmission (i.e., reliability) is p_i, for all $i \in \{1, \ldots, M\}$. Evaluation of $c_i(t)$ is given in (2).

$$c_i(t) = \begin{cases} 1 & \text{if the channel is ON ,} \\ 0 & \text{if the channel is OFF .} \end{cases} \tag{2}$$

To have a successful transmission in frame t, the receiver i must be selected for transmission, and the channel status of that receiver must be available for transmission. Let $u_i(t)$ denote the overall result of the transmission to receiver i at frame t (3). Evaluation of $u_i(t)$ is given in (3).

$$u_i(t) = \begin{cases} 1 & \text{if } c_i(t)a_i(t) = 1, \\ 0 & \text{otherwise .} \end{cases} \tag{3}$$

We also define $f_i(t)$ as complementary of $u_i(t)$ for simplification of some equations throughout the paper, that is, $f_i(t) = 1 - u_i(t)$.

The instantaneous AoI of receiver i at the beginning of the tth frame is denoted by $\Delta_i(t)$. Note that $\Delta_i(t)$ drops to one if the transmission to receiver i succeeds and increases by 1 if receiver i is not selected for transmission or fails to successfully receive a packet. Evaluation of $\Delta_i(t)$ is given in (4).

$$\Delta_i(t+1) = \begin{cases} 1 & \text{if } u_i(t) = 1 \\ \Delta_i(t) + 1 & \text{otherwise} \end{cases} \tag{4}$$

$d_i(t)$ indicates the query presence. If a query arrives to the receiver i at frame t, $d_i(t)$ will be equal to one. Otherwise, $d_i(t)$ will be equal to zero. Evaluation of $d_i(t)$ is given in (5).

$$d_i(t) = \begin{cases} 1 & \text{if a query arrives to the receiver} \\ 0 & \text{otherwise} \end{cases} \tag{5}$$

The instantaneous query age of receiver i at the beginning of the tth frame is $\Delta_{q_i}(t)$. The evaluation of the $\Delta_{q_i}(t)$ varies with the adopted query response scenario within the system model. In the scope of this study, we assume that query arrival to the receiver and receiver's response will happen at the beginning of the frame. We denote this query response scenario as the "instantaneous serving" scenario. Evaluation of $\Delta_{q_i}(t)$ for instantaneous serving scenario is given in (6).

$$\Delta_{q_i}(t) = d_i(t)\Delta_i(t) \tag{6}$$

An alternative query response scenario called "proactive serving" is defined in the literature in [29]. In proactive serving, the response to the query may be delayed by at most one frame. The purpose of this delay is to put the newest information into the query if the receiver acquires a packet within the queried frame. Nevertheless, unless stated otherwise, the instantaneous serving strategy will be adopted throughout this study. The overall system model is illustrated in Figure 1.

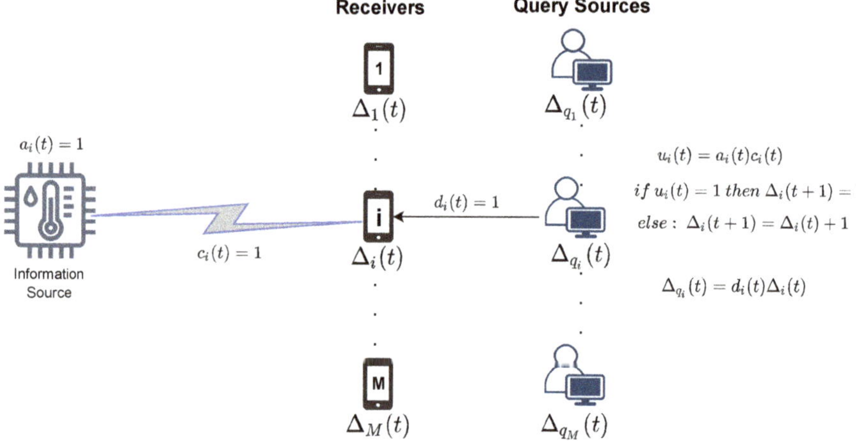

Figure 1. The architecture of the system model.

Next, we formally define the AoI and QAoI minimization problems in Sections 3.1 and 3.2, respectively.

3.1. AoI Minimization Problem

The analytical expressions for the AoI minimization problem have been previously studied in [5]. The objective of the scheduling policy is to minimize the average AoI in the network. Average AoI is calculated for M receivers across T frames. The objective is to find a stationary scheduling policy $\pi \in \Pi$ that minimizes the long-term average AoI, which is defined in (7).

$$\min_{\pi \in \Pi} \lim_{T \to \infty} \mathbb{E}[J_A(\pi)], \text{ where } J_A(\pi) = \frac{1}{TM} \sum_{t=1}^{T} \sum_{i=1}^{M} \Delta_i(t) \tag{7}$$

3.2. QAoI Minimization Problem

For the QAoI problem, the main objective of the scheduling policy is to minimize the average age of the query sources in the network. This problem differs from the AoI problem since the query sources do not require fresh data at every instant but only at the queried frames. The difference between the two problem statements has also been previously investigated in [28,30].

There are two major approaches in the literature to calculate the average ages of the users at query instants in pull-based communication systems. In the first approach, the sum of the ages at query instants is divided by the total number of frames. This method is followed by [27,29] to develop age-aware scheduling policies, and the metric is called *effective age of information (EAoI)*. Note that [28] also follows a similar approach in the discounted setting for single-user pull-based communication.

The objective function obtained by utilizing this approach is given in (8).

$$\min_{\pi \in \Pi} \lim_{T \to \infty} \mathbb{E}[J_E(\pi)], \text{ where } J_E(\pi) = \frac{1}{TM} \left[\sum_{i=1}^{M} \sum_{t=1}^{T} \Delta_{q_i}(t) \right] \tag{8}$$

The second approach divides the sum of all query ages by the total number of query arrivals. This approached is used by [30] for the average query age calculation. The objective function obtained by utilizing this approach is given in (9).

$$\min_{\pi \in \Pi} \lim_{T \to \infty} \mathbb{E}[J_Q(\pi)], \text{ where } J_Q(\pi) = \frac{1}{M} \sum_{i=1}^{M} \frac{1}{N_i(T)} \left[\sum_{t=1}^{T} \Delta_{q_i}(t) \right], \tag{9}$$

where $N_i(T)$ denotes the total number of queries arrived at receiver i throughout T frames.

Throughout this study, we refer to the metric aligned with the first approach as the *effective age of information (EAoI)*, following its definition in [27]. We call the metric evaluated with the second approach the *query age of information (QAoI)*.

In the average EAoI calculation, the query age of the frames for which the query is not present is taken as zero and included in the average. This calculation method may lead to misleading results for measuring the average AoI of the query sources. This is because even if the AoI of a rarely queried receiver is very high at the time of query, it remains low on average due to the inclusion of unqueried frames. Similarly, for a frequently queried receiver, since the number of unqueried frames is low, the number of zeros included in the calculation of the average EAoI will be low. Therefore, the EAoI of this receiver will tend to be higher than the rarely queried receiver. The effect of the scheduling policy becomes less apparent as the query frequency decreases. Therefore, to measure the performance of scheduling policies, comparing average EAoI values of two individual systems with different query arrival frequencies would provide inconsistent results. When the same problem is analyzed from the QAoI perspective, the effect of the scheduling policy becomes more decisive, as the unqueried frames are discarded in the average query age calculation.

To examine the QAoI problem, we first consider the case where the query generation is an independent Bernoulli process. Note that [28,30] indicates that, to see a difference between QAoI and AoI metrics, the query arrival process must be non-stationary. For the Bernoulli query arrival case, the QAoI problem converges to the AoI problem. On the other hand, EAoI can yield results different than AoI even under the Bernoulli-arrival scheme.

4. Age-Aware Downlink Scheduling Policies

In this section, we define scheduling policies to minimize age-aware metrics. We describe AoI-aware, EAoI-aware, and QAoI-aware scheduling policies in Section 4.1, Section 4.2, Section 4.3, respectively.

4.1. Scheduling Policies for AoI Minimization

AoI-aware scheduling policies have been previously investigated in [5,7,8,10]. The WI and MW policies proposed in [5,7] utilize the instantaneous ages of the receivers and the reliabilities of the corresponding links to calculate the expected costs $\{C_i\}$ associated with each receiver. To maximize the cost reduction, the scheduling policy selects the receiver with the highest C_i at each frame.

The max-weight policy is an adaptation of the Lyapunov optimization technique to the AoI minimization problem. Lyapunov optimization provides a method for penalty minimization while maintaining the queue stability [31]. The objective of the MW policy is to minimize Lyapunov drift in the network with the appropriate scheduling decision. Lyapunov drift measures the expected cost increase between two consecutive frames. In each frame, the policy calculates the expected Lyapunov drift of the receivers. Then, the policy selects the receiver with the highest Lyapunov drift. With this decision, the policy aims to minimize the overall cost. The calculation of expected costs for the MW policy is given in (10). At each frame, the scheduling policy selects the receiver with the highest C_i.

$$C_i(\Delta_i(t)) = p_i \Delta_i(t)(\Delta_i(t) + 2) \tag{10}$$

The WI policy has been presented in [5,7,10] by formulating the AoI minimization problem in (7) as a restless multi-armed bandit (MAB) problem. The MAB problem in general aims to optimize the reward in an unknown environment through a series of trials where the decision-maker can activate only one of the arms and each arm has an immediate reward (or penalty for the minimization problem case) associated with it. The closed-form costs (indexes) for the WIP are given in (11). At each frame, the scheduling policy transmits to the receiver with the highest C_i.

$$C_i(\Delta_i(t)) = p_i \Delta_i(t)\left[\Delta_i(t) + \frac{2 - p_i}{p_i}\right] \tag{11}$$

In our study, we implement AoI-aware MW and WI policies on the USRP testbed and compare their performances with round-robin and greedy policies.

4.2. Scheduling Policies for EAoI Minimization

In the USRP testbed, we implement and evaluate the performance of the EAoI aware WI policy that was previously proposed in [29]. In addition, we propose an EAoI-aware MW policy by modifying the AoI-aware max-weight policy previously proposed in [5] and compare their performances.

EAoI-aware WI in [29] is given in (12). In each frame, the policy chooses the receiver with the highest C_i.

$$C_i(t) = q_i(p_i \Delta_i(t) + 2)(\Delta_i(t) - 1) \tag{12}$$

We can derive the max-weight policy for the pull-based instantaneous serving scenario: First, we calculate the Lyapunov drift of the instantaneous EAoI's between consecutive frames. Then, in line with [5], we select quadratic Lyapunov function to calculate the Lyapunov drift.

Lemma 1. *In each frame, EAoI-MW policy selects the receiver with highest $C_i(t)$, which can be computed as in* (13).

$$C_i(t) = q_i p_i \left(\Delta_i^2(t) + 2\Delta_i(t) \right). \tag{13}$$

Derivation of the EAoI-MW Policy can be found in Appendix A.

4.3. Scheduling Policies for QAoI Minimization

For the QAoI metric investigation, we evaluate the cases where the query arrival process forms a Markov chain. Within the Markov chain, we designate one state as the "Query" state and other states as "non-query" states. When the current state of the Markov chain reaches the query state, a query arrives.

For QAoI minimization, we propose a max-weight-based scheduling policy, following similar steps as in [5]. To adapt this policy to the QAoI model, we utilize the main features of the Markov chain, which determines the query process. In the first step, we calculate the future AoI $\Delta_i(t + K)$ in terms of current AoI $\Delta_i(t)$. The evaluation of AoI between consecutive frames is given in (14).

$$\begin{aligned}
\Delta_i(t+1) &= u_i(t) + (1 - u_i(t))(\Delta_i(t) + 1) \\
&= a_i(t)c_i(t) + (1 - a_i(t)c_i(t))(\Delta_i(t) + 1) \\
&= 1 + f_i(t)\Delta_i(t)
\end{aligned} \tag{14}$$

Repeating this approach multiple times enables us to obtain the future AoI in terms of current AoI. The result is given in (15).

$$\begin{aligned}
\Delta_i(t+1) &= 1 + f_i(t)\Delta_i(t) \\
\Delta_i(t+2) &= 1 + f_i(t+1) + f_i(t+1)f_i(t)\Delta_i(t) \\
\Delta_i(t+3) &= 1 + f_i(t+2) + f_i(t+2)f_i(t+1) + f_i(t+2)f_i(t+1)f_i(t)\Delta_i(t)
\end{aligned} \tag{15}$$

In the following equations, we indicate the future time frames as $\hat{t}$. Although it may lead to suboptimal results, for computational convenience, we assume that future decisions $a_i(\hat{t})$ are independent variables and stationary through time with a fixed expected value. Based on this assumption, we can argue that $f_i(\hat{t})$ is also stationary. Therefore, we define f_i as the stationary version of the $f_i(\hat{t})$ as shown in Equation (16).

$$\mathbb{E}\left[f_i(\hat{t})\right] = \mathbb{E}[f_i(t+1)] = \mathbb{E}[f_i(t+2)] = \mathbb{E}[f_i(t+K)] = f_i \tag{16}$$

Then, we define the closed-form version the future AoI with current AoI in (17).

$$\Delta_i(t+K) = \left[\sum_{k=1}^{K} f_i^{k-1}\right] + f_i^{K-1} f_i(t)\Delta_i(t) \tag{17}$$

To simplify the notation, we define $F_s(K)$ and $F_m(K)$ as in (18) and (19).

$$F_s(K) = \sum_{k=1}^{K} f_i^{k-1} \tag{18}$$

$$F_m(K) = f_i^{K-1} \tag{19}$$

We then rewrite the simplified version of Equation (17) in Equation (20).

$$\Delta_i(t+K) = F_s(K) + F_m(K)f_i(t)\Delta_i(t) \tag{20}$$

We proceed with the max-weight policy derivation steps by the definition of Lyapunov function and Lyapunov drifts. Similar to [5], we use the quadratic Lyapunov function as given in Equation (21). However, rather than calculating the Lyapunov drift between consecutive frames, we calculate the Lyapunov drift $Y_i(t)$ between the current frame t

and the expected query-arrival frame $t + K$. Calculation of Lyapunov drift is given in Equation (22).

$$L(t) = \frac{1}{M} \sum_{i=1}^{M} \Delta_{q_i}^2(t) \tag{21}$$

$$
\begin{aligned}
Y_i(t) &= \mathbb{E}\left[\Delta_i^2(t+K) - \Delta_i^2(t)\right] \\
&= \mathbb{E}\left[F_s(K)^2 + 2f_i(t)F_s(K)F_m(K)\Delta_i(t) + f_i(t)F_m^2(K)\Delta_i^2(t) - \Delta_i^2(t)\right] \\
&= \mathbb{E}\left[F_s(K)^2 - \Delta_i^2(t) + (1 - a_i(t)c_i(t))\left[2F_s(K)F_m(K)\Delta_i(t) + F_m^2(K)\Delta_i^2(t)\right]\right]
\end{aligned}
\tag{22}
$$

In Equation (22), $a_i(t)$ is the only decision variable from which the scheduling policy can choose its value. For simplification, we ignore terms in the calculation of $Y_i(t)$ that are not affected by the decision $a_i(t)$. We denote the modifiable part of the Lyapunov drift with $a_i(t)$ decision as $\hat{Y}_i(t)$.

$$
\begin{aligned}
\hat{Y}_i(t) &= -\mathbb{E}[c_i(t)]\mathbb{E}[a_i(t)]\mathbb{E}\left[2F_s(K)F_m(K)\Delta_i(t) + F_m^2(K)\Delta_i^2(t)\right] \\
&= \mathbb{E}[a_i(t)]p_i\mathbb{E}\left[2F_s(K)F_m(K)\Delta_i(t) + F_m^2(K)\Delta_i^2(t)\right]
\end{aligned}
\tag{23}
$$

At each frame, the main objective of the scheduling policy is to minimize the Lyapunov drift. Therefore, the scheduling policy must eliminate the receiver with the highest $C_i(t)$ to cause maximum reduction to Lyapunov drift.

Lemma 2. *For each frame, QAoI-aware max-weight (Q-MW) policy selects the receiver with highest immediate cost $C_i(t)$. Calculation of immediate cost is given in (24).*

$$C_i(t) = p_i F_m(K)\Delta_i(t)(F_m(K)\Delta_i(t) + 2F_s(K)) \tag{24}$$

To emphasize what our system model corresponds to in practice and depict the difference between AoI- and QAoI-aware policies, we can consider a simple IoT network as an example. This network consists of sensors, microprocessors, a base station, and individual users. In the network, sensor devices generate time-sensitive data about their current status. Nevertheless, the sensors cannot process this data, and they have to transfer it over a wireless network to remote microprocessors. The sensors send the data to a base station, and the base station transmits this data over the wireless network. However, the transmission capacity of the base station is limited, and it cannot simultaneously transmit data to multiple processors.

There is a dedicated microprocessor for each sensor. Microprocessors use the sensor data and generate status reports. Each microprocessor is tracked by an individual user that queries the processor to obtain the freshest status report about the sensor. Query arrivals to each microprocessor are independent of each other and occur infrequently.

QAoI-aware policies come to the fore if the requirement in the system precedes the query source's request for timely information. For the system model given in this example, AoI-aware policies concentrate on the AoI at the microprocessors, and the QAoI-aware policies focus on the AoI at the individual users. The impact of the QAoI-aware policy is shown in Figure 2. The figure examines the instantaneous AoI of a receiver (microcontroller in our example) in a multi-user network. A query source (individual user in our example) generates queries at the 41st, 81st, and 121st frames. From the query source's perspective, freshness is only important at query instants. In line with the query source's demands, the QAoI-aware policy aims to minimize the AoI of the receiver at the 41st, 81st, and 121st frames. Since there is no need for AoI minimization in all frames, the transmission constraint in the system can be relieved, and transmission attempts can be utilized more efficiently.

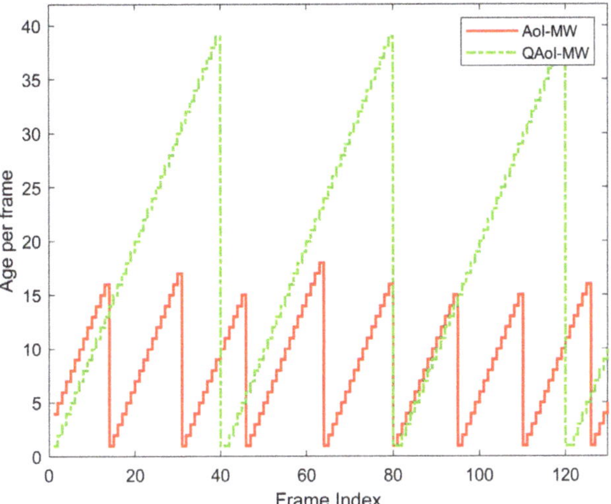

Figure 2. Instantaneous AoI of a receiver in a multi-user network.

5. Implementation

In this section, we describe our implementation work on USRPs. We firstly share detailed information about the implementation environment. Then, we describe the packet interface that we use to transmit time-sensitive information in Section 5.2, and we explain the runtime of our setup in Section 5.3.

Software-based radios, also called "software radios" in pioneering studies, are radios that allow the user to change main parameters of communication systems such as center frequency, bandwidth, and coding of the communication system only by changing the software [32,33]. With SDRs, all layers of the communication system, from the physical layer to the application layer, can be changed only by software modifications. These radios play an important role in the development of today's technologies that require rapid prototyping of various parameters, protocols, and standards, because software-based radios reduce the burden of extra hardware production for test and development studies and provide significant improvement in terms of time and cost.

For the AoI testbed implementation, we use one Ettus USRP N210, one NI USRP 2930, and two NI USRP 2930 SDR devices. General specifications of the devices are available in the devices' datasheets [34,35]. Both USRPs have independent transmit and receive modules. For this reason, these devices can operate as a transmitter and a receiver simultaneously. Nevertheless, it is not possible to run two transmission operations simultaneously.

The host computer runs a LabVIEW application that interfaces with the USRP devices. USRP communicates with the host computer via a 1 Gb Ethernet link. Signals are fragmented to in-phase and quadrant components and carried over in the Ethernet packets. Each transmitter and receiver module contains an amplifier that is controllable through software. In the experiments, we often use these amplifiers to change channel reliabilities.

The LabVIEW environment contains useful built-in functions for system implementation. We use them frequently in our study. We also benefited from the examples regarding the PSK-modulated communication system and packet-based digital link tutorials and examples provided by LabVIEW and the LabVIEW community [36].

5.1. Setup

Among four USRPs, one USRP is configured as the base station, and the other three are the receiver modules. The setup configuration for the implementation is given in Table 1. An overview of the USRP testbed is given in Figure 3.

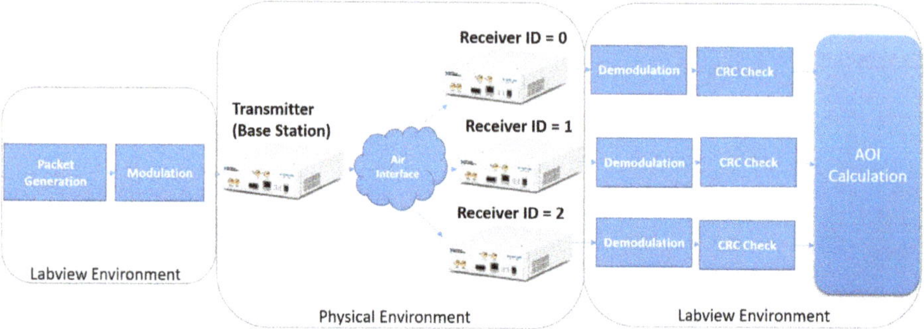

Figure 3. Overview of the implementation environment.

System time is discretized in 50 ms duration frames. The LabVIEW application keeps track of the frame number, that is, the total number of frames that have passed since the experiment began. The frame number is the system's reference clock. All radios run in separate threads over a single LabVIEW application running on the host computer. Thus, difficulties related to synchronization are reduced, as all USRPs are managed from a single host.

We use QPSK modulation in the air interface. The maximum operating frequency of the USRP-2920 is 2.2 GHz [35], and we choose a center frequency of 1.9 GHz for all receivers. We prefer the high center frequency of the carrier signal to induce higher path loss since we have a limited area in the test environment.

The sampling rate of the USRP is configurable via the LabVIEW application. Detailed information about this configuration is described in the USRP documentation [35]. NI specifies that the I/Q ratio must be multiplied by 0.8 to convert to the sample rate [37]. In the implementation, we use the I/Q ratio 500k samples/s, which corresponds to a sampling rate of 400k samples/sec or bandwidth of 200 kHz. This bandwidth meets the requirements of our target application. Selecting higher I/Q rates increases the bandwidth. However, increasing the sample rate causes more data to be processed and transported. Therefore, more data would put a higher load on the USRP and Ethernet connection and eventually induce higher delay. Since timeliness is the primary concern in AoI calculations, we keep the I/Q Ratio low to achieve a more stable operating point without overloading the USRP and Ethernet.

Table 1. Overview of parameters.

Modulation:	Quadrature Phase-Shift Keying (QPSK)
Center Frequency:	1.9 GHz
I/Q Rate:	500k Samples/s
Sample Rate:	400k Samples/s
Bandwidth:	200 kHz
Bits Per Symbol:	2
Samples Per Symbol:	8
Duration of one Frame:	50 ms

5.2. Packet Interface

In the implementation, time-sensitive information is carried through the packets. The structure of the packet interface is summarized in Figure 4. There are six guard bits at the beginning of a packet. These bits are placed to prevent the pulse shaping filter from damaging the message content. The synchronization bit field starts after the guard bits. A 30-bit synchronization sequence is known in advance by both the sending and receiving modules. This sequence is created by a LabVIEW function that generates pseudo-random bits in the Galois domain. Receivers that continuously acquire data from the air interface use the synchronization sequence to detect the beginning of the packet.

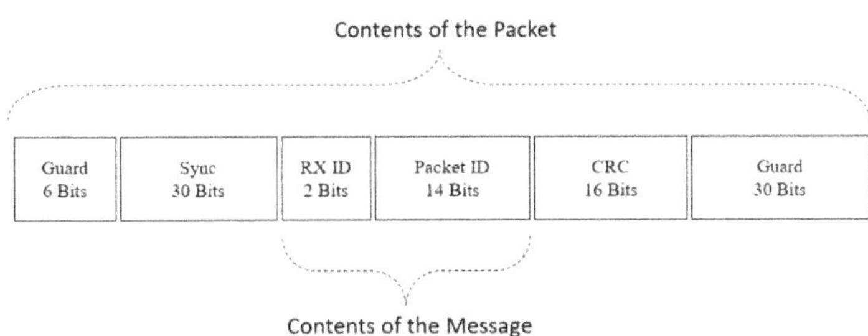

Figure 4. Packet content in the air interface.

The message field contains time-sensitive information. The message consists of Receiver ID (RX ID) and Packet ID fields. RX ID field is a 4-bit address that is used to distinguish receivers. Each receiver has a unique RX ID. When a receiver obtains a packet, it locates the RX ID field in the packet content and compares it with its RX ID. If the RX ID of the packet and the receiver do not match, the receiver discards the packet, and $\Delta_i(t)$ for that receiver increases by one for the next frame.

The frame number is the reference clock of the entire system. It initially starts from one at the beginning of the experiment and increases by one for each frame. Upon the generation of a packet, the Packet ID field is filled with the frame number of the system. Thus, the Packet ID field operates as the packet timestamp. Since the receiver also knows the current frame number, the difference between the packet's creation frame (contained in the Packet ID field) and the current frame gives the instant information age $\Delta_i(t)$.

Packets sent over an unreliable channel may suffer corruption due to noise. The receiver should discard packets containing incorrect information since processing this data may lead to incorrect AoI measurements. To detect errors, we use cyclic redundancy check (CRC). Within the packet generation process, we pass the message field through the 16-bit CRC and write the result to the CRC field. When a receiver obtains a packet, it first calculates the CRC of the message field of the packet and compares the result with the CRC field in the packet. If both CRCs are equal, we consider the message to be error-free. We track the number of successful CRC checks for each receiver, thereby dynamically measuring the reliability of the channel. In the implementation, we dynamically estimate channel reliability throughout the experiment. Accurate calculation of the channel reliability values is essential, as MW policy and WIP take this value as input. We pre-run the setup to initialize the channel reliabilities. During the pre-run, we discard AoI calculations.

5.3. Runtime

The LabVIEW program allows multi-threading, which allows us to execute processes independently in different threads. We implement the Receiver, Transmitter, and Logging modules as separate threads in the program. In this way, we were able to perform these operations simultaneously. Moreover, the LabVIEW program has the feature of providing synchronization between threads. With the activation of this feature, it has been possible to organize processes running in different threads and following each other. The runtime of the system can be described step by step as follows:

1. The new frame starts with incrementing the frame number.
2. The scheduling policy performs the receiver selection for the new frame. AoIs of the receivers, query arrival status, and channel statistics are the inputs of the scheduling policy.
3. In the meantime, receiver threads start acquiring a signal from receiver USRPs. The acquired signal is demodulated using LabVIEW's built-in demodulation function. We synchronize the transmit and receive threads using the synchronization function of the LabVIEW program. Moreover, we keep the receiver thread's acquisition duration

long enough compared to the transmission thread's time to complete its task so that the receiver can acquire the signal sent by the transmitter.

4. A time-sensitive packet is constructed in the transmitter thread. The ID of the selected receiver is inserted into the new packet. The current frame number is also inserted in the "Package ID" field.

5. The constructed packet is modulated using LabVIEW's Modulation function and prepared for transmission. The transmitter thread transmits this packet to the USRP over an Ethernet connection.

6. The transmitter USRP converts this packet to an RF signal and broadcasts it on the air interface.

7. The receiver threads demodulate the signal in the air interface and try to catch the transmitted packet. The demodulator tries to detect synchronization bits to find the beginning of the packet. If the demodulator finds the synchronization bits, it returns a bit field that contains the packet.

8. At the next stage, the receiver thread checks the CRC value of the acquired packet. The CRC field in the packet content is extracted. Next, the receiver thread passes the first 16 bits of the packet through the CRC. Then, the receiver compares this CRC result with the extracted CRC field in the packet. If both CRCs are equal, the receiver thread concludes that the packet is valid. Otherwise, the receiver discards the packet.

9. The total number of successful CRC checks for each receiver is used to calculate channel reliabilities.

10. After the CRC check, the receiver thread checks the Receiver ID field of the packet. If the Receiver ID field in the packet is different from the ID of the receiver, the receiver discards the packet again. If the Receiver IDs align, the acquired packet is assumed to be successfully received.

11. Receiver threads that have finished their processes are set to idle for a while. The receiver thread will start listening to the air interface again before the new frame starts to activate the receiving process before the transmission occurs.

12. Results obtained within a frame are passed to the logging thread. The main task of this thread is to calculate the average AoI, EAoI, and QAoI based on the result of the experiment. In addition, channel reliability and other statistics about the experiment are also calculated in this thread.

13. Before the frame ends, the transmitter thread calculates instantaneous AoIs of the receivers with the data received within the frame. In the next frame, AoIs of the receivers and channel statistics will be used as input to the scheduling policy.

This experimental procedure is repeated at each frame. After the overall experiment is finished, results are saved to a text file.

6. Experiments and Results

Throughout this section, we share the results that we obtained in the USRP environment and MATLAB simulations. We share the performances of AoI-aware policies in Section 6.1, EAoI-aware policies in Section 6.2, and QAoI-aware policies in Section 6.3.

6.1. Evaluation of AoI-Aware Scheduling Policies

In this section, we share the results of the experiments conducted in the SDR network. We evaluate the performances of AoI-aware scheduling policies, and compare their AoI performances with round-robin and greedy policies. Round-robin policy activates all links sequentially, one per frame, regardless of any prior knowledge obtained about receivers. Greedy policy uses the AoIs of the receivers and selects the receiver with the highest age for packet transmission.

We evaluate the scheduling policies under various conditions by changing the channel reliabilities of the receivers among experiments. To change channel reliabilities, we manipulate the gains of the receiver and transmitter USRPs. LabVIEW allows configuring the signal gains of USRPs. Moreover, we locate receiver USRPs with different distances to the transmitter USRP to induce diverse path losses to receiver USRPs. When the signal power of the transmitter USRP increases, all receivers in the network acquire stronger signals. Therefore, the channel reliabilities of all receivers increase. Throughout the experiments, we also adjust the power gains of the receivers to alter the channel reliabilities. The receiver's power gain is directly proportional to its channel reliability. Increasing the signal gain of a receiver reduces the error probability for that receiver and increases the channel reliability.

In the experiments, we run scheduling policies multiple times at each power gain level and take the average of the obtained results. We compare the scheduling policies in terms of the average AoI and the throughput of the network.

6.1.1. Adjusting the Gain of an Individual Receiver

In this case, we increase the input signal gain of an individual receiver USRP. Throughout the experiments, we test the policies ten times at each transmitter gain level and average the results of redundant experiments. In each experiment, the frame length is $K = 7500$, and $M = 3$ receivers are available in the network. Results of the experiments in terms of AoI and throughput are given in Figure 5. Average channel reliabilities for each USRP gain level are given in Table 2.

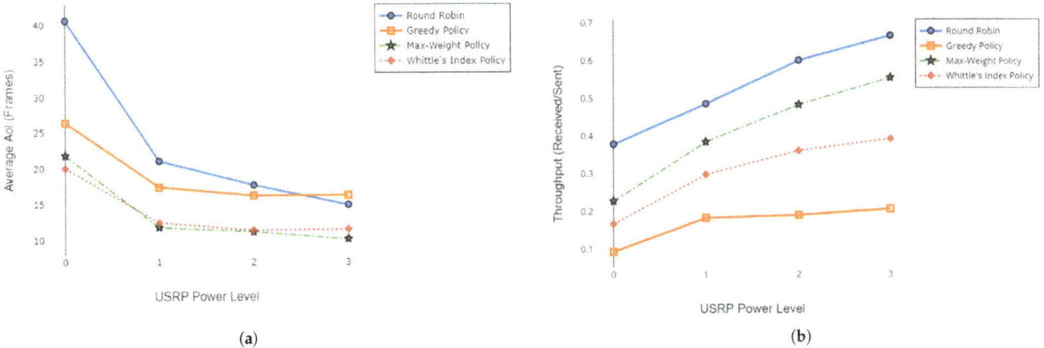

Figure 5. Evaluation of average AoI J_A (**a**) and throughput (**b**) with varying receiver gain (SDR testbed).

Table 2. Channel statistics in the first experiment set.

Experiment Index	p_1	p_2	p_3	Coefficient of Variation C_V among Channels
0	0.9997	0.0517	0.0779	0.84
1	0.9997	0.3698	0.078	1.16
2	0.9997	0.7135	0.0747	1.337
3	0.9997	0.9139	0.0795	1.361

6.1.2. Adjusting the Gain of the Base Station

In this case, we increase the output signal gain of the transmitter USRP (base station). Throughout the experiments, we test the policies five times at each transmitter gain level and average the results of redundant experiments. In each experiment, the frame length is $K = 7500$, and $M = 3$ receivers are active in the network. Average AoI and throughput of age-aware policies are illustrated in Figure 6, and the channel statistics for the experimental setup are given in Table 3.

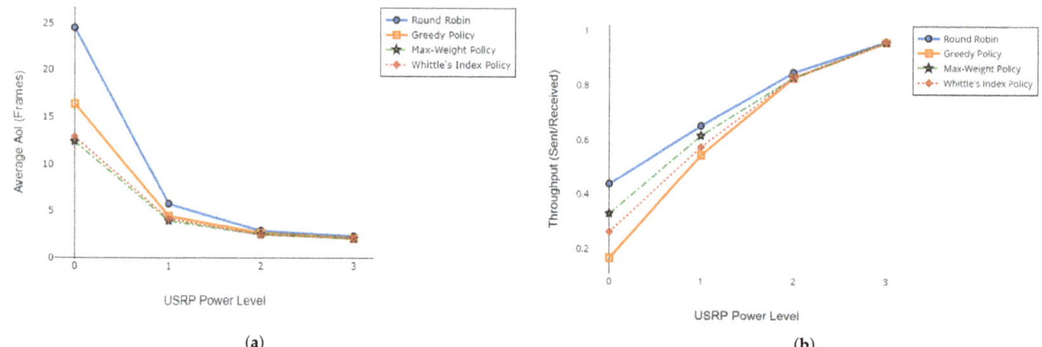

Figure 6. Evaluation of average AoI J_A (**a**) and throughput (**b**) with varying BS output gain (SDR testbed).

Table 3. Channel statistics in the second experiment set.

Experiment Index	p_1	p_2	p_3	Coefficient of Variation C_V among Channels
0	0.9997	0.0814	0.2317	0.988
1	0.9997	0.3566	0.5891	0.496
2	0.9997	0.6811	0.8587	0.196
3	0.9997	0.9055	0.9733	0.051

6.1.3. Comparison of SDR Testbed Results with Simulations

In this section, we share the results of the comparison between simulation and realization. We use the results of the experiment mentioned in Section 6.1.2 as a reference to the simulation. We use the same channel reliabilities from Table 3 for the simulation environment and evaluate the policies. Results of the comparison in terms of average AoI and throughput are given in Figure 7.

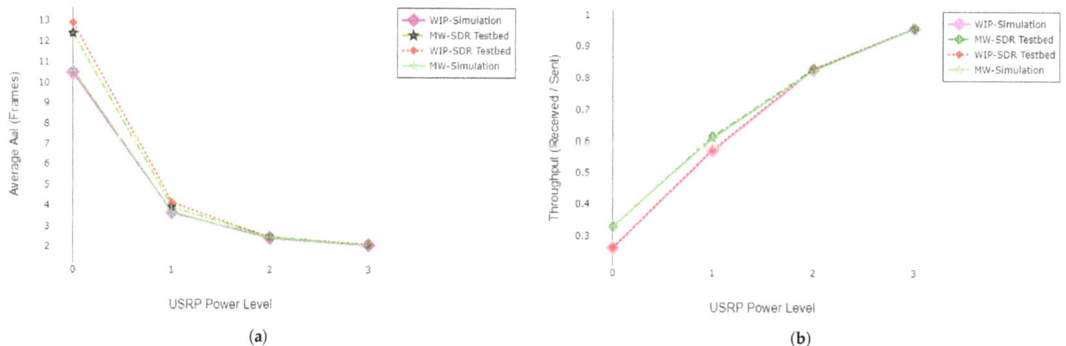

Figure 7. Comparison of simulation and implementation in terms of average AoI J_A (**a**) and throughput (**b**).

6.1.4. Interpretation of the Results

As channel reliability decreases, the performances of MW and WIP differ positively from the others. MW and WIP policies take channel reliability into account in the scheduling decision. This information enables more efficient use of transmission attempts. On the other hand, greedy policy does not utilize channel reliability information. If a receiver with a very low-quality channel is present in the network, the greedy policy may block the network by continuous unsuccessful update attempts to that receiver. This results in an increase in the average AoI of the network. For the first experiment set with the results illustrated in Figure 5, the performance degradation of the greedy algorithm is more apparent. Greedy policy always tries to send an update packet to the receiver with the worst channel condition. However, that receiver rarely receives packets successfully, and the base station gets stuck in that receiver until a successful packet reception. On the other hand, since the round-robin policy proceeds by transmitting to all receivers one by one without using any information about whether the packet is successfully received or not, the starvation problem does not occur. In both experiments, we also observe that as channel reliability values of receivers improve and asymmetry of channels decreases, greedy policy performs better than round-robin. As the channel conditions improve and the asymmetry among the channels decreases, performances of both policies converge to the optimal. As the channel reliability rises to 100%, all scheduling policies behave like round-robin and transmit to all receivers in a cyclic order.

For the SDR testbed simulation comparison case, we use the same average channel reliabilities in both experiments. We do not observe any significant difference in throughput, as expected. However, in terms of AoI, we found that the simulation results yield lower AoI than the SDR implementation. In the simulation environment, the channel status is a Bernoulli random variable. However, in the SDR implementation, the channel status is formed by realistic conditions and doesn't have to be stationary or follow the Bernoulli distribution. The regularity of the packet arrivals is an essential factor for low AoI. Even if the channel reliabilities over time are equal for SDR realization and simulation environments, the imperfections of the realistic channel may reduce the update regularity more drastically than the Bernoulli-distributed channel.

6.2. Evaluation of EAoI-Aware Scheduling Policies

In this section, we compare the EAoI-aware policies with the traditional policies. Traditional policies do not utilize query statistics for scheduling decisions, and we aim to observe the outcomes of using query statistics. We evaluate the policies in the SDR environment and use EAoI as the primary performance indicator. We also share results about AoI and throughput metrics. Throughout the experiments, query presences at each frame are implemented as i.i.d. Bernoulli random variables. In each experiment, the frame length is $K = 7500$, and $M = 3$ receivers are active in the network. We use the proactive serving method as the query response scenario.

6.2.1. Adjusting the Gain of an Individual Receiver

In this case, we increase the output signal gain of an individual receiver USRP. Throughout the experiments, we test the policies ten times for each gain level and average the results of redundant experiments. Evaluation of EAoI and AoI throughout the experiments are given in Figure 8. Channel statistics corresponding to USRP gain levels are given in Table 4 and query statistics are given in Table 5.

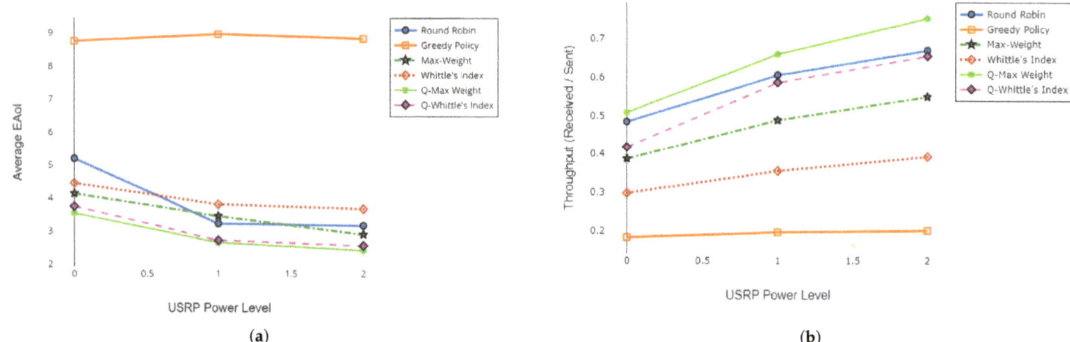

Figure 8. Evaluation of effective AoI J_E (**a**) and throughput (**b**) with varying input gain of second receiver (SDR testbed).

Table 4. Channel statistics.

Gain	p_1	p_2	p_3
0	0.999	0.363	0.078
1	0.999	0.735	0.076
2	0.999	0.930	0.077

Table 5. Query statistics.

q_1	q_2	q_3
0.9	0.9	0.1

6.2.2. Adjusting the Gain of the Base Station

In this case, we increase the output signal gain of the transmitter USRP (base station). Throughout the experiments, we test the policies at least five times for each gain level and average the results of redundant experiments. Evaluation of EAoI and AoI throughout the experiments are given in Figure 9. Channel statistics corresponding to USRP gain levels are given in Table 6 and query statistics are given in Table 7.

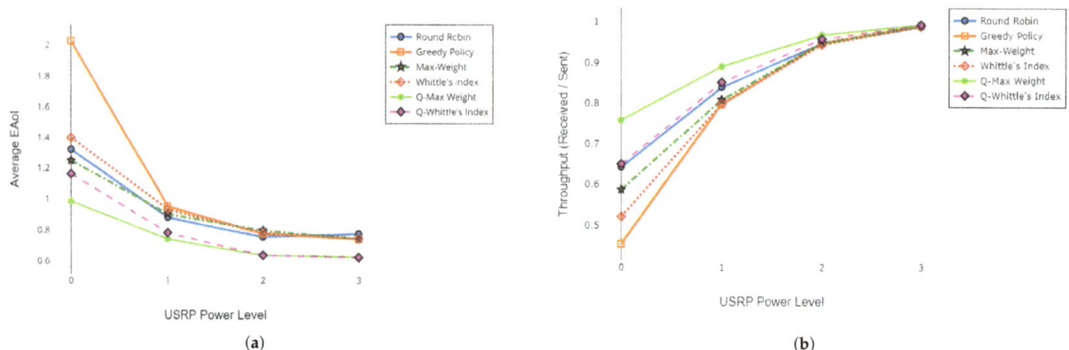

Figure 9. Evaluation of effective AoI J_E (**a**) and throughput (**b**) with varying output gain of BS (SDR testbed).

Table 6. Channel statistics.

Gain	p_1	p_2	p_3
0	0.9997	0.6652	0.2439
1	0.9997	0.9073	0.5983
2	0.9997	0.9830	0.8651
3	0.9997	0.9980	0.9736

Table 7. Query statistics.

q_1	q_2	q_3
0.9	0.1	0.1

6.2.3. Interpretation of the Results

For the EAoI minimization objective, the EAoI-MW and EAoI-WI policies outperform the policies that do not utilize query information. Moreover, experimental results show that EAoI-MW surpasses the EAoI-WIP in terms of throughput. For EAoI-aware scheduling policies, whether the policy is derived for the instantaneous serving or the proactive serving scenario does not cause a significant difference in EAoI performance. Rather than utilizing the exact timings of the query arrivals, EAoI-aware policies weight receivers according to their long-term query arrival statistics. Since there is no significant difference between the proactive response and instant response scenarios in the long-term query arrival statistics, there is no significant difference between the performances of the policies. As can be seen from Figures 8 and 9, the EAoI performances of EAoI-MW derived for the instantaneous response scenario and the EAoI-WIP derived for the proactive response scenario are very close to each other.

6.3. Evaluation of QAoI-Aware Scheduling Policies

In this section, we share the results of our experiments. We conducted the experiments in the simulation environment and the SDR environment. Throughout the experiments, we evaluated the performance of the QAoI-aware MW policy in terms of QAoI and AoI, and we used the AoI-aware MW policy as a benchmark.

6.3.1. Results from Simulation Environment

We conducted four experiments in the MATLAB environment. In each experiment, the frame length was $K = 1,100,000$, and $M = 10$ receivers were active in the network. Within the experiments, we adjusted the query period of the receivers and observed the result of this increment from the AoI and QAoI perspectives. We initialized query periods to prevent the overlap of the query frames for each receiver. We assume $a_i(t)$ is stationary through time by taking advantage of non-overlapping queries, and we calculate f_i as $f_i = 1 - p_i$ in Q-MW policy. Channel reliabilities (long-term average packet success rates) measured in the experiments are summarized in Table 8. Average QAoI and AoI obtained by Q-MW policy for each experiment are given in Figures 10 and 11, respectively.

Table 8. Channel statistics for simulations.

Experiment Index	p_1	p_2	p_3	p_4	p_5	p_6	p_7	p_8	p_9	p_{10}
1	0.9	0.9	0.9	0.9	0.9	0.9	0.9	0.9	0.9	0.9
2	0.9	0.9	0.9	0.9	0.9	0.1	0.1	0.1	0.1	0.1
3	0.1	0.1	0.1	0.1	0.1	0.1	0.1	0.1	0.1	0.1
4	0.1	0.2	0.3	0.4	0.5	0.6	0.7	0.8	0.9	1

(**a**) Experiment-1

(**b**) Experiment-2

(**c**) Experiment-3

(**d**) Experiment-4

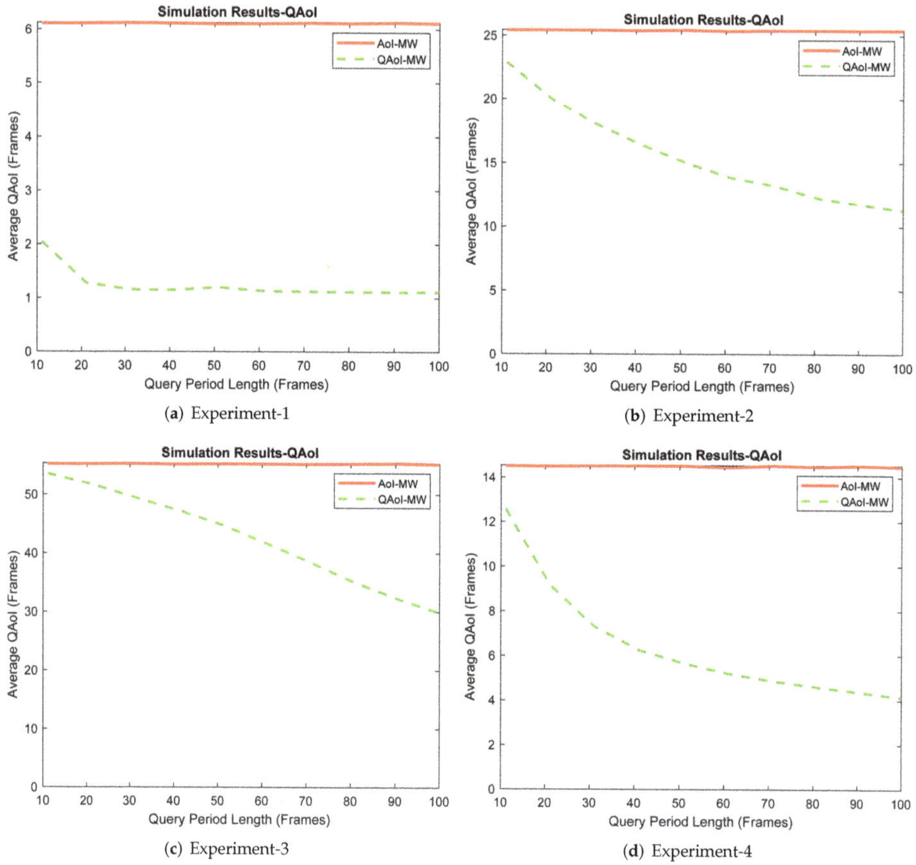

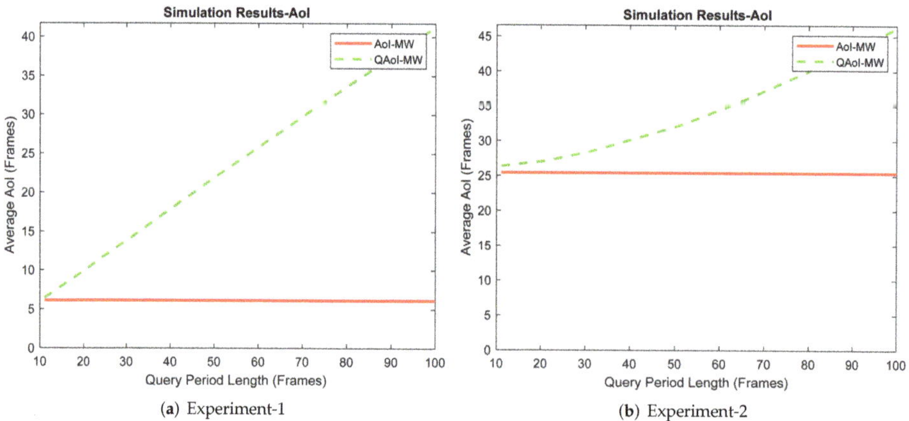

Figure 10. Evaluation of Q-MW policy in terms of average QAoI (J_Q).

(**a**) Experiment-1

(**b**) Experiment-2

Figure 11. *Cont.*

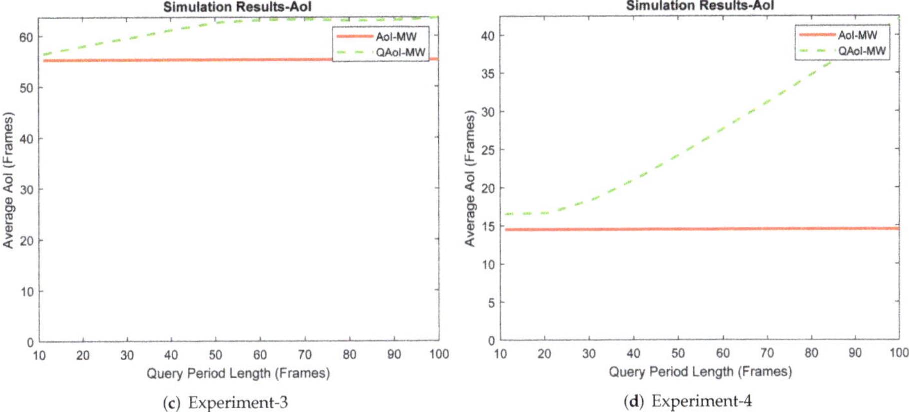

(c) Experiment-3   (d) Experiment-4

Figure 11. Evaluation of Q-MW policy in terms of average AoI (J_A).

6.3.2. Results from the USRP Testbed

We conducted two experiment sets in the SDR testbed. In both experiment sets, we increased the output signal gain of the transmitter USRP (base station) to observe the effects of various channel reliabilities (i.e., packet success rates). For each signal gain level, we test the policies at least ten times and average the results of redundant experiments. The frame length of each test was $K = 7500$, and there were $M = 3$ receivers in the network. In the first experiment set, the query period of receivers was 25, and in the second experiment set, the query period of the receivers was 5. In both experiment sets, we initialize the query periods to prevent the arrival of multiple queries within the same frame. We assume $a_i(t)$ is stationary through time by taking advantage of non-overlapping queries, and we calculate f_i as $f_i = 1 - p_i$ in Q-MW policy.

For the first experiment set, evaluation of QAoI and AoI are given in Figures 12 and 13, respectively. Channel reliabilities corresponding to USRP gain levels are given in Table 9. For the second experiment set, evaluation of QAoI and AoI are given in Figures 14 and 15, respectively. Channel reliabilities corresponding to USRP gain levels are given in Table 10.

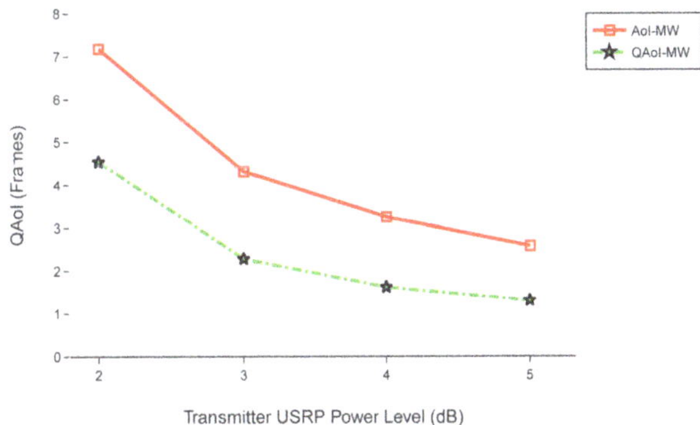

Figure 12. Evaluation of QAoI (J_Q) for varying power levels of transmitter USRP, 25 frames length query period.

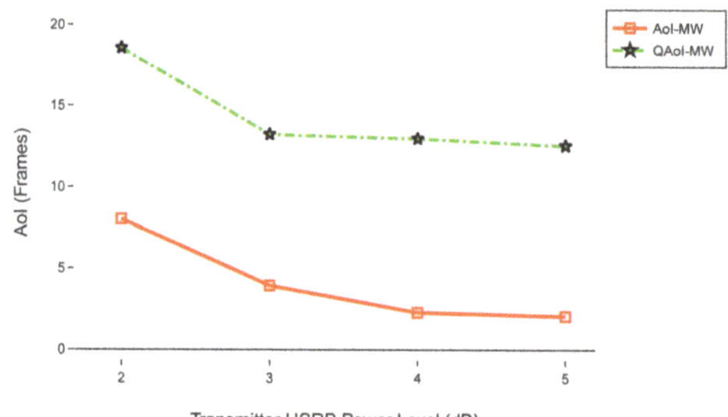

Figure 13. Evaluation of AoI (J_A) for varying power levels of transmitter USRP, 25 frames length query period.

Table 9. Channel statistics.

USRP Power Level	p_1	p_2	p_3
2	0.9997	0.2331	0.1681
3	0.9997	0.4203	0.3175
4	0.9997	0.5567	0.4775
5	0.9997	0.7245	0.6589

In the second experiment, we increase the output signal gain of the transmitter USRP (base station). Throughout the experiments, we test the policies at least ten times for each gain level and average the results of redundant experiments. In each experiment, the frame length is $K = 7500$, and $M = 3$ receivers are active in the network. In this experiment, the query period for each receiver is 5 frames. We initialize the query periods such that the queried frames of receivers do not overlap at the same frame.

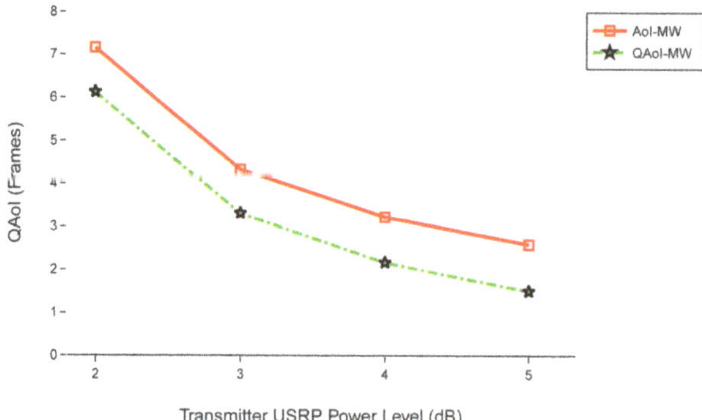

Figure 14. Evaluation of QAoI for varying power levels of transmitter USRP.

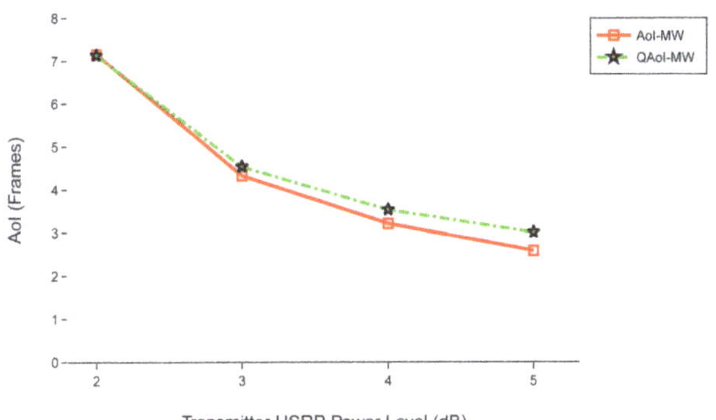

Figure 15. Evaluation of AoI for varying power levels of transmitter USRP.

Table 10. Channel statistics.

USRP Power Level	p_1	p_2	p_3
2	0.9997	0.2380	0.1687
3	0.9997	0.4109	0.3167
4	0.9997	0.5721	0.4796
5	0.9996	0.7299	0.6635

6.3.3. Comparison of SDR Testbed Results with Simulations

In this section, we share the results of the comparison between simulation and realization. We use the results of the experiment illustrated in Figure 14 as a reference for the simulation. We use the same channel reliabilities from Table 10 for the simulation environment. Results of the comparison in terms of QAoI are given in Figure 16.

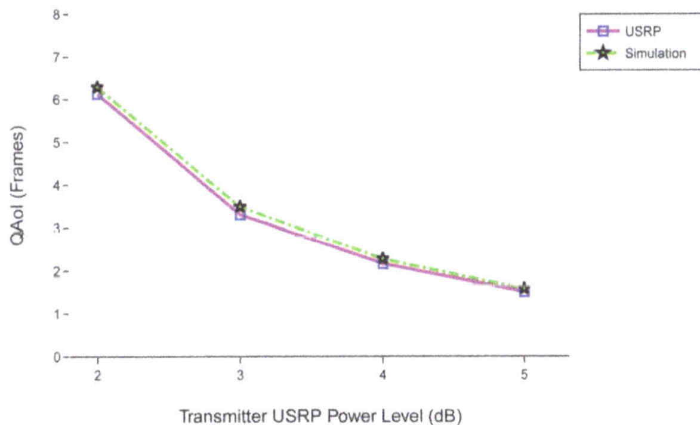

Figure 16. Comparison of simulation and realization results.

6.3.4. Interpretation of the Results

Within the scope of the experiments, we studied the case where query arrivals are periodic. According to the results of both SDR realization and simulations, the Q-MW policy outperforms the AoI-MW policy for the QAoI minimization objective. By utilizing the query arrival information, the Q-MW scheduling policy can select receivers more efficiently, and thus it can exhibit superior QAoI performance compared to AoI-MW.

Throughout the simulations, we investigated Q-MW in networks with various channel reliabilities. In the first experiment, we considered ten receivers with good channel reliability. According to the results of this experiment, in cases where the query periods of the receivers do not overlap, the QAoI policy can reduce the average QAoI in the network to approximately one, which is the lowest possible limit. In the first experiment, the expected number of attempts to update a receiver is close to one. Having a reduced number of attempts enables the scheduling policy to distribute scheduling decisions more effectively and eases the alignment of the scheduling decisions with the query periods. In the second experiment, all receivers have poor channel qualities,and the number of attempts needed to update a receiver is high. As the channel reliabilities decrease, the expected number of attempts to update a receiver increases, and aligning the scheduling decisions with the query periods become more challenging. In this case, the performance of query-aware policies is reduced. As the query period increases, queried frames of the receivers become more distant from each other, which positively affects the performance of query-aware policies.

The fact that the transmission can only be allocated to a single receiver in each frame is one of the most fundamental limitations of the network. The QAoI-aware MW policy we recommend, on the other hand, reduces the need for packet transmission in the network by taking into account the query periods of receivers' timely information requests and eases the transmission allocation constraint in line with the query periods.

7. Conclusions and Future Work

Within the scope of the paper, we have examined the AoI, EAoI, and QAoI, which are semantic communication metrics that prioritize information freshness. We implemented a multi-user wireless network with SDRs to examine these metrics in real-world scenarios. We investigated the performance of AoI-aware scheduling policies by comparing their AoI performance with traditional scheduling policies. The emulation results reveal that the WI and MW policies are superior to the round-robin and greedy policies, as they exploit the information on the link reliabilities and AoIs of the receivers. Experimental results obtained in the SDR testbed are close to simulation results when packet drops are rare, but as the link reliabilities decrease, they begin to show some slight discrepancies. We attribute this to the following: the AoI-aware policies adopted in this work were derived under Bernoulli-distributed packet drops; however, as channels get poorer, the sequence of packet drops tends to acquire a memory.

We have also studied the Effective AoI and Query AoI metrics to examine the freshness of information from the perspective of the query source in pull-based networks. For the EAoI domain, we proposed EAoI-MW policy by leveraging the formerly defined AoI-aware policies. We implemented and tested the EAoI-MW and EAoI-WI policies on the SDR Network. Experiment results show that utilizing the statistical information about the query process significantly improves EAoI performance. We observed that EAoI-MW policy exhibits a comparable performance with EAoI-WI and yields better results throughput. For the Query AoI metric, we have proposed a scheduling policy by adapting the max-weight policy to the QAoI case for multi-user pull-based network scenarios. We tested the resulting Q-MW policy in simulation and SDR implementation environments. Results reveal that utilizing the Q-MW policy can reduce QAoI significantly compared to AoI-aware policies.

In future studies, we seek to examine different semantic metrics beyond AoI. To this end, we want to expand the scope of our work on the QAoI. We also aim to investigate and optimize the Q-MW policy for the stochastic query arrival scenarios.

Author Contributions: Conceptualization, T.K.O., E.T.C., E.U., and T.G.; methodology, T.K.O., E.T.C., and E.U.; software, T.K.O.; validation, T.K.O.; formal analysis, T.K.O., E.T.C., and E.U.; investigation, T.K.O., E.T.C., E.U., and T.G.; resources, T.K.O., E.T.C., E.U., and T.G.; data curation, T.K.O., E.T.C.; writing—original draft preparation, T.K.O.; writing—review and editing, E.T.C., E.U., and T.G.; visualization, T.K.O., E.T.C., E.U., and T.G.; supervision, E.T.C., E.U., and T.G.; project administration, E.U.; funding acquisition, E.U. All authors have read and agreed to the published version of the manuscript.

Funding: This work has been supported by TUBITAK-BIDEB Grant 119C028.

Institutional Review Board Statement: Not applicable.

Informed Consent Statement: Not applicable.

Data Availability Statement: Not applicable.

Conflicts of Interest: The authors declare no conflict of interest.

Abbreviations

The following abbreviations are used in this manuscript:

AoI	Age of information
WIP	Whittle's index policy
EAoI-WIP	EAoI-aware Whittle's index policy
MWP	Max-weight policy
EAoI-MW	EAoI-aware max-weight policy
Q-MW	QAoI-aware max-weight policy
QAoI	Query age of information
EAoI	Effective age of information
MAB	Multi-armed bandit
RMAB	Restless multi-armed bandit
MDP	Markov decision process
SDR	Software-defined radio
USRP	Universal software radio peripheral
PSK	Phase Shift Keying

Appendix A

The framework of the Lyapunov optimization aims to minimize the Lyapunov function at every time instant t. Lyapunov function at time t is given in (A1).

$$L(t) = \frac{1}{M} \sum_{i=1}^{M} \Delta_{q_i}^2(t) \tag{A1}$$

At frame transitions, each receiver causes a drift in L. For the EAoI concept, we focus on the drift between consecutive frames. The drift is associated with the receiver's AoI evolution. For each receiver, calculation of the Lyapunov drift $Y_i(t)$ between the frames t and $t+1$ is given in (A2).

$$Y_i(t) = \mathbb{E}\left[\Delta_{q_i}^2(t+1) - \Delta_{q_i}^2(t)\right] = \mathbb{E}\left[d_i(t)\Delta_i^2(t+1) - d_i(t)\Delta_i^2(t)\right] \tag{A2}$$

Since we assume that policies are non-anticipative, which means the policies do not have information about future channel or query status, we can argue that $\Delta_i(t)$ and $d_i(t)$ are independent.

$$\mathbb{E}[d_i(t)\Delta_i(t)] = \mathbb{E}[d_i(t)]\mathbb{E}[\Delta_i(t)] \tag{A3}$$

We rewrite the Lyapunov drift using $\mathbb{E}[d_i(t)] = q_i$ and $\mathbb{E}[d_i(t+1)] = q_i$.

$$Y_i(t) = \mathbb{E}\left[q_i\Delta_i^2(t+1) - q_i\Delta_i^2(t)\right] \tag{A4}$$

After this modification, the derivation process becomes identical with [5]. We write the transition of $\Delta_i(t)$ between consecutive frames in (A5).

$$\begin{aligned}\Delta_i(t+1) &= u_i(t) + (1 - u_i(t))(\Delta_i(t) + 1) \\ &= a_i(t)c_i(t) + (1 - a_i(t)c_i(t))(\Delta_i(t) + 1)\end{aligned} \tag{A5}$$

Then, we rewrite the Lyapunov drift by expressing $\Delta_i(t+1)$ in terms of $\Delta_i(t)$.

$$Y_i(t) = \mathbb{E}\left[q_i[u_i(t) + (1 - u_i(t))(\Delta_i(t) + 1)]^2 - q_i\Delta_i^2(t)\right] \tag{A6}$$

Since $u_i(t)$ is 0-or-1 variable, we can argue that $u_i^2(t) = u_i(t)$, $(1 - u_i(t))^2 = (1 - u_i(t))$ and $u_i(t)(1 - u_i(t)) = 0$. With these simplifications, we can rewrite $Y_i(t)$.

$$
\begin{aligned}
Y_i(t) =& \mathbb{E}\left[q_i\left[u_i(t) + (1 - u_i(t))(\Delta_i(t) + 1)^2\right] - q_i\Delta_i^2(t)\right] \\
=& \mathbb{E}\left[q_i\left[\Delta_i^2(t) + 2\Delta_i(t) + 1 - u_i(t)\Delta_i^2(t) - 2u_i(t)\Delta_i(t)\right] - q_i\Delta_i^2(t)\right] \\
=& \mathbb{E}\left[q_i\left[2\Delta_i(t) + 1 - u_i(t)\Delta_i^2(t) - 2u_i(t)\Delta_i(t)\right]\right] \\
=& q_i\left[2\Delta_i(t) + 1 - \mathbb{E}[u_i(t)]\Delta_i^2(t) - 2\mathbb{E}[u_i(t)]\Delta_i(t)\right] \\
=& \left[2q_i\Delta_i(t) + q_i - q_ip_i\mathbb{E}[a_i(t)]\Delta_i^2(t) - 2q_ip_i\mathbb{E}[a_i(t)]\Delta_i(t)\right]
\end{aligned}
\tag{A7}
$$

The $a_i(t)$ is the decision variable that we can select zero or one. We only aim to investigate the effect of changing $a_i(t)$. Since the results of other terms in $Y_i(t)$ do not change as $a_i(t)$ change, we omit them and focus on the terms that have $a_i(t)$ as a coefficient.

$$
\begin{aligned}
\hat{Y}_i(t) =& - q_ip_i\mathbb{E}[a_i(t)]\Delta_i^2(t) - 2q_ip_i\mathbb{E}[a_i(t)]\Delta_i(t) \\
=& -\mathbb{E}[a_i(t)]\left(q_ip_i\left(\Delta_i^2(t) + 2\Delta_i(t)\right)\right) \\
=& -\mathbb{E}[a_i(t)]C_i(t)
\end{aligned}
\tag{A8}
$$

$$
C_i(t) = q_ip_i\left(\Delta_i^2(t) + 2\Delta_i(t)\right)
\tag{A9}
$$

The max-weight policy aims to minimize the average Lyapunov drift by selecting the receiver i with maximum $C_i(t) = q_ip_i\left(\Delta_i^2(t) + 2\Delta_i(t)\right)$ at every frame t.

References

1. Uysal, E.; Kaya, O.; Ephremides, A.; Gross, J.; Codreanu, M.; Popovski, P.; Assaad, M.; Liva, G.; Munari, A.; Soleymani, T.; et al. Semantic Communications in Networked Systems. *arXiv* **2021**, arXiv:2103.05391.
2. Kaul, S.; Gruteser, M.; Rai, V.; Kenney, J. Minimizing age of information in vehicular networks. In Proceedings of the IEEE Communications Society Conference on Sensor, Mesh and Ad Hoc Communications and Networks (SECON), Rome, Italy, 6–9 July 2021; pp. 350–358.
3. Yates, R.D. Lazy is timely: Status updates by an energy harvesting source. In Proceedings of the 2015 IEEE International Symposium on Information Theory (ISIT), Hong Kong, China, 14–19 June 2015; pp. 3008–3012. [CrossRef]
4. Sun, Y.; Uysal-Biyikoglu, E.; Yates, R.D.; Koksal, C.E.; Shroff, N.B. Update or Wait: How to Keep Your Data Fresh. *IEEE Trans. Inf. Theory* **2017**, *63*, 7492–7508. [CrossRef]
5. Kadota, I.; Sinha, A.; Uysal-Biyikoglu, E.; Singh, R.; Modiano, E. Scheduling Policies for Minimizing Age of Information in Broadcast Wireless Networks. *IEEE/ACM Trans. Netw.* **2018**, *26*, 2637–2650. [CrossRef]
6. He, Q.; Yuan, D.; Ephremides, A. Optimal Link Scheduling for Age Minimization in Wireless Systems. *IEEE Trans. Inf. Theory* **2018**, *64*, 5381–5394. [CrossRef]
7. Kadota, I.; Uysal-Biyikoglu, E.; Singh, R.; Modiano, E. Minimizing Age of Information in Broadcast Wireless Networks. In Proceedings of the 54th Annual Allerton Conference on Communication, Control, and Computing, Monticello, IL, USA, 28–30 September 2016.
8. Talak, R.; Kadota, I.; Karaman, S.; Modiano, E. Scheduling Policies for Age Minimization in Wireless Networks with Unknown Channel State. In Proceedings of the 2018 IEEE International Symposium on Information Theory (ISIT), Vail, CO, USA, 17–22 June 2018; pp. 2564–2568. [CrossRef]
9. Beytur, H.B.; Uysal, E. Age Minimization of Multiple Flows using Reinforcement Learning. In Proceedings of the 2019 International Conference on Computing, Networking and Communications (ICNC), Honolulu, HI, USA, 18–21 February 2019; pp. 339–343. [CrossRef]
10. Ceran, E.T.; Gündüz, D.; György, A. A Reinforcement Learning Approach to Age of Information in Multi-User Networks with HARQ. *IEEE J. Sel. Areas Commun.* **2021**, *39*, 1412–1426. [CrossRef]
11. Kadota, I.; Modiano, E. Minimizing the Age of Information in Wireless Networks with Stochastic Arrivals. *IEEE Trans. Mob. Comput.* **2021**, *20*, 1173–1185. [CrossRef]
12. Kaul, S.K.; Yates, R.D.; Gruteser, M. Status updates through queues. In Proceedings of the 2012 46th Annual Conference on Information Sciences and Systems (CISS), Princeton, NJ, USA, 21–23 March 2012; pp. 1–6. [CrossRef]
13. Yates, R.D.; Kaul, S.K. The Age of Information: Real-Time Status Updating by Multiple Sources. *IEEE Trans. Inf. Theory* **2016**, *65*, 1807–1827. [CrossRef]

14. Sert, E.; Sönmez, C.; Baghaee, S.; Uysal-Biyikoglu, E. Optimizing age of information on real-life TCP/IP connections through reinforcement learning. In Proceedings of the 2018 26th Signal Processing and Communications Applications Conference (SIU), Izmir, Turkey, 2–5 May 2018; pp. 1–4. [CrossRef]
15. Sönmez, C.; Baghaee, S.; Ergişi, A.; Uysal-Biyikoglu, E. Age-of-Information in Practice: Status Age Measured Over TCP/IP Connections Through WiFi, Ethernet and LTE. In Proceedings of the 2018 IEEE International Black Sea Conference on Communications and Networking (BlackSeaCom), Batumi, Georgia, 4–7 June 2018; pp. 1–5. [CrossRef]
16. Beytur, H.B.; Baghaee, S.; Uysal, E. Towards AoI-aware Smart IoT Systems. In Proceedings of the 2020 International Conference on Computing, Networking and Communications (ICNC), Big Island, HI, USA, 17–20 February 2020; pp. 353–357. [CrossRef]
17. Guloglu, U.; Baghaee, S.; Uysal, E. Evaluation of Age Control Protocol (ACP) and ACP+ on ESP32. In Proceedings of the 2021 17th International Symposium on Wireless Communication Systems (ISWCS), Berlin, Germany, 6–9 September 2021; pp. 1–6. [CrossRef]
18. Kadota, I.; Rahman, M.S.; Modiano, E. WiFresh: Age-of-Information from Theory to Implementation. In Proceedings of the 2021 International Conference on Computer Communications and Networks (ICCCN), Athens, Greece, 19–22 July 2021.
19. Ayan, O.; Özkan, H.; Kellerer, W. An Experimental Framework for Age of Information and Networked Control via Software-Defined Radios. In Proceedings of the IEEE International Conference on Communications (ICC), Montreal, QC, Canada, 14–23 June 2021; pp. 1–6.
20. Han, Z.; Liang, J.; Gu, Y.; Chen, H. Software-Defined Radio Implementation of Age-of-Information-Oriented Random Access. In Proceedings of the IECON 2020 The 46th Annual Conference of the IEEE Industrial Electronics Society, Singapore, 18–21 October 2020; pp. 4374–4379. [CrossRef]
21. Oğuz, T.K.; Uysal, E.; Girici, T. Implementation and Evaluation of Age-Aware Downlink Scheduling Policies. In Proceedings of the 2021 IEEE International Black Sea Conference on Communications and Networking, Bucharest, Romania, 24–28 May 2021; pp. 1–3. [CrossRef]
22. Kam, C.; Kompella, S.; Ephremides, A. Experimental evaluation of the age of information via emulation. In Proceedings of the MILCOM 2015—2015 IEEE Military Communications Conference, Tampa, FL, USA, 26–28 Octpber 2015; pp. 1070–1075. [CrossRef]
23. Uysal, E.; Kaya, O.; Baghaee, S.; Beytur, H.B. Age of Information in Practice. *arXiv* **2021**, arXiv:2106.02491.
24. Maatouk, A.; Assaad, M.; Ephremides, A. The Age of Incorrect Information: An Enabler of Semantics-Empowered Communication. *arXiv* **2021**, arXiv:2012.13214.
25. Kam, C.; Kompella, S.; Ephremides, A. Age of Incorrect Information for Remote Estimation of a Binary Markov Source. In Proceedings of the IEEE INFOCOM 2020—IEEE Conference on Computer Communications Workshops (INFOCOM WKSHPS), Toronto, ON, Canada, 6–9 July 2020; pp. 1–6. [CrossRef]
26. Maatouk, A.; Kriouile, S.; Assaad, M.; Ephremides, A. The Age of Incorrect Information: A New Performance Metric for Status Updates. *IEEE/ACM Trans. Netw.* **2020**, *28*, 2215–2228. [CrossRef]
27. Li, F.; Sang, Y.; Liu, Z.; Li, B.; Wu, H.; Ji, B. Waiting But Not Aging: Optimizing Information Freshness Under the Pull Model. *IEEE/ACM Trans. Netw.* **2021**, *29*, 465–478. [CrossRef]
28. Chiariotti, F.; Holm, J.; Kalør, A.E.; Soret, B.; Jensen, S.K.; Pedersen, T.B.; Popovski, P. Query Age of Information: Freshness in Pull-Based Communication. *IEEE Trans. Commun.* **2022**, *70*, 1606–1622. [CrossRef]
29. Yin, B.; Zhang, S.; Cheng, Y.; Cai, L.X.; Jiang, Z.; Zhou, S.; Niu, Z. Only Those Requested Count: Proactive Scheduling Policies for Minimizing Effective Age-of-Information. In Proceedings of the IEEE INFOCOM 2019—IEEE Conference on Computer Communications, Paris, France, 29 April–2 May 2019; pp. 109–117. [CrossRef]
30. Ildiz, M.E.; Yavascan, O.T.; Uysal, E.; Kartal, O.T. Pull or Wait: How to Optimize Query Age of Information. *arXiv* **2021**, arXiv:2111.02309.
31. Neely, M. *Stochastic Network Optimization with Application to Communication and Queueing Systems*; Morgan & Claypool: San Rafael, CA, USA, 2010.
32. Mitola, J. The software radio architecture. *IEEE Commun. Mag.* **1995**, *33*, 26–38. [CrossRef]
33. Cruz, P.; Carvalho, N.B.; Remley, K.A. Designing and Testing Software-Defined Radios. *IEEE Microw. Mag.* **2010**, *11*, 83–94. [CrossRef]
34. NI. *What Is NI USRP Hardware?* White Papers; Technical Report; National Instruments. Available online: https://www.ni.com/en-tr/innovations/white-papers/11/what-is-ni-usrp-hardware-.html (accessed on 22 March 2022).
35. NI. USRP-2920 Software Defined Radio Device Specifications. Available online: https://www.ni.com/pdf/manuals/375839c.pdf (accessed on 22 March 2022).
36. Packet-Based Digital Link. National Instruments Community. Available online: https://forums.ni.com/t5/Software-Defined-Radio/Packet-based-Digital-Link/ta-p/3509866 (accessed on 22 March 2022).
37. NI. *Acquiring an Analog Signal: Bandwidth, Nyquist Sampling Theorem, and Aliasing*; White Papers; Technical Report; National Instruments. Available online: https://www.ni.com/en-tr/innovations/white-papers/06/acquiring-an-analog-signal--bandwidth--nyquist-sampling-theorem-.html (accessed on 22 March 2022).

 *entropy*

Article

Age of Information Minimization for Radio Frequency Energy-Harvesting Cognitive Radio Networks

Juan Sun [1], Shubin Zhang [1], Changsong Yang [2] and Liang Huang [1,*]

[1] College of Computer Science and Technology, Zhejiang University of Technology, Hangzhou 310014, China; sunjuan@zjut.edu.cn (J.S.); zhangshubin@zjut.edu.cn (S.Z.)
[2] Guangxi Key Laboratory of Cryptography and Information Security, Guilin University of Electronic Technology, Guilin 541004, China; csyang@guet.edu.cn
* Correspondence: lianghuang@zjut.edu.cn

Abstract: The Age of Information (AoI) measures the freshness of information and is a critic performance metric for time-sensitive applications. In this paper, we consider a radio frequency energy-harvesting cognitive radio network, where the secondary user harvests energy from the primary users' transmissions and opportunistically accesses the primary users' licensed spectrum to deliver the status-update data pack. We aim to minimize the AoI subject to the energy causality and spectrum constraints by optimizing the sensing and update decisions. We formulate the AoI minimization problem as a partially observable Markov decision process and solve it via dynamic programming. Simulation results verify that our proposed policy is significantly superior to the myopic policy under different parameter settings.

Keywords: Age of Information; RF energy-harvesting; cognitive radio network; dynamic programming

Citation: Sun, J.; Zhang, S.; Yang, C.; Huang, L. Age of Information Minimization for Radio Frequency Energy-Harvesting Cognitive Radio Networks. *Entropy* **2022**, *24*, 596. https://doi.org/10.3390/e24050596

Academic Editors: Anthony Ephremides and Yin Sun

Received: 23 February 2022
Accepted: 21 April 2022
Published: 24 April 2022

Publisher's Note: MDPI stays neutral with regard to jurisdictional claims in published maps and institutional affiliations.

1. Introduction

To cope with both the spectrum scarcity and the energy shortage challenges in future wireless networks, radio frequency (RF) energy-harvesting in cognitive radio networks (CRN) has been increasingly attractive. Cognitive radio technology allows secondary users (SUs) to opportunistically access the primary users' (PUs) licensed spectrum, based on the condition that the SUs transmission must not cause harmful interference to PUs [1–4]. Meanwhile, the RF energy-harvesting technique conquers the intermittency and uncontrollability of the conventional charging techniques absorbing energy from renewable energy sources [5–7]. Hence, it can simultaneously improve energy efficiency and spectral efficiency, where the SUs can both capture energy and spectrum [8].

While existing works mainly investigated throughput of the RF energy-harvesting CRN, many emerging applications require timely status-update delivery [9–15], i.e., health monitoring, environment monitoring, smart building, vehicle-to-vehicle networking, and so on. For example, in health monitoring, the sensors continuously measure and update blood pressure and heartbeat to the health monitoring platform, which implies the importance of the freshness and timeliness of status-update. The Age of Information (AoI) as a recently proposed performance metric can be used to quantify the freshness and timeliness of status-update [16–23]. It is defined as the time elapsed since the generation time of the latest successfully received status-update at the destination.

Some innovative efforts have been devoted to the AoI of CRN [24–28]. In [24], the authors considered a cognitive wireless sensor network with a cluster of SUs, where the authors proposed a joint and scheduling strategy that optimized energy efficiency of a communication system subject to the expected AoI. The authors in [25] considered an overlay CRN where the SU acted as a relay. The SU forwarded the PU's packets or transmitted its own packets. The optimal policy for status-update and packet relaying was investigated to minimize the average AoI and energy efficiency. In [26], the authors analyzed the average peak AoI of the PU and SU for both overlay and underlay schemes. The asymptotic expressions of the average peak AoI were derived when the PU operated at high signal-to-noise ratio. Considering that it is difficult for PU keeping time-slotted synchronization with SU, the authors in [27] investigated AoI minimization in CRN with an unslotted PU. The closed-form expression was derived by conducting a Markov chain analysis. In [28], the authors considered AoI minimization for energy-harvesting CRN. They assumed that the SU harvests energy from ambient energy sources and derived the optimal sensing and update policies for both perfect and imperfect spectrum sensing.

Overall, the aforementioned research efforts rarely address AoI minimization for RF energy-harvesting CRN. Motivated by this, this article attempts to minimize the average AoI by adaptively making sensing and updating decisions subject to the energy causality and spectrum constraints with imperfect spectrum sensing. The system consists of one PU and one SU. Different from [28], the SU harvests RF energy from PU transmissions instead of ambient energy sources, which is further used to generate and deliver the status-update data pack when the PU is idle. The SU utilizes the harvested energy to perform spectrum sensing and updating. The main contributions of this paper are summarized as follows:

- We study the average AoI minimization for RF energy-harvesting CRN where the SU harvests energy from PU transmissions. In each time slot, the SU adaptively makes sensing and updating decisions based on the channel state information, the AoI value, the available energy, and the belief of PU's spectrum.
- We formulate the decision-making problem as a framework of a partially observable Markov decision process (POMDP) with finite state and action spaces. Then we use dynamic programming to obtain the optimal policy.
- We demonstrate through extensive simulations that the proposed policy can essentially improve the system performance compared to the myopic policy under different system parameter settings.

The remaining part of the paper is organized as follows. In Section 2, we review the works on RF energy-harvesting CRN in the literature. Section 3 describes the studied system model for RF energy-harvesting CRN. Section 4 first formulates the AoI minimization problem as a POMDP framework and then solves it through the dynamic programming. Section 5 presents simulation results and discussions. Finally, Section 6 concludes this paper.

2. Related Works on RF Energy-Harvesting CRN

Recently, cognitive radio technology has drawn significant attention as a promising solution to overcome the licensed spectrum severe scarcity. Cognitive radio allows SUs to opportunistically access PUs' licensed spectrum, based on the condition that the SUs transmission must not cause harmful interference to PUs [1–3]. Spectrum sensing is an important functionality in the cognitive radio system [29], by which the SUs decide whether the spectrum is occupied by the PUs. It can be performed by a single SU or in cooperation with multiple SUs. The SUs can only transmit data when the PUs are idle [30]. Various spectrum-sensing approaches have been developed based on employing different features of the PUs' signal [31], such as coherent detection [32], energy detection [33], and feature detection [34].

On the other hand, energy shortage is also a challenge in future wireless networks. Over the last past years, the RF energy-harvesting technique has emerged as a candidate method for charging low-power wireless devices, which can conquer the intermittency and uncontrollability of the conventional charging techniques absorbing energy from renewable energy sources [5–7]. In [35], the authors proposed the harvest-then-transmit (HTT) protocol as one of the important transmission strategies of RF energy-harvesting technology, where the users first harvest energy from the hybrid access point (HAP) and then use the captured energy to transmit information to the HAP. There have been some related works before. In [36], the authors investigated the wireless-powered network (WPCN) where one HAP coordinated the wireless information and energy transmissions to a set of nodes, where the transmission completion time minimization subject to the throughput requirement per node was considered. Furthermore, the authors studied a similar WPCN scenario in [37], where they focused on energy provision minimization for two physical-layer protocols, non-orthogonal multiple access (NOMA) and time-division multiple access (TDMA). Different from the common WPCN with a fixed HAP, the transmission completion time minimization was investigated in aerial vehicle-enabled WPCN in [38].

To jointly solve the aforementioned two challenges including spectrum scarcity and energy shortage, introducing RF energy-harvesting in CRN has been increasingly attractive due to the fact that it can simultaneously improve energy efficiency and spectral efficiency, where the SUs can both capture energy and spectrum [8]. The timely-delivery probability of data packs for the RF energy-harvesting SU was derived in [39], where the SU opportunistically accesses the spectrum vacated by the PU to deliver real-time data packs and harvests RF energy when the PU is active. Unlike the traditional RF energy-harvesting CRN system where the SU keeps synchronization with the PU, the authors in [40] considered unslotted PU. The sensing intervals were derived to balance between energy harvesting and spectrum access. However, both [39,40] focused on a simple CRN consisting of one PU and one SU. The authors in [41–43] considered a more general scenario where there were multiple SUs or multiple PUs. In [41], the multiple selection strategy was proposed for RF energy-harvesting CRN to maximize the SUs' average throughput. In [42], the authors studied a hybrid energy-harvesting SU that can capture energy from both renewable sources and ambient radio frequency signals. The asymptotic activity behavior of a single SU was analyzed by deriving the theoretical upper bound on sensing and transmission opportunities. In [43], the authors investigated the end-to-end throughput maximization by jointly optimizing the transmit power and time allocation for multiple SUs.

3. System Model

As illustrated in Figure 1, we investigated AoI minimization for a RF energy-harvesting CRN, where the system consists of one PU, one SU, and one CBS communicating with the SU. The SU is a wireless sensor node that monitors the physical process and randomly generates status updates to the CBS. It has no embedded power supply available and harvests RF energy from PU transmissions. Additionally, it opportunistically accesses the PU's licensed spectrum. We considered a time-slotted system with a time interval of T time slots. The duration of each time slot is sufficient for the SU to generate one status-update data packet and receive it successfully at the CBS. Without loss of generality, we assume that the time slot duration is 1 s. The important notations are summarized in Table 1.

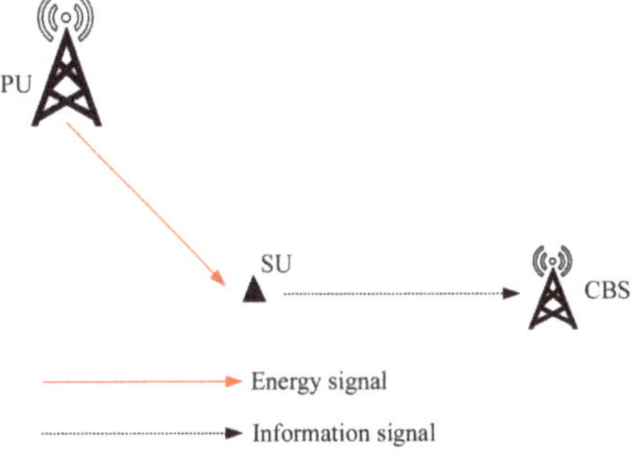

Figure 1. System model. In each time slot, the SU can harvest energy from the PU transmissions and can deliver the status-update date pack to the CBS when the channel is idle.

Table 1. List of notations used in this paper.

Notation	Definition
p_{ai}	The transition probability from the active state to the idle state
p_{ii}	The transition probability from the idle state to the idle state
p_f	The false alarm probability
p_d	The detection probability
ϕ_t	The sensing decision at time slot t
θ_t	The update decision at time slot t
$\hat{q}_t$	The sensing result
δ	The energy consumption on sensing spectrum
τ_s	The time consumption on sensing spectrum
$E_{T,t}$	The energy consumption on update
τ_t	The time consumption on update
S	The size of status-update data pack
a_t	The AoI at time slot t
ρ_t	The belief probability
b_{max}	The maximum battery energy level
g_{max}	The maximum channel power gain level from the SU to the CBS
h_{max}	The maximum channel power gain level from the PU to the SU
Φ	The belief space
η	The energy-harvesting efficiency
σ_2	The noise power
B_{max}	The battery capacity of the SU
A_{max}	The upper of AoI
s_t	The current state
x_t	The action at time slot t

3.1. Primary User Model

The occupancy of a channel by the PU is modeled as a two-state continuous-time Markov chain [44], i.e., active (A) and idle (I) states. In each time slot, the PU either stays in the idle state or occupies the spectrum in an active state. The two-state (active/idle) Markov chain model for modeling PU activity has been verified to be an appropriate model to characterize spectrum occupancy in the time domain [45]. Let $q_t \in \{A, I\}$ denote the state of the PU for $t = 0, 1, \ldots, T - 1$. The transition probabilities of the two-state Markov chain are expressed as p_{ai} and p_{ii}, which represent transitioning from the active state to the idle state, and still staying in the idle state, respectively. For $t = 0, 1, \ldots, T - 1$, we have

$$p_{ai} \triangleq \mathbb{P}(q_{t+1} = I | q_t = A), \tag{1}$$

$$p_{ii} \triangleq \mathbb{P}(q_{t+1} = I | q_t = I) \tag{2}$$

The transition probabilities are known to SU, which can be obtained by long-term measurements.

3.2. Secondary User Model

We considered the SU time-slotted synchronization with the PU. At the beginning of each time slot, the SU needs to decide whether to sense the PU's spectrum. If it decides not to sense the spectrum, it takes the entire time slot to harvest energy from the PU transmissions. That is, energy can be harvested when the PU is active; otherwise, no energy is harvested. We assume the imperfect sensing outcome for the SU [46]. We denote the probability of a false alarm by p_f (i.e., the probability of deciding the spectrum is occupied by the PU while it is not). The probability of detection is denoted by p_d (i.e., the probability of deciding PU is active when it is active). Then, we have

$$p_f = \mathbb{P}(\hat{q}_t = A | q_t = I), \tag{3}$$

$$p_d = \mathbb{P}(\hat{q}_t = I | q_t = I). \tag{4}$$

The SU will take two actions after obtaining the sensing result. When the PU is sensed to be active, the SU will not deliver the status-update data pack. This means that it can harvest energy when the PU is actually active. On the other hand, if the sensing result is that the spectrum is vacated by the PU, the SU needs to further decide whether to update. If an update package is delivered, the SU will receive a 1-bit feedback signal from the CBS to determine whether the update is successful or not. When the sensing result $\hat{q}_t = I$ is correct, the update is successful. This happens with probability $1 - p_f$. Update failure occurs if the PU is active despite the SU declaring it idle. This happens with probability $1 - p_d$. The SU aims to minimize the average AoI by making the optimal sensing and update decisions over time slot $t = 0, 1, \ldots, T - 1$. We denote the decision of time slot t by $x_t = (\phi_t, \theta_t)$, where $\phi_t \in \{0(\text{not sense}), 1(\text{sense})\}$ and $\theta_t \in \{0(\text{not update}), 1(\text{update})\}$ denote the sensing and update decisions, respectively. The optimal sensing and update decisions are based on the SU's states and its statistical knowledge of the PU activity.

(1) Belief model: The SU observes the availability of the PU spectrum by adaptively detecting and accessing the spectrum. The belief state of the PU spectrum can be obtained based on the SU's action and observation history. That is, at the beginning of each time slot t, the SU forms the belief ρ_t. The belief ρ_t is the conditional probability that the PU is in an idle state given the SU's action and observation history.

(2) Channel model: Denote the channel power gains from the PU to the SU and from the SU to the CBS by h_t and g_t over time slot t. We consider a quasi-static channel model based on one time slot by assuming that the channel state information is constant in a single time slot and variable in different time slots. Especially, as is commonly assumed in the works about the wireless communication system, the channel state information of the current time slot can be perfectly obtained.

(3) RF energy-harvesting model: The batter-free SU harvests energy from the occupied spectrum by the PU. For the SU, the HTT protocol is employed. That is, the SU first captures energy from the PU transmissions and then utilizes the harvested energy to sense spectrum and transmit data. Overall, there are two cases where energy can be harvested over time slots: (1) The not sensing decision is made, and the PU is inactive, and (2) the sensing decision is made, and the sensing result $\hat{q}_t = A$ is correct. The energy captured by the SU is expressed as

$$E_{H,t}^m = \eta\tau P h_t, \tag{5}$$

for $t = 0, 1, \ldots, T - 1$ and $m = 1, 2$, where η, τ and P denote the energy-harvesting efficiency, energy-harvesting time and transmit power at the PU, respectively. The superscript m denotes the two cases of energy-harvesting mentioned above. The captured energy is used to perform sensing spectrum and update. Denote the energy and time consumption on sensing spectrum by δ and τ_s, respectively. Similarly, let $E_{T,t}$ and τ_t denote the energy and time consumption on update, respectively. Energy consumption $E_{T,t}$ is time-varying, which is related to the channel state information g_t from the SU to the CBS. According to Shannon's formula [47], the transmission rate $\frac{S}{\tau_t}$ can be expressed as $\frac{S}{\tau_t} = W \log_2(1 + \frac{E_{T,t}g_t}{\tau_t\sigma^2})$, where σ^2 is the noise power at the CBS, S is the size of status-update data pack, and W is the bandwidth. Reorganizing the expression, we obtain the energy consumption, $E_{T,t}$, as

$$E_{T,t} = \frac{\sigma^2\tau_t}{g_t}\left(2^{\frac{S}{\tau_t W}} - 1\right). \tag{6}$$

Since the size of the status-update data pack is fixed, $E_{T,t}$ is only related to the channel state information from the SU to the CBS. Although the update decision can reduce the AoI to one, when the channel quality is poor, it may be better not to deliver the status-update data pack to conserve energy. Note that update failure occurs if the sensing result $\hat{q}_t = I$ is incorrect. In this case, the SU will consume all its available energy. Let $B_{\max}$ denote the battery capacity of the SU. In time slot t, the battery state is b_t, which evolves as

$$b_{t+1} = \min\{b_t + E_{H,t}^m - \phi_t\delta - \theta_t E_{T,t}, B_{\max}\}, t = 0, 1, \ldots, T - 1. \tag{7}$$

Hence, for the SU, the energy causality constraint should satisfy

$$\phi_t \delta + \theta_t E_{T,t} \le b_t, \ t = 0, 1, \ldots, T-1. \tag{8}$$

(4) AoI model: We consider a linear model for the AoI [16], where the AoI is defined as the time elapsed from the moment when the most recently received update was generated to the present. Let the AoI at time slot t denote by $a_t \in \mathcal{A} \triangleq \{1, 2, \ldots, A_{\max}\}$. Here $A_{\max}$ is the upper of the AoI and is defined as

$$A_{\max} = a_0 + T. \tag{9}$$

In the considered system, the SU adopts the generate-at-will scheme. That is, the SU generates and delivers a status-update data pack after making an update decision. At each time slot t, the size of the data packet S is small enough to be generated and updated immediately and received by the end of the current time slot when the update decision is made and the sensing result $\hat{q}_t = I$ is correct. If the update is received at the CBS, AoI decreases to one; otherwise, it increases by one. We consider an error-free channel through which the status-update data pack can be successfully received at the CBS when the update decision is made and the sensing result $\hat{q}_t = I$ is correct. The average AoI for an interval of T time slots is expressed as

$$\overline{A} = \frac{1}{T} \sum_{t=0}^{T-1} a_t, \ t = 0, 1, \ldots, T-1. \tag{10}$$

4. POMDP for AoI Minimization

In this section, we formulate the AoI minimization as a finite-horizon POMDP problem and solve for the optimal solutions via dynamic programming.

4.1. POMDP Formulation

We use a POMDP framework to model the optimal sensing and update decisions for the SU's AoI minimization. The components of POMDP are described as follows.

- Actions: At the beginning of each time slot t, the SU needs to decide whether to sense the spectrum. If it decides not to sense the spectrum, then it captures energy from the PU transmissions and does not update, i.e., $x_t = (0, 0)$. If it decides to sense the spectrum and finds that the PU is idle, it further decides whether to update based on the available energy, the AoI value, the channel state information from the SU to the CBS and from the PU to the SU, i.e., $x_t = (1, 0)$ and $x_t = (1, 1)$. Thus, the action for each time slot t is $x_t = (\phi_t, \theta_t) \in \mathcal{X} \triangleq \{(0, 0), (1, 0), (1, 1) : b_t > \phi_t \delta + \theta_t F_{T,t}\}$, where $\psi_t \in \Gamma_\phi \triangleq \{0, 1 : b_t \ge \phi_t \delta\}$ and $\theta_t \in \Gamma_\theta \triangleq \{0, 1 : b_t \ge \delta + \theta_t E_{T,t}\}$.

- Observations and beliefs: Let $\hat{q}_t \in \{A, I\}$ denote the observation of the PU's state. The belief $\rho_t \in [0, 1]$ is a condition probability that the spectrum is vacated by the PU. The belief is updated according to the following cases.

  Case 1: The SU does not sense the spectrum; the new belief is given by

$$\rho_{t+1} = \Lambda_0(\rho_t) = \rho_t p_{ii} + (1 - \rho_t) p_{ai}. \tag{11}$$

  Case 2: If the PU is sensed to be active, the SU harvests energy in the remaining time of the current time slot, i.e., the battery energy increases. This implies the true state of the PU is $q_t = A$. The belief is updated as

$$\rho_{t+1} = p_{ai}. \tag{12}$$

Case 3: If the PU is sensed to be active, the SU does not harvest energy; i.e., the battery energy does not change and is lower than $B_{\max}$. This implies the true state of the PU is $q_t = I$. The new belief is expressed as

$$\rho_{t+1} = p_{ii}. \tag{13}$$

Case 4: If the PU is sensed to be active, the battery energy is $B_{\max}$ at time slot t. The new belief is given by

$$\rho_{t+1} = \Lambda_{1A}(\rho_t) = \zeta_t p_{ii} + (1 - \zeta_t) p_{ai}, \tag{14}$$

where

$$\zeta_t \triangleq \mathbb{P}(q_t = I | \hat{q}_t = A) = \frac{\rho_t(1 - p_f)}{\rho_t p_f + (1 - \rho_t)(1 - p_d)}. \tag{15}$$

Case 5: If the PU is sensed to be idle, the SU does not update. The belief is updated as

$$\rho_{t+1} = \Lambda_{1I}(\rho_t) = \bar{\zeta}_t p_{ii} + (1 - \bar{\zeta}_t) p_{ai}, \tag{16}$$

where

$$\bar{\zeta}_t \triangleq \mathbb{P}(q_t = I | \hat{q}_t = I) = \frac{\rho_t(1 - p_f)}{\rho_t(1 - p_f) + (1 - \rho_t)(1 - p_d)}. \tag{17}$$

Case 6: If the PU is sensed to be idle, the SU updates successfully. This implies that the true state of the PU is $q_t = I$. Then, we have

$$\rho_{t+1} = p_{ii}. \tag{18}$$

Case 7: If the PU is sensed to be idle, the SU update fails. This implies that the true state of the PU is $q_t = A$. Then, we have

$$\rho_{t+1} = p_{ai}. \tag{19}$$

Although (11)–(19) cover seven cases from case one to case seven, the new beliefs in both case two and case seven are denoted as p_{ai}, and the new beliefs in both case three and case six are denoted as p_{ii}. Hence, the SU can only transit to five beliefs. That is, the number of possible beliefs is finite over T time slots. Thus, for the length of T time slots, the belief space Φ is a finite set.

- States: Denote the discrete battery energy level of the SU at the beginning of time slot t by $b_t' \in \mathcal{B} \triangleq \{0, 1, 2, \ldots, b_{\max}\}$, where $b_{\max}$ is the maximum battery energy level that can be stored inside the battery of the SU. Then, each energy quantum of the SU's battery contains $\frac{B_{\max}}{b_{\max}}$ Joules. In this case, we use $b_t' = \left\lfloor \frac{b_t b_{\max}}{B_{\max}} \right\rfloor$ to convert the continuous battery energy of the SU to the discrete battery energy level, by which a lower bound to the AoI of the original continuous system is obtained. Similarly, divide continuous channel power gain into finite number of intervals according to fading probability density function (PDF). Thus, the discrete channel power gain levels from the SU to the CBS and from the PU to the SU are expressed as $g_t' \in \mathcal{G} \triangleq (0, 1, 2, \ldots, g_{\max})$ and $h_t' \in \mathcal{H} \triangleq (0, 1, 2, \ldots, h_{\max})$, respectively. Here, $g_{\max}$ and $h_{\max}$ denote the corresponding maximum channel power gain levels. At each time slot t, the completely observable states include channel state from the PU to the SU, channel state from the SU to the CBS, the AoI state, and battery state, denoted by $s_t \triangleq (h_t', g_t', a_t, b_t')$. Note that the state space, i.e., $\mathcal{S} \triangleq \mathcal{H} \times \mathcal{G} \times \mathcal{A} \times \mathcal{B}$, is finite. Due to imperfect sensing, an update may be unsuccessful when the sensing result is $\hat{q}_t = I$ and the update decision is $\theta_t = 1$. Thus,

$$a_{t+1} = \begin{cases} 1, & \text{when } x_t = (1, 1) \text{ and } \hat{q}_t = q_t, \\ a_t + 1, & \text{otherwise,} \end{cases} \tag{20}$$

for $t = 1, 2, \ldots, T$. Alternatively, we can express $a_{t+1} = (1 - \theta_t)a_t + 1$ when the sensing result $\hat{q}_t = I$ is correct. Additionally, the PU's spectrum state is only partially observable, which is described by the belief ρ_t. Thus, for each time slot t, the complete system state is denoted by (s_t, ρ_t). Since $\mathcal{S}$ and Φ are finite, the SU experiences a finite number of possible system states $(s_t, \varrho_t) \in \mathcal{S} \times \Phi$.

- Transition probabilities: For time slot t, given the current state $s_t = (h'_t, g'_t, a_t, b'_t)$ and the action $x_t = (\phi_t, \theta_t)$, the transition probability to the next state $s_{t+1} = (h'_{t+1}, g'_{t+1}, a_{t+1}, b'_{t+1})$ is denoted by $p_{x_t}(s_{t+1}|s_t)$. Since the captured energy and the channel power gains are independently and identically distributed (i.i.d), the transition probabilities for taking actions other than $x_t = (1, 1)$ are given as follows.

$$p_{x_t}(s_{t+1}|s_t) = \mathbb{P}(a_{t+1}|a_t, x_t)\mathbb{P}(b'_{t+1}|b'_t, g'_t, h'_t, x_t)\mathbb{P}(g'_{t+1})\mathbb{P}(h'_{t+1}), \tag{21}$$

where

$$\mathbb{P}(a_{t+1}|a_t, x_t) = \begin{cases} 1, & \text{when } a_{t+1} = (1 - \theta_t)a_t + 1, \\ 0, & \text{otherwise,} \end{cases} \tag{22}$$

$$\mathbb{P}(b'_{t+1}|b'_t, g'_t, h'_t, x_t) = \begin{cases} 1, & \text{when } \phi_t = 0 \text{ and } b_{t+1} = \min\{b_t + E^1_{H,t}, B_{\max}\}, \\ 1, & \text{when } \phi_t = 0 \text{ and } b_{t+1} = b_t, \\ 1, & \text{when } \phi_t = 1, \theta_t = 0, \text{ and } b_{t+1} = \min\{b_t - \delta + E^2_{H,t}, B_{\max}\}, \\ 1, & \text{when } \phi_t = 1, \theta_t = 0, \text{ and } b_{t+1} = b_t - \delta, \\ 0, & \text{otherwise.} \end{cases} \tag{23}$$

For the action $x_t = (1, 1)$, the transition probability is expressed as follows.

$$p_{x_t}(s_{t+1}|s_t, \hat{q}_t, q_t) = \mathbb{P}(a_{t+1}|a_t, x_t \hat{q}_t, q_t) \times \mathbb{P}(b'_{t+1}|b'_t, g'_t, h'_t, x_t) \times \mathbb{P}(g'_{t+1})\mathbb{P}(h'_{t+1}), \tag{24}$$

where

$$\mathbb{P}(a_{t+1}|a_t, x_t) = \begin{cases} \bar{\zeta}, & \text{when } a_{t+1} = 1 \text{ and } q_t = \hat{q}_t, \\ 1 - \bar{\zeta}, & \text{when } a_{t+1} = a_t + 1 \text{ and } q_t \leq \hat{q}_t, \\ 0, & \text{otherwise,} \end{cases} \tag{25}$$

and

$$\mathbb{P}(b'_{t+1}|b'_t, g'_t, h'_t, x_t) = \begin{cases} 1, & \text{when } \phi_t = 1, \theta_t = 1, b_{t+1} = b_t - \delta - E_{T,t}, \text{ and } \hat{q}_t = q_t, \\ 1, & \text{when } \phi_t = 1, \theta_t = 1, b_{t+1} = 0, \hat{q}_t = I, \text{ and } q_t = A, \\ 0, & \text{otherwise.} \end{cases} \tag{26}$$

- Cost: Let the immediate cost at state s_t denoted by $C(s_t)$, which is the accumulated AoI at time slot t. Then, we have

$$C(s_t) = a_t, t = 0, 1, \ldots, T - 1. \tag{27}$$

- Policy: The policy is expressed as $\pi = \{\vartheta_0, \vartheta_1, \ldots, \vartheta_{T-1}\}$, where ϑ_t is the deterministic decision rule that maps a system state $(s_t, \rho_t) \in \mathcal{S} \times \Phi$ into an action $x_t \in \mathcal{X}$, i.e., $x_t = \vartheta_t(s_t, \rho_t)$. In this paper, let Π denote the set of all deterministic decision policies.

  Given the SU's initial state s_0 and belief ρ_0 of PU's spectrum, the average AoI of T time slots under the policy π is given by

$$\overline{A}^\pi(s_0, \rho_0) = \frac{1}{T}\mathbb{E}\left[\sum_{t=0}^{T-1} C(s_t)|s_0, \rho_0\right], \tag{28}$$

where the expectation is caused by policy π. Based on the above analysis, minimize the average AoI by finding the optimal sensing and update policy corresponds to solving

$$\min_{\pi \in \Pi} \overline{A}^\pi(s_0, \rho_0). \tag{29}$$

Given T, (29) is a finite-state MDP with total cost. Based on (28) and (29), to minimize the average AoI, the SU should sense the spectrum and deliver the status-update data pack as long as it has sufficient energy. However, considering the channel state information, the belief of PU's spectrum, and the battery energy available, preferring the spectrum sensing and status-update may not be the best decision. As a result, there is an optimal decision scheduling problem.

4.2. POMDP Solution

In this section, we use dynamic programming to solve total cost minimization of T time slots in (29) [48]. At a time slot t, the successive actions $\{x_k\}_{k=t}^{T-1}$ affect the states s_k along with the accumulated AoI $C(s_k)$ for all $k = t, t+1, \ldots, T-1$. Let $V_t(s_t, \rho_t)$ denote the state-value function, which is given by

$$V_t(s_t, \rho_t) \triangleq \min_{\{x_k\}_{k=t}^{T-1}} \mathbb{E}\left[\sum_{k=t}^{T-1} C(s_k)|s_t, \rho_t\right]. \tag{30}$$

It is the minimum expected cost accumulated from time slot t to $T-1$ given state (s_t, ρ_t). Thus, denote the minimum AoI in (29) by $A^* = V_0(s_0, \rho_0)/T$. Additionally, given (s_t, ρ_t) and sensing action ϕ_t, let $Q_t^{\phi_t}(s_t, \rho_t)$ represent the action-value function or Q-function, which is the minimum expected cost for taking sensing action ϕ_t at state (s_t, ρ_t). The Q-function includes two parts: the immediate cost of taking action at the current state and the expected sum of the state-value functions from the next time slot.

Overall, the formulated MDP problem can be solved recursively by dynamic programming as follows. For $t = 0, 1, \ldots, T-1$,

$$V_t(s_t, \rho_t) = \min_{\phi_t \in \Gamma_\phi} Q_t^{\phi_t}(s_t, \rho_t), \tag{31}$$

When $t = T - 1$, we have

$$Q_{T-1}^0(s_{T-1}, \rho_{T-1}) = C(s_{T-1}) + C(s_T), \tag{32}$$

$$Q_{T-1}^1(s_{T-1}, \rho_{T-1}) = (1 - \Delta_{T-1})C(s_{T-1}) + \rho_{T-1} \times \Delta_{T-1} \min_{\phi_{T-1} \in \Gamma_\phi} C(s_{T-1}) + C(s_T). \tag{33}$$

When $t = 0, 1, \ldots, T - 2$, we have

$$Q_t^0(s_t, \rho_t) = C(s_t) + \sum_{s_{t+1}} p_{00}(s_{t+1}|s_t)V_{t+1}(s_{t+1}, \Lambda_0(\varrho_t)), \tag{34}$$

$$Q_t^1(s_t, \rho_t) = (1 - \Delta_t)Q_t^{1A}(s_t, \rho_t) + \Delta_t \min_{\theta_t \in \Gamma_\phi} Q_t^{1\phi_t}(s_t, \rho_t), \tag{35}$$

$$Q_t^{1A}(s_t, \rho_t) = C(s_t) + \sum_{s_{t+1}} p_{10}(s_{t+1}|s_t)V_{t+1}(s_{t+1}, \Lambda_{1A}(\varrho_t)), \tag{36}$$

$$Q_t^{10}(s_t, \varrho_t) = C(s_t) + \sum_{s_{t+1}} p_{10}(s_{t+1}|s_t)V_{t+1}(s_{t+1}, \Lambda_{1I}(\varrho_t)), \tag{37}$$

$$\begin{aligned}Q_t^{11}(s_t, \rho_t) = &C(s_t) + \sum_{s_{t+1}} p_{11}(s_{t+1}|s_t, \hat{q}_t = q_t)V_{t+1}(s_{t+1}, \Lambda_I(\varrho_t)) \\ &+ \sum_{s_{t+1}} p_{11}(s_{t+1}|s_t, \hat{q}_t \leq q_t)V_{t+1}(s_{t+1}, \Lambda_A(\varrho_t)),\end{aligned} \tag{38}$$

where Δ_t represents the probability of observing PU idle. That is

$$\Delta_t = P(\hat{q}_t = I) = \rho_t(1 - p_f) + (1 - \rho_t)(1 - p_d). \tag{39}$$

Especially, $Q_t^{1A}(s_t, \rho_t)$ in (36) represents the minimum expected cost by adopting sensing action $\phi_t = 1$ and sensing result $\hat{q}_t = A$, i.e., $x_t = (1,0)$. In (37) and (38), given the sensing action $\phi_t = 1$ and sensing result $\hat{q}_t = I$, $Q_t^{10}(s_t, \varrho_t)$ and $Q_t^{11}(s_t, \varrho_t)$ denote the minimum expected costs by adopting update action $\theta_t = 0$ and $\theta_t = 1$, respectively. Then, by recursion in (31)–(38), the optimal policies for sensing and update are given by

$$\phi_t^*(s_t, \rho_t) \in \underset{\phi_t \in \Gamma_\phi}{\operatorname{argmin}} Q_t^{\phi_t}(s_t, \varrho_t), \tag{40}$$

$$\theta_t^*(s_t, \rho_t) \in \underset{\phi_t \in \Gamma_\theta}{\operatorname{argmin}} Q_t^{1\theta_t}(s_t, \varrho_t). \tag{41}$$

5. Numerical Results

In this section, we evaluate the performance of our proposed optimal policy through comparing it with the myopic policy and the random policy. At the beginning of time slot t, for the myopic policy, the SU senses the spectrum if it has enough energy. When the sensing result is $\hat{q}_t = I$, the SU generates and delivers a status-update data pack if the energy available is sufficient. For the random policy, the SU randomly chooses to deliver the status-update data pack or harvest energy with a probability. Taking into account the protection of the PU's transmission, the probability of harvesting energy is set to be 90%, and the probability of delivering the status-update data pack is set to be 10%. If the SU chooses to deliver the status-update data pack while the spectrum is occupied by the PU, the status-update fails, and the AoI increases by one. The PU's state transition probabilities are $p_{ii} = 0.8$ and $p_{ai} = 0.5$. The probability of detecting an active PU is $p_d = 0.8$. The channel power gains from the PU to the SU and from the SU to the CBS are modeled as $h = Y\Psi^2 d_1^{-\kappa}$ and $g = Y\Psi^2 d_2^{-\kappa}$, where d_1 and d_2 denote the distances from the PU to the SU, and the SU to the CBS, respectively. Y represents a signal power gain at a 1 m's reference distance, $\Psi \sim \exp(1)$ denotes the small-scale fading gain, and $d_1^{-\kappa}$ and $d_2^{-\kappa}$ are standard power law path-loss with exponent κ. In the simulations, the system parameter values are set as follows: $\eta = 0.5$, $\sigma^2 = -95$ dBm, $W = 1$ MHz, $Y = 0.2$, $\kappa = 2$, $b_{\max} = 5$, $g_{\max} = h_{\max} = 10$, $\rho_0 = p_{ii}$, $\tau_s = 0.2$ s, and $A_{\max} = 13$.

Figure 2 shows one sample path of the AoI by the optimal policy. The transmit power of the PU is 35 dBm, the energy consumption is one energy quantum, the distance from the the SU to the CBS is 20 m, the distance from the PU to the SU is 25 m, the size of the status-update data pack is 14 Mbits, and the battery capacity is 0.5 mJoules. The trend of the AoI over time slots is clearly observed. In the simulations, we found the SU did not perform sensing spectrum even the remaining energy was enough, which verifies the foresight of the optimal policy compared with the myopic policy.

Figure 3 shows the size of the status-update data packet versus the AoI, where the simulation setup is similar as in Figure 3. It is clear that our proposed policy is superior to the other policies. For the random policy, the AoI is obviously high due to the low probability of delivering the status-update data pack. For the random policy, the AoI is greater than 5.57, due to the low probability of delivering the status-update data pack. Considering the poor AoI performance of the random policy, we only compare our algorithm with the myopic algorithm in the following numerical evaluations. We can observe that the AoI increases with the size of the status-update data packet. The reason is that the increase in the size of the status-update data packet will result in increasing the energy needed to deliver one status-update data pack. This decreases the possibility that the SU will have enough energy to update, and hence the AoI is increased.

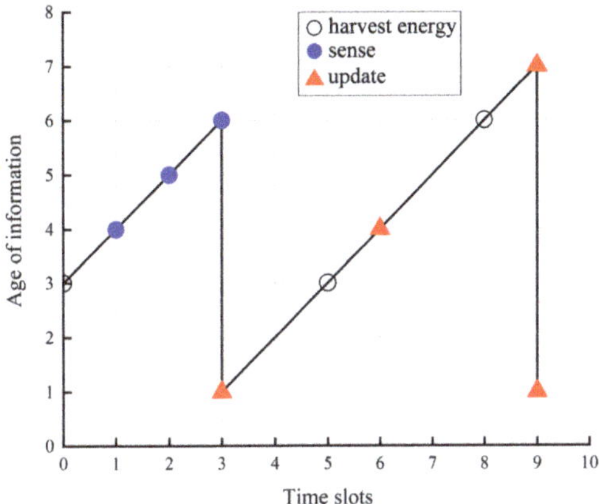

Figure 2. One sample path of the AoI by the optimal policy.

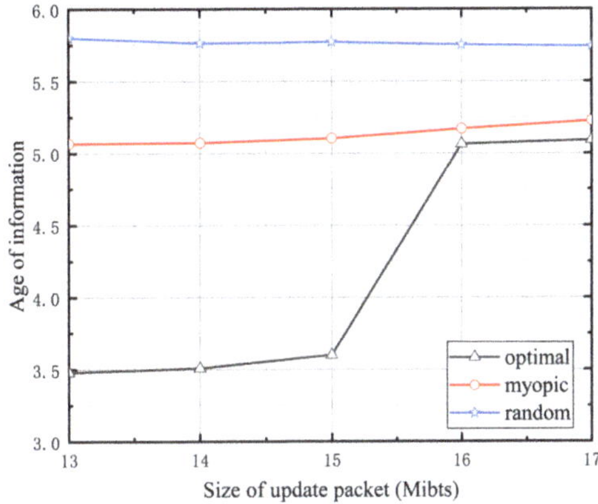

Figure 3. The size of status-update data packet versus the AoI when $T = 10$.

Figure 4 shows the transmit power of the PU versus the AoI, where the capacity of battery is 0.2 mJoules, the distance from the PU to the SU is 5 m, the distance from SU to the CBS is 25 m, the size of status-update data pack is 15 Mbits. We can observe from Figure 4 that the average AoI increases with the transmit power of PU. The reason is that the SU will harvest more energy as the transmit power of PU increases, which allows the SU to store more energy in the battery. This increases the possibility that the SU will have enough energy to update, and hence the AoI is decreased.

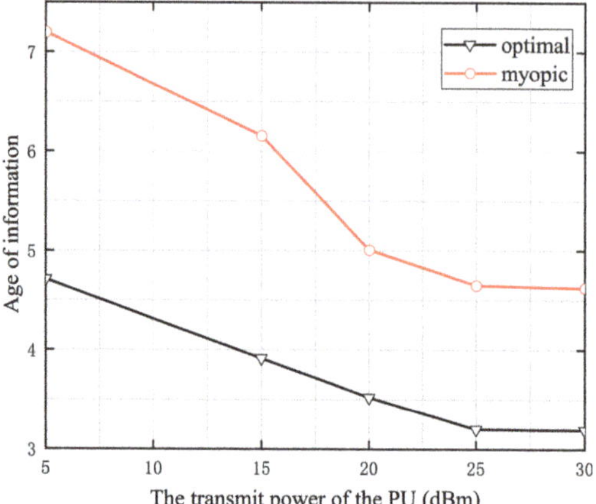

Figure 4. The transmit power of PU versus the AoI when $T = 10$.

Figure 5 shows the battery capacity versus the AoI, where the size of the status-update data pack is 15 Mbits, the energy consumption on the sensing spectrum is one energy quantum, the transmit power of the PU is 35 dBm, the distance from the SU to the CBS is 10 m, and the distance from the PU to the SU is 5 m. It is clearly observed that the proposed policy essentially improves the AoI as compared to the myopic policy. We can also observe the average AoI decreases with the battery capacity. The reason is that increasing the battery capacity allows more harvested energy to be stored inside the battery. Thus, the SU will have enough energy to perform an update, and hence the AoI is reduced.

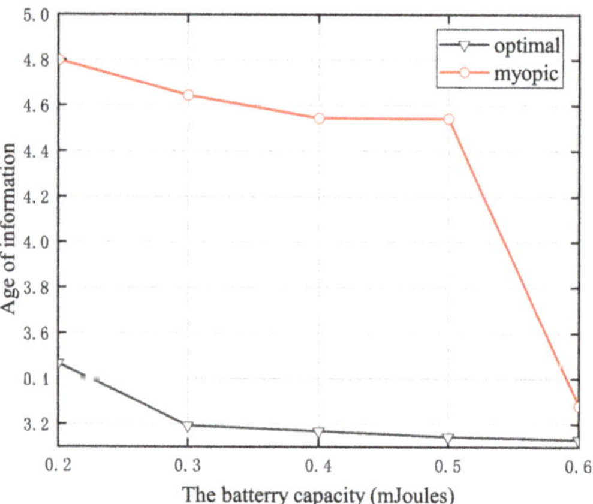

Figure 5. The battery capacity versus the AoI when $T = 10$.

Figure 6 shows the energy consumption on sensing spectrum versus the AoI. The simulation setup is the similar as in the Figure 5. It is observed that the average AoI increases with the energy consumption on sensing action. The reason is that increasing the energy consumption on sensing spectrum can result in less energy remaining inside the battery. This, in turn, decreases the possibility that the SU will have enough energy to deliver status-update data packet, and hence the AoI is increased.

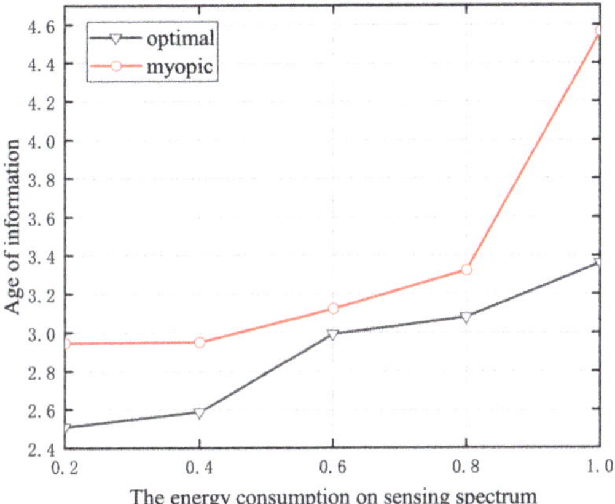

Figure 6. The energy consumption on sensing spectrum versus the AoI when $T = 10$.

6. Conclusions

In this paper, we investigated RF energy-harvesting CRN with the aim of AoI minimization subject to the energy causality and spectrum constraints. We first used POMDP to formulate this average AoI minimization based on the AoI value, the channel state information, the energy available, and the PU's spectrum belief, and then dynamic programming was adopted to find the optimal sensing and update decisions. Numerical results showed the influence of system parameters on the AoI, and demonstrated that the proposed policy significantly outperform the myopic policy.

Author Contributions: Conceptualization, all co-authors; methodology, all co-authors; software, not applicable; validation, all co-authors; formal analysis, all co-authors; investigation, all co-authors; resources, all co-authors; data curation, S.J.; writing—original draft preparation, J.S. and S.Z.; writing—review and editing, all co-authors; visualization, J.S. and C.Y.; supervision, S.Z.; project administration, L.H.; funding acquisition, L.H. All authors have read and agreed to the published version of the manuscript.

Funding: This research was funded in part by the National Natural Science Foundation of China under grant number 62072410 and the Zhejiang Provincial Natural Science Foundation of China under grant number LQ22F020009.

Institutional Review Board Statement: Not applicable.

Informed Consent Statement: Not applicable.

Data Availability Statement: Not applicable.

Conflicts of Interest: The authors declare no conflict of interest.

References

1. He, Y.; Xue, J.; Ratnarajah, T.; Sellathurai, M.; Khan, F. On the Performance of Cooperative Spectrum Sensing in Random Cognitive Radio Networks. *IEEE Syst. J.* **2018**, *12*, 881–892. [CrossRef]
2. Zheng, K.; Liu, X.; Liu, X.; Zhu, Y. Hybrid overlay-underlay cognitive radio networks with energy harvesting. *IEEE Trans. Commun.* **2019**, *67*, 4669–4682. [CrossRef]
3. Liu, X.; Zheng, K.; Chi, K.; Zhu, Y. Cooperative spectrum sensing optimization in energy-harvesting cognitive radio networks. *IEEE Trans. Wirel. Commun.* **2020**, *19*, 7663–7676. [CrossRef]
4. Yu, H.; Zikria, Y.B. Cognitive Radio Networks for Internet of Things and Wireless Sensor Networks. *Sensors* **2020**, *20*, 5288. [CrossRef]
5. Bi, S.; Zeng, Y.; Zhang, R. Wireless powered communication networks: An overview. *IEEE Wirel. Commun.* **2016**, *23*, 10–18. [CrossRef]
6. Zhang, S.; Kong, S.; Chi, K.; Huang, L. Energy Management for Secure Transmission in Wireless Powered Communication Networks. *IEEE Internet Things J.* **2022**, *9*, 1171–1181. [CrossRef]

7. Guo, H.; Li, J.; Liu, J.; Na, T.; Kato, N. A Survey on Space-Air-Ground-Sea Integrated Network Security in 6G. *IEEE Commun. Surv. Tutor.* **2021**, *24*, 53–87. [CrossRef]
8. Zhang, Y.; Han, W.; Li, D.; Zhang, P.; Cui, S. Power versus spectrum 2-D sensing in energy harvesting cognitive radio networks. *IEEE Trans. Signal Process.* **2015**, *63*, 6200–6212. [CrossRef]
9. Khan, A.A.; Rehmani, M.H.; Rachedi, A. When cognitive radio meets the Internet of Things. In Proceedings of the International Wireless Communications & Mobile Computing Conference (IWCMC 2016), Paphos, Cyprus, 5–9 September 2016; pp. 469–474.
10. Perera, C.; Liu, C.H.; Jayawardena, S. The emerging internet of things marketplace from an industrial perspective: A survey. *IEEE Trans. Emerg. Top. Comput.* **2015**, *3*, 585–598. [CrossRef]
11. Guo, H.; Huang, W.; Liu, J.; Wang, Y. Inter-Server Collaborative Federated Learning for Ultra-Dense Edge Computing. *IEEE Trans. Wirel. Commun.* 2021, *in press*. [CrossRef]
12. Sun, W.; Lei, S.; Wang, L.; Liu, Z.; Zhang, Y. Adaptive Federated Learning and Digital Twin for Industrial Internet of Things. *IEEE Trans. Ind. Inform.* **2021**, *17*, 5605–5614. [CrossRef]
13. Bi, S.; Huang, L.; Zhang, Y.-J.A. Joint Optimization of Service Caching Placement and Computation Offloading in Mobile Edge Computing Systems. *IEEE Trans. Wirel. Commun.* **2020**, *19*, 4947–4963. [CrossRef]
14. Sun, W.; Xu, N.; Wang, L.; Zhang, H.; Zhang, Y. Dynamic Digital Twin and Federated Learning with Incentives for Air-Ground Networks. *IEEE Trans. Netw. Sci. Eng.* **2022**, *9*, 321–333. [CrossRef]
15. Khan, A.A.; Rehmani, M.H.; Rachedi, A. Cognitive-radio-based internet of things: Applications, architectures, spectrum related functionalities, and future research directions. *IEEE Wirel. Commun.* **2017**, *24*, 17–25. [CrossRef]
16. Kaul, S.; Yates, R.; Gruteser, M. Real-time status: How often should one update? In Proceedings of the IEEE International Conference on Computer Communications (INFOCOM), Orlando, FL, USA, 25–30 March 2012; pp. 2731–2735.
17. Kam, C.; Kompella, S.; Ephremides, A. Effect of message transmission diversity on status age. In Proceedings of the IEEE International Symposium on Information Theory (ISIT), Honolulu, HI, USA, 29 June–4 July 2014; pp. 2411–2415.
18. Costa, M.; Codreanu, M.; Ephremides, A. Age of information with packet management. In Proceedings of the IEEE International Symposium on Information Theory (ISIT), Seoul, Korea, 29 June–4 July 2014; pp. 1583–1587.
19. Sun, Y.; Uysal-Biyikoglu, E.; Yates, R.; Koksal, C.E.; Shroff, N.B. Update or wait: How to keep your data fresh. In Proceedings of the IEEE International Conference on Computer Communications (INFOCOM), San Francisco, CA, USA, 10–14 April 2016; pp. 1–9.
20. Huang, L.; Qian, L.P. Age of Information for Transmissions over Markov Channels. In Proceedings of the IEEE Global Communications Conference (GLOBECOM), Singapore, 4–8 December 2017; pp. 1–9.
21. Abd-Elmagid, M.A.; Pappas, N.; Dhillon, H.S. On the role of age of information in the Internet of Things. *IEEE Commun. Mag.* **2019**, *57*, 72–77. [CrossRef]
22. Nguyen, G.; Kompella, S.; Kam, C.; Wieselthier, J.; Ephremides, A. Impact of hostile interference on information freshness: A game approach. In Proceedings of the 15th International Symposium on Modeling and Optimization in Mobile, Ad Hoc, and Wireless Networks (WiOpt), Paris, France, 15–19 May 2017.
23. Garnaev, A.; Zhang, W.; Zhong, J.; Yates, R.D. Maintaining information freshness under jamming. In Proceedings of the IEEE Conference on Computer Communications Workshops (INFOCOM WKSHPS), Paris, France, 29 April–2 May 2019; pp. 90–95.
24. Valehi, A.; Razi, A. Maximizing Energy Efficiency of Cognitive Wireless Sensor Networks With Constrained Age of Information. *IEEE Trans. Cogn. Commun. Netw.* **2017**, *3*, 643–654. [CrossRef]
25. Zhao, Y.; Zhou, B.; Saad, W.; Luo, X. Age of information analysis for dynamic spectrum sharing. In Proceedings of the Global Conference on Signal and Information Processing (GlobalSIP), Ottawa, ON, Canada, 11–14 November 2019; pp. 1–5.
26. Gu, Y.; Chen, H.; Zhai, C.; Li, Y.; Vucetic, B. Minimizing age of information in cognitive radio-based IoT systems: Underlay or Overlay? *IEEE Internet Things J.* **2019**, *6*, 10273–10288. [CrossRef]
27. Wang, Q.; Chen, H.; Gu, Y.; Li, Y.; Vucetic, B. Minimizing the Age of Information of cognitive radio-based IoT systems under a collision constraint. *IEEE Trans. Wirel. Commun.* **2020**, *19*, 8054–8067. [CrossRef]
28. Leng, S.; Yener, A. Age of information minimization for an energy harvesting cognitive radio. *IEEE Trans. Cognit. Commun. Netw.* **2019**, *5*, 427–439. [CrossRef]
29. Yilmaz, Y.; Moustakides, G.V.; Wang, X. Cooperative Sequential Spectrum Sensing Based on Level-Triggered Sampling. *IEEE Trans. Signal Process.* **2012**, *60*, 4509–4524. [CrossRef]
30. Zhao, W.; Ali, S.S.; Jin, M.; Cui, G.; Zhao, N.; Yoo, S.-J. Extreme Eigenvalues-Based Detectors for Spectrum Sensing in Cognitive Radio Networks. *IEEE Trans. Commun.* **2022**, *70*, 538–551. [CrossRef]
31. Ma, J.; Li, G.Y.; Juang, B.H. Signal processing in cognitive radio. *Proc. IEEE* **2009**, *97*, 805–823.
32. Tang, H. Some physical layer issues of wide-band cognitive radio systems. In Proceedings of the 1st IEEE International Symposium on New Frontiers in Dynamic Spectrum Access Networks (DySPAN), Baltimore, MD, USA, 8–11 November 2005; pp. 151–159.
33. Akyildiz, I.F.; Lo, B.F.; Balakrishnan, R. Cooperative spectrum sensing in cognitive radio networks: A survey. *Phys. Commun.* **2011**, *4*, 40–62. [CrossRef]
34. Ghasemi, A.; Sousa, E.S. Collaborative spectrum sensing for opportunistic access in fading environments. In Proceedings of the 1st IEEE International Symposium on New Frontiers in Dynamic Spectrum Access Networks (DySPAN), Baltimore, MD, USA, 8–11 November 2005; pp. 131–136.

35. Ho, C.K.; Zhang, R. Optimal Energy Allocation for Wireless Communications with Energy Harvesting Constraints. *IEEE Trans. Signal Process.* **2012**, *60*, 4808–4818. [CrossRef]
36. Chi, K.; Zhu, Y.; Li, Y.; Huang, L.; Xia, M. Minimization of Transmission Completion Time in Wireless Powered Communication Networks. *IEEE Internet Things J.* **2017**, *4*, 1671–1683. [CrossRef]
37. Chi, K.; Chen, Z.; Zheng, K.; Zhu, Y.-H.; Liu, J. Energy Provision Minimization in Wireless Powered Communication Networks With Network Throughput Demand: TDMA or NOMA? *IEEE Trans. Commun.* **2019**, *67*, 6401–6414. [CrossRef]
38. Chen, Z.; Chi, K.; Zheng, K.; Dai, G.; Shao, Q. Minimization of transmission completion time in UAV-enabled wireless powered communication networks. *IEEE Internet Things J.* **2020**, *7*, 1245–1259. [CrossRef]
39. Bae, Y.H.; Baek, J.W. Performance analysis of delay-constrained traffic in a cognitive radio network with RF energy harvesting. *IEEE Commun. Lett.* **2019**, *23*, 2177–2181. [CrossRef]
40. Li, K.H.; Teh, K.C. Optimal spectrum access and energy supply for cognitive radio systems with opportunistic RF energy harvesting. *IEEE Trans. Veh. Technol.* **2017**, *66*, 7114–7122.
41. Xu, M.; Jin, M.; Guo, Q.; Li, Y. Multichannel Selection for Cognitive Radio Networks With RF Energy Harvesting. *IEEE Wirel. Commun. Lett.* **2018**, *7*, 178–181. [CrossRef]
42. Celik, A.; Alsharoa, A.; Kamal, A.E. Hybrid energy harvesting- Based cooperative spectrum sensing and access in heterogeneous cognitive radio networks. *IEEE Tran. Cognit. Commun. Netw.* **2017**, *3*, 37–48. [CrossRef]
43. Xu, C.; Zheng, M.; Liang, W.; Yu, H.; Liang, Y. End-to-End Throughput Maximization for Underlay Multi-Hop Cognitive Radio Networks with RF Energy Harvesting. *IEEE Trans. Wirel. Commun.* **2017**, *16*, 3561–3572. [CrossRef]
44. Masonta, M.T.; Mzyece, M.; Ntlatlapa, N. Spectrum decision in cognitive radio networks: A survey. *IEEE Commun. Surv. Tutor.* **2013**, *15*, 1088–1107. [CrossRef]
45. López-Benítez, M.; Casadevall, F. Empirical time-dimension model of spectrum use based on a discrete-time Markov chain with deterministic and stochastic duty cycle models. *IEEE Trans. Veh. Technol.* **2011**, *60*, 2519–2533. [CrossRef]
46. Unnikrishnan, J.; Veeravalli, V.V. Cooperative sensing for primary detection in cognitive radio. *IEEE J. Select. Top. Signal Process.* **2006**, *2*, 18–27. [CrossRef]
47. Shannon, C.E. A mathematical theory of communication. *ACM SIGMOBILE Mob. Comput. Commun. Rev.* **2001**, *5*, 3–55. [CrossRef]
48. Bertsekas, D.P. *Dynamic Programming and Optimal Control*; Athena Scientific: Belmont, MA, USA, 2005; Volume 1–2.

Article

On the Age of Information in a Two-User Multiple Access Setup

Mehrdad Salimnejad * and Nikolaos Pappas *

Department of Science and Technology, Linköping University, SE-60174 Norrköping, Sweden
* Correspondence: mehrdad.salimnejad@liu.se (M.S.); nikolaos.pappas@liu.se (N.P.)

Abstract: This work considers a two-user multiple access channel in which both users have Age of Information (AoI)-oriented traffic with different characteristics. More specifically, the first user has external traffic and cannot control the generation of status updates, and the second user monitors a sensor and transmits status updates to the receiver according to a generate-at-will policy. The receiver is equipped with multiple antennas and the transmitters have single antennas; the channels are subject to Rayleigh fading and path loss. We analyze the average AoI of the first user for a discrete-time first-come-first-served (FCFS) queue, last-come-first-served (LCFS) queue, and queue with packet replacement. We derive the AoI distribution and the average AoI of the second user for a threshold policy. Then, we formulate an optimization problem to minimize the average AoI of the first user for the FCFS and LCFS with preemption queue discipline to maintain the average AoI of the second user below a given level. The constraints of the optimization problem are shown to be convex. It is also shown that the objective function of the problem for the first-come-first-served queue policy is non-convex, and a suboptimal technique is introduced to effectively solve the problem using the algorithms developed for solving a convex optimization problem. Numerical results illustrate the performance of the considered optimization algorithm versus the different parameters of the system. Finally, we discuss how the analytical results of this work can be extended to capture larger setups with more than two users.

Keywords: age of information; multiple access channels; multiple-input multiple-output Rayleigh fading channel; discrete-time Markov chain; convex optimization

Citation: Salimnejad, M.; Pappas, N. On the Age of Information in a Two-User Multiple Access Setup. *Entropy* **2022**, *24*, 542. https://doi.org/10.3390/e24040542

Academic Editor: Song-Nam Hong

Received: 1 February 2022
Accepted: 11 April 2022
Published: 12 April 2022

Publisher's Note: MDPI stays neutral with regard to jurisdictional claims in published maps and institutional affiliations.

1. Introduction

Age of Information (AoI) is considered to be a metric for characterizing the timeliness and freshness of data [1–4]. AoI was first introduced in [4], and it is defined as the time difference between the current time and the time that the latest status update was successfully received by a destination. In [4–9], the authors derived the average AoI for systems with different availability of resources using different queuing models. The $M/M/1$, $M/D/1$, and $D/M/1$ queues were studied under the first-come-first served (FCFS) queue management protocols in [4]. In [5–9], the authors considered the last-come-first-served (LCFS) queue protocols with or without the ability to preempt the packet in service. Recently, different types of traffic associated with different source nodes have been considered in which some nodes generate time-sensitive status updates and other nodes strive to achieve high throughput. The performance of a multiple access channel with heterogeneous traffic has been investigated in [10] where one user has bursty arrivals of regular data packets while another AoI-oriented sensor has energy-harvesting capabilities.

The interplay between delay guarantees and information freshness in a two-user multiple access channel with multi-packet reception (MPR) capability at the receiver and heterogeneous traffic is studied in [11]. Motivated by [11], in [12] the interplay of deadline-constrained traffic and the average AoI in a two-user random-access channel with MPR reception capabilities was investigated. The authors obtained analytical expressions for the throughput and drop rate of a user with external bursty traffic, which is the deadline-constrained and analytical expression for the average AoI of a user monitoring the sensor.

In [13], the authors presented the analysis of the average AoI with and without packet management at the transmission queue of the source nodes. In the proposed system, each source node has a buffer of infinite capacity to store incoming bursty traffic in the form of packets.

A small average AoI corresponds to having fresh information, which is a key requirement in various applications, including Internet of Things (IoT) scenarios, wireless sensor networks, industrial control, and vehicular networks. The problem of optimizing the process, i.e., of sending status updates from a user to minimize the average AoI, was studied in [14–24]. The works [14,25] consider real-time IoT monitoring systems, where IoT devices sample a physical process and transmit status updates to a remote monitor to minimize the average AoI. In [15], the worst-case average AoI and average peak AoI from a sensor in a system where all other sensors have a saturated queue are analyzed. In [16], a randomized policy, a MaxWeight policy, and a Whittle's Index policy have been proposed to minimize the AoI subject to minimum throughput requirements. In [17], the problem of minimizing AoI in various continuous-time and discrete-time queuing systems, such as the FCFS G/G/1, the LCFS G/G/1, and the G/G/∞, has been studied. In [18], the age-optimal scheduling policies in a network with general interference constraints have been studied. In [19], the authors considered an energy-harvesting sensor and determined the optimal status update policy to minimize the average AoI. In [20], several methods have been proposed for solving an AoI minimizing problem with throughput constraints. In [20–24], the authors developed the Drift-Plus-Penalty (DPP) policy from the Lyapunov optimization theory which is often used for solving stochastic network optimization problems with stability constraints. In [23], the authors applied the Lyapunov DPP method to minimize the average AoI total transmit power of sensors under constraints on the maximum average AoI and the maximum power of each sensor. In [24], the authors proposed the probabilistic random-access (PRA) and DPP methods for solving an optimization problem that aims to minimize the average AoI of the energy-harvesting node subject to the queue stability constraint of the grid-connected node. Recently, the performance of AoI has been investigated in Multiple-Input Multiple-Output (MIMO) systems [26–30]. In [26,27], the user scheduling problem has been investigated to minimize AoI in a multiuser MIMO status update system where multiple single-antenna devices send their information over a common wireless uplink channel to a multiple-antenna access point. In [28], a novel MIMO broadcast setting is studied to minimize the sum average AoI through precoding and transmission scheduling. The age-limited capacity through MIMO setup was investigated in [29], where a random subset of users are active in any transmission period. In [30], the authors analyzed and optimized the performance of AoI in a grant-free random-access system with massive MIMO.

Contributions

Motivated by [12,24,29] in this paper we consider a multiple access channel (MAC) with two users that have AoI-oriented traffic with different characteristics. The receiver has multiple antennas, and the communication channels are subject to Rayleigh fading and path loss, as depicted in Figure 1. The key contributions of this paper are:

1. We consider a MIMO Rayleigh fading channel in which the receiver has multiple antennas. Most of the literature assumes single-antenna setups.
2. In a two-user multiple access setup, we study the performance of the average AoI of the first user for the cases of FCFS, LCFS with preemption, and queue with packet replacement when external status update packets are arriving according to a Bernoulli process. The case of FCFS can be useful when we do not allow out-of-order transmissions, since we intend to also monitor the evolution of a process in addition to the freshness. The policies of LCFS with preemption and the queue with replacement are more relevant when we do not care for the evolution of the process; thus, we can allow packet dropping.

3. Furthermore, we consider a threshold for the AoI of the second user which affects the sampling and transmission frequency, and we conduct a detailed analysis for the AoI of the second user. In particular, we derive exact analytical expressions for

   - The distribution of the AoI of the second user;
   - The probability that the AoI of the second user is greater than a threshold;
   - The average AoI of the second user;
   - The average AoI of the first user for the LCFS with preemption policy by assuming the threshold for the AoI of the second user.

4. We formulate and analyze a constrained optimization problem where the objective function is the average AoI of the first user for the FCFS queue discipline, with a constraint on the average AoI for the second user, which should be less than a threshold. We show that the problem is not convex in general. Then, we propose a suboptimal approach to solve the problem. Furthermore, we solve the optimization problem where the objective function is the average AoI of the first user for the LCFS scheme with preemption.

The considered setup is expected to occur in several scenarios in wireless industrial automation (Industry 4.0, Industrial IoT), in which several processes are coexisting by sharing the same network resources, and sensing the states of a set of systems is essential.

The remainder of this paper is organized as follows. In Section 2, the system model is introduced. In Section 3, we analyze the average AoI of the first and second users; we formulate an optimization problem and propose a convex optimization algorithm to minimize the average AoI of the first user under the constraint on the average AoI for the second user. In Section 4, we present the numerical and simulation results to evaluate the performance of the proposed optimization method. Conclusions are drawn in Section 6.

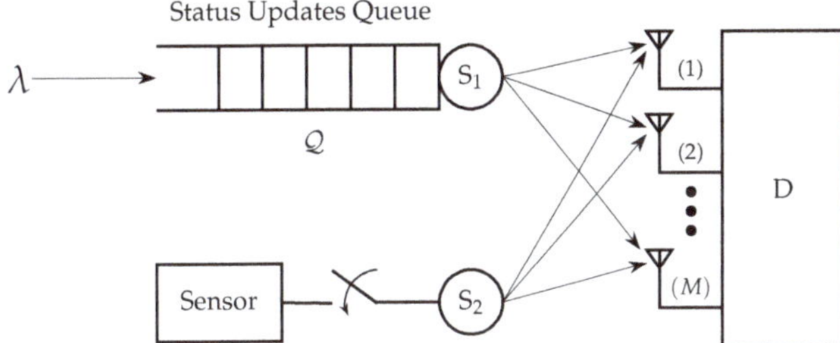

Figure 1. User 1 has AoI-oriented external bursty traffic with probability λ, user 2 has also AoI-oriented traffic but it can control the generation of status updates.

2. System Model

We consider a time-slotted MAC with two users equipped with a single antenna transmitting their information in the form of packets over a MIMO Rayleigh fading channel to a common receiver with M antenna, as shown in Figure 1. We assume that both users have AoI-oriented traffic, but with different characteristics. One of the main differences between the users is that the first one does not have control over the generation of the status update packets, but they are externally generated according to a Bernoulli process with a probability λ, while the second user can control the generation of status update packets. Let $Q(t)$ denote the status update queue of the first user in time slot t, which has infinite capacity. When the queue of the first user is not empty, it attempts to transmit its status update packets with a probability q_1. Additionally, it is assumed that the second user samples and transmits its status updates with probability q_2 based on a generate-at-will policy. Note that in Section 3.5, we will consider the case where the second user can adjust its sampling and transmission probability based on an AoI threshold.

2.1. Physical Layer Model

We assume a quasi-static Rayleigh fading model for the duration of the timeslot in which $h_i \in \mathbb{C}^M$ denotes the $M \times 1$ channel vector between the user i $(i = 1, 2)$ and the receiver and reads

$$h_i = \sqrt{\beta_i} g_i, \tag{1}$$

where $g_i \in \mathbb{C}^M$ denotes the fast-fading coefficients between user i and receiver antenna, and β_i models path loss where $\beta_i = r_i^{-\alpha}$. Please note that r_i is the distance between user i and the receiver and α is the path-loss exponent that $2 < \alpha < 7$. At each time slot, the received signal-to-noise ratio (SNR) at the receiver when only user i transmits and the received signal to interference and noise ratio (SINR) at the receiver when both users transmit are given by

$$\text{SNR}_i = \frac{P_{t,i}\beta_i \|g_i\|^2}{\sigma^2},$$

$$\text{SINR}_i = \frac{P_{t,i}\beta_i \|g_i\|^2}{\sigma^2 + P_{t,j}\beta_j |\tilde{g}_i^{\mathrm{H}} g_j|^2}, \tag{2}$$

where $P_{t,i}$ is the transmitted power by user i and σ^2 is the variance of the complex additive white Gaussian noise (AWGN) at the receiver. Please note that $\|g_i\|^2$ follows a gamma distribution with shape parameter M, and scale parameter 1 (i.e., $\|g_i\|^2 \sim \Gamma(M, 1)$). Additionally, it is shown in [31] that $\tilde{g}_i^{\mathrm{H}} g_j \sim \mathcal{CN}(0, 1) \forall i, j$, where $\tilde{g}_i^{\mathrm{H}} = g_i^{\mathrm{H}} / \|g_i\|$ and they are mutually independent and independent of $\|g_i\|^2$. In this paper, we assume MPR capability at the receiver, which means that the receiver can correctly decode packets from multiple simultaneous transmissions that are interfering with each other. It is assumed that a packet is successfully transmitted from the user i if the received SNR or SINR at the receiver exceeds a certain threshold. The success transmission probability for user i when only user i transmits $p_{i/i}$ and when both users transmit $p_{i/i,j}$ can be obtained as [31]

$$p_{i/i} = \Pr\{\text{SNR}_i > \gamma\} = \int_{\frac{\gamma\sigma^2}{P_{t,i}\beta_i}}^{\infty} \frac{z^{M-1} e^{-z}}{(M-1)!} dz = \frac{\Gamma\left[M, \frac{\gamma\sigma^2}{P_{t,i}\beta_i}\right]}{(M-1)!}, i = \{1, 2\} \tag{3a}$$

$$p_{i/i,j} = \Pr\{\text{SINR}_i > \gamma\} = \int_0^{\infty} \int_{\left(\frac{\gamma\sigma^2}{P_{t,i}\beta_i} + \frac{P_{t,j}\beta_j}{P_{t,i}\beta_i}\gamma t\right)}^{\infty} \frac{z^{M-1} e^{-(z+t)}}{(M-1)!} dz\, dt, i = \{1, 2\}, j \neq i. \tag{3b}$$

Note that for the special case of $M = 1$, (3a) and (3b) can be written as

$$p_{i/i} = \Pr\{\text{SNR}_i > \gamma\} = \exp\left(-\frac{\gamma\sigma^2}{P_{t,i}\beta_i}\right) \tag{4a}$$

$$p_{i/i,j} = \Pr\{\text{SINR}_i > \gamma\} = \exp\left(-\frac{\gamma\sigma^2}{P_{t,i}\beta_i}\right)\left(1 + \gamma\frac{P_{t,j}\beta_j}{P_{t,i}\beta_i}\right)^{-1}, i = \{1, 2\}, j \neq i. \tag{4b}$$

The results presented in this work are general and can also be applied to other types of wireless channels as long as we can calculate the aforementioned success probabilities.

2.2. The Service Probability

The service probability of a user is defined as the probability of successful transmission in a timeslot. The service probability for the first user is given by

$$\mu_1 = q_1(1 - q_2)p_{1/1} + q_1 q_2 p_{1/1,2}. \tag{5}$$

To obtain the service probability of the second user, three cases are considered as follows.

1. When the queue of the first user is empty.
2. When the queue of the first user is not empty, and it does not transmit a status update to the receiver.

3. When the queue of the first user is not empty, and it transmits a status update to the receiver with probability q_1.

The service probability of the second user can be written as

$$\mu_2 = \Pr\{Q = 0\}q_2 p_{2/2} + \Pr\{Q \neq 0\}(1 - q_1)q_2 p_{2/2} + \Pr\{Q \neq 0\}q_1 q_2 p_{2/1,2}$$

$$= q_2\left(1 - q_1 \Pr\{Q \neq 0\}\right)p_{2/2} + q_2 q_1 \Pr\{Q \neq 0\}p_{2/1,2}. \tag{6}$$

The status updates at the first user are arriving according to a Bernoulli process with a probability λ. When the status update queue of the first user is stable $\lambda < \mu_1$, the probability that the queue of user 1 is not empty can be written as

$$\Pr\{Q \neq 0\} = \frac{\lambda}{\mu_1}. \tag{7}$$

Now, using (7), the expression (6) can be written as

$$\mu_2 = q_2\left(p_{2/2} - \frac{\lambda(p_{2/2} - p_{2/1,2})}{p_{1/1} - q_2(p_{1/1} - p_{1/1,2})}\right). \tag{8}$$

3. Analysis of the Age of Information and Problem Formulation

In this section, we analyze the average AoI of the first and second users, and formulate an optimization problem to minimize the average AoI of the first user. In the following subsections, we first derive the average AoI of the first user for a discrete-time FCFS queue, LCFS queue with preemption, and queue with packet replacement policies. Then, we obtain the AoI and average AoI of the second user for a case threshold policy.

3.1. Average Age of Information of the First User

The AoI of the first user at the receiver is defined as a random process $\Delta_t = t - G(t)$, where $G(t)$ is the time slot when the latest successfully received a status update from the first user. The evolution of AoI of the first user is illustrated in Figure 2. In this figure, we assume that all packets need to be delivered to the destination regardless of the freshness of the status update information. Therefore, we consider that jth status update is generated at time slot t_j, and received by the receiver at time slot t'_j. Then, we denote $T_j = t'_j - t_j$ and $Y_j = t_j - t_{j-1}$ as the system time of update j and the interarrival time of update j, respectively. Without loss of generality, the average AoI of the first user for an interval of observation $(0, \tau)$ is defined as

$$\Delta_\tau = \frac{1}{\tau}\sum_{t=0}^{N(\tau)}\Delta_t, \tag{9}$$

where $N(\tau)$ is the number of samples during the observation interval. Using Figure 2, Equation (9) can be calculated as the area under Δ_t. Starting from $t = 0$, the area is decomposed into the areas $J_1, J_2, \ldots, J_{N(\tau)}$, and the area of width T_n over the time interval (t_n, t'_n) that is denoted by $\tilde{J}$. Therefore, one can write the average AoI of the first user as a sum of disjoint geometric parts as

$$\Delta_\tau = \frac{1}{\tau}\left(J_1 + \tilde{J} + \sum_{j=2}^{N(\tau)} J_j\right) = \frac{J_1 + \tilde{J}}{\tau} + \frac{N(\tau) - 1}{\tau}\frac{1}{N(\tau) - 1}\sum_{j=2}^{N(\tau)} J_j. \tag{10}$$

Now, the average AoI of the first user is given by

$$\bar{A}_1 = \lim_{\tau \to \infty}\Delta_\tau, \tag{11}$$

we can write the expression given in (11) as [5]

$$\bar{A}_1 = \frac{1}{\mathbb{E}[Y]}\left(\mathbb{E}[YT] + \frac{\mathbb{E}[Y^2]}{2} + \frac{\mathbb{E}[Y]}{2}\right). \tag{12}$$

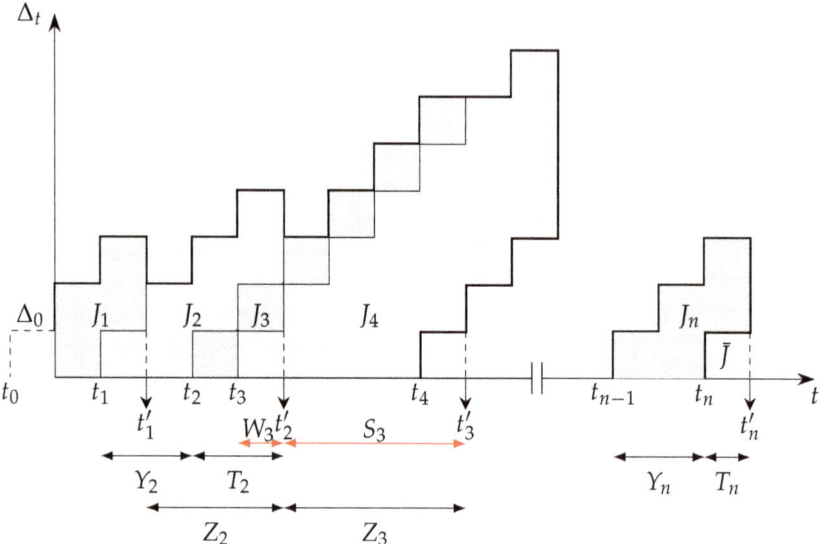

Figure 2. An example of the age evolution of user 1 at the receiver.

3.2. The FCFS Geo/Geo/1 Queue

In this section, we obtain the average AoI of the first user for a discrete-time Geo/Geo/1 queue discipline of FCFS. When status update packets are arriving according to the Bernoulli process with a probability λ, the interarrival times Y_j are i.i.d. sequences that follow a geometric distribution with probability mass function (PMF) as

$$\Pr\{Y_j = y\} = \lambda(1 - \lambda)^{y-1}, y = 1, 2, \ldots. \tag{13}$$

Thus, we can obtain $\mathbb{E}[Y]$ and $\mathbb{E}[Y^2]$ in Equation (12) as

$$\mathbb{E}[Y] = \frac{1}{\lambda}, \ \mathbb{E}[Y^2] = \frac{2 - \lambda}{\lambda^2}. \tag{14}$$

Also, the expression $\mathbb{E}[YT]$ can be obtained as [13]

$$\mathbb{E}[YT] = \frac{\lambda(1 - \mu_1)}{(\mu_1 - \lambda)\mu_1^2} + \frac{1}{\lambda\mu_1}. \tag{15}$$

Now, using Equations (14) and (15), the expression given in (12) can be written as

$$\bar{A}_1 = \frac{1}{\lambda} + \frac{1 - \lambda}{\mu_1 - \lambda} - \frac{\lambda}{\mu_1^2} + \frac{\lambda}{\mu_1}. \tag{16}$$

3.3. The Preemptive LCFS Geo/Geo/1 Queue

In this section, we consider a discrete-time LCFS Geo/Geo/1 queue with preemptive service, where a newly generated packet is given priority for service immediately. It is assumed that status update packets are arriving according to the Bernoulli process with probability λ. The probability distribution of interarrival time between the jth and $(j+1)$th status update packet is assumed to be geometric with mean $\mathbb{E}[Y] = 1/\lambda$, and the probability distribution time until successful delivery is assumed to be geometric distribution with mean $\mathbb{E}[S] = 1/\mu_1$, in which μ_1 denotes the service probability of the first user. In [32], it is shown that the PMF of AoI for a discrete-time LCFS Geo/Geo/1 queue is given by

$$\Pr\{A_1 = x\} = \frac{\lambda\mu_1\left[(1-\lambda)^{x-1} - (1-\mu_1)^{x-1}\right]}{\mu_1 - \lambda}. \tag{17}$$

Now, we can write the average AoI of the first user as

$$\bar{A}_1 = \sum_{x=1}^{\infty} x\Pr\{A_1 = x\} = \frac{1}{\lambda} + \frac{1}{\mu_1}. \tag{18}$$

3.4. Queue with Replacement

In this section, we derive the average AoI of the first user for a queue with replacement. In this case, it is assumed that a newly generated packet discards the packet waiting in the queue. As shown in Figure 2, we express the areas J_j with respect to the random variables Z_j as follows

$$
\begin{aligned}
J_j &= \sum_{m=1}^{T_{j-1}+Z_j} m - \sum_{m=1}^{T_j} m \\
&= \frac{(T_{j-1} + Z_j)(T_{j-1} + Z_j + 1)}{2} - \frac{T_j(T_j + 1)}{2}.
\end{aligned} \tag{19}
$$

We use the fact that in the steady state T_{j-1} and T_j are identically distributed. Therefore, the average AoI of the first user for a queue with packet replacement is given by [13]

$$\bar{A}_1 = \lambda_e\left(\mathbb{E}[ZT] + \frac{\mathbb{E}[Z^2]}{2} + \frac{\mathbb{E}[Z]}{2}\right), \tag{20}$$

where one can obtain λ_e, $\mathbb{E}[Z]$, $\mathbb{E}[Z^2]$ and $\mathbb{E}[ZT]$ as [13]

$$
\begin{aligned}
\lambda_e &= \lambda - \frac{\lambda^3(1-\mu_1)}{\lambda^2(1-\mu_1) + \lambda(1-\mu_1)\mu_1 + \mu_1^2} \\
\mathbb{E}[Z] &= \frac{\lambda^2(1-\mu_1) + \lambda(1-\mu_1)\mu_1 + \mu_1^2}{\lambda\mu_1(\lambda + \mu_1 - \lambda\mu_1)} \\
\mathbb{E}[Z^2] &= \frac{(2\lambda^2 + 2\lambda\mu_1 - \lambda^2\mu_1 + 2\mu_1^2 - \lambda\mu_1^2)(\mu_1 - \lambda\mu_1)}{\lambda^2\mu_1^2(\lambda + \mu_1 - \lambda\mu_1)} + \frac{\lambda(2 - \mu_1)}{\mu_1^2(\lambda + \mu_1 - \lambda\mu_1)} \\
\mathbb{E}[ZT] &= \frac{1}{\mu_1^2} + \frac{1-\lambda}{\lambda\mu_1} - \frac{1+\lambda}{(\lambda + \mu_1 - \lambda\mu_1)^2} + \frac{1+2\lambda}{\lambda + \mu_1 - \lambda\mu_1} + \frac{\lambda(1 - 2\mu_1 + \lambda(3\mu_1 - 2))}{\lambda^2(1-\mu_1)^2 + \lambda\mu_1(1 - 2\mu_1) + \mu_1^2}.
\end{aligned} \tag{21}
$$

3.5. Age of Information and the Average Age of Information of the Second User

We assume $A_2(t)$ be a positive integer that represents the AoI associated with the second user at the receiver. The AoI evolution between two consecutive time slots at the receiver can be written as

$$A_2(t+1) = \begin{cases} 1, & \text{successful packet reception at time slot t} \\ A_2(t) + 1, & \text{otherwise.} \end{cases} \tag{22}$$

According to (22), the AoI drops to one when there is a successful reception of a status update at the receiver. Otherwise, it increases by one. We can model the evolution of the AoI of the second user as a Discrete-Time Markov Chain (DTMC). The DTMC is shown in Figure 3, in which when $A_2(t) < \kappa$ (κ is the threshold of the AoI of the second user), a packet is transmitted with the probability q_2. Also, a packet is transmitted with the probability q_2' when $A_2(t) \geqslant \kappa$. The service probability of the first user can be written as

$$\mu_1' = q_1(1 - q_2')p_{1/1} + q_1 q_2' p_{1/1,2}. \tag{23}$$

According to the DTMC described in Figure 3, we can obtain the steady-state probabilities of the AoI of the second user as follows

$$\pi_i = \begin{cases} (1 - \mu_2)^{i-1}\pi_1 & , i < \kappa \\ \left(\frac{1-\mu_2}{1-\mu_2'}\right)^{\kappa-1}(1 - \mu_2')^{i-1}\pi_1 & , i \geqslant \kappa \end{cases} \tag{24}$$

where for $\lambda < \mu_1'$, μ_2' is given by

$$\mu_2' = q_2'\left(p_{2/2} - \frac{\lambda(p_{2/2} - p_{2/1,2})}{p_{1/1} - q_2'(p_{1/1} - p_{1/1,2})}\right). \tag{25}$$

Additionally, we can obtain π_1 as

$$\pi_1 = \begin{cases} \mu_2' & , \kappa = 1 \\ \frac{\mu_2 \mu_2'}{\mu_2' + (\mu_2 - \mu_2')(1 - \mu_2)^{\kappa-1}} & , \kappa \geqslant 2. \end{cases} \tag{26}$$

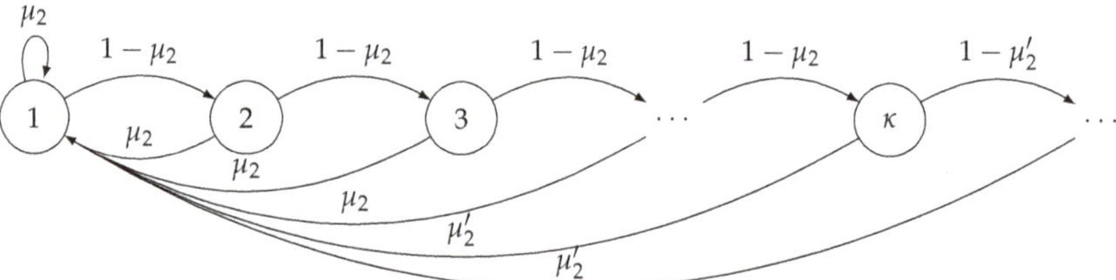

Figure 3. The DTMC, which models the evolution of AoI of the second user.

Using Equations (24) and (26), we can write the probability that the AoI of the second user is smaller than a threshold κ, as follows

$$\Pr\{A_2 < \kappa\} = \sum_{i=1}^{\kappa-1}(1 - \mu_2)^{i-1}\pi_1 = \frac{(1 - (1 - \mu_2)^{\kappa-1})\pi_1}{\mu_2}. \tag{27}$$

Furthermore, one can write the probability that the AoI of the second user is greater than a threshold, κ, as follows

$$\Pr\{A_2 \geqslant \kappa\} = \sum_{i=\kappa}^{\infty}\left(\frac{1-\mu_2}{1-\mu_2'}\right)^{\kappa-1}(1 - \mu_2')^{i-1}\pi_1$$
$$= \frac{(1 - \mu_2)^{\kappa-1}\pi_1}{\mu_2'}. \tag{28}$$

Now, using Equations (27) and (28), the average AoI of the second user is described as

$$\bar{A}_2 = \sum_{i=1}^{\infty} i\pi_i = \sum_{i=1}^{\kappa-1} i(1 - \mu_2)^{i-1}\pi_1 + \sum_{i=\kappa}^{\infty} i\left(\frac{1-\mu_2}{1-\mu_2'}\right)^{\kappa-1}(1 - \mu_2')^{i-1}\pi_1$$
$$= \frac{1 - \mu_2 - (1 - \mu_2)^{\kappa}[1 - \mu_2 + \kappa\mu_2]}{(1 - \mu_2)\mu_2^2}\pi_1 + \frac{(1 - \mu_2)^{\kappa}[1 - \mu_2' + y\mu_2']}{(1 - \mu_2)\mu_2'^2}\pi_1, \tag{29}$$

where μ_2, μ_2', and π_1 are given by (8), (25) and (26), respectively. Additionally, when $\kappa \to \infty$ and $\lambda < \mu_1$, Equation (29) can be written as

$$\bar{A}_2 = \frac{1}{\mu_2}.$$ (30)

3.6. The Average AoI of S_1 for the Preemptive LCFS Geo/Geo/1 Queue for the Threshold-Based Policy of S_2

By considering the threshold-based policy explained in Section 3.5, the average AoI of S_1 for the preemptive LCFS queue discipline given in (17) can be written as

$$\Pr\{A_1 = x\} = \Pr\{A_1 = x | A_2 < \kappa\}\Pr\{A_2 < \kappa\} + \Pr\{A_1 = x | A_2 \geqslant \kappa\}\Pr\{A_2 \geqslant \kappa\},$$ (31)

where $\Pr\{A_2 < \kappa\}$, $\Pr\{A_2 \geqslant \kappa\}$ are given by (27) and (28). Using Equation (17), the first and second conditional probabilities given in (31) can be written as

$$\Pr\{A_1 = x | A_2 < \kappa\} = \frac{\lambda\mu_1\left[(1-\lambda)^{x-1} - (1-\mu_1)^{x-1}\right]}{\mu_1 - \lambda}$$ (32a)

$$\Pr\{A_1 = x | A_2 \geqslant \kappa\} = \frac{\lambda\mu_1'\left[(1-\lambda)^{x-1} - (1-\mu_1')^{x-1}\right]}{\mu_1' - \lambda},$$ (32b)

where μ_1' is given by (23). Now, we can write the average AoI of S_1 for threshold-based policy of the AoI of S_2 as

$$\begin{aligned}
\bar{A}_1 &= \sum_{x=1}^{\infty} x\Pr\{A_1 = x\} \\
&= \left(\frac{1}{\lambda} + \frac{1}{\mu_1}\right)\left(\frac{[1-(1-\mu_2)^{\kappa-1}]\pi_1}{\mu_2}\right) + \left(\frac{1}{\lambda} + \frac{1}{\mu_1'}\right)\left(\frac{(1-\mu_2)^{\kappa-1}\pi_1}{\mu_2'}\right),
\end{aligned}$$ (33)

where π_1 is given by (26).

3.7. Optimizing the Average AoI of S_1 subject to AoI constraints on S_2

3.7.1. Using the Average AoI of the FCFS as the objective function

In this section, our objective is to minimize the average AoI of user 1 for a discrete-time Geo/Geo/1 queue discipline of FCFS with a constraint on the average AoI for user 2, which should be less than a threshold. Let $A_{\max}$ be a strictly positive real value that represents the maximum average AoI of user 2. Thus, the optimization problem is formulated as follows

$$\text{minimize} \quad \bar{A}_1$$ (34a)
$$\text{subject to} \quad \bar{A}_2 < A_{\max}.$$ (34b)

Using the expressions given in Equations (16) and (30), one can write Equation (34) as follows

$$\underset{q_1, q_2, \lambda}{\text{minimize}} \quad \frac{1}{\lambda} + \frac{1-\lambda}{\mu_1 - \lambda} - \frac{\lambda}{\mu_1^2} + \frac{\lambda}{\mu_1}$$ (35a)

$$\text{subject to} \quad \frac{1}{\mu_2} < A_{\max},$$ (35b)

$$0 \leqslant \lambda < \mu_1,$$ (35c)

$$q_1, q_2 \in [0, 1].$$ (35d)

where the constraint in (35c) ensures that the queue of the first user is stable. To solve this optimization problem, we first note that for $\lambda < \mu_1$ when the service probability of the first user increases, the objective function given in (35a) decreases. Hence, to minimize the objective function, we must obtain the maximum value of μ_1. The service probability of the first user given in (5) can be simplified as

$$
\begin{aligned}
\mu_1 &= q_1(1 - q_2)p_{1/1} + q_1 q_2 p_{1/1,2} \\
&= q_1 \left[p_{1/1} - q_2(p_{1/1} - p_{1/1,2}) \right].
\end{aligned}
\tag{36}
$$

According to Equation (36), μ_1 has its maximum value when q_1 is maximum. Therefore, by selecting $q_1 = 1$, we can maximize the service probability of the first user and minimize the average AoI of the first user as an objective function. Therefore, the optimal value of q_1 is given by

$$
q_1^* = 1.
\tag{37}
$$

Now, using Equations (8), (36) and (37), we can write the optimization problem given in (35) as

$$
\underset{q_2, \lambda}{\text{minimize}} \quad \frac{1}{\lambda} + \frac{1 - \lambda}{p_{1/1} - q_2(p_{1/1} - p_{1/1,2}) - \lambda} - \frac{\lambda \left(1 - p_{1/1} + q_2(p_{1/1} - p_{1/1,2}) \right)}{\left(p_{1/1} - q_2(p_{1/1} - p_{1/1,2}) \right)^2}
\tag{38a}
$$

$$
\text{subject to} \quad \frac{p_{1/1} - q_2(p_{1/1} - p_{1/1,2})}{q_2 \left(p_{1/1} p_{2/2} - q_2 p_{2/2}(p_{1/1} - p_{1/1,2}) - \lambda(p_{2/2} - p_{2/1,2}) \right)} - A_{\max} < 0, \tag{38b}
$$

$$
\lambda - p_{1/1} + q_2(p_{1/1} - p_{1/1,2}) < 0, \tag{38c}
$$

$$
\lambda, q_2 \in [0, 1]. \tag{38d}
$$

By definition, an optimization problem is convex when its objective function and the inequality constraints are convex, and its equality constraints are affine, see Chapter 4.2 in [33]. We can show that the Hessian matrix of the objective function given in (38a) is positive semi-definite for some parameters of λ and q_2 and for some others is not positive semi-definite and therefore it is not a convex function. Additionally, it can be verified that the Hessian matrices of the inequality constraints (38b) and (38c) are positive semi-definite for different values of λ and q_2. Therefore, this optimization problem is not a convex optimization problem, a trivial solution does not exist for this problem, and finding the optimal solution is computationally involved. Hence, to find the optimal values of λ and q_2, a suboptimal technique is proposed to effectively solve the problem using an algorithm developed for solving convex optimization problems. This approach is known as the bilevel optimization algorithm and is used when optimization parameters are interdependent, and the optimization problem is convex with respect to each of the optimization parameters when other parameters are fixed [34].

3.7.2. Bilevel Convex Optimization

Using the procedure explained in Appendix A, it can be verified that the objective function given in (38a) is a convex function of λ when q_2 is fixed and $\lambda < \mu_1$. Therefore, the optimization problem can be solved for λ by assuming that q_2 is fixed. Then, substituting for λ in (38a) from the previous stage and assuming that this parameter is fixed, we can solve the optimization problem for q_2. This procedure continues until the convergence condition is satisfied (for example, the change in the objective function in two successive iterations is lower than a small threshold).

In this paper, an interior-point method is used to solve the optimization problem in each iteration of the bilevel optimization algorithm. The iteration complexity of this method is shown in Chapter 3.4.3 in the work of den Hertog [35] to be $\mathcal{O}(\nu(c\sqrt{n}))$, where ν denotes the number of iterations, n is the number of constraints and c is a constant, which depends on system parameters such as tolerance.

3.7.3. Using the Average AoI of the LCFS with Preemption as an Objective Function

In this section, our objective is to minimize the average AoI of user 1 for a discrete-time preemptive LCFS Geo/Geo/1 queue discipline with a constraint on the average AoI for user 2, which should be less than a threshold. Using the expressions given in Equations (18) and (30), the optimization problem is formulated as follows

$$\underset{q_1,q_2,\lambda}{\text{minimize}} \quad \frac{1}{\lambda} + \frac{1}{\mu_1} \tag{39a}$$

$$\text{subject to} \quad \frac{1}{\mu_2} < A_{\max}, \tag{39b}$$

$$0 \leqslant \lambda < \mu_1, \tag{39c}$$

$$q_1, q_2 \in [0,1]. \tag{39d}$$

Using the procedure explained in Section 3.7.1, the $q_1^* = 1$ and the optimization problem given in (39) is simplified as

$$\underset{q_2,\lambda}{\text{minimize}} \quad \frac{1}{\lambda} + \frac{1}{p_{1/1} - q_2(p_{1/1} - p_{1/1,2})} \tag{40a}$$

$$\text{subject to} \quad \frac{p_{1/1} - q_2(p_{1/1} - p_{1/1,2})}{q_2\left(p_{1/1}p_{2/2} - q_2 p_{2/2}(p_{1/1} - p_{1/1,2}) - \lambda(p_{2/2} - p_{2/1,2})\right)} - A_{\max} < 0, \tag{40b}$$

$$\lambda - p_{1/1} + q_2(p_{1/1} - p_{1/1,2}) < 0, \tag{40c}$$

$$\lambda, q_2 \in [0,1]. \tag{40d}$$

We can prove that the Hessian matrix of the objective function given in (40a) is positive semi-definite and the optimization problem is convex (see Appendix B). Therefore, this optimization problem can be solved using an algorithm developed for solving convex optimization problems such as the interior-point method.

4. Numerical Results and Discussion

In this section, we illustrate our analytical results presented in Section 3 and we verify them by means of computer simulation. Simulation results are obtained using 10^6 independent realizations of the system. Additionally, we evaluate the performance of the proposed interior-point algorithm presented in Section 3. It is assumed that the users are located at a distance $r_i = 30$ m ($i = 1, 2$) from the receiver. The receiver noise power is assumed to be $\sigma^2 = -100$ dBm, and the path-loss exponent is $\alpha = 4$. Additionally, the assumed transmit powers are $P_{t,1} = P_{t,2} = 5$ mW, and the transmission channels between the users and receiver are subject to Rayleigh fading model and we use the expressions for the success probabilities that were presented in Section 2. Furthermore, the initial point for the interior-point algorithm is zero, i.e., $\left(\lambda^{(0)}, q_2^{(0)}\right) = 0$.

Figure 4 shows the average AoI of the first user for the FCFS Geo/Geo/1 queue, preemptive LCFS Geo/Geo/1 queue, and queue with replacement as a function of λ, $\gamma = -5$ dB, $M = 1$, $q_1 = 0.8$, and $q_2 = 0.2$. As seen in this figure, the preemptive LCFS Geo/Geo/1 queue outperforms the FCFS Geo/Geo/1 queue, and queue with replacement. Additionally, note that the average AoI of the first user for the FCFS Geo/Geo/1 queue we plot for $\lambda < 0.7$ to satisfy the stability requirements. Figure 4 also shows that the simulation results match the analytical results.

The average AoI of S_1 and S_2 are shown in Figure 5 as a function of κ, for $\lambda = 0.5$, $q_1 = 1$, $q_2 = 0.2$, $q_2' = 0.5$, $\gamma = 3$ dB, and various values of M. As seen in this figure, as κ increases, the slope of the average AoI of S_2 and S_1 decreases because when κ becomes larger, the average AoI of S_2 and S_1 tends to $1/\mu_2$ and $1/\lambda + 1/\mu_1$, respectively, and becomes independent of q_2'. Therefore, by changing q_2' the average AoI of S_2 and S_1 will not change. Furthermore, when κ increases, the average AoI of S_2 increases and the average AoI of S_1 decreases. This is because for smaller values of κ, a packet is transmitted with probability q_2' that $q_2' > q_2$ sooner than larger values of κ. Therefore, the average AoI of the first and second users has larger and smaller values, respectively, for smaller values of κ. Additionally, note that when M increases, the average AoI of the first and second users decreases because for a larger value of M the service probabilities of the first and second users have larger values such that they decrease the average AoI of S_1 and S_2.

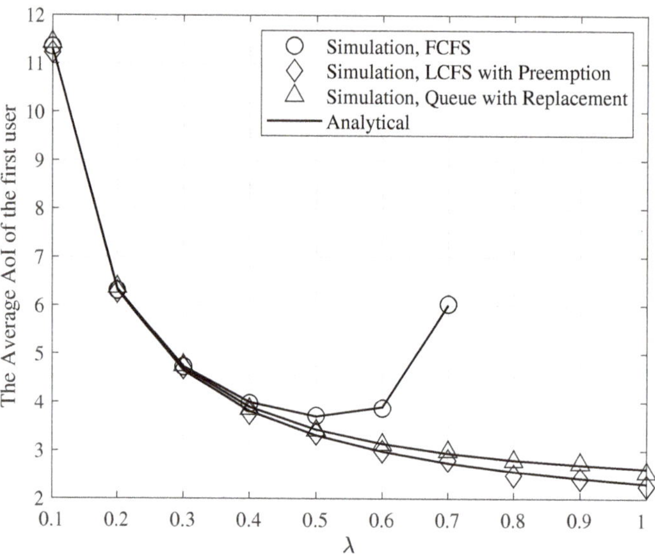

Figure 4. The average AoI of the first user for the FCFS Geo/Geo/1 queue, preemptive LCFS Geo/Geo/1 queue, and queue with replacement for $\gamma = -5$ dB, $M = 1$, $q_1 = 0.8$, and $q_2 = 0.2$, and various values of λ.

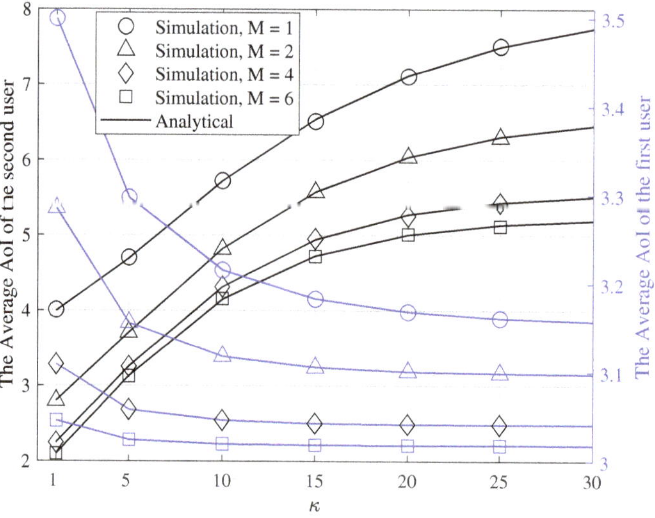

Figure 5. The average AoI of the S_1 and S_2 for $\gamma = 3$ dB, $\lambda = 0.5$, $q_1 = 1$, $q_2 = 0.2$, $q_2' = 0.5$, $\kappa = 1, 5, 10, \ldots, 30$, and various values of M.

Figure 6 shows the probability that the AoI of the second user to be greater than a threshold as a function of λ for $q_1 = 1$, $q_2 = 0.2$, $q_2' = 0.5$, $\gamma = 3$ dB, $x = 5$, and selected values of M. As seen in this figure, when λ increases the probability $\Pr\{A_2 \geqslant x\}$ increases. This is because when λ increases the service probability of the second user decreases and therefore it increases the AoI of the second user. Furthermore, by increasing M, the success transmission probabilities increase, and the service probability of the second user increases. As a result, the AoI of the second user decreases and the probability $\Pr\{A_2 \geqslant x\}$ has lower values. Note, importantly, that when $M = 1$, the probability $\Pr\{A_2 \geqslant x\}$ does not have a value for $\lambda > 0.6$. This is because for $\lambda > 0.6$, λ becomes larger than μ_1 and thus the AoI of the second user does not have a value.

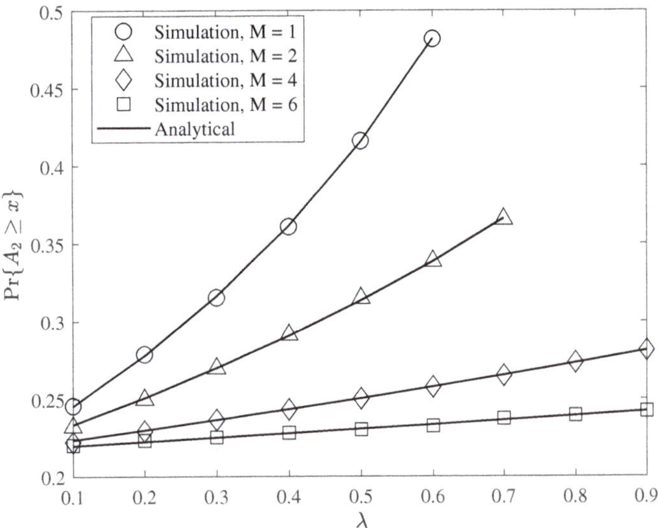

Figure 6. The probability the AoI of the S_2 at the receiver greater than a threshold, $x = 5$, for $\gamma = 3$ dB, $q_1 = 1$, $q_2 = 0.2$, $q_2' = 0.5$, $\lambda = 0.1, 0.2, \ldots, 1$, and various values of M.

Figure 7 shows the average service time of the first user, $1/\mu_1$, as a function of $P_{t,1}$ ($P_{t,1} = P_{t,2}$) for $\sigma^2 = -50$ dBm, $\alpha = 4$, $r_i = 30$ m ($i = 1, 2$), $\gamma = 0$ dB, $q_1 = 0.8$, $q_2 = 0.4$, and selected values of M. As seen in this figure, when transmitted power increases, the average service time decreases. This is because by increasing the transmitted power, the success transmission probabilities increases and thus the average service time decreases. Furthermore, when M increases, the average service time decreases. This is because when M increases, the service probability increases such that it decreases the average service time.

The average service time is illustrated in Figure 8 as a function of r_1 ($r_1 = r_2$) for $\sigma^2 = -50$ dBm, $\alpha = 4$, $P_{t,1} = P_{t,2} = 5$ mW, $\gamma = 0$ dB, $q_1 = 0.8$, $q_2 = 0.4$, and various values of M. As seen in this figure, the average service time increases by increasing r_1. This is because when r_1 increases, the service probability of the first user decreases, which results in decreasing in the average service time. Moreover, by increasing M, the average service time decreases. This is because when M increases, the success probabilities increase and therefore the average service time decreases.

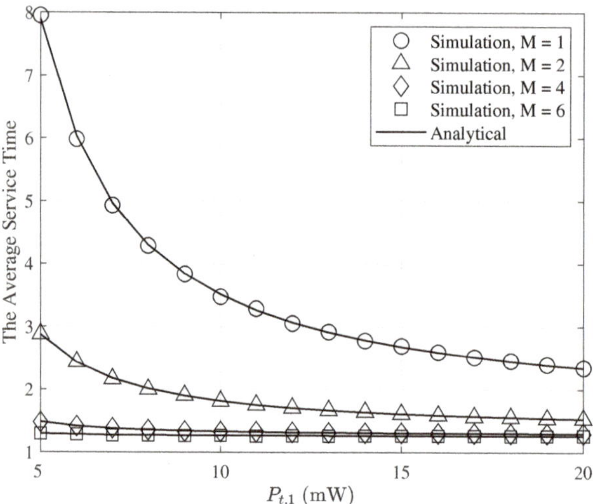

Figure 7. The average service time of the first user as a function of $P_{t,1}$ for $\sigma^2 = -50$ dBm, $\alpha = 4$, $r_1 = 30$ m $(i = 1, 2)$, $\gamma = 0$ dB, $q_1 = 0.8$, $q_2 = 0.4$, and various values of M.

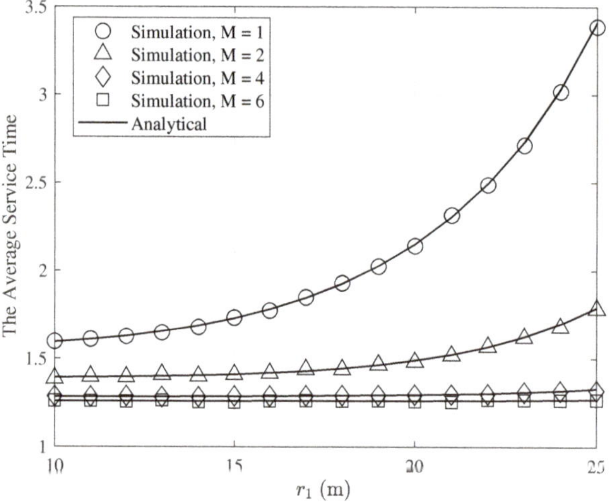

Figure 8. The average service time of the first user as a function of r_1 for $\sigma^2 = -50$ dBm, $\alpha = 4$, $P_{t,1} = P_{t,2} = 5$ mW, $\gamma = 0$ dB, $q_1 = 0.8$, $q_2 = 0.4$, and various values of M.

The minimum average AoI of the first user for the case where $M = 1$ as a function of γ and selected values of $A_{\max}$ is illustrated in Figure 9. As seen in this figure and Tables 1–4, the minimum average AoI of the first user has a larger value when the SNR threshold γ is larger. This is because a higher γ gives lower success probabilities and therefore increases the minimum average AoI. An important observation is that as $A_{\max}$ increases, the average AoI of the first user does not depend on $A_{\max}$. This is because when $A_{\max}$ increases, the constraint on the average AoI of the second user becomes independent of $A_{\max}$. Hence, when $A_{\max}$ increases, the optimal value of transmit probability q_2, λ and therefore the minimum average AoI of the first user will not change.

Table 1. The minimum average AoI of the first user and the optimal values of λ^*, q_1^*, and q_2^* for $A_{\max} = 2$.

γ	λ^*	q_1^*	q_2^*	The Minimum Average AoI of the First User
-5 dB	0.6159	1	0.6047	3.10
-3 dB	0.5264	1	0.6443	3.54
-1 dB	0.4263	1	0.6859	4.20
1 dB	0.3264	1	0.7175	5.16
3 dB	0.2425	1	0.7289	6.46
5 dB	0.1832	1	0.7239	8.02

Table 2. The minimum average AoI of the first user and the optimal values of λ^*, q_1^*, and q_2^* for $A_{\max} = 5$.

γ	λ^*	q_1^*	q_2^*	The Minimum Average AoI of the First User
-5 dB	0.7539	1	0.2477	2.59
-3 dB	0.6947	1	0.2684	2.78
-1 dB	0.6260	1	0.2935	3.02
1 dB	0.5519	1	0.3196	3.32
3 dB	0.4806	1	0.3416	3.67
5 dB	0.4208	1	0.3560	4.04

Table 3. The minimum average AoI of the first user and the optimal values of λ^*, q_1^*, and q_2^* for $A_{\max} = 10$.

γ	λ^*	q_1^*	q_2^*	The Minimum Average AoI of the First User
-5 dB	0.8246	1	0.1257	2.39
-3 dB	0.7815	1	0.1376	2.50
-1 dB	0.7303	1	0.1531	2.64
1 dB	0.6730	1	0.1708	2.81
3 dB	0.6147	1	0.1880	3.01
5 dB	0.5622	1	0.2019	3.19

Table 4. The minimum average AoI of the first user and the optimal values of λ^*, q_1^*, and q_2^* for $A_{\max} = 15$.

γ	λ^*	q_1^*	q_2^*	The Minimum Average AoI of the First User
-5 dB	0.8563	1	0.0844	2.31
-3 dB	0.8206	1	0.0929	2.40
-1 dB	0.7777	1	0.1043	2.51
1 dB	0.7288	1	0.1179	2.63
3 dB	0.6775	1	0.1320	2.78
5 dB	0.6297	1	0.1441	2.92

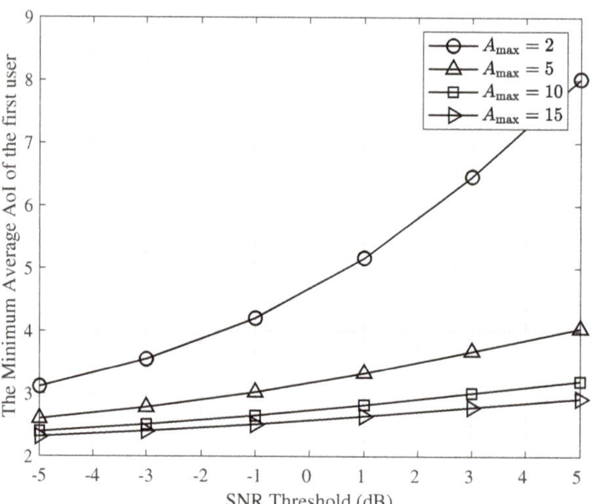

Figure 9. The minimum average AoI of user 1, for $M = 1$, $A_{\max} = 2, 5, 10, 15$, and various values of γ.

Figure 10 shows the interplay between the average AoI of the first user for a discrete-time Geo/Geo/1 queue discipline of FCFS and the average AoI of the second user when $y \to \infty$ as a function of q_2 and selected values of γ. In this figure, we consider the weak/strong MPR capabilities. We denote that the strong and weak MPR capability of a receiver corresponds to $K = \frac{p_{1/1,2}}{p_{1/1}} + \frac{p_{2/1,2}}{p_{2/2}} > 1$ and $K = \frac{p_{1/1,2}}{p_{1/1}} + \frac{p_{2/1,2}}{p_{2/2}} < 1$, respectively [24]. When $M = 1$ for $\gamma = -5$ dB and $\gamma = -3$ dB, $K = 1.51$ and $K = 1.33$, respectively. Therefore, the receiver has strong MPR capabilities. Furthermore, for $\gamma = 1$ dB and $\gamma = 3$ dB, $K = 0.88$ and $K = 0.66$, respectively; thus, the receiver has weak MPR capabilities. Additionally, when $M = 2$ for $\gamma \in \{-5, -3, 1, 5\}$ dB, $K \in \{1.88, 1.77, 1.37, 1.11\}$, respectively. Moreover, when $M = 4$ for $\gamma \in \{-5, -3, 1, 5\}$ dB, $K \in \{1.99, 1.97, 1.80, 1.60\}$, respectively. Therefore, when $M > 1$ the receiver has strong MPR capabilities for selected values of γ. In Figure 10, we consider a different scenario from the optimization problem scenario in (34). Here we intend to find transmission probabilities q_1 and q_2 that both users transmit at the same time to keep the average AoI of the first and second users below a threshold $A_{1\max 1}$ and $A_{2\max}$ respectively. Observe that when the receiver has strong MPR capabilities, we have $\bar{A}_1 < A_{1\max}$ and $\bar{A}_2 < A_{2\max}$ with a high value of transmit probability q_2. Therefore, in this case, both users can transmit at the same time with a high probability. For example, we assume the thresholds for the average AoI of the first and second users are equal to $A_{1\max} = 6$ and $A_{2\max} = 6$, respectively. As seen in this figure when M equals 1, and $\gamma = -5$ dB (strong MPR capability), we can achieve our purpose with $q_1 = 0.6$ and $q_2 = 0.5$ while for $\gamma = 1$ dB (weak MPR capability), we cannot find a value of q_2 to achieve our goal. In this case, both users cannot transmit at the same time. Additionally, observe that in Figure 10b, when $\gamma = -5$ dB the first and second users can transmit at the same time with the probabilities of $q_1 = 0.6$ and $q_2 = 1$. Furthermore, when $\gamma = 1$ dB, the transmit probability of q_1 and q_2 are equal to $q_1 = 0.6$ and $q_2 = 0.4$. In addition, as shown in Figure 10c, when $\gamma = -5$ dB and $\gamma = 1$ dB, both users can transmit at the same time with the probabilities of $q_1 = 0.6$ and $q_2 = 1$. This reflects the fact that when the number of receiver antenna increases, the receiver has a strong MPR capability for higher values of γ and thus both users can transmit at the same time with a high probability.

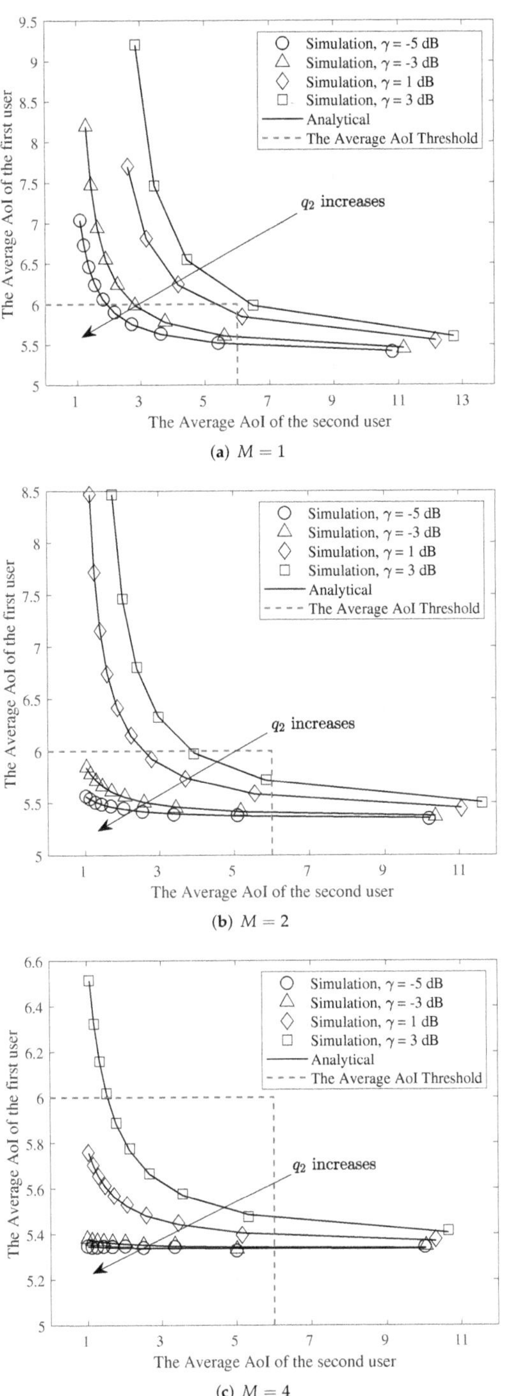

(**a**) $M = 1$

(**b**) $M = 2$

(**c**) $M = 4$

Figure 10. The interplay between the average AoI of the first and second users for $q_1 = 0.6$, $\lambda = 0.3$, $q_2 = 0.1, 0.2, \ldots, 1$, (**a**) $M = 1$, (**b**) $M = 2$, and (**c**) $M = 4$.

5. Discussion on Larger Topology

In this section, we discuss how this work can be extended to capture more than two users. However, detailed analysis and optimization for more than two users is left for a future publication. Below, we provide some details for a setup with two users with external traffic and two users with control over the generation of the status updates as depicted in Figure 11. More specifically, it is assumed that the users S_1 and S_2 do not have control over the generation of status update packets, and they are externally generated according to Bernoulli processes with probabilities λ_1 and λ_2, respectively. When the queue of S_i, $i = 1, 2$ is not empty, S_i attempts to transmit with probability q_i. We also consider that users S_3 and S_4 can control the generation of status update packets; thus, they sample and transmit with probabilities q_3 and q_4, respectively, based on a generate-at-will policy. Using the same approach presented in Section 2.2, we derive the service probabilities of S_1, and S_3 as

$$
\begin{aligned}
\mu_1 =\ & q_1 \Pr\{Q_2 = 0\}(1 - q_3)(1 - q_4)p_{1/1} + q_1 \Pr\{Q_2 = 0\}q_3(1 - q_4)p_{1/1,3} \\
& + q_1 \Pr\{Q_2 = 0\}(1 - q_3)q_4 p_{1/1,4} + q_1 \Pr\{Q_2 = 0\}q_3 q_4 p_{1/1,3,4} \\
& + q_1 \Pr\{Q_2 \neq 0\}(1 - q_2)(1 - q_3)(1 - q_4)p_{1/1} + q_1 \Pr\{Q_2 \neq 0\}(1 - q_2)q_3(1 - q_4)p_{1/1,3} \\
& + q_1 \Pr\{Q_2 \neq 0\}(1 - q_2)(1 - q_3)q_4 p_{1/1,4} + q_1 \Pr\{Q_2 \neq 0\}(1 - q_2)q_3 q_4 p_{1/1,3,4} \\
& + q_1 \Pr\{Q_2 \neq 0\}q_2(1 - q_3)(1 - q_4)p_{1/1,2} + q_1 \Pr\{Q_2 \neq 0\}q_2 q_3(1 - q_4)p_{1/1,2,3} \\
& + q_1 \Pr\{Q_2 \neq 0\}q_2(1 - q_3)q_4 p_{1/1,2,4} + q_1 \Pr\{Q_2 \neq 0\}q_2 q_3 q_4 p_{1/1,2,3,4} \\
=\ & q_1(1 - q_3)(1 - q_4)\left[p_{1/1} - q_2 \Pr\{Q_2 \neq 0\}(p_{1/1} - p_{1/1,2})\right] \\
& + q_1 q_3(1 - q_4)\left[p_{1/1,3} - q_2 \Pr\{Q_2 \neq 0\}(p_{1/1,3} - p_{1/1,2,3})\right] \\
& + q_1 q_4(1 - q_3)\left[p_{1/1,4} - q_2 \Pr\{Q_2 \neq 0\}(p_{1/1,4} - p_{1/1,2,4})\right] \\
& + q_1 q_3 q_4\left[p_{1/1,3,4} - q_2 \Pr\{Q_2 \neq 0\}(p_{1/13,4} - p_{1/1,2,3,4})\right]
\end{aligned}
\tag{41}
$$

$$
\begin{aligned}
\mu_3 =\ & \Pr\{Q_1 = 0, Q_2 = 0\}q_3(1 - q_4)p_{3/3} + \Pr\{Q_1 = 0, Q_2 = 0\}q_3 q_4 p_{3/3,4} \\
& + \Pr\{Q_1 \neq 0, Q_2 = 0\}q_3(1 - q_1)(1 - q_4)p_{3/3} + \Pr\{Q_1 \neq 0, Q_2 = 0\}q_3 q_4(1 - q_1)p_{3/3,4} \\
& + \Pr\{Q_1 \neq 0, Q_2 = 0\}q_1 q_3(1 - q_4)p_{3/1,3} + \Pr\{Q_1 \neq 0, Q_2 = 0\}q_1 q_3 q_4 p_{3/1,3,4} \\
& + \Pr\{Q_1 = 0, Q_2 \neq 0\}q_3(1 - q_2)(1 - q_4)p_{3/3} + \Pr\{Q_1 = 0, Q_2 \neq 0\}q_3(1 - q_2)q_4 p_{3/3,4} \\
& + \Pr\{Q_1 = 0, Q_2 \neq 0\}q_2 q_3(1 - q_4)p_{3/2,3} + \Pr\{Q_1 = 0, Q_2 \neq 0\}q_2 q_3 q_4 p_{3/2,3,4} \\
& + \Pr\{Q_1 \neq 0, Q_2 \neq 0\}q_3(1 - q_1)(1 - q_2)(1 - q_4)p_{3/3} \\
& + \Pr\{Q_1 \neq 0, Q_2 \neq 0\}q_2 q_3(1 - q_1)(1 - q_4)p_{3/2,3} \\
& + \Pr\{Q_1 \neq 0, Q_2 \neq 0\}q_3 q_4(1 - q_1)(1 - q_2)p_{3/3,4} \\
& + \Pr\{Q_1 \neq 0, Q_2 \neq 0\}q_1 q_3(1 - q_2)(1 - q_4)p_{3/1,3} \\
& + \Pr\{Q_1 \neq 0, Q_2 \neq 0\}q_2 q_3 q_4(1 - q_1)p_{3/2,3,4} + \Pr\{Q_1 \neq 0, Q_2 \neq 0\}q_1 q_2 q_3(1 - q_4)p_{3/1,2,3} \\
& + \Pr\{Q_1 \neq 0, Q_2 \neq 0\}q_1 q_3 q_4(1 - q_2)p_{3/1,3,4} + \Pr\{Q_1 \neq 0, Q_2 \neq 0\}q_1 q_2 q_3 q_4 p_{3/1,2,3,4}
\end{aligned}
\tag{42}
$$

Similarly, we can write the service probabilities for S_2 and S_4. As we can observe, we see that the service probability of S_1 depends on the state of the queue of S_2 and vice versa. Thus, the queues are coupled, which is a known problem and closed form solutions cannot be obtained for more than three users. Furthermore, the service probabilities of S_3 and S_4 depend on the joint PDF of the queues of S_1 and S_2.

A way to bypass the difficulty due to coupling among the queues is to assume independence as in [13], or if we further assume that the queues have finite capacity then we can use semi-analytical methods from queuing theory. In the first case, we can approximate the performance and that approximation is tight for higher values of the arrival probabilities. After characterizing the service probabilities for each node, we can use the provided analysis in the earlier sections. In a scenario where we have only one user with a queue and N users with the generate-at-will policy, the extension becomes a trivial exercise of our analytical results since one can use the expressions directly.

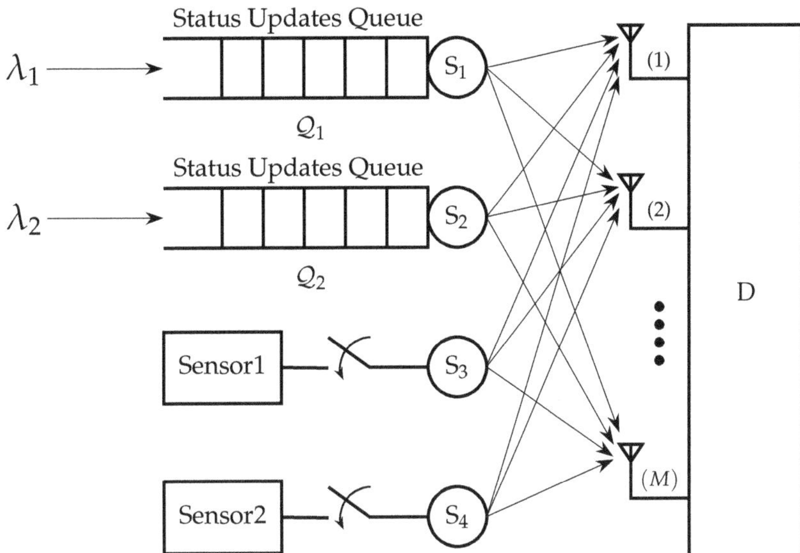

Figure 11. S_1 and S_2 have AoI-oriented external bursty traffic, S_3 and S_4 have also AoI-oriented traffic but they can control the generation of status updates.

6. Conclusions

In this work, we considered a two-user multiple access channel in which both users have AoI-oriented traffic, but with different characteristics. All transmission channels were assumed to be subject to path loss and fading. We have investigated the performance of the average AoI of the first user for the FCFS, LCFS Geo/Geo/1 with preemption, and queue with replacement. Additionally, we have derived the AoI and the average AoI of the second user by considering a threshold for the AoI of the second user. Then, we have formulated an optimization problem to minimize the average AoI of the first user with a constraint on the average AoI of the second user. To solve the proposed optimization problem, we used the interior-point method. Numerical results showed the performance of the proposed algorithm for the different parameters of the system and the impact of multiple antennas.

Future extensions of this work include larger topologies, as discussed in Section 5. Furthermore, an interesting extension is to consider more elaborate schemes at the physical layer such as the MMSE receiver or zero-forcing. Another interesting direction is to consider power control schemes with dynamic programming methodologies such as Markov Decision Processes or stochastic optimization.

Author Contributions: Conceptualization, N.P.; Formal analysis, M.S.; Supervision, N.P.; Writing—original draft, M.S.; Writing—review & editing, M.S. and N.P. All authors have read and agreed to the published version of the manuscript.

Funding: This work was supported by the Swedish Research Council (VR), ELLIIT, and CENIIT.

Institutional Review Board Statement: Not applicable.

Informed Consent Statement: Not applicable.

Data Availability Statement: Not applicable.

Conflicts of Interest: The authors declare no conflict of interest.

Appendix A

To prove the convexity of the objective function, we first define the expression given in (38a) as

$$\mathcal{A} = \frac{1}{\lambda} + \frac{1-\lambda}{p_{1/1} - q_2(p_{1/1} - p_{1/1,2}) - \lambda} - \frac{\lambda\left(1 - p_{1/1} + q_2(p_{1/1} - p_{1/1,2})\right)}{\left(p_{1/1} - q_2(p_{1/1} - p_{1/1,2})\right)^2}. \tag{A1}$$

Now, by taking the second derivative $\frac{\mathrm{d}^2 \mathcal{A}}{\mathrm{d}\lambda^2}$ we have

$$\frac{\mathrm{d}^2 \mathcal{A}}{\mathrm{d}\lambda^2} = \frac{2}{\lambda^3} + \frac{2(1-X)}{(X-\lambda)^3} \tag{A2}$$

where $X = p_{1/1} - q_2(p_{1/1} - p_{1/1,2})$, and using the expression given in (38c), we have $\lambda < X \leqslant 1$. Therefore, for all values of X and λ, the second derivative $\frac{\mathrm{d}^2 \mathcal{A}}{\mathrm{d}\lambda^2}$ is positive, and thus the objective function is a convex function of λ when q_2 is fixed. Similarly, by taking the second derivative $\frac{\mathrm{d}^2 \mathcal{A}}{\mathrm{d}q_2^2}$ when λ is fixed we have

$$\frac{\mathrm{d}^2 \mathcal{A}}{\mathrm{d}q_2^2} = (p_{1/1} - p_{1/1,2})^2 \left[\frac{-2\lambda(3-X)}{X^4} + \frac{2(1-\lambda)}{(X-\lambda)^3} \right] \tag{A3}$$

where $X = p_{1/1} - q_2(p_{1/1} - p_{1/1,2})$ and $\lambda < X \leqslant 1$. It can be readily shown that for all values of X and λ, $\frac{\mathrm{d}^2 \mathcal{A}}{\mathrm{d}q_2^2}$ is positive and therefore the objective function $\mathcal{A}$ is a convex function of q_2 when λ is fixed.

Appendix B

In order to prove the convexity of the objective function given in (40a), we must verify that the Hessian matrix of the objective function is positive semi-definite. We first consider the objective function as

$$\mathcal{B} = \frac{1}{\lambda} + \frac{1}{p_{1/1} - q_2(p_{1/1} - p_{1/1,2})}. \tag{A4}$$

To obtain the Hessian matrix, we need to derive $\frac{\mathrm{d}^2 \mathcal{B}}{\mathrm{d}\lambda^2}$, $\frac{\mathrm{d}^2 \mathcal{B}}{\mathrm{d}\lambda \mathrm{d}q_2}$, and $\frac{\mathrm{d}^2 \mathcal{B}}{\mathrm{d}q_2^2}$.

$$\frac{\mathrm{d}^2 \mathcal{B}}{\mathrm{d}\lambda^2} = \frac{2}{\lambda^3} \geqslant 0$$

$$\frac{\mathrm{d}^2 \mathcal{B}}{\mathrm{d}\lambda \mathrm{d}q_2} = 0$$

$$\frac{\mathrm{d}^2 \mathcal{B}}{\mathrm{d}q_2^2} = \frac{2(p_{1/1} - p_{1/1,2})}{(p_{1/1} - q_2(p_{1/1} - p_{1/1,2}))^3} \geqslant 0. \tag{A5}$$

According to the stability constraint in (40c), $\lambda < p_{1/1} - q_2(p_{1/1} - p_{1/1,2}) \leqslant 1$ and therefore $\frac{\mathrm{d}^2 \mathcal{B}}{\mathrm{d}q_2^2} \geqslant 0$. Since $\frac{\mathrm{d}^2 \mathcal{B}}{\mathrm{d}\lambda^2} \times \frac{\mathrm{d}^2 \mathcal{B}}{\mathrm{d}q_2^2} - \left(\frac{\mathrm{d}^2 \mathcal{B}}{\mathrm{d}\lambda \mathrm{d}q_2} \right)^2 > 0$, the Hessian matrix is positive semi-definite and the objective function is convex.

References

1. Kosta, A.; Pappas, N.; Angelakis, V. Age of information: A new concept, metric, and tool. *Found. Trends Netw.* **2017**, *12*, 162–259. [CrossRef]
2. Sun, Y.; Kadota, I.; Talak, R.; Modiano, E. Age of information: A new metric for information freshness. *Synth. Lect. Commun. Netw.* **2019**, *12*, 1–224. [CrossRef]

3. Kaul, S.; Gruteser, M.; Rai, V.; Kenney, J. Minimizing age of information in vehicular networks. In Proceedings of the 2011 8th Annual IEEE Communications Society Conference on Sensor, Mesh and Ad Hoc Communications and Networks, Salt Lake City, UT, USA, 27–30 June 2011; pp. 350–358.

4. Kaul, S.; Yates, R.; Gruteser, M. Real-time status: How often should one update? In Proceedings of the 2012 IEEE INFOCOM, Orlando, FL, USA, 25–30 March 2012; pp. 2731–2735.

5. Kaul, S.K.; Yates, R.D.; Gruteser, M. Status updates through queues. In Proceedings of the 2012 46th Annual Conference on Information Sciences and Systems (CISS), Princeton, NJ, USA, 21–23 March 2012; pp. 1–6.

6. Najm, E.; Nasser, R. Age of information: The gamma awakening. In Proceedings of the 2016 IEEE International Symposium on Information Theory (ISIT), Barcelona, Spain, 10–15 July 2016; pp. 2574–2578.

7. Bedewy, A.M.; Sun, Y.; Shroff, N.B. Optimizing data freshness, throughput, and delay in multi-server information-update systems. In Proceedings of the 2016 IEEE International Symposium on Information Theory (ISIT), Barcelona, Spain, 10–15 July 2016; pp. 2569–2573.

8. Yates, R.D. Age of information in a network of preemptive servers. In Proceedings of the IEEE INFOCOM 2018—IEEE Conference on Computer Communications Workshops (INFOCOM WKSHPS), Honolulu, HI, USA, 15–19 April 2018; pp. 118–123.

9. Najm, E.; Telatar, E. Status updates in a multi-stream M/G/1/1 preemptive queue. In Proceedings of the IEEE Infocom 2018—IEEE Conference On Computer Communications Workshops (Infocom Wkshps), Honolulu, HI, USA, 15–19 April 2018; pp. 124–129.

10. Chen, Z.; Pappas, N.; Björnson, E.; Larsson, E.G. Age of information in a multiple access channel with heterogeneous traffic and an energy harvesting node. In Proceedings of the IEEE INFOCOM 2019—IEEE Conference on Computer Communications Workshops (INFOCOM WKSHPS), Paris, France, 29 April–2 May 2019; pp. 662–667.

11. Pappas, N.; Kountouris, M. Delay violation probability and age of information interplay in the two-user multiple access channel. In Proceedings of the 2019 IEEE 20th International Workshop on Signal Processing Advances in Wireless Communications (SPAWC), Cannes, France, 2–5 July 2019; pp. 1–5.

12. Fountoulakis, E.; Charalambous, T.; Nomikos, N.; Ephremides, A.; Pappas, N. Information Freshness and Packet Drop Rate Interplay in a Two-User Multi-Access Channel. *arXiv* **2020**, arXiv:2006.01515.

13. Kosta, A.; Pappas, N.; Ephremides, A.; Angelakis, V. Age of information performance of multiaccess strategies with packet management. *J. Commun. Netw.* **2019**, *21*, 244–255. [CrossRef]

14. Zhou, B.; Saad, W. Joint status sampling and updating for minimizing age of information in the Internet of Things. *IEEE Trans. Commun.* **2019**, *67*, 7468–7482. [CrossRef]

15. Moltafet, M.; Leinonen, M.; Codreanu, M. Worst case age of information in wireless sensor networks: A multi-access channel. *IEEE Wirel. Commun. Lett.* **2019**, *9*, 321–325. [CrossRef]

16. Kadota, I.; Sinha, A.; Modiano, E. Optimizing age of information in wireless networks with throughput constraints. In Proceedings of the IEEE INFOCOM 2018—IEEE Conference on Computer Communications, Honolulu, HI, USA, 16–19 April 2018; pp. 1844–1852.

17. Talak, R.; Karaman, S.; Modiano, E. Can determinacy minimize age of information? *arXiv* **2018**, arXiv:1810.04371.

18. Talak, R.; Karaman, S.; Modiano, E. Optimizing information freshness in wireless networks under general interference constraints. *IEEE/ACM Trans. Netw.* **2019**, *28*, 15–28. [CrossRef]

19. Wu, X.; Yang, J.; Wu, J. Optimal status update for age of information minimization with an energy harvesting source. *IEEE Trans. Green Commun. Netw.* **2017**, *2*, 193–204. [CrossRef]

20. Kadota, I.; Sinha, A.; Modiano, E. Scheduling algorithms for optimizing age of information in wireless networks with throughput constraints. *IEEE/ACM Trans. Netw.* **2019**, *27*, 1359–1372. [CrossRef]

21. Fountoulakis, E.; Pappas, N.; Codreanu, M.; Ephremides, A. Optimal sampling cost in wireless networks with age of information constraints. In Proceedings of the IEEE INFOCOM 2020—IEEE Conference on Computer Communications Workshops (INFOCOM WKSHPS), Toronto, ON, Canada, 6–9 July 2020; pp. 918–923.

22. Neely, M.J. Stochastic network optimization with application to communication and queueing systems. *Synth. Lect. Commun. Netw.* **2010**, *3*, 1–211. [CrossRef]

23. Moltafet, M.; Leinonen, M.; Codreanu, M.; Pappas, N. Power minimization in wireless sensor networks with constrained AoI using stochastic optimization. In Proceedings of the 2019 53rd Asilomar Conference on Signals, Systems, and Computers, Pacific Grove, CA, USA, 3–6 November 2019; pp. 406–410.

24. Chen, Z.; Pappas, N.; Björnson, E.; Larsson, E.G. Optimizing information freshness in a multiple access channel with heterogeneous devices. *IEEE Open J. Commun. Soc.* **2021**, *2*, 456–470. [CrossRef]

25. Stamatakis, G.; Pappas, N.; Traganitis, A. Optimal Policies for Status Update Generation in an IoT Device with Heterogeneous Traffic. *IEEE Internet Things J.* **2020**, *7*, 5315–5328. [CrossRef]

26. Chen, H.; Wang, Q.; Dong, Z.; Zhang, N. Multiuser scheduling for minimizing age of information in uplink MIMO systems. In Proceedings of the 2020 IEEE/CIC International Conference on Communications in China (ICCC), Chongqing, China, 9–11 August 2020; pp. 1162–1167.

27. Zhu, Z.; Yu, B.; Cai, Y. Status Update Performance in Uplink Massive MU-MIMO Short-packet Communication Systems. In Proceedings of the 2020 International Conference on Wireless Communications and Signal Processing (WCSP), Nanjing, China, 21–23 October 2020; pp. 115–119.

28. Feng, S.; Yang, J. Precoding and Scheduling for AoI Minimization in MIMO Broadcast Channels. *arXiv* **2020**, arXiv:2009.00171.
29. Tadele, B.; Shyianov, V.; Bellili, F.; Mezghani, A.; Hossain, E. Age-limited capacity of Massive MIMO. *arXiv* **2020**, arXiv:2007.05071.
30. Yu, B.; Cai, Y. Age of Information in Grant-Free Random Access with Massive MIMO. *IEEE Wirel. Commun. Lett.* **2021**, *10*, 1429–1433. [CrossRef]
31. Shah, A.; Haimovich, A.M. Performance analysis of maximal ratio combining and comparison with optimum combining for mobile radio communications with cochannel interference. *IEEE Trans. Veh. Technol.* **2000**, *49*, 1454–1463. [CrossRef]
32. Kosta, A.; Pappas, N.; Ephremides, A.; Angelakis, V. The Age of Information in a Discrete Time Queue: Stationary Distribution and Non-Linear Age Mean Analysis. *IEEE J. Sel. Areas Commun.* **2021**, *39*, 1352–1364. [CrossRef]
33. Boyd, S.; Boyd, S.P.; Vandenberghe, L. *Convex Optimization*; Cambridge University Press: Cambridge, UK, 2004.
34. Colson, B.; Marcotte, P.; Savard, G. An overview of bilevel optimization. *Ann. Oper. Res.* **2007**, *153*, 235–256. [CrossRef]
35. Den Hertog, D. *Interior Point Approach to Linear, Quadratic and Convex Programming: Algorithms and Complexity*; Springer Science & Business Media: New York, NY, USA, 2012; Volume 277.

Article

On the Value of Information in Status Update Systems †

Peng Zou and Suresh Subramaniam *

Department of Electrical and Computer Engineering, George Washington University, Washington, DC 20052, USA; pzou94@gwu.edu
* Correspondence: suresh@gwu.edu
† Part of this paper was published and presented in the IEEE WCNC 2020, Seoul, Korea, 25–28 May 2020.

Abstract: The age of information (AoI) is now well established as a metric that measures the freshness of information delivered to a receiver from a source that generates status updates. This paper is motivated by the inherent value of packets arising in many cyber-physical applications (e.g., due to precision of the information content or an alarm message). In contrast to AoI, which considers all packets are of equal importance or value, we consider status update systems with update packets carrying values as well as their generated time stamps. A status update packet has a random initial value at the source and a deterministic deadline after which its value vanishes (called ultimate staleness). In our model, the value of a packet either remains constant until the deadline or decreases in time (even after reception) starting from its generation to the deadline when it vanishes. We consider two metrics for the value of information (VoI) at the receiver: *sum VoI* is the sum of the current values of all packets held by the receiver, whereas *packet VoI* is the value of a packet at the instant it is delivered to the receiver. We investigate various queuing disciplines under potential dependence between value and service time and provide closed form expressions for both average sum VoI and packet VoI at the receiver. Numerical results illustrate the average VoI for different scenarios and relations between average sum VoI and average packet VoI.

Keywords: age of information; status update system; value of information

Citation: Zou, P.; Subramaniam, S. On the Value of Information in Status Update Systems. *Entropy* **2022**, *24*, 449. https://doi.org/10.3390/e24040449

Academic Editors: Anthony Ephremides and Yin Sun

Received: 2 March 2022
Accepted: 21 March 2022
Published: 24 March 2022

Publisher's Note: MDPI stays neutral with regard to jurisdictional claims in published maps and institutional affiliations.

1. Introduction

In many cyber-physical applications, the need for *real-time* communication of information packets involves not only maintaining information freshness but is also accompanied by the need to preserve the importance or *value* of those packets. Examples of such cases include autonomous cars and general vehicular networks [1–3], sensor networks [4–6], tactical networks [7] and other systems making decisions in *real-time* [8,9]. In this context, the value of information is another crucial dimension in addition to the notion of timeliness associated with information. In this paper, we address this issue in a queuing system carrying status update packets.

Status update systems with the age of information (AoI) metric measuring end-to-end freshness of packets have received extensive interest recently. Pioneered by the analysis in [10,11] motivated from vehicular status update systems, the AoI metric has been found to be useful in various scenarios such as single server queuing systems [12–14], energy harvesting systems [15–20], single and multi-hop networks [21–25], cognitive radio [26,27] and vehicular communication networks [28]. The AoI metric provides exclusive meaning to the timing of packets and connects a packet's usefulness at the receiver with how long the packet spends before its reception. As such, each packet is assumed to be created with the same value starting at generation. The current literature on status update system abstractions is focused mostly on information freshness and does not consider real-time communication of information packets involving a (time-varying) value associated with its content as well as timing, with some attempts in [29–33] being exceptions. In particular, different packets may have different values with respect to the application at the receiver using it. In such scenarios, the AoI metric falls short of capturing all the dimensions of the problem, and a separate *value of information (VoI)* metric has to be introduced.

In this paper, we abstract out the VoI of a status update packet as a time-varying quantity with a random initial value which becomes zero after a deterministic deadline (identical over all packets) inspired by the AoI metric. Packets are assumed to be useless after the deadline, which we term as *ultimate staleness*. We also assume a functional dependence between the initial value of an information packet and its service time to capture the relation between value and data size (e.g., packets carrying higher resolution information are more valuable but larger in size), the growth rate of processes to be monitored (e.g., state estimation in cyber-physical systems) and the content of packets regarding an alarming event. We propose two definitions for VoI. The *sum VoI* is the sum of the current values of all packets held by the receiver, which is reminiscent of throughput. Note that the value of a packet continues to decay after it is received until ultimate staleness. On the other hand, the *packet VoI* is simply the instantaneous value of a packet at the moment it is delivered to the receiver. By comparing the initial value and the packet value, we aim to understand the effect of communication on the lost value.

We note that the use of deadlines has been a topic of research in earlier works in the literature on AoI, motivating us to further explore it in the context of value of information updates. Reference [34] shows how packet deadlines, buffer sizes and packet replacement influence average AoI. Closed-form expressions for average AoI with deadline are derived in [35,36]. Reference [37] studies AoI in a status update system with random packet deadlines and infinite buffer capacity.

Previous works in [29–33] have components related to our view on value of information. For example, references [29,32] consider the quality of information associated with the distortion observed at the receiving end and [38] considers partial updates. Similarly, [31,39] relate the timeliness of observations with the correctness of information. The author of [30] considers age and the value of information with a notion of value taking into account the non-linear costs regarding information updates in various queuing disciplines. The work in [33] evaluates the value of information in addition to age of information in uplink/downlink transmissions in network control systems. The authors of [40] study the performance of VoI and AoI in a first responders' health monitoring system; their VoI metric is very closely related to our VoI metric originally presented in [41]. In the current paper, we propose a new notion of VoI where a packet's inherent properties at the time of generation determine its value, in contrast to a value evaluated after processing at the receiver as in previous work. We investigate VoI in M/GI/1/1, M/GI/1/2, M/GI/1/2* and M/GI/1/1* queuing disciplines and provide closed-form expressions for average sum VoI and packet VoI.

The work in this paper is a significantly extended version of our conference paper [41]. In particular, we include the following:

1. We propose and analyze a second VoI metric (average packet VoI) in addition to the average sum VoI analyzed in [41].
2. We add the case of constant value over time until deadline to our analysis on top of the previous work on linear value descent over time until deadline.
3. We analyze the performance of a new queuing scheme, which is M/GI/1/1* in the server. This extended analysis enables us to study the possible use for the value of status update packets in different kinds of systems.
4. We present more numerical results on the two VoI metrics and the four queuing schemes that enable the reader to obtain a clear picture of the various trade-offs involved.

2. System Model

We consider a point-to-point communication system with a single transmitter sending status updates from a source to a receiver, as shown in Figure 1. The update packets arrive at the transmitter as a Poisson process with arrival rate λ at instants t_i. A packet may be discarded in the queuing phase; those that are not discarded enter the server. A packet may also be preempted and discareded while undergoing service; otherwise, it is received by the receiver after system time T_i at $t_i' = t_i + T_i$. In this paper, we cover M/GI/1/1, M/GI/1/2, M/GI/1/2* and M/GI/1/1* queuing schemes. In M/GI/1/1, there are no buffer and packets arriving in the server-busy state that are discarded. In M/GI/1/2, there is a single data buffer with a first come first serve discipline so that an arriving packet that finds the buffer occupied will be discarded. In M/GI/1/2*, there is a single data buffer but, in this case, an arriving packet will preempt the packet stored in the buffer. In M/GI/1/1*, there are no buffer and packets arriving in the server-busy state that will preempt the current packet in service. For the two no-buffer schemes M/GI/1/1 and M/GI/1/1*, $T_i = S_i$ where S_i is the service time for the ith packet, which is independent and identically distributed with $f_S(s)$. For the two schemes with buffer M/GI/1/2 and M/GI/1/2*, $T_i = S_i + W_i$ where W_i is the waiting time for the ith packet. We derive T_i for different schemes in Section 3. We focus on these four queuing systems because previous research has shown that excessive queuing in large buffer systems can adversely impact AoI, and limited-buffer systems with packet management can improve AoI [12,34]. Since the value also potentially becomes worse with time, a similar behavior is expected for VoI.

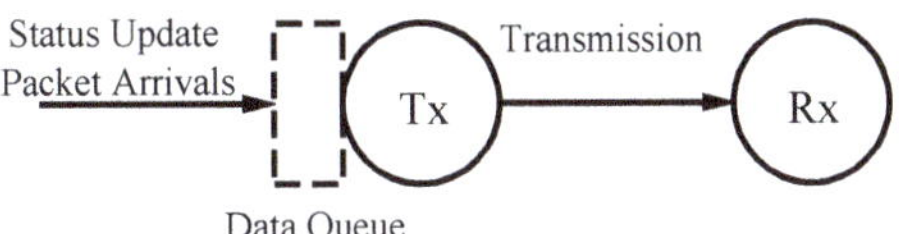

Figure 1. System model with status update packets arriving at a single server transmission queue.

2.1. Value of a Packet

The ith update packet has initial value $V_{0,i}$ at the generation instant. This is a random sequence independent over different i. $V_{0,i}$ has the identical general distribution $f_V(v)$ with mean value $\mathbb{E}[V]$. This initial value represents the importance of a packet for an application. It could be related to the precision of a measurement, proximity of the sensor to the measured object or it could indicate an alarm event. Each packet has a deterministic lifetime D after which it reaches ultimate staleness. Hence, after a fixed time period D from packet generation, the packet has no value for the receiver. We use $V_{r,i}$ to denote the instantaneous value of the ith update packet when it is delivered to the receiver and $\rho_i = \frac{V_{r,i}}{V_{0,i}}$ to denote the fraction of the initial value of the ith update packet that is delivered to the receiver.

Motivated by various applications of sensor networking and the value of information in them [1–6], in our model, we assume that packet i's value can decrease from its time of generation at t_i until it hits the deadline at $t_i + D$. The value $V_i(\tau) = h_i(V_{0,i}, \tau)$ for the ith packet decreases with $\tau = t - t_i$, representing the time passed after generation at the transmitter. This value keeps on decreasing (even after a packet is received) until it becomes zero. We have $h_i(V_{0,i}, 0) = V_{0,i}$ and $h_i(V_{0,i}, D) = 0$. In this paper, we consider two different *descend functions* $h(.)$ for the value: (i) constant value and (ii) linear descend. The former models the case where the packet's value does not change with time as long as it is delivered by the deadline, while the latter models the case where a packet that is delivered earlier has a higher value. In the constant value case, we have the following.

$$V_i(\tau) = h_i(V_{0,i}, \tau) = \begin{cases} V_{0,i} & (\tau < D) \\ 0 & (\tau > D). \end{cases} \tag{1}$$

In the linear case, since $h_i(V_{0,i}, 0) = V_{0,i}$ and $h_i(V_{0,i}, D) = 0$, we have a linear descend function.

$$V_i(\tau) = h_i(V_{0,i}, \tau) = \begin{cases} -\frac{V_{0,i}}{D}\tau + V_{0,i} & (\tau < D) \\ 0 & (\tau > D). \end{cases} \tag{2}$$

Then we have the following:

$$V_{r,i} = h_i(V_{0,i}, T_i), \tag{3}$$

$$\rho_i = \frac{h_i(V_{0,i}, T_i)}{V_{0,i}}, \tag{4}$$

for packets that are deliverd to the receiver. We set $V_{r,i} = 0$, $\rho_i = 0$, for packets that are not delivered to the receiver.

2.2. Value-Dependent Service Times

We consider two possibilities for a packet's service time. In one model, the service times are independent of the initial value of a packet. In another model, the service time of a packet depends on the initial value of the packet through a non-decreasing function g.

$$S_i = g(V_{0,i}). \tag{5}$$

In this case, the distribution function of S_i is $f_S(s) = f_V(g^{-1}(s))\frac{dg^{-1}(s)}{ds}$ where $g^{-1}(.)$ is the inverse function of $g(.)$, and the mean service time is $\mathbb{E}[S] = \mathbb{E}[g(V)]$. Corresponding to the general distribution, we have the moment generating function (MGF) evaluated at $-\gamma$ for $\gamma \geq 0$:

$$M_S(\gamma) \triangleq \mathbb{E}[e^{-\gamma S}].$$

This monotonic relation reflects the fact that a larger packet takes longer time to transmit and its reception yields more value. This relation causes an interesting tradeoff between value and age as a larger value is obtained at the receiver by paying a longer service time.

In this paper, we consider two definitions for VoI. The first one is Y_{sum}, which denotes the sum VoI, i.e., the sum of the current values of all packets received by the receiver (cf. [4–6] where the additive nature of VoI is discussed in various wireless sensor networks). Hence, $Y_{\text{sum}}(t)$ is as follows:

$$Y_{\text{sum}}(t) = \sum_{j=1}^{i_t} V_j(t) \tag{6}$$

where $i_t = \max\{i : t_i' \leq t\}$. The time average of $Y_{\text{sum}}(t)$ is the following.

$$\mathbb{E}[Y_{\text{sum}}] = \lim_{T \to \infty} \frac{1}{T} \int_{t=0}^{T} Y_{\text{sum}}(t). \tag{7}$$

Another definition is Y_{packet}, which measures the instantaneous value of a packet at the moment it is delivered to the receiver (if it is delivered). Packets that are dropped are assumed to have zero value. The average packet VoI is then defined as follows.

$$\mathbb{E}[Y_{\text{packet}}] = \mathbb{E}[V_{r,i}]. \tag{8}$$

$\mathbb{E}[\rho_i]$ is the expected fraction of the initial value that is delivered to the receiver, which illustrates the amount of value received by the receiver compared to the generated initial value at the source. We reiterate that $\mathbb{E}[V_{r,i}]$ and $\mathbb{E}[\rho_i]$ are expectations over *all* packets; dropped packets contribute zero received value.

We illustrate the evolution of value with an example. In Figures 2 and 3, the evolution of value for specific packets generated over time is shown in an M/GI/1/1 system with constant value and linearly descending values, respectively. We use X_i to denote the inter-arrival period between two packets $i - 1$ and i. Therefore, X_i is an exponentially distributed random variable with rate parameter λ. Packet 1 finds the server idle and begins service at t_1; service ends at t_1'. Packet 2 arrives between t_1 and t_1', and it is discarded. The service of packet 1 finishes at t_1' before the deadline of packet 1, $D_1 = t_1 + D$. The value of packet 1 at t_1', when received by the receiver, is non-zero, and it becomes zero at D_1. Packets 3, 4 and 5 arrive to the system during the idle period, and they are received at t_3', t_4' and t_5'. Note that when packet 4 is received, packet 3 has a non-zero value; thus, the sum VoI, which is shown with a solid red line, is the sum of the values of these packets.

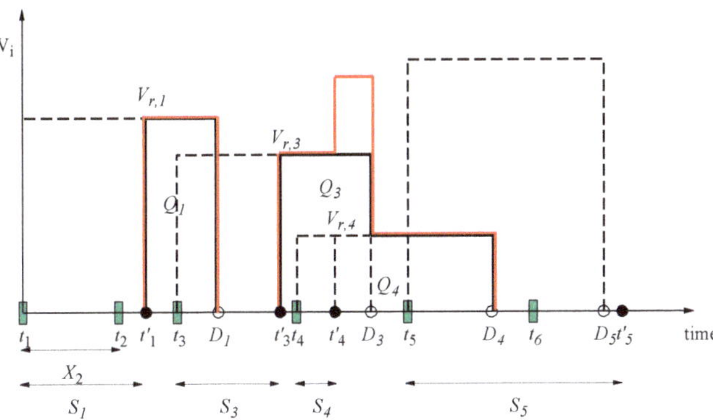

Figure 2. Evolution of value in M/GI/1/1 system when the value remains constant until deadline.

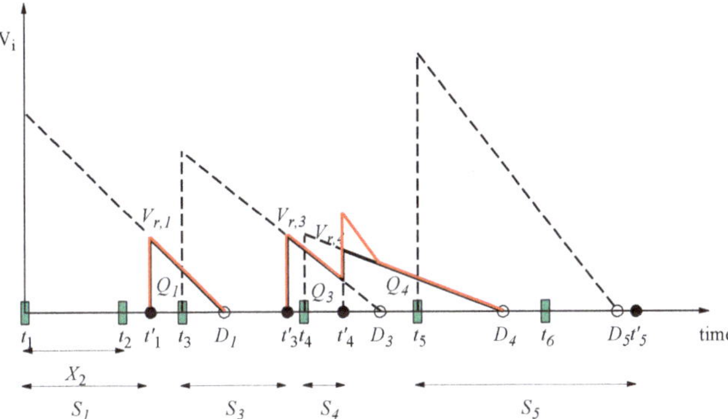

Figure 3. Evolution of thevalue in the M/GI/1/1 system with linearly descending values.

We define areas Q_i under the rectangular regions of the curve shown in Figure 2 or the triangular regions of the curve shown in Figure 3, and we set $Q_i = 0$ for packets discarded in the queuing phase. Then, the expected sum VoI at the receiver is as follows:

$$\mathbb{E}[Y_{\text{sum}}] = \lambda \mathbb{E}[Q_i], \tag{9}$$

where λ is the arrival rate of packets at the transmitter.

3. Evaluating Value of Information

In this section, we derive closed-form expressions for $\mathbb{E}[V_{r,i}]$, $\mathbb{E}[Q_i]$ and $\mathbb{E}[\rho]$ for the various queuing systems. $\mathbb{E}[Y_{\text{packet}}]$ and $\mathbb{E}[Y_{\text{sum}}]$ can then be obtained by using Equations (8) and (9).

3.1. Average VoI for M/GI/1/1

In the M/GI/1/1 queueing system, there is a single server and no buffer. Packets that arrive in the idle period are taken to service immediately and those arriving in busy period are dropped. In view of the renewal structure, we have the following stationary probabilities for each state:

$$p_I = \frac{1}{\lambda T_{\text{cycle}}}, \quad p_B = \frac{\mathbb{E}[S]}{T_{\text{cycle}}}, \tag{10}$$

where $T_{\text{cycle}} = \frac{1}{\lambda} + \mathbb{E}[S]$ is the expected length of one renewal cycle; and I and B indicate the idle and busy states. In the M/GI/1/1 system, packets are delivered to the receiver if they arrive when the server is idle. Recall that if the total time spent by the packet before reaching the receiver is larger than D, its value vanishes. Since a packet that is taken to service spends service time S_i in the queue before reaching the receiver, the packet's value vanishes if S_i is larger than D. Hence, we just need to consider condition $S_i < D$ and i arriving in idle states. Based on the two time-dependent functions for the value shown in (1) and (2) and the relationship shown in (3)–(5), we have the following:

$$\mathbb{E}[V_{r,i}] = p_I \int_0^{\tilde{V}} h_i(v, g(v)) f_V(v) dv, \tag{11}$$

$$\mathbb{E}[\rho_i] = p_I \int_0^{\tilde{V}} \frac{h_i(v, g(v))}{v} f_V(v) dv, \tag{12}$$

$$\mathbb{E}[Q_i] = p_I \int_0^{\tilde{V}} \int_{g(v)}^{D} h_i(v, \tau) f_V(v) d\tau dv, \tag{13}$$

where $\tilde{V} = g^{-1}(D)$ denotes the corresponding initial value when the related service time is equal to the deadline.

3.2. Average VoI for M/GI/1/2

In the M/GI/1/2 queueing system, there is a single buffer. The server is in either idle or busy states. Packets that arrive in the idle period are served immediately; those that arrive in the busy period are stored in the buffer if there is no other packet in it and they are discarded otherwise. In view of the renewal structure, we have the following stationary probabilities for each state of the server:

$$p_I = \frac{1}{\lambda T_{\text{cycle}}}, \quad p_B = \frac{\mathbb{E}[S]}{T_{\text{cycle}} M_S(\lambda)}, \tag{14}$$

where we use $M_S(\lambda)$ to denote the moment generating function of the service distribution evaluated at $-\lambda$:

$$M_S(\lambda) = \mathbb{E}[e^{-\lambda S}], \tag{15}$$

where $T_{\text{cycle}} = \frac{1}{\lambda} + \frac{\mathbb{E}[S]}{M_S(\lambda)}$ is the expected length of one renewal cycle. Next, we evaluate $\mathbb{E}[V_{r,i}]$ and $\mathbb{E}[Q_i|(s)]$ for $s \in \mathcal{S}_{M/GI/1/2} = \{I, B\}$ and conditioning is on the server state observed by packet i. Due to the PASTA property, $\Pr[P_i = (s)] = p_s$, where $p_s, s \in \mathcal{S}_{M/GI/1/2}$ are as in (14).

3.2.1. Idle State Analysis

As a packet arriving in the idle state is served immediately, we have the following.

$$\mathbb{E}[V_{r,i}|I] = \int_0^{\tilde{V}} h_i(v, g(v)) f_V(v) dv, \tag{16}$$

$$\mathbb{E}[\rho_i|I] = \int_0^{\tilde{V}} \frac{h_i(v, g(v))}{v} f_V(v) dv, \tag{17}$$

$$\mathbb{E}[Q_i|I] = \int_0^{\tilde{V}} \int_{g(v)}^D h_i(v,\tau) f_V(v) d\tau dv.$$

(18)

3.2.2. Busy State Analysis

Since only the first packet that arrives during the busy period is served and the others are discarded, we introduce a lemma for the probability that an arriving packet is the first one that arrives in the busy state. To do so, we first define states B_1 and B_2 as the busy states of the server with zero and one packet waiting in the queue, respectively. The renewal cycle is as follows. After the idle period, an arrival happens and the system turns to B_1 state. Now, a time duration of service S starts and if during the service period another arrival occurs, the system turns to B_2 state. This back-and-forth between B_1 and B_2 states continues until no packet arrives in one service time. We provide an example in Figure 4 for the three states in the M/GI/1/2 scheme. At time t_0, packet 1 arrives and finds the system idle. Packet 2 finds the system in B_1 state at t_1 and is stored in the buffer. Packet 3 finds the system in B_2 state at t_2 and is dropped.

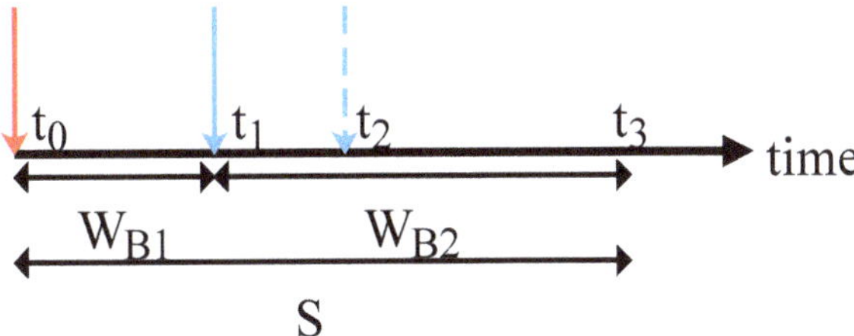

Figure 4. Three states that can be observed by packets in M/GI/1/2 scheme.

This renewal structure yields the following result.

Lemma 1. *In the M/GI/1/2 scheme, the waiting time of a packet in the buffer conditioned on its arrival in B_1 state is as follows*

$$\mathbb{E}[W_{B_2}] = \mathbb{E}[S - X | X < S] Pr[X < S]$$
$$= \mathbb{E}[S] + \frac{1}{\lambda} M_S(\lambda) - \frac{1}{\lambda}.$$

The stationary probability of B_2 state is as follows:

$$p_{B_2} = p_B \frac{\mathbb{E}[W_{B_2}]}{\mathbb{E}[S]} = p_B \left(1 + \frac{M_S(\lambda) - 1}{\lambda \mathbb{E}[S]}\right),$$

and the probability of B_1 state is $p_{B_1} = p_B - p_{B_2}$.

Then, we have $\mathbb{E}[Q_i|B] = \mathbb{E}[Q_i|B_1]$ and we provide the probability distribution function for the conditional residual service time W' under the condition that the packet arrives in the B_1 state:

$$\mathbb{P}[W' > w] = \mathbb{P}[S - X > w | X < S]$$
$$= \frac{\int_w^\infty \int_0^{s-w} f_S(s) f_X(x) dx ds}{\mathbb{P}[X < S]}$$
$$= \frac{\int_w^\infty f_S(s)(1 - e^{-\lambda(s-w)}) ds}{1 - M_S(\lambda)},$$

and we have the following.

$$f_{W'}(w) = \frac{d(1 - \mathbb{P}[W' > w])}{dw}. \tag{19}$$

Then, we have the following.

$$\mathbb{E}[V_{r,i}|B_1] = \int_0^{\tilde{V}} \int_0^{D-g(v)} h_i(v, g(v) + w) f_{W'}(w) f_V(v) dw dv, \tag{20}$$

$$\mathbb{E}[\rho_i|B_1] = \int_0^{\tilde{V}} \int_0^{D-g(v)} \frac{h_i(v, g(v) + w)}{v} f_{W'}(w) f_V(v) dw dv, \tag{21}$$

$$\mathbb{E}[Q_i|B_1] = \int_0^{\tilde{V}} \int_0^{D-g(v)} \int_{g(v)+w}^{D} h_i(v, \tau) f_{W'}(w) f_V(v) d\tau dw dv. \tag{22}$$

Therefore, we have $\mathbb{E}[V_{r,i}] = \mathbb{E}[V_{r,i}|I]p_I + \mathbb{E}[V_{r,i}|B_1]p_{B_1}$, $\mathbb{E}[\rho_i] = \mathbb{E}[\rho_i|I]p_I + \mathbb{E}[\rho_i|B_1]p_{B_1}$ and $\mathbb{E}[Q_i] = \mathbb{E}[Q_i|I]p_I + \mathbb{E}[Q_i|B_1]p_{B_1}$.

3.3. Average VoI for M/GI/1/2*

The M/GI/1/2* queueing system is the same as M/GI/1/2 except that we use a last-come first-serve order with packet discarding in the buffer. The latest packet arriving in a busy period takes the place of the old packet in the buffer. Therefore, we have the same stationary probabilities for each state as the M/GI/1/2 system in (14). Additionally, the expressions for $\mathbb{E}[V_{r,i}|I]$, $\mathbb{E}[\rho_i|I]$ and $\mathbb{E}[Q_i|I]$ are the same as in (16)–(18) separately. We now derive expressions for $\mathbb{E}[Q_i|B]$ and $\mathbb{E}[V_{r,i}|B]$.

Busy State Analysis

If the ith packet arrives to the server during the busy period, it will be transmitted to the receiver conditioned on event $\{X_i > W_{i-1}\}$, which means the next packet arrives for the server after the current service finishes. W is the general residual service time for all packets arriving in the busy state, and we have the following: $f_W(w) = \frac{\mathbb{P}[S > w]}{\mathbb{E}[S]}$. Then, the following is the case.

$$\mathbb{E}[V_{r,i}|B] = \int_0^{\tilde{V}} \int_0^{D-g(v)} \int_w^{\infty} h_i(v, g(v) + w) f_X(x)$$
$$f_W(w) f_V(v) dx dw dv, \tag{23}$$

$$\mathbb{E}[\rho_i|B] = \int_0^{\tilde{V}} \int_0^{D-g(v)} \int_w^{\infty} \frac{h_i(v, g(v) + w)}{v} f_X(x)$$
$$f_W(w) f_V(v) dx dw dv, \tag{24}$$

$$\mathbb{E}[Q_i|B] = \int_0^{\tilde{V}} \int_0^{D-g(v)} \int_w^{\infty} \int_{g(v)+w}^{D} h_i(v, \tau) f_X(x)$$
$$f_W(w) f_V(v) d\tau dx dw dv. \tag{25}$$

Therefore, we have $\mathbb{E}[V_{r,i}] = \mathbb{E}[V_{r,i}|I]p_I + \mathbb{E}[V_{r,i}|B_1]p_{B_1}$, $\mathbb{E}[\rho_i] = \mathbb{E}[\rho_i|I]p_I + \mathbb{E}[\rho_i|B_1]p_{B_1}$ and $\mathbb{E}[Q_i] = \mathbb{E}[Q_i|I]p_I + \mathbb{E}[Q_i|B_1]p_{B_1}$.

3.4. Average VoI for M/GI/1/1*

In the M/GI/1/1* queueing system, there is no buffer and a new packet that arrives during busy state will preempt the current packet in service. Since the arrival process is a

Poisson with rate λ, p_e, the probability that a packet is delivered to the receiver is given by the following:

$$p_e = \mathbb{P}[S_i < X_{i+1}] = M_S(\lambda), \tag{26}$$

which means, in preemption scheme, only the packet that has a service time less than the upcoming inter-arrival period is delivered to the receiver. We use relation $f_{G|G<F}(t) = f_G(t)\frac{\mathbb{P}(F>t)}{\mathbb{P}(G<F)}$ from [13] where G and F are arbitrary random variables. Since $\mathbb{P}(G < F) = M_G(\lambda)$ and $\mathbb{P}(F > t) = e^{-t\lambda}$, we have the probability density function for conditional service time.

$$f_{S|S<X}(s) = f_S(s)\frac{e^{-s\lambda}}{M_S(\lambda)}. \tag{27}$$

We use S' to denote the conditional service time S; therefore, we have $f_{S'}(s) = f_{S|S<X}(s)$. In this case, we rewrite Equation (1) as follows:

$$h_i(g^{-1}(s),\tau) = \begin{cases} g^{-1}(S_i') & (\tau < D) \\ 0 & (\tau > D) \end{cases} \tag{28}$$

and Equation (2) as the following.

$$h_i(g^{-1}(s),\tau) = \begin{cases} -\frac{g^{-1}(S_i')}{D}\tau + g^{-1}(S_i') & (\tau < D) \\ 0 & (\tau > D) \end{cases} \tag{29}$$

Then, we have the following.

$$\mathbb{E}[V_{r,i}] = p_e \int_0^D h_i(g^{-1}(s),s) f_{S'}(s)ds, \tag{30}$$

$$\mathbb{E}[\rho_i] = p_e \int_0^D \frac{h_i(g^{-1}(s),s)}{g^{-1}(s)} f_{S'}(s)ds, \tag{31}$$

$$\mathbb{E}[Q_i] = p_e \int_0^D \int_s^D h_i(g^{-1}(s),\tau) f_{S'}(s)d\tau ds, \tag{32}$$

4. Numerical Results

In this section, we provide numerical results for average VoI for various cases. We also perform packet-based queue simulations offline for 10^6 packets as verification of the analytical results. An example of our simulation results is shown in Figure 5. We use $g(V) = V$ as the relation between service time and value to model the case where the value is directly proportional to the packet size. Results are presented for three different distributions for the initial value of packets.

4.1. Uniformly Distributed Initial Value

First, we assume that the initial value of each packet is uniformly distributed between V_{min} and V_{max} and the value follows the linear descend function. In Appendix A, we provide closed-form expressions for $\mathbb{E}[Y_{sum}]$ and $\mathbb{E}[Y_{packet}]$ in various systems with linearly descending value.

We show a comparison of average Y_{sum} and average Y_{packet} in Figure 5. In Figure 5a, we show average Y_{sum} versus arrival rate λ for the four queuing schemes. We observe that M/GI/1/1 and M/GI/1/2* perform better than M/GI/1/1* as λ increases. In particular, due to the linear relation between time and value, keeping a packet in the buffer to keep the server busy turns out to yield smaller value at the receiver with respect to keeping none and serving only the freshest packets. For M/GI/1/1* and M/GI/1/2, on the other hand, there is an optimal value of λ after which average Y_{sum} drops. For M/GI/1/2, it is due to undesired increases in waiting times in the data buffer while for M/GI/1/1*, it is due to undesired decrease in the number of delivered packets.

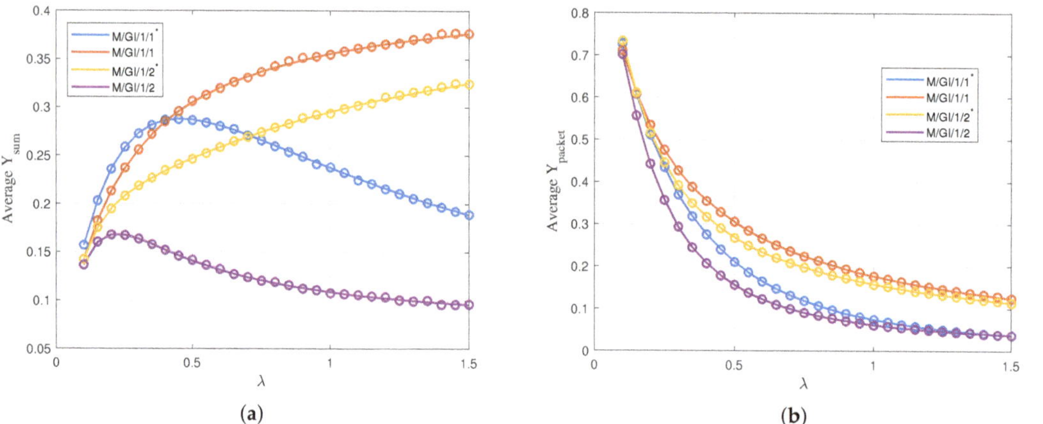

(a)   (b)

Figure 5. (**a**) Average Y_{sum} and (**b**) average Y_{packet} for uniformly distributed initial value with linear descend function versus λ; $V_{min} = 0$, $V_{max} = 10$, $D = 8$. Circles are simulation results.

In Figure 5b, we show Y_{packet} versus arrival rate λ for the four queuing schemes. Again, we observe that M/GI/1/1 performs better than the other three. We observe that as λ increases, $\mathbb{E}[Y_{packet}]$ decreases in all four queuing schemes due to the fact that most of the generated packets are discarded in the queuing phase and have zero value for the receiver.

In Figure 6, we show $\mathbb{E}[\rho]$, which denotes the average ratio of the received value compared to the generated values over all the generated packets. We observe that as λ increases, $\mathbb{E}[\rho]$ decreases in all four queuing schemes, which matches the result for $\mathbb{E}[Y_{packet}]$. However, interestingly, M/GI/1/1* scheme performs best for $\mathbb{E}[\rho]$. This is because as λ increases, even though there will be more packets dropped, the packets delivered to the receiver have smaller service times, which increases the ratio of the delivered value to the initial value.

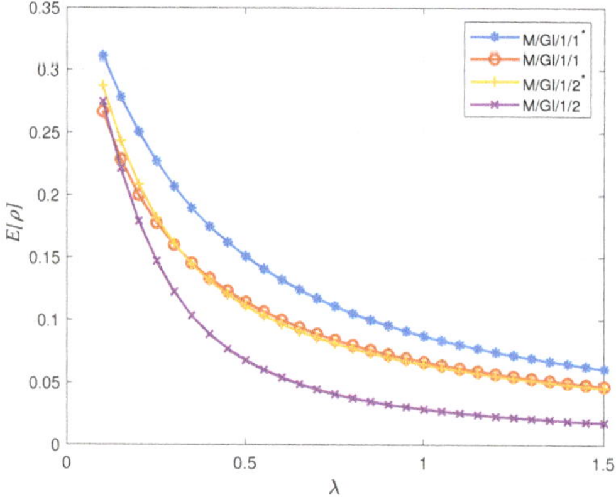

Figure 6. $\mathbb{E}[\rho]$ for uniformly distributed initial value with linear descend function versus λ; $V_{min} = 0$, $V_{max} = 10$, $D = 8$.

Next, we consder the case when the service times are independent of the initial values and are exponentially distributed with service rate μ. In Figure 7, we show the average Y_{sum} versus arrival rate λ for the four queuing schemes. We observe that M/GI/1/1* performs better than the other three. This is because the service time is independent of the initial value, and large-valued packets may have small service times. In particular, due to the linear relation between time and value, keeping a packet in the buffer to keep the server busy turns out to yield smaller values at the receiver compared to keeping none and serving only the freshest packets.

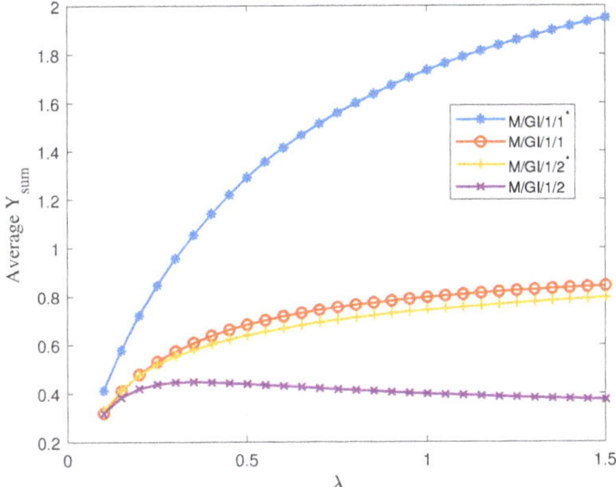

Figure 7. Average Y_{sum} for uniformly distributed initial value with linear descend function and exponential independent service time versus λ; $V_{min} = 0$, $V_{max} = 10$, $D = 8$, $\mu = 0.2$.

Finally, in Figure 8, we show the average Y_{sum} versus service rate μ for the four queuing schemes when the service times are independent of the initial values and are exponentially distributed. We observe that M/GI/1/2 and M/G/1/2* perform better than M/GI/1/1* as μ increases. This is because, as the average service time deceases, fewer packets will expire, i.e., reach ultimate staleness, during the waiting period in the buffer, and in this case, having a buffer to store the packets turns out to yield larger value at the receiver with respect to dropping the packets in the server.

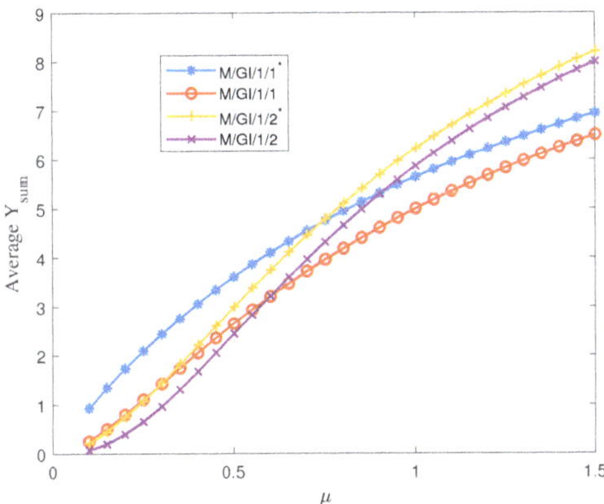

Figure 8. Average Y_{sum} for uniformly distributed initial value with linear descend function and exponential independent service time versus μ; $V_{min} = 0$, $V_{max} = 10$, $D = 8$, $\lambda = 1$.

4.2. Exponentially Distributed Initial Value

Next, we consider $f_V(v) = \mu_v e^{-\mu_v v}$ with constant value. In this case, we have service rate $\mu = \mu_v$ due to $g(V) = V$. We compare average AoI with average sum VoI for the same schemes as both of them are time-average metrics over all the packets. In Appendix B, we provide closed-form expressions for $\mathbb{E}[Y_{sum}]$ and $\mathbb{E}[Y_{packet}]$ in various systems for constant values.

In Figure 9a, we plot the average Y_{sum} with respect to λ for various schemes. We observe that M/M/1/2* always performs better than the others. This is connected to the fact that when the value of packet is constant over time, all packets received within the deadline contribute their full initial value. Since Y_{sum} is the accumulated value of received packet values, the total value is higher if a packet is stored in the buffer instead of dropping it. At the same time, we observe that M/M/1/1* performs the worst in terms of value since the dependence between service time and value causes higher value packets to be preempted in this system, resulting in no contribution to VoI at the receiver.

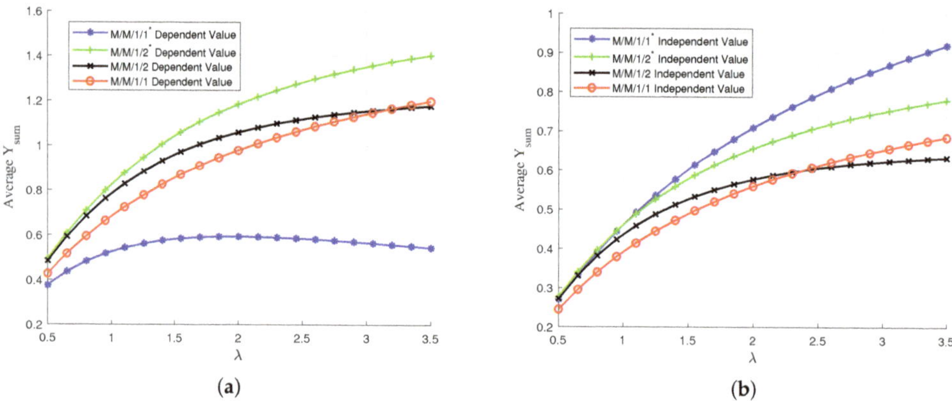

Figure 9. Average Y_{sum} for exponentially distributed service time with constant value versus λ for M/M/1/1, M/M/1/2, M/M/1/2* and M/M/1/1* schemes with $\mu_v = 1.5$ and $D = 3$. (**a**) Dependent Value. (**b**) Independent Value.

Next in Figure 9b, we show average Y_{sum} for independent initial value and service time under the same marginal distributions. We observe that, with independent service time, the M/M/1/1* scheme becomes the best case while it is the worst case with dependent service time. The other three schemes yield higher values as the adverse relation between initial value and service rate is removed.

Finally, in Figure 10, we show $\mathbb{E}[\rho]$ versus deadline D for the four queuing schemes. We observe that, as D increases, $\mathbb{E}[\rho]$ for all queuing schemes increases, but never reaches threshold 1 due to the fact that some packets are discarded in the queuing phase.

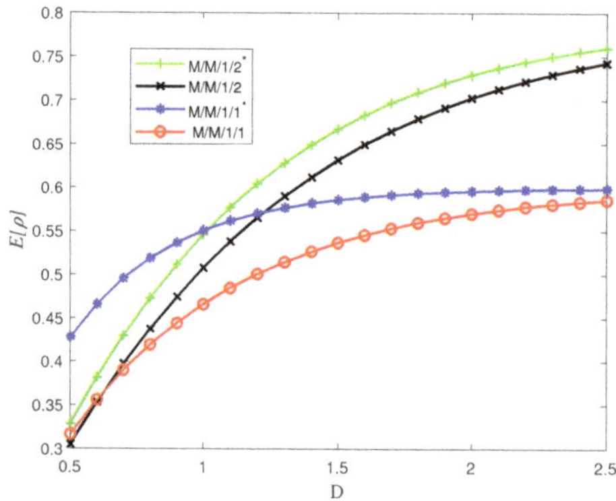

Figure 10. $\mathbb{E}[\rho]$ for exponentially distributed service time with constant value versus D for $\lambda = 1$.

4.3. Binary Distributed Initial Value

We finally consider binary distributed initial value for two classes of update packets. Class 1 and class 2 packets have $V_{0,i} = V_1$ and $V_{0,i} = V_2$. Each packet is independently chosen to be in class 1 or 2 with probability p and $(1 - p)$, respectively. This situation models the case when a packet of one class contains a message about an alarming event yielding high value once received, whereas the other class of packets are assumed to be regular status updates.

In Figure 11, we set $V_1 = 1.33$, $V_2 = 0.4$ and $p = 0.2$. We compare plots showing average Y_{sum} versus λ for three different service policies in an M/M/1/1 system. The first policy serves all packets without regard to the value, the second policy involves serving only class 1 packets, and the third policy serves only class 2 packets. Note that if the service time is dependent on the value, class 1 packets will have exponentially distributed service time with mean $E[S] = E[V_1]$, and similarly, class 2 packets will have exponentially distributed service time with mean $E[S] = E[V_1]$. If the service time is independent of the value, both class packets will have exponentially distributed service time with $\mu = 1.5$. Our numerical results show that when service time is independent of value, always serving the high-value packet will yield the highest average value. On the other hand, in the dependent case when arrival rate becomes large, serving the packet with low value but smaller service time and high probability will benefit the average Y_{sum} compared to serving all the packets or serving the high-value packets with larger service time and low probability.

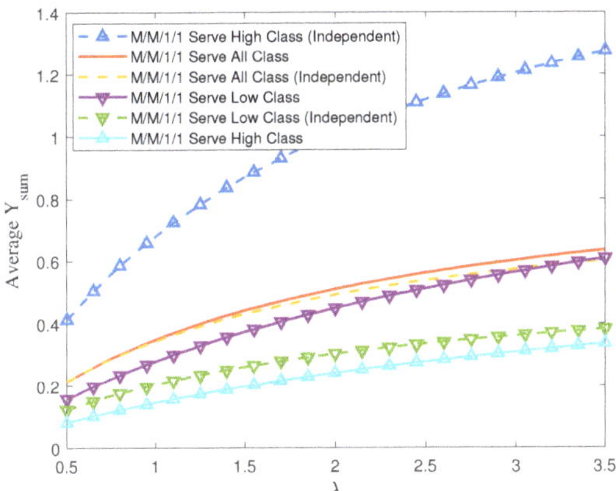

Figure 11. Exponentially distributed service time dependent on or independent of the binary value in M/M/1/1 scheme.

5. Conclusions

Age of information (AoI) is a well-known metric that quantifies the freshness of information at a receiver in status update systems. This metric ignores the potential differences in the importance of various update packets. In this paper, we consider the value of information in status update systems wherein packets have various initial values upon generation. We investigate various queuing disciplines with initial-value-dependent packet service times and obtain closed-form expressions for two different VoI metrics. Our numerical results illustrate the trade-off between the two VoI metrics and the contrast between these two metrics. We show average sum VoI and average packet VoI for different scenarios and the fraction of received value comparing to the inital value for different systems.

Author Contributions: Conceptualization, P.Z. and S.S.; formal analysis, P.Z.; investigation, P.Z.; supervision, S.S.; validation, P.Z.; writing—original draft, P.Z.; writing—review and editing, P.Z. and S.S. All authors have read and agreed to the published version of the manuscript.

Funding: This research received no external funding.

Institutional Review Board Statement: Not applicable.

Informed Consent Statement: Not applicable.

Data Availability Statement: Not applicable.

Conflicts of Interest: The authors declare no conflict of interest.

Appendix A. $\mathbb{E}[Y_{sum}]$ and $\mathbb{E}[Y_{packet}]$ for Uniformly Distributed Initial Value with Linear Descend Function

In uniform case, we have $f_V(v) = \frac{1}{u}$, where $u = V_{max} - V_{min}$, and we assume $g(V) = V$. Thus, we have the mean service time.

$$\mathbb{E}[S] = \mathbb{E}[V] = \frac{V_{max} + V_{min}}{2}.$$

Appendix A.1. M/GI/1/1

From (10), we have the following.

$$p_I = \frac{1}{1 + \lambda \mathbb{E}[S]}.$$

We calculate $\mathbb{E}[V_{r,i}]$ from (2) and (11) and we have the following.

$$\mathbb{E}[V_{r,i}] = p_I \int_0^{\tilde{V}} (v - \frac{v}{D}g(v))f_V(v)dv.$$

Here, $\tilde{V} = D$. Define $V_{up} = \tilde{V}$ if $\tilde{V} < V_{max}$ and $V_{up} = V_{max}$ otherwise. Then, we have the following.

$$\begin{aligned}
\mathbb{E}[V_{r,i}] &= p_I \int_{V_{min}}^{V_{up}} (v - \frac{v}{D}g(v))f_V(v)dv \\
&= \frac{p_I}{u} \int_{V_{min}}^{V_{up}} (v - \frac{v^2}{D})dv \\
&= \frac{p_I}{u} \left(\frac{1}{2}(V_{up}^2 - V_{min}^2) - \frac{1}{3D}(V_{up}^3 - V_{min}^3) \right).
\end{aligned}$$

Then, we calculate $\mathbb{E}[Q_i]$ from (2) and (13) and we have the following.

$$\begin{aligned}
\mathbb{E}[Q_i] &= p_I \int_0^{\tilde{V}} \int_{g(v)}^{D} (v - \frac{v}{D}\tau)f_V(v)d\tau dv \\
&= \frac{p_I}{2} \int_{V_{min}}^{V_{up}} \frac{v}{D}(D - v)^2 f_V(v)dv \\
&= \frac{p_I}{2Du} \int_{V_{min}}^{V_{up}} (D^2 v - 2Dv^2 + v^3)dv \\
&= \frac{p_I}{2Du} \left(\frac{D^2}{2}(V_{up}^2 - V_{min}^2) - \frac{2D}{3}(V_{up}^3 - V_{min}^3) \right. \\
&\quad \left. + \frac{1}{4}(V_{up}^4 - V_{min}^4) \right).
\end{aligned}$$

Finally, we have $\mathbb{E}[Y_{packet}] = \frac{1}{p_I}\mathbb{E}[V_{r,i}]$ and $\mathbb{E}[Y_{sum}] = \lambda \mathbb{E}[Q_i]$.

Appendix A.2. M/GI/1/2

From (15), we have the following.

$$M_S(\lambda) = \frac{1}{u\lambda}(e^{-\lambda V_{\min}} - e^{-\lambda V_{\max}}).$$

Then, from (14), we have the following.

$$p_I = \frac{M_S(\lambda)}{M_S(\lambda) + \lambda\mathbb{E}[S]},$$

$$p_B = \frac{\lambda\mathbb{E}[S]}{M_S(\lambda) + \lambda\mathbb{E}[S]}.$$

From Lemma 1, we have the following.

$$p_{B_1} = \frac{1 - M_S(\lambda)}{\lambda\mathbb{E}[S]}p_B,$$

$$\mathbb{P}[W' > w] = \frac{\lambda(V_{\max} - w) + e^{\lambda(w - V_{\max})} - 1}{u\lambda(1 - M_S(\lambda))}.$$

Then, from (19), we have the following.

$$f_{W'}(w) = \frac{-1 + \lambda e^{\lambda(w - V_{\max})}}{u\lambda(M_S(\lambda) - 1)}.$$

For the idle case, from (2), (16) and (18), we have the following.

$$\mathbb{E}[V_{r,i}|I] = \frac{1}{u}\left(\frac{1}{2}(V_{\text{up}}^2 - V_{\min}^2) - \frac{1}{3D}(V_{\text{up}}^3 - V_{\min}^3)\right),$$

$$\mathbb{E}[Q_i|I] = \frac{1}{2Du}\left(\frac{D^2}{2}(V_{\text{up}}^2 - V_{\min}^2) - \frac{2D}{3}(V_{\text{up}}^3 - V_{\min}^3) + \frac{1}{4}(V_{\text{up}}^4 - V_{\min}^4)\right).$$

For the busy case, since waiting time W' has the same domain of definition as initial value $V_{0,i}$, there are three conditions: $D < V_{\max}$, $V_{\max} < D < 2V_{\max}$ and $D < 2V_{\max}$. We show the expression for the condition $D < V_{\max}$, which corresponds to our parameter setting in numerical results. Then, from (2), (20) and (22), we have the following.

$$\begin{aligned}
\mathbb{E}[V_{r,i}|B_1] &= \frac{1}{u}\int_{V_{\min}}^{D}\int_{V_{\min}}^{D-v}\left(v - \frac{v}{D}(v + w)\right)f_{W'}(w)\,dw\,dv \\
&= \frac{1}{u^2\lambda(M_S(\lambda) - 1)}\left(\frac{D^2 V_{\min}}{6} - \frac{D^3}{24} - \frac{5V_{\min}^3}{6}\right. \\
&\quad + \frac{17V_{\min}^4}{24D} + \left(-\frac{D^2}{6} + \frac{V_{\min}^2}{2} - \frac{5V_{\min}^3}{6D} + \frac{DV_{\min}}{2}\right. \\
&\quad - \frac{D}{2\lambda} + \frac{V_{\min}^2}{2D\lambda}\bigg)e^{\lambda(V_{\min} - V_{\max})} \\
&\quad \left. - \frac{e^{-D\lambda}(D\lambda + 1) - e^{-\lambda V_{\min}}(\lambda V_{\min} + 1)}{D\lambda^3}e^{D\lambda}e^{-\lambda V_{\max}}\right),
\end{aligned}$$

$$\mathbb{E}[Q_i|B_1] = \frac{1}{u}\int_{V_{\min}}^{D}\int_{V_{\min}}^{D-v}\frac{v}{D}(D-(v+w))^2 f_{W'}(w)\,dw\,dv$$

$$= \frac{1}{u^2\lambda(M_S(\lambda)-1)}\left(\frac{D^3 V_{\min}}{12} - \frac{2DV_{\min}^3}{3} - \frac{D^4}{60}\right.$$

$$- \frac{2e^{-\lambda V_{\max}}}{\lambda^3} + \frac{17V_{\min}^4}{12} - \frac{49V_{\min}^5}{60D} - \frac{2e^{-\lambda V_{\max}}}{D\lambda^4}$$

$$+ (-\frac{D^3}{12} - \frac{5V_{\min}^3}{3} - \frac{D^2}{3\lambda} + \frac{17V_{\min}^4}{12D} + \frac{V_{\min}^2}{\lambda} - \frac{D}{\lambda^2}$$

$$+ \frac{D^2 V_{\min}}{3} + \frac{DV_{\min}}{\lambda} - \frac{5V_{\min}^3}{3D\lambda} + \frac{V_{\min}^2}{D\lambda^2})e^{\lambda(V_{\min}-V_{\max})}$$

$$+ (\frac{2}{D\lambda^4} + \frac{2V_{\min}}{D\lambda^3})e^{\lambda(D-V_{\min}-V_{\max})}\bigg).$$

Finally, we have the following: $\mathbb{E}[V_{r,i}] = \mathbb{E}[V_{r,i}|I]p_I + \mathbb{E}[V_{r,i}|B_1]p_{B_1}$ and $\mathbb{E}[Q_i] = \mathbb{E}[Q_i|I]p_I + \mathbb{E}[Q_i|B_1]p_{B_1}$.

Then, $\mathbb{E}[Y_{\text{packet}}] = \frac{1}{p_I+p_{B_1}}\mathbb{E}[V_{r,i}]$ and $\mathbb{E}[Y_{\text{sum}}] = \lambda\mathbb{E}[Q_i]$.

*Appendix A.3. M/GI/1/2**

For M/GI/1/2* system, we have the same p_I, p_B, $\mathbb{E}[V_{r,i}|I]$ and $\mathbb{E}[Q_i|I]$ as in the M/GI/1/2 system. Next, we calculate the $\mathbb{E}[V_{r,i}|B]$ and $\mathbb{E}[Q_i|B]$. We have the following.

$$f_W(w) = \frac{\mathbb{P}[S>w]}{\mathbb{E}[S]} = \frac{V_{\max}-w}{u\mathbb{E}[S]}.$$

Then, we consider the condition $D < V_{\max}$ and from (2), (23) and (25); we have the following.

$$\mathbb{E}[V_{r,i}|B] = \frac{1}{u}\int_{V_{\min}}^{D}\int_{V_{\min}}^{D-v}(v-\frac{v}{D}(v+w))e^{-\lambda w}f_W(w)\,dw\,dv$$

$$= \frac{1}{u^2\mathbb{E}[S]}\left(\frac{V_{\max}}{\lambda^3} - \frac{3}{\lambda^4} + \frac{4}{D\lambda^5} - \frac{V_{\max}}{D\lambda^4} + (\frac{D}{\lambda^3}\right.$$

$$- \frac{D^2}{6\lambda^2} + \frac{V_{\min}^3}{2\lambda} + \frac{V_{\min}^2}{2\lambda^2} - \frac{V_{\min}^2 V_{\max}}{2\lambda} - \frac{5V_{\min}^4}{6D\lambda}$$

$$- \frac{4V_{\min}^3}{3D\lambda^2} - \frac{V_{\min}^2}{D\lambda^3} + \frac{D(2V_{\min}-V_{\max})}{2\lambda^2} + \frac{DV_{\min}^2}{2\lambda}$$

$$+ \frac{D^2(V_{\max}-V_{\min})}{6\lambda} + \frac{5V_{\min}^3 V_{\max}}{6D\lambda} + \frac{V_{\min}^2 V_{\max}}{2D\lambda^2}$$

$$- \frac{DV_{\min}V_{\max}}{2\lambda})e^{-\lambda V_{\min}} + (\frac{4V_{\min}}{D\lambda^4} + \frac{V_{\max}}{D\lambda^4} - \frac{V_{\min}^2}{D\lambda^3}$$

$$- \frac{V_{\min}V_{\max}}{D\lambda^3} - \frac{4}{\lambda^4} + \frac{V_{\min}}{\lambda^3} - \frac{4}{D\lambda^5})e^{\lambda(V_{\min}-D)}\bigg),$$

$$\mathbb{E}[Q_i|B] = \frac{1}{u}\int_{V_{\min}}^{D}\int_{V_{\min}}^{D-v}\frac{v}{D}(D-(v+w))^2 e^{-\lambda w}f_W(w)\,dw\,dv$$

$$= \frac{1}{u^2\mathbb{E}[S]}\left(\frac{8}{\lambda^5} - \frac{2V_{\max}}{\lambda^4} - \frac{10}{D\lambda^6} + \frac{2V_{\max}}{D\lambda^5}\right.$$

$$\left(-\frac{3D}{\lambda^4} + \frac{2D^2}{3\lambda^3} - \frac{D^3}{12\lambda^2} - \frac{5V_{\min}^4}{3\lambda} - \frac{8V_{\min}^3}{3\lambda^2} - \frac{2V_{\min}^2}{\lambda^3}\right.$$

$$+ \frac{5V_{\min}^3 V_{\max}}{3\lambda} + \frac{V_{\min}^2 V_{\max}}{\lambda^2} + \frac{D^2 V_{\min}^2}{3\lambda} + \frac{17V_{\min}^5}{12D\lambda}$$

$$+ \frac{37V_{\min}^4}{12D\lambda^2} + \frac{13V_{\min}^3}{3D\lambda^3} + \frac{3V_{\min}^2}{D\lambda^4} - \frac{3DV_{\min}}{\lambda^3} + \frac{DV_{\max}}{\lambda^3}$$

$$- \frac{DV_{\min}^2}{\lambda^2} + \frac{2D^2 V_{\min}}{3\lambda^2} - \frac{D^3 V_{\min}}{12\lambda} - \frac{D^2 V_{\max}}{3\lambda^2} + \frac{D^3 V_{\max}}{12\lambda}$$

$$+ \frac{DV_{\min}V_{\max}}{\lambda^2} - \frac{D^2 V_{\min}V_{\max}}{3\lambda} - \frac{17V_{\min}^4 V_{\max}}{12D\lambda}$$

$$\left.- \frac{5V_{\min}^3 V_{\max}}{3D\lambda^2} - \frac{V_{\min}^2 V_{\max}}{D\lambda^3}\right)e^{-\lambda V_{\min}} + \left(-\frac{10V_{\min}}{D\lambda^5}\right.$$

$$- \frac{2V_{\max}}{D\lambda^5} + \frac{2V_{\min}^2}{D\lambda^4} + \frac{2V_{\min}V_{\max}}{D\lambda^4} + \frac{2}{\lambda^5} - \frac{2V_{\min}}{\lambda^4}$$

$$\left.\left.+ \frac{10}{D\lambda^6}\right)e^{\lambda(V_{\min}-D)}\right).$$

Finally, we have the following: $\mathbb{E}[V_{r,i}] = \mathbb{E}[V_{r,i}|I]p_I + \mathbb{E}[V_{r,i}|B]p_B$ and $\mathbb{E}[Q_i] = \mathbb{E}[Q_i|I]p_I + \mathbb{E}[Q_i|B]p_B$.

Then, $\mathbb{E}[Y_{\text{packet}}] = \frac{1}{p_I+p_{B_1}}\mathbb{E}[V_{r,i}]$ and $\mathbb{E}[Y_{\text{sum}}] = \lambda\mathbb{E}[Q_i]$.

*Appendix A.4. M/GI/1/1**

Since we have $M_S(\lambda) = \frac{1}{u\lambda}(e^{-\lambda V_{\min}} - e^{-\lambda V_{\max}})$, from (26) we have $p_e = M_S(\lambda)$, and from (27), we have the following.

$$f_{S'}(s) = \frac{e^{-\lambda s}}{uM_S(\lambda)}.$$

Note that due to $g(V) = V$, conditional service time S' has the same domain of definition as the initial value $V_{0,i}$. Then, we calculate $\mathbb{E}[V_{r,i}]$ from (29) and (30), and we have the following.

$$\mathbb{E}[V_{r,i}] = p_e\int_{V_{\min}}^{V_{\text{up}}}(s - \frac{s}{D}s)f_{S'}(s)\,ds$$

$$= \frac{p_I}{uM_S(\lambda)}\left(\frac{e^{-\lambda V_{\min}}(\lambda V_{\min}+1)}{\lambda^2} - \frac{e^{-\lambda V_{\text{up}}}(\lambda V_{\text{up}}+1)}{\lambda^2}\right.$$

$$- \frac{e^{-\lambda V_{\min}}(\lambda^2 V_{\min}^2 + 2\lambda V_{\min}+2)}{D\lambda^3}$$

$$\left.- \frac{e^{-\lambda V_{\text{up}}}(\lambda^2 V_{\text{up}}^2 + 2\lambda V_{\text{up}}+2)}{D\lambda^3}\right).$$

From (29) and (32), we have the following.

$$\mathbb{E}[Q_i] = \frac{p_e}{2} \int_{V_{\min}}^{V_{up}} \frac{s}{D}(D-s)^2 f_{S'}(s)\,ds$$

$$= \frac{p_I}{2uM_S(\lambda)D\lambda^4}\left(e^{-\lambda V_{\min}}(D^2\lambda^3 V_{\min} + D^2\lambda^2 - 2D\lambda^3 V_{\min}^2\right.$$
$$- 4D\lambda^2 V_{\min} - 4D\lambda + \lambda^3 V_{\min}^3 + 3\lambda^2 V_{\min}^2 + 6\lambda V_{\min} + 6)$$
$$- e^{-\lambda V_{up}}(D^2\lambda^3 V_{up} + D^2\lambda^2 - 2D\lambda^3 V_{up}^2$$
$$\left. - 4D\lambda^2 V_{up} - 4D\lambda + \lambda^3 V_{up}^3 + 3\lambda^2 V_{up}^2 + 6\lambda V_{up} + 6)\right).$$

Finally, we have $\mathbb{E}[Y_{packet}] = \frac{1}{p_e}\mathbb{E}[V_{r,i}]$ and $\mathbb{E}[Y_{sum}] = \lambda\mathbb{E}[Q_i]$.

Appendix B. $\mathbb{E}[Y_{sum}]$ and $\mathbb{E}[Y_{packet}]$ for Constant Value with Exponentially Distributed Initial Value

For an exponentially distributed initial value, we have $f_V(v) = \mu e^{-\mu v}$, $\mathbb{E}[S] = \mathbb{E}[V] = \frac{1}{\mu}$ and $\tilde{V} = D$.

Appendix B.1. M/M/1/1

From (10), we have the following.

$$p_I = \frac{\mu}{\lambda + \mu}.$$

Next, we calculate $\mathbb{E}[V_{r,i}]$ from (1) and (11) and we have the following.

$$\mathbb{E}[V_{r,i}] = p_I \int_0^D v f_V(v)\,dv$$

$$= p_I \int_0^D v f_V(v)\,dv$$

$$= p_I\left(-De^{-\mu D} - \frac{1}{\mu}(e^{-\mu D} - 1)\right).$$

Then, we calculate $\mathbb{E}[Q_i]$ from (1) and (13) and we have the following.

$$\mathbb{E}[Q_i] = p_I \int_0^D \int_{g(v)}^D v f_V(v)\,d\tau dv$$

$$= p_I \int_0^D v(D-v)f_V(v)\,dv$$

$$= p_I \int_0^D (Dv - v^2)f_V(v)\,dv$$

$$= p_I\left(-D^2 e^{-\mu D} - \frac{D}{\mu}(e^{-\mu D} - 1) + D^2 e^{-\mu D}\right.$$
$$\left. + \frac{2}{\mu}De^{-\mu D} + \frac{2}{\mu^2}(e^{-\mu D} - 1)\right).$$

Finally we have $\mathbb{E}[Y_{packet}] = \frac{1}{p_I}\mathbb{E}[V_{r,i}]$ and $\mathbb{E}[Y_{sum}] = \lambda\mathbb{E}[Q_i]$.

Appendix B.2. M/M/1/2

From (15), we have the following.

$$M_S(\lambda) = \frac{\mu}{\lambda + \mu}.$$

Then, from (14), we have the following.

$$p_I = \frac{\mu^2}{\mu^2 + \mu\lambda + \lambda^2},$$

$$p_B = \frac{\mu\lambda + \lambda^2}{\mu^2 + \mu\lambda + \lambda^2}.$$

From Lemma 1, we have the following.

$$p_{B_1} = \frac{\mu\lambda}{\mu^2 + \mu\lambda + \lambda^2},$$

$$\mathbb{P}[W' > w] = e^{-\mu w}.$$

Then, from (19), we have the following.

$$f_{W'}(w) = \mu e^{-\mu w}.$$

For the idle case, from (1), (16) and (18), we have the following.

$$\mathbb{E}[V_{r,i}|I] = -De^{-\mu D} - \frac{1}{\mu}(e^{-\mu D} - 1),$$

$$\mathbb{E}[Q_i|I] = - D^2 e^{-\mu D} - \frac{D}{\mu}(e^{-\mu D} - 1) + D^2 e^{-\mu D}$$
$$+ \frac{2}{\mu}De^{-\mu D} + \frac{2}{\mu^2}(e^{-\mu D} - 1).$$

For the busy case, from (1), (20) and (22), we have the following.

$$\mathbb{E}[V_{r,i}|B_1] = \int_0^D \int_0^{D-v} v f_V(v) f_{W'}(w)\,dw\,dv$$
$$= \frac{1}{\mu} - \frac{e^{-D\mu}(D\mu + 1)}{\mu} - \frac{D^2 \mu e^{-D\mu}}{2}.$$

$$\mathbb{E}[Q_i|B_1] = \int_0^D \int_0^{D-v} v(D - (v + w)) f_V(v) f_{W'}(w)\,dw\,dv$$
$$= \frac{e^{-D\mu}}{2\mu^2}\left(4D\mu - 6e^{D\mu} + D^2\mu^2 + 2D\mu e^{D\mu} + 6\right).$$

Finally, we have the following: $\mathbb{E}[V_{r,i}] = \mathbb{E}[V_{r,i}|I]p_I + \mathbb{E}[V_{r,i}|B_1]p_{B_1}$ and $\mathbb{E}[Q_i] - \mathbb{E}[Q_i|I]p_I + \mathbb{E}[Q_i|B_1]p_{B_1}$.
Then, $\mathbb{E}[Y_{\text{packet}}] = \frac{1}{p_I + p_{B_1}}\mathbb{E}[V_{r,i}]$ and $\mathbb{E}[Y_{\text{sum}}] = \lambda\mathbb{E}[Q_i]$.

*Appendix B.3. M/GI/1/2**

For M/GI/1/2* system, we have the same p_I, p_B, $\mathbb{E}[V_{r,i}|I]$ and $\mathbb{E}[Q_i|I]$ as in the M/GI/1/2 system. Next, we calculate $\mathbb{E}[V_{r,i}|B]$ and $\mathbb{E}[Q_i|B]$. We have the following.

$$f_W(w) = \frac{\mathbb{P}[S > w]}{\mathbb{E}[S]} = \mu e^{-\mu w}.$$

Then, from (1), (23) and (25), we have the following.

$$E[V_{r,i}|B] = \int_0^D \int_0^{D-v} ve^{-\lambda w} f_V(v) f_W(w)\,dw\,dv$$

$$= \frac{1 - e^{-D\mu}(D\mu + 1)}{\lambda + \mu} - \frac{\mu^2 e^{-D\mu}(e^{-D\lambda} + D\lambda - 1)}{\lambda^2(\lambda + \mu)},$$

$$E[Q_i|B] = \int_0^D \int_0^{D-v} v(D - (v + w))e^{-\lambda w} f_V(v) f_W(w)\,dw\,dv$$

$$= \frac{e^{-D(\lambda+\mu)}}{\lambda^2\mu(\lambda + \mu)^2}(2\lambda^3 e^{D\lambda} - \mu^3 e^{D\lambda} + \mu^3 + 3\lambda^2\mu e^{D\lambda}$$

$$- 2\lambda^3 e^{D(\lambda+\mu)} + D\lambda\mu^3 e^{D\lambda} + D\lambda^3\mu e^{D\lambda}$$

$$- 3\lambda^2\mu e^{D(\lambda+\mu)} + 2D\lambda^2\mu^2 e^{D\lambda} + D\lambda^2\mu^2 e^{D(\lambda+\mu)}$$

$$+ D\lambda^3\mu e^{D(\lambda+\mu)}).$$

Finally, we have the following: $E[V_{r,i}] = E[V_{r,i}|I]p_I + E[V_{r,i}|B]p_B$ and $E[Q_i] = E[Q_i|I]p_I + E[Q_i|B]p_B$.

Then, $E[Y_{\text{packet}}] = \frac{1}{p_I + p_{B_1}} E[V_{r,i}]$ and $E[Y_{\text{sum}}] = \lambda E[Q_i]$.

*Appendix B.4. M/GI/1/1**

Since we have $M_S(\lambda) = \frac{\mu}{\lambda+\mu}$, from (26) we have $p_e = \frac{\mu}{\lambda+\mu}$ and from (27), we have the following.

$$f_{S'}(s) = (\lambda + \mu)e^{-(\lambda+\mu)s}.$$

Note that since $g(V) = V$, we calculate $E[V_{r,i}]$ from (29) and (30), and we have the following.

$$E[V_{r,i}] = p_e \int_0^D s f_{S'}(s)\,ds$$

$$= p_e\left(\frac{1}{\lambda + \mu} - De^{-D(\lambda+\mu)} - \frac{e^{-D(\lambda+\mu)}}{\lambda + \mu}\right).$$

From (29) and (32), we have the following.

$$E[Q_i] = p_e \int_0^D s(D - s) f_{S'}(s)\,ds$$

$$- \frac{p_e}{(\lambda + \mu)^2}(e^{-D(\lambda+\mu)}(D\lambda + D\mu + 2) + D\lambda + D\mu - 2).$$

Finally, we have $E[Y_{\text{packet}}] = \frac{1}{p_e} E[V_{r,i}]$ and $E[Y_{\text{sum}}] = \lambda E[Q_i]$.

References

1. Giordani, M.; Zanella, A.; Higuchi, T.; Altintas, O.; Zorzi, M. Investigating value of information in future vehicular communications. In Proceedings of the 2019 IEEE 2nd Connected and Automated Vehicles Symposium (CAVS), Honolulu, HI, USA, 22–23 September 2019; pp. 1–5.
2. Higuchi, T.; Giordani, M.; Zanella, A.; Zorzi, M.; Altintas, O. Value-anticipating v2v communications for cooperative perception. In Proceedings of the 2019 IEEE Intelligent Vehicles Symposium, Paris, France, 9–12 June 2019; pp. 1947–1952.
3. Giordani, M.; Higuchi, T.; Zanella, A.; Altintas, O.; Zorzi, M. A framework to assess value of information in future vehicular networks. In Proceedings of the 1st ACM MobiHoc Workshop on Technologies, Models, and Protocols for Cooperative Connected Cars, Catania, Italy, 2 July 2019; pp. 31–36.
4. Bölöni, L.; Turgut, D.; Basagni, S.; Petrioli, C. Scheduling data transmissions of underwater sensor nodes for maximizing value of information. In Proceedings of the 2013 IEEE Global Communications Conference (GLOBECOM), Atlanta, GA, USA, 9–13 December 2013; pp. 438–443.

5. Turgut, D.; Bölöni, L. IVE: Improving the value of information in energy-constrained intruder tracking sensor networks. In Proceedings of the 2013 IEEE International Conference on Communications (ICC), Budapest, Hungary, 9–13 June 2013; pp. 6360–6364.

6. Bidoki, N.H.; Baghdadabad, M.B.; Sukthankar, G.; Turgut, D. Joint value of information and energy aware sleep scheduling in wireless sensor networks: A linear programming approach. In Proceedings of the 2018 IEEE International Conference on Communications (ICC), Kansas City, MO, USA, 13 May 2018.

7. Suri, N.; Benincasa, G.; Lenzi, R.; Tortonesi, M.; Stefanelli, C.; Sadler, L. Exploring value-of-information-based approaches to support effective communications in tactical networks. *IEEE Commun. Mag.* **2015**, *53*, 39–45. [CrossRef]

8. Kamar, E.; Horvitz, E. Light at the end of the tunnel: A monte carlo approach to computing value of information. In Proceedings of the International Conference on Autonomous Agents and Multi-Agent Systems, St. Paul, MN, USA, 6–10 May 2013; pp. 571–578.

9. Rosenthal, S.; Bohus, D.; Kamar, E.; Horvitz, E. Look versus leap: Computing value of information with high-dimensional streaming evidence. In Proceedings of the Twenty-Third International Joint Conference on Artificial Intelligence, Beijing, China, 3–9 August 2013.

10. Kaul, S.; Yates, R.; Gruteser, M. Real-time status: How often should one update? In Proceedings of the 2012 Proceedings IEEE INFOCOM, Orlando, FL, USA, 25–30 March 2012; pp. 2731–2735.

11. Kaul, S.; Gruteser, M.; Rai, V.; Kenney, J. Minimizing age of information in vehicular networks. In Proceedings of the IEEE Conference on Sensor, Mesh and Ad Hoc Communications and Networks, Secon, Japan, 27–30 June 2011; pp. 350–358.

12. Costa, M.; Codreanu, M.; Ephremides, A. On the age of information in status update systems with packet management. *IEEE Trans. Inf. Theory* **2016**, *62*, 1897–1910. [CrossRef]

13. Najm, E.; Nasser, R. Age of information: The gamma awakening. In Proceedings of the 2016 IEEE International Symposium on Information Theory (ISIT), Barcelona, Spain, 10–15 July 2016; pp. 2574–2578.

14. Inoue, Y.; Masuyama, H.; Takine, T.; Tanaka, T. A general formula for the stationary distribution of the age of information and its application to single-server queues. *IEEE Trans. Inf. Theory* **2019**, *65*, 2574–2578. [CrossRef]

15. Baknina, A.; Ozel, O.; Yang, J.; Ulukus, S.; Yener, A. Sending information through status updates. In Proceedings of the 2018 IEEE International Symposium on Information Theory (ISIT), Vail, CO, USA, 17–22 June 2018.

16. Jia, X.; Cao, S.; Xie, M. Age of Information of Dual-Sensor Information Update System With HARQ Chase Combining and Energy Harvesting Diversity. *IEEE Wirel. Commun. Lett.* **2021**, *10*, 2027–2031. [CrossRef]

17. Yates, R. Lazy is timely: Status updates by an energy harvesting source. In Proceedings of the 2015 IEEE International Symposium on Information Theory (ISIT), Hong Kong, China, 14–19 June 2015.

18. Gindullina, E.; Badia, L.; Gündüz, D. Age-of-Information With Information Source Diversity in an Energy Harvesting System. *IEEE Trans. Green Commun. Netw.* **2021**, *5*, 1529–1540. [CrossRef]

19. Koukoutsidis, I. A Fluid Reservoir Model for the Age of Information through Energy-Harvesting Transmitters. In Proceedings of the 2021 International Symposium on Performance Evaluation of Computer and Telecommunication Systems (SPECTS), Fairfax, WV, USA, 19–21 July 2021; pp. 1–8.

20. Arafa, A.; Yang, J.; Ulukus, S.; Poor, H.V. Timely Status Updating Over Erasure Channels Using an Energy Harvesting Sensor: Single and Multiple Sources. *IEEE Trans. Green Commun. Netw.* **2022**, *6*, 6–19. [CrossRef]

21. Talak, R.; Karaman, S.; Modiano, E. Minimizing age-of-information in multi-hop wireless networks. In Proceedings of the Communication, Control, and Computing (Allerton), 2017 55th Annual Allerton Conference on, Monticello, IL, USA, 3–6 October 2017; pp. 486–493.

22. Liu, S.; Jiao, J.; Ni, Z.; Wu, S.; Zhang, Q. Age-Optimal NC-HARQ Protocol for Multi-hop Satellite-based Internet of Things. In Proceedings of the 2021 IEEE Wireless Communications and Networking Conference (WCNC), Nanjing, China, 29 March 2021; pp. 1–6.

23. Yates, R.D. The age of information in networks: Moments, distributions, and sampling. *IEEE Trans. Inf. Theory* **2020**, *66*, 5712–5728. [CrossRef]

24. Maatouk, A.; Assaad, M.; Ephremides, A. The age of updates in a simple relay network. In Proceedings of the 2018 IEEE Information Theory Workshop (ITW), Guangzhou, China, 25–29 November 2018.

25. Ding, J.; Jiao, J.; Liu, S.; Wu, S.; Zhang, Q. Freshness-Critical Transmission Scheme with IR-HARQ over Multi-Hop Satellite-IoT. In Proceedings of the 2021 IEEE 94th Vehicular Technology Conference (VTC2021-Fall), Virtual, 27 September–28 October 2021; pp. 1–5.

26. Leng, S.; Ni, X.; Yener, A. Age of information for wireless energy harvesting secondary users in cognitive radio networks. In Proceedings of the IEEE International Conference on Mobile Ad Hoc and Sensor Systems, Shenzhen, China, 11–13 December 2019; pp. 353–361.

27. Leng, S.; Yener, A. Age of information minimization for an energy harvesting cognitive radio. *IEEE Trans. Cogn. Commun. Netw.* **2019**, *5*, 427–439. [CrossRef]

28. Alabbasi, A.; Aggarwal, V. Joint information freshness and completion time optimization for vehicular networks. *IEEE Trans. Serv. Comput.* **2020**. [CrossRef]

29. Rajaraman, N.; Vaze, R.; Reddy, G. Not just age but age and quality of information. *IEEE J. Sel. Areas Commun.* **2021**, *39*, 1325–1338. [CrossRef]

30. Kosta, A.; Pappas, N.; Ephremides, A.; Angelakis, V. Age and value of information: Non-linear age case. In Proceedings of the 2017 IEEE International Symposium on Information Theory(ISIT), Aachen, Germany, 25–30 June 2017; pp. 326–330.
31. Maatouk, A.; Kriouile, S.; Assaad, M.; Ephremides, A. The age of incorrect information: A new performance metric for status updates. *IEEE/ACM Trans. Netw.* **2020**, *28*, 2215–2228. [CrossRef]
32. Bastopcu, M.; Ulukus, S. Age of information for updates with distortion. In Proceedings of the 2019 IEEE Information Theory Workshop (ITW), Visby, Sweden, 25–28 August 2019.
33. Ayan, O.; Vilgelm, M.; Klügel, M.; Hirche, S.; Kellerer, W. Age-of-information vs. value-of-information scheduling for cellular networked control systems. In Proceedings of the ACM/IEEE International Conference on Cyber-Physical Systems, Montreal, QC, Canada, 16–18 April 2019; pp. 109–117.
34. Kam, C.; Kompella, S.; Nguyen, G.D.; Wieselthier, J.E.; Ephremides, A. Controlling the age of information: Buffer size, deadline, and packet replacement. In Proceedings of the MILCOM 2016—2016 IEEE Military Communications Conference, Baltimore, MD, USA, 1–3 November 2016; pp. 301–306.
35. Kam, C.; Kompella, S.; Nguyen, G.D.; Wieselthier, J.E.; Ephremides, A. Age of information with a packet deadline. In Proceedings of the 2016 IEEE International Symposium on Information Theory ISIT, Barcelona, Spain, 10–15 July 2016; pp. 2564–2568.
36. Kam, C.; Kompella, S.; Nguyen, G.D.; Wieselthier, J.E.; Ephremides, A. On the age of information with packet deadlines. *IEEE Trans. Inf. Theory* **2018**, *64*, 6419–6428. [CrossRef]
37. Inoue, Y. Analysis of the age of information with packet deadline and infinite buffer capacity. In Proceedings of the 2018 IEEE International Symposium on Information Theory (ISIT), Vail, CO, USA, 17–22 June 2018; pp. 2639–2643.
38. Bastopcu, M.; Ulukus, S. Minimizing age of information with soft updates. *J. Commun. Netw.* **2019**, *21*, 233–243. [CrossRef]
39. Kam, C.; Kompella, S.; Ephremides, A. Content based status updates. In Proceedings of the IEEE INFOCOM AoI Workshop, Beijing, China, 27 April 2020.
40. Ali, S.M.; Stefanović, Č. A Study on Value-of-Information and Age-of-Information in a First Responders Health Monitoring System. In Proceedings of the 2021 International Balkan Conference on Communications and Networking (BalkanCom), Novi Sad, Serbia, 20–22 September 2021; pp. 133–137.
41. Zou, P.; Ozel, O.; Subramaniam, S. On age and value of information in status update systems. In Proceedings of the 2020 IEEE Wireless Communications and Networking Conference (WCNC), Seoul, Korea, 25–28 May 2020; pp. 1–6.

Article

Age of Information of Parallel Server Systems with Energy Harvesting

Josu Doncel

Mathematics Department, University of the Basque Country, UPV/EHU, 48940 Leioa, Spain; josu.doncel@ehu.eus

Abstract: Motivated by current communication networks in which users can choose different transmission channels to operate and also by the recent growth of renewable energy sources, we study the average Age of Information of a status update system that is formed by two parallel homogeneous servers and such that there is an energy source that feeds the system following a random process. An update, after getting service, is delivered to the monitor if there is energy in a battery. However, if the battery is empty, the status update is lost. We allow preemption of updates in service and we assume Poisson generation times of status updates and exponential service times. We show that the average Age of Information can be characterized by solving a system with eight linear equations. Then, we show that, when the arrival rate to both servers is large, the average Age of Information is one divided by the sum of the service rates of the servers. We also perform a numerical analysis to compare the performance of our model with that of a single server with energy harvesting and to study in detail the aforementioned convergence result.

Keywords: parallel servers; energy harvesting; Age of Information

Citation: Doncel, J. Age of Information of Parallel Server Systems with Energy Harvesting. *Entropy* **2021**, *23*, 1549. https://doi.org/10.3390/e23111549

Academic Editors: Anthony Ephremides and Yin Sun

Received: 2 November 2021
Accepted: 19 November 2021
Published: 21 November 2021

Publisher's Note: MDPI stays neutral with regard to jurisdictional claims in published maps and institutional affiliations.

1. Introduction

1.1. Motivation

The Age of Information is a recent metric of the performance of systems and it measures the freshness of the information that a monitor has about the status of a remote process of interest. There is a wide range of applications in which information about a source must be as recent as possible. An example of this is given in autonomous driving systems since the location of the vehicles must be known as soon as possible. Or, in other words, obsolete information about the traffic might lead to bad consequences (traffic accidents, for instance) to the users.

Status update systems are formed by sources of generation status updates, a transmission channel and a monitor that receives the updates. The transmission channel takes care of sending the status updates from the source to the destination. It is clear that the devices of the transmission channel require energy to work. Therefore, it is important to consider energy consumption in the modeling of the transmission channel. Furthermore, there has been recently an increasing amount of different types of renewable energy sources that feed the energy network. Some examples are solar or wind energy sources, which are clearly very volatile. As a consequence, the randomness of the generation of energy also needs to be taken into account in the modeling of the transmission channel.

Current communication networks are very complex and often allow users to operate using different transmission channels. This is the case, for instance, when a user is a part of an overlay network (i.e., when it belongs to a set of nodes that are located in different spots over the Internet and collaborate with each other to forward data between any pair of nodes with minimum delay). In fact, in this instance, the user can choose the transmission channel that provides the IP protocol or through the overlay network. Therefore, in this work, we study the average Age of Information in a status update system with energy harvesting. That is, we consider that the transmission channel is formed by parallel servers that do not interchange information and a battery that can store energy that can be used to send status updates after getting service in the servers.

1.2. Related Work

The Age of Information has been introduced in [1,2] as a metric to measure the freshness of the information about the state of a remote system. Since its introduction, there has been many researcher of different areas that has been interested in analyzing this metric. In the first works following the seminal papers, the goal has been to characterize the average Age of Information of status update systems where the transmission channel is modeled as a single queue. For instance, the authors in [3] characterize the average Age of Information of a single server (i.e., a queue without buffer) and a single source. Regarding optimality, the authors in [4] show that the preemptive Last Generated First Served policy minimizes the Age of Information. Unfortunately, the characterization of the average Age of Information of many models is known to be an extremely difficult task. Therefore, some authors has been interested in other similar metric of performance such as the Peak Age of Information [5] or the Age of Incorrect Information [6]. We refer to the following surveys on this topic for full details of these metrics and their properties [7–9].

Let us now discuss the work of some authors that have been interested in analyzing the Age of Information of a system with energy harvesting. In [10,11] it is considered a system with Poisson arrivals of energy and that there is no losses of packets. Their goal is to find the optimal status updates policy such that the battery is not empty upon an arrival of a status update. The authors in [12–14] generalize the model of [10,11] by allowing status update losses and also focus on optimal policies for generation updates, with or without knowledge (or feedback) whether the status updates are delivered successfully. Our model, that has been in inspired by the Energy Packet Networks [15,16], is different from these models for different reasons. First, we do not impose the presence of energy to receive a status update. Another difference is that the generation of status updates follows a Poisson process in our model, which is not the case in these works. Finally, our goal is different since we are interested in characterizing the average Age of Information and studying its properties and, hence, we do not aim to find the optimal policy.

1.3. Contribution

We consider a system with two parallel homogeneous servers and one battery that stores energy packets. Energy packets model a certain amount of energy and are necessary to send the status updates (or data packets) to the monitor after ending service. This means that a data packet is sent to the battery when it ends service and, if the battery is empty, the data packet is lost, whereas if battery is not empty the data packet is delivered to the monitor and one energy packet disappears. We consider that arrivals of data packets and energy packets follow a Poisson process and the queues that handle data packets and energy packets do not have buffer. We allow preemption of data packets, i.e., when a data packet arrives to a server that is busy, the incoming packet replaces the packet in service.

The first contribution of this work is to characterize the average Age of Information of the above status update system using the Stochastic Hybrid System technique [17]. More specifically, we show that the average Age of Information can be computed by solving a system of 8 linear equations. We then consider the regime where the arrival rate of data packets to both servers tends to infinity and we show that the average Age of Information is one divided by two times the service rate of data packets.

The model we study here generalizes

- the work of Section IIIA of [18] where it is studied the Age of Information of two parallel servers. In our work, we consider energy harvesting in their model. In fact, when in our model the arrival rate of energy packets is very large, it coincides with the model of [18].

- the work of [19] where it is analyzed a system with a single server and energy harvesting. In our work, we consider the same energy harvesting model, but with two parallel servers.

We go beyond the above presented analytical results with a numerical work that we describe next. First, we aim to compare the performance of a single servers with two parallel servers with energy harvesting. For this purpose, we consider the following systems: (i) a single server with arrival rate $\lambda/2$ and service rate μ, (ii) a single server with arrival rate λ and service rate 2μ and (iii) two parallel servers with service rate μ, each of them handling an arrival rate of $\lambda/2$. Let us note that the ratio of the arrival rate over the service rate coincide in all the servers of the systems under consideration. This comparison has been previously done in Section IIIA of [18], but they do not consider energy harvesting. Our first finding is that, when the arrival rate of energy packets is very large, we obtain the plot as Figure 4 of [18] and, therefore, their conclusions follow in our model as well (i.e., the system with double service rate and a single server minimizes the average Age of Information). We then investigate whether the conclusions of [18] also hold when the arrival rate of energy packets is not large. We observe that the average Age of Information is smaller for the system with two parallel servers and this difference increases when we decrease the arrival rate of energy packets. Finally, we study how the average Age of Information converges, when the arrival rate to the servers increases, to the value obtained in our analytical part. We conclude that the average Age of Information is not monotone with respect to the arrival rate of energy packets when the arrival rate to both servers is small. However, the average Age of Information does not depend on the arrival rate of energy packets when the arrival rate of packets to both servers is very large.

Potential applications of this model include systems in which two different transmission channels can be chosen to send updates. This is the case, for instance, when the source that generates status updates is part of an overlay network (i.e., when it belongs to a set of nodes that are located in different spots over the Internet and collaborate with each other to forward data between any pair of nodes with minimum delay) and it can choose to send the status updates through the path the provides the IP protocol or through the overlay routing.

1.4. Organization

The rest of the paper is organized as follows. First, in Section 2, we describe the model we study in this article. The average Age of Information analysis of this model is presented in Section 3. In Section 4, we focus on our numerical work and, finally, in Section 5, we draw the main conclusions of this work.

2. Model Description

2.1. Age of Information

We study the transmission of status updates (or data packets) to a monitor. We consider that data packet i is generated at time t_i and that it is delivered to the monitor at time t_i'. We denote by $N(t)$ the index of the last successfully delivered data packet to the monitor at time t, i.e.,

$$N(t) = \max\{i|t_i' \leq t\}.$$

Taking into account that the generation time of the last received data packet before time t is $t_{N(t)}$, we define the Age of Information at time t as follows:

$$\Delta(t) = t - t_{N(t)},$$

that is, the Age of Information at time t is the time elapsed since the generation of the last delivered packet to the monitor. We show in Figure 1 an example of $\Delta(t)$.

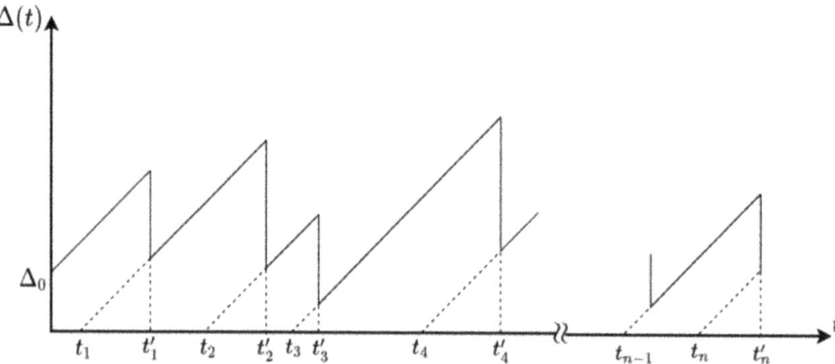

Figure 1. An example of $\Delta(t)$.

Assuming that the updating system is stable, the average Age of Information can be computed as the area below a "saw-tooth" shaped curve with teeth at the times at which the data packets are delivered (see Figure 1). Hence, if we denote by Δ the average Age of Information, we have that

$$\Delta = \lim_{\tau \to \infty} \frac{1}{\tau} \int_0^\tau \Delta(t)dt.$$

In this article, we are interested in calculating the average Age of Information in an energy harvesting model. In the following section, we describe the model we analyze.

2.2. Energy Harvesting Model

In our model, we represent energy by packets of discrete units called energy packets that model a certain quantity of energy (energy packets) measured in Joules, whereas the status updates of a process of interest are represented by packets that we call data packets. We consider an energy harvesting model formed by two parallel queues that store data packets (data queues) and a single queue that stores the energy packets. We show in Figure 2 the model under consideration in this work.

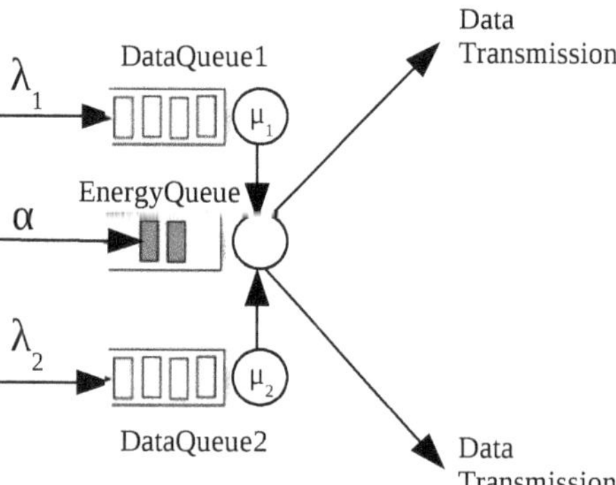

Figure 2. The energy harvesting model with two parallel data queues and a single energy queue. Energy packets are depicted with gray and data packets with white.

Energy packets arrive to the system according to a Poisson process of rate α and data packets (or workload packets) with rate λ. Upon arrival, a packet is dispatched to data queue 1 with probability $p > 0$ and to data queue 2 with probability $1 - p > 0$. Therefore, the arrival rate to data queue 1 is $\lambda_1 = \lambda p$ and to data queue 2 is $\lambda_2 = \lambda(1 - p)$.

Remark 1. *The probability p can be seen as the willingness of a source to use an alternative path (for instance, the path of an overlay network) rather than the usual transmission channel.*

We consider that the service rate of jobs of data queue i is exponentially distributed with rate μ, i = 1, 2

In this model, we consider that data packets, i.e., the packets of the data queues, start the transfer to a single energy queue. This means that, when a data packet gets served (in data queue 1 or data queue 2), it is sent to the energy queue. If the energy queue is empty upon arrival of a data packet, the data packet is lost. However, if there are energy packets when a data packet arrives to the energy queue, the data packet is transferred successfully to the monitor and one energy packet disappears.

Remark 2. *Our model considers a single energy queue. This models that the destination requires energy to receive status updates. This occurs, for instance, when there is a wireless antenna in charge of receiving the status updates at the destination (indeed, in absence of energy the antenna cannot deliver packets to the monitor).*

Here, we assume that the energy queue and the data queues do not have buffer. Therefore, the number of packets in each queue is, at most, one. Besides, energy packets that arrive when the energy queue is full are dropped, whereas when a data packet arrives to a full data queue, it replaces the job in execution.

3. Average Age of Information Analysis

In this section, we aim to analyze the average Age of Information of a system formed by two parallel queues with energy harvesting. We will use the Stochastic Hybrid System method to characterize the average Age of Information of the system under consideration. The Stochastic Hybrid System is formed by two values: the state of a continuous time Markov Chain and a vector containing the generation times of all the packets in the system as well of the current Age of Information. The Markov chain we consider is presented in Figure A1.

Let $s_0\,s_1\ldots s_7$ be the solution of the following system of equations:

$$0 = -s_0(\lambda + \alpha) + s_2\mu + s_3\mu + s_4\mu + s_5\mu \tag{1a}$$
$$0 = s_0\alpha - s_1\lambda \tag{1b}$$
$$0 = s_0\lambda(1 - p) - s_2(\lambda p + \mu + \alpha) + s_6\mu + s_7\mu \tag{1c}$$
$$0 = s_1\lambda(1 - p) + s_2\alpha - s_3(\lambda p + \mu) \tag{1d}$$
$$0 = s_0\lambda p - s_4(\lambda(1 - p) + \alpha + \mu) + s_6\mu + s_7\mu \tag{1e}$$
$$0 = s_1\lambda p + s_4\alpha - s_5(\lambda(1 - p) + \mu) \tag{1f}$$
$$0 = s_2\lambda p + s_4\lambda(1 - p) - s_6(\alpha + 2\mu) \tag{1g}$$
$$0 = s_3\lambda p + s_5\lambda(1 - p) + s_6\alpha - s_7\,2\mu \tag{1h}$$

that satisfies that $\sum_{i=0}^{7} s_i = 1$. As we will see in Appendix B, the solution of the above system of equations provides the steady-state distribution of the Markov chain of Figure A1.

We also define $x_1, x_2, \ldots, x_{16}$ as the solution of the following system of equations:

$$-s_0 = -x_1(\lambda + \alpha) + \mu x_7 + \mu x_{10} + \mu x_3 + \mu x_6 \tag{2a}$$

$$-s_1 = -x_2(\lambda + \alpha) + \alpha x_1 + \alpha x_2 \tag{2b}$$

$$-s_2 = -x_3(\lambda + \alpha + \mu) + \lambda(1-p)x_1 + \lambda(1-p)x_3$$
$$+ \mu x_{15} \tag{2c}$$

$$-s_2 = -x_4(\lambda + \alpha + \mu) + \mu x_{16} \tag{2d}$$

$$-s_3 = -x_5(\lambda + \alpha + \mu) + \alpha x_3 + \alpha x_5 + \lambda(1-p)x_2$$
$$+ \lambda(1-p)x_5 \tag{2e}$$

$$-s_3 = -x_6(\lambda + \alpha + \mu) + \alpha x_4 + \alpha x_6 \tag{2f}$$

$$-s_4 = -x_7(\lambda + \alpha + \mu) + \lambda p x_1 + \lambda p x_7 + \mu x_{16} \tag{2g}$$

$$-s_4 = -x_8(\lambda + \alpha + \mu) + \mu x_{15} \tag{2h}$$

$$-s_5 = -x_9(\lambda + \alpha + \mu) + \lambda p x_2 + \lambda p x_9 + \alpha x_9 + \alpha x_7 \tag{2i}$$

$$-s_5 = -x_{10}(\lambda + \alpha + \mu) + \alpha x_{10} + \alpha x_8 \tag{2j}$$

$$-s_6 = -x_{11}(\lambda + 2\mu + \alpha) + \lambda p x_3 + \lambda p x_{11}$$
$$+ \lambda(1-p)x_7 + \lambda(1-p)x_{11} \tag{2k}$$

$$-s_6 = -x_{12}(\lambda + 2\mu + \alpha) + \lambda(1-p)x_8 + \lambda(1-p)x_{12} \tag{2l}$$

$$-s_6 = -x_{13}(\lambda + 2\mu + \alpha) + \lambda p x_4 + \lambda p x_{13} \tag{2m}$$

$$-s_7 = -x_{14}(\lambda + 2\mu) + \lambda p x_5 + \lambda p x_{14} + \lambda(1-p)x_9$$
$$+ \lambda(1-p)x_{14} + \alpha x_{11} \tag{2n}$$

$$-s_7 = -x_{15}(\lambda + 2\mu) + \lambda(1-p)x_{10} + \lambda(1-p)x_{15}$$
$$+ \alpha x_{12} \tag{2o}$$

$$-s_7 = -x_{16}(\lambda + 2\mu) + \lambda p x_6 + \lambda p x_{16} + \alpha x_{13}, \tag{2p}$$

where $s_0, s_1, \ldots, s_7$ are given in Equation (1a–h). As we explain in Appendix B, the values $x_1, \ldots, x_{16}$ coincide with the generation time of all the packets in the system for all the possible states of the Markov chain.

In the following result, we use the Stochastic Hybrid System technique [17] to characterize the average Age of Information of this system and we show that it can be done by solving the above system of equations.

Proposition 1. *The average Age of Information of a system with two parallel servers with the energy harvesting is given by*

$$x_1 + x_2 + x_3 + x_5 + x_7 + x_9 + x_{11} + x_{14},$$

where $x_1, x_2, \ldots, x_{16}$ are the solution of Equation (2a–p).

Proof. See Appendix B. $\square$

In Proposition 1, we show that the computation of the average Age of Information of the system under study requires to solve Equation (2a–p), which is a system of 16 linear equations with 16 variables. Now, we aim to show that this system of equations has a special structure and how it can be used to obtain a method to compute the average Age of Information by solving a simpler system. Let us first present the following auxiliary results.

Lemma 1. *The Equation (2d,f,m,p), form a system of 4 linear equations with 4 variables (x_4, x_6, x_{13} and x_{16}). Let*

$$c = \frac{\lambda p \alpha}{\lambda + \alpha + \mu}\left(\frac{1}{\lambda + \mu} + \frac{1}{\lambda(1-p) + 2\mu + \alpha}\right). \tag{3}$$

We have that

$$x_{16} = \frac{s_7 - \frac{\lambda p s_3}{\lambda + \mu} - \frac{\alpha s_6}{\lambda(1-p)+2\mu+\alpha} - c\, s_2}{c\mu - (\lambda(1-p)+2\mu)}, \tag{4}$$

and

$$x_4 = \frac{\mu x_{16} + s_2}{\lambda + \alpha + \mu}. \tag{5}$$

as well as

$$x_6 = \frac{\alpha x_4 + s_3}{\lambda + \mu}. \tag{6}$$

Proof. See Appendix C. $\square$

Lemma 2. *The Equation (2h,j,l,o), form a system of 4 linear equations with 4 variables (x_8, x_{10}, x_{12} and x_{15}). Let*

$$d = \frac{\lambda(1-p)\alpha}{\lambda + \alpha + \mu}\left(\frac{1}{\lambda + \mu} + \frac{1}{\lambda p + 2\mu + \alpha}\right). \tag{7}$$

We have that

$$x_{15} = \frac{s_7 - \frac{\lambda p s_5}{\lambda + \mu} - \frac{\alpha s_6}{\lambda(1-p)+2\mu+\alpha} - d\, s_4 +}{d\mu - (\lambda(1-p)+2\mu)}, \tag{8}$$

and

$$x_8 = \frac{\mu x_{15} + s_4}{\lambda + \alpha + \mu}. \tag{9}$$

as well as

$$x_{10} = \frac{\alpha x_8 + s_5}{\lambda + \mu}. \tag{10}$$

Proof. The proof is symmetric to the proof of Lemma 1 and, therefore, we omit it for clarity of the presentation. $\square$

We now writ Equation (2a–p) that have not been analyzed in the previous lemmas:

$$-s_0 = -x_1(\lambda + \alpha) + \mu x_7 + \mu x_{10} + \mu x_3 + \mu x_6 \tag{11a}$$
$$-s_1 = -x_2\lambda + \alpha x_1 \tag{11b}$$
$$-s_2 = -x_3(\lambda p + \alpha + \mu) + \lambda(1-p)x_1 + \mu x_{15} \tag{11c}$$
$$-s_3 = -x_5(\lambda p + \mu) + \alpha x_3 + \lambda(1-p)x_2 \tag{11d}$$
$$-s_4 = -x_7(\lambda(1-p) + \alpha + \mu) + \lambda p x_1 + \mu x_{16} \tag{11e}$$
$$-s_5 = -x_9(\lambda(1-p) + \mu) + \lambda p x_2 + \alpha x_7 \tag{11f}$$
$$-s_6 = -x_{11}(2\mu + \alpha) + \lambda p x_3 + \lambda(1-p)x_7 \tag{11g}$$
$$-s_7 = -x_{14}2\mu + \lambda p x_5 + \lambda(1-p)x_9 + \alpha x_{11} \tag{11h}$$

which is a system of 8 equations with 12 variables (the variables are $x_1, x_2, x_3, x_5, x_6, x_7, x_8, x_9,$ $x_{10}, x_{11}, x_{14}, x_{15}$ and x_{16}). However, an explicit expression of x_6 and x_{16} have been obtained in Lemma 1 and an explicit explicit expression of x_{10} and x_{15} in Lemma 2. Therefore, the next result follows.

Proposition 2. *Let $x_1, x_2, x_3, x_5, x_7, x_9, x_{11}, x_{14}$ be the solution of Equation (11a–h) (recall that x_6 and x_{16} are given in Lemma 1 and x_{10} and x_{15} in Lemma 2). Therefore, the average Age of Information of a two parallel servers system with energy harvesting is given by*

$$x_1 + x_2 + x_3 + x_5 + x_7 + x_9 + x_{11} + x_{14}.$$

3.1. Analysis When λ Tends to ∞

We now consider the asymptotic regime where $\lambda \to \infty$ when the parameters α and μ are finite.

We first focus on the solution of Equation (1a–h).

Lemma 3. *When $\lambda \to \infty$ and $\max(\alpha, \mu) < 0$, the solution of Equation (1a–h) satisfies that $s_0 = s_1 = s_2 = s_3 = s_4 = s_5 = 0$.*

Proof. See Appendix D. $\square$

From this result, we conclude that, in the asymptotic regime under study, $s_6 + s_7 = 1$ and that Equation (11a–h) can be written as

$$0 = -x_1(\lambda + \alpha) + \mu x_7 + \mu x_{10} + \mu x_3 + \mu x_6 \tag{12a}$$
$$0 = -x_2\lambda + \alpha x_1 \tag{12b}$$
$$0 = -x_3(\lambda p + \alpha + \mu) + \lambda(1 - p)x_1 + \mu x_{15} \tag{12c}$$
$$0 = -x_5(\lambda p + \mu) + \alpha x_3 + \lambda(1 - p)x_2 \tag{12d}$$
$$0 = -x_7(\lambda(1 - p) + \alpha + \mu) + \lambda p x_1 + \mu x_{16} \tag{12e}$$
$$0 = -x_9(\lambda(1 - p) + \mu) + \lambda p x_2 + \alpha x_7 \tag{12f}$$
$$-s_6 = -x_{11}(2\mu + \alpha) + \lambda p x_3 + \lambda(1 - p)x_7 \tag{12g}$$
$$-s_7 = -x_{14}2\mu + \lambda p x_5 + \lambda(1 - p)x_9 + \alpha x_{11} \tag{12h}$$

If we replace the last equation by the sum the last two equations, we get the following equivalent system:

$$0 = -x_1(\lambda + \alpha) + \mu x_7 + \mu x_{10} + \mu x_3 + \mu x_6 \tag{13a}$$
$$0 = -x_2\lambda + \alpha x_1 \tag{13b}$$
$$0 = -x_3(\lambda p + \alpha + \mu) + \lambda(1 - p)x_1 + \mu x_{15} \tag{13c}$$
$$0 = -x_5(\lambda p + \mu) + \alpha x_3 + \lambda(1 - p)x_2 \tag{13d}$$
$$0 = -x_7(\lambda(1 - p) + \alpha + \mu) + \lambda p x_1 + \mu x_{16} \tag{13e}$$
$$0 = -x_9(\lambda(1 - p) + \mu) + \lambda p x_2 + \alpha x_7 \tag{13f}$$
$$-s_6 = -x_{11}(2\mu + \alpha) + \lambda p x_3 + \lambda(1 - p)x_7 \tag{13g}$$
$$-1 = -(x_{14} + x_{11})2\mu + \lambda p x_5 + \lambda(1 - p)x_9$$
$$+ \lambda p x_3 + \lambda(1 - p)x_7 \tag{13h}$$

We now analyze the solution of Equation (13a–h) for large λ.

Lemma 4. *When $\lambda \to \infty$ and $\max(\alpha, \mu) < 0$, the solution of Equation (13a–h) satisfies that $x_1 = x_2 = x_3 = x_5 = x_7 = x_9 = 0$.*

Proof. The proof uses the same arguments than those of the proof of Lemma 3 and, therefore, we omit it. $\square$

From the above results, we conclude that the average Age of Information of this system is given by $x_{11} + x_{14}$. Furthermore, using that $x_3 = x_5 = x_7 = x_9 = 0$ and from (13h), we obtain that $x_{11} + x_{14} = \frac{1}{2\mu}$, which gives the following result:

Proposition 3. *When $\lambda \to \infty$ and $\max(\alpha, \mu) < \infty$, the average Age of Information of a two parallel servers system with energy harvesting is given by $\frac{1}{2\mu}$.*

It is important to remark that the average Age of Information of the system under study in the considered asymptotic regime does not depend on the arrival rate of energy packets, i.e., on α.

3.2. Limitations to Analyze More Complex Models

We have tried to extend the results presented in this section to more complex systems and we have noticed that this task is extremely difficult. The main reason for this that the Markov chain to be considered (and, as a consequence, the number of equations to be solved) increases at a very high rate with the complexity of the system. This suggests that the analysis of the average Age of Information of more complex systems requires to consider other techniques such as simulations or approximation techniques.

4. Performance Evaluation

In the previous section, we have obtained an explicit expression of the average Age of Information of a system with two parallel servers and energy harvesting. Now, we aim to evaluate the obtained expression to analyze its main properties. We have performed a large number of simulations changing the values of the parameters and the illustrations of this section are illustrative of the general pattern. (The plots of this section can be reproduced using the code of https://github.com/josudoncel/AioParallelEnergy, accessed on 18 November 2021).

4.1. Parallel Servers vs. Single Server

We aim to compare the average Age of Information of the model under study in this paper with that of a system with a single server with energy harvesting (the average Age of Information of the latter model has been studied in [19]). For this purpose, we consider the following systems: (i) a single server with arrival rates λ and α of data packets and energy packets, respectively, and service rate 2μ (which is represented with a dotted line); (ii) a single server with arrival rates $\lambda/2$ and α of data packets and energy packets, respectively, and service rate 2μ (which is represented with a solid line); and (iii) two parallel servers with $p = 0.5$, i.e., the arrival rate to each server is $\lambda/2$, the service rate is μ and the arrival rate of energy packets is α (which is represented with a dashed line).

We first consider that the arrival rate of energy packet is very large. For this instance, there is always energy to transmit the data packet when it ends service, or in other words, the data packets are never lost because there is no energy. We note that, when this occurs, the comparison study we carry out here coincides with the analysis of Section III2 of [18]. In Figure 3, we consider $\mu = 1$ and $\alpha = 10^3$ and we plot the evolution of the average Age of Information of the systems under study in this section when λ varies from 0.1 to 10^3. We observe that the obtained plot coincides with Figure 4 of [18]. From this illustration, the authors in [18] conclude that some properties of the classical performance metrics of queuing theory, such as mean delay, are verified for the average Age of Information (the system that minimizes performance is the single server with service 2μ), but other properties do not (the mean delay of a system with two parallel servers with arrival rate to each equal to $\lambda/2$ is the same as that of a single server with arrival rate $\lambda/2$, but this is not the case for the average Age of Information). This illustration shows that, as expected, these conclusions also hold for our model when α is large.

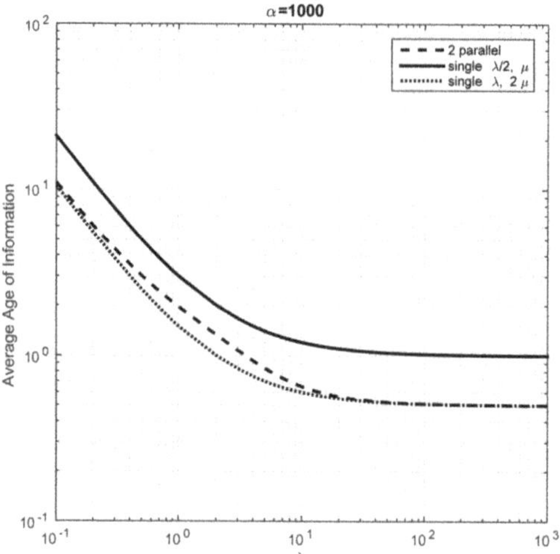

Figure 3. Average Age of Information of the three systems under comparison when λ changes from 0.1 to 10^3. $\alpha = 1000$ and $\mu = 1$.

We now aim to compare the performance of these systems when the arrival rate is not large. Thus, we fix the parameters equal to the previous plot and we consider different values of α. First, we consider $\alpha = 1$ and in Figure 4, we observe that the average Age of Information of all the systems do not change substantially with respect to the previous plot when λ is small. However, as λ grows, the average Age of Information of the systems with a single server decreases less than that of two parallel servers. We have also seen that, when λ is large, the average Age of Information of two parallel servers is 0.5, of a single server with double service rate 1.5 and of a single server with half traffic is 1.67. The difference on the average Age of Information between these system is even larger if we consider a smaller value of α. For instance, in Figure 5, we illustrate that the system with the smallest average Age of Information is the system with two parallel servers for almost all the values of λ we have considered.

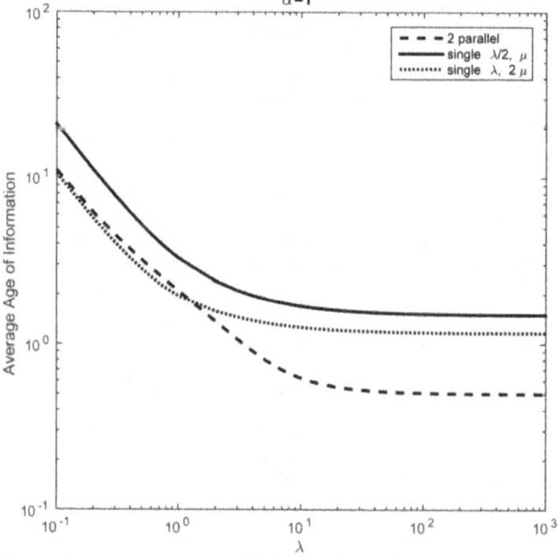

Figure 4. Average Age of Information of the three systems under comparison when λ changes from 0.1 to 10^3. $\alpha = 1$ and $\mu = 1$.

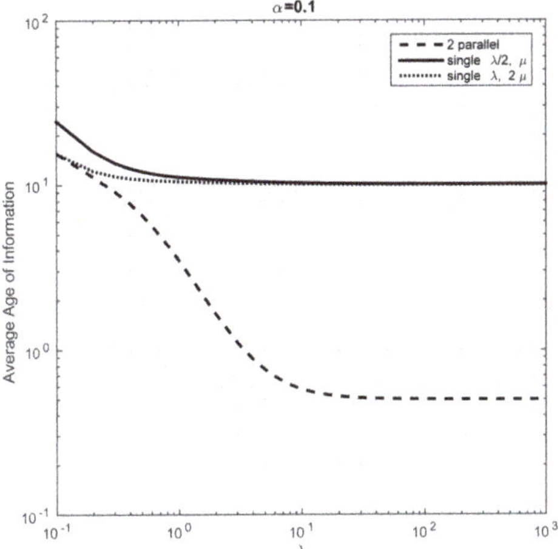

Figure 5. Average Age of Information of the three systems under comparison when λ changes from 0.1 to 10^3. $\alpha = 0.1$ and $\mu = 1$.

4.2. Convergence Analysis of Proposition 3

In Proposition 3, we show that, when $\lambda \to \infty$, the average Age of Information tends to $\frac{1}{2\mu}$, i.e., it does not depend on α or p. In this part of the article, we aim to study this convergence. We consider $\mu = 1$ and $p = 0.1$ and we vary α from 0.1 to 10^3. We plot the evolution of the average Age of Information for different values of λ in Figure 6 and we observe that, as λ increases, the obtained values tend to 0.5, which confirms the result of Proposition 3. We consider $p = 0.45$ in Figure 7 to study how this convergence depends on p and we observe that it seems to converge at the same rate to 0.5 than in the previous case.

From these illustrations, we also conclude that the average Age of Information is not monotone with respect to α (note that in [19] it has been shown that the average Age of Information of a system with a single server with energy harvesting is monotone with respect to α). For instance, we see that, when $p = 0.45$ and $\lambda = 10$, the average Age of Information increases when λ is small, then decreases and finally decreases again.

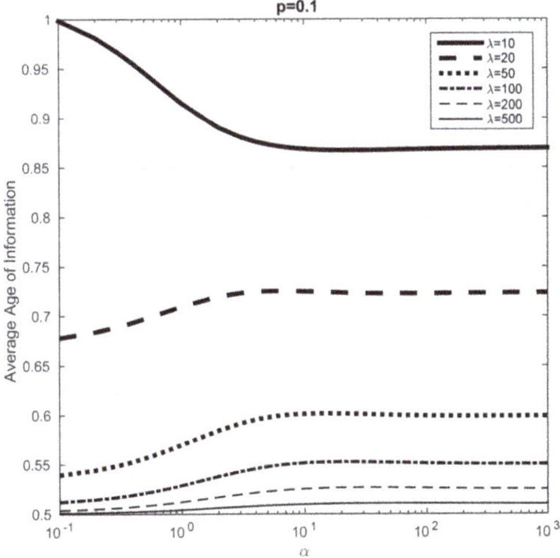

Figure 6. Average Age of Information with respect to α for different values of λ, when α varies from 0.1 to 10^3. $\mu = 1$ and $p = 0.1$.

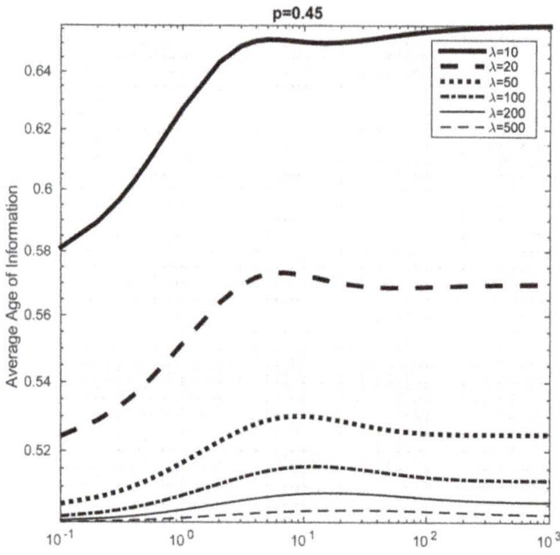

Figure 7. Average Age of Information with respect to α for different values of λ, when α varies from 0.1 to 10^3. $\mu = 1$ and $p = 0.45$.

4.3. Analysis of ther Parameter p

We now focus on the parameter p that determines the proportion of the total incoming arrival rate is sent to each of the servers. In Figure 8, we consider $\alpha = 1$ and we plot the average Age of Information when p varies from 0.01 to 0.99 for different values of λ and μ. We observe that, in all the considered instances, the average Age of Information first decreases with p and then increases. This suggests that the minimum of the average Age of Information for p is achieved when this parameter is close to 0.5, i.e., in the symmetric case that has been studied in Section 4.1.

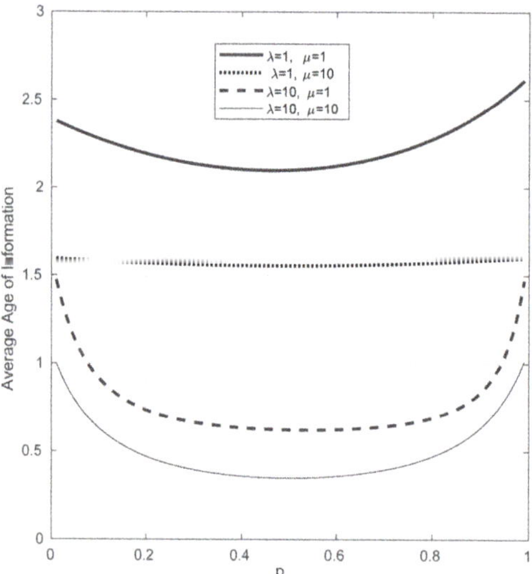

Figure 8. Average Age of Information with respect to p for different values of λ and μ, when p varies from 0.01 to 0.99. $\alpha = 1$.

5. Conclusions

We consider a system with two transmission channels that do not communicate and a energy source that generates energy following a random process. We model this system as a system with Poisson arrivals of status updates to two parallel servers and of energy packets to a battery. We study the average Age of Information of system using the Stochastic Hybrid System method. We first show that the average Age of Information of this system can be computed by solving a system of 8 linear equations (Proposition 2). We then consider that the arrival rate tends to infinity and, in this case, we show that the average Age of Information is equal to one divided by the sum of the service rate of the servers. This implies that, in this regime, the average Age of Information does not depend on the probability to dispatch jobs to the server and on the arrival rate of energy packets. We then study numerically the performance of our model with single server systems with energy harvesting and the same load of data packets as in our model. We conclude that, when the arrival rate of energy packets tends to infinity, the same conclusions of [18] follow in our model (i.e., the average Age of Information does not satisfy the same properties that other performance measures used in queuing theory such as number of packets in the queue). However, when the arrival rate of energy packets is not large, we conclude that the parallel server system outperforms the single server systems.

For future work, we would like to analyze the average Age of Information with a larger number of servers and with buffer for the energy packets and the data packets. Furthermore, we would like to study optimality of this model for some parameters such as p. We would also like to extend this model to include other parameters of the system such as transmit power or channel statistics to address real-life problems. Finally, we are also interested in exploring the performance of this model when it requires energy not only to send a status update to the monitor after getting service, but also to receive data packets from the source.

Funding: Josu Doncel has received funding from the Department of Education of the Basque Government through the Consolidated Research Group MATHMODE (IT1294-19), from the Marie Sklodowska-Curie grant agreement No 777778 and from the Spanish Ministry of Science and Innovation with reference PID2019-108111RB-I00 (FEDER/AEI).

Institutional Review Board Statement: Not applicable.

Informed Consent Statement: Not applicable.

Data Availability Statement: The code to reproduce the results of this article are available at https://github.com/josudoncel/AioParallelEnergy, accessed on 18 November 2021.

Conflicts of Interest: The authors declare no conflict of interest.

Appendix A. Average Age of Information and SHS

In the SHS, the system is modeled as a hybrid state $(q(t), \mathbf{x}(t))$, where $q(t)$ a state of a continuous time Markov Chain and $\mathbf{x}(t)$ is a vector whose components is the generation time of each of the updates. We assume that in the monitor there is the update with the latest generation time.

A link l of the Markov Chain represents a transition between two states, which occurs with rate λ^l. In each transition l, the vector $\mathbf{x}$ changes to $\mathbf{x}'$ using a transformation matrix $\mathbf{A}_l$, that is, $\mathbf{x}' = \mathbf{x}\mathbf{A}_l$. Therefore, $\mathbf{x}(t)$ is a piece-wise function.

For each state q of the Markov Chain, we define $\mathbf{b}_q$ as the vector whose elements are zero or one. One values represent the packets that are present in the system and therefore that the time from their generation increases at unit rate. On the other hand, zero values represent the updates that are not in the system.

We assume the Markov Chain is ergodic and we denote by π_q the stationary distribution of state q. We denote by $\mathcal{L}_q$ the set of outgoing links of state q and $\mathcal{L}'_q$ the set of links that get into state q. We now present the following theorem that will be used to characterize the average Age of Information:

Theorem A1 ([17], Thm 4). *Let $v_q(i)$ denote the i-th element of the vector $\mathbf{v}_q$. For each state q, if $\mathbf{v}_q$ is a non-negative solution of the following system of equations*

$$\mathbf{v}_q \sum_{l \in \mathcal{L}_q} \lambda^l = \mathbf{b}_q \pi_q + \sum_{l \in \mathcal{L}_q'} \lambda^l \mathbf{v}_{q_l} \mathbf{A}_l, \tag{A1}$$

then the average Age of Information is $\Delta = \sum_q v_q(0)$.

Appendix B. Proof of Proposition 1

We use the SHS technique to compute the average age of information of this system. We denote by $\mathbf{x} = [x_0(t)\ x_1(t)\ x_2(t)]$, where $x_0(t)$ represents the age of the information at time t and $x_i(t)$ the age of information if a data packet that is getting service in data queue i is sent successfully to the monitor, $i = 1, 2$.

The Markov Chain of this model is presented in Figure A1. We denote by ijk the state where there are i data packets in data queue 1, j data packets in data queue 2 and k energy packets in the energy queue. We now describe each of the transitions of this model.

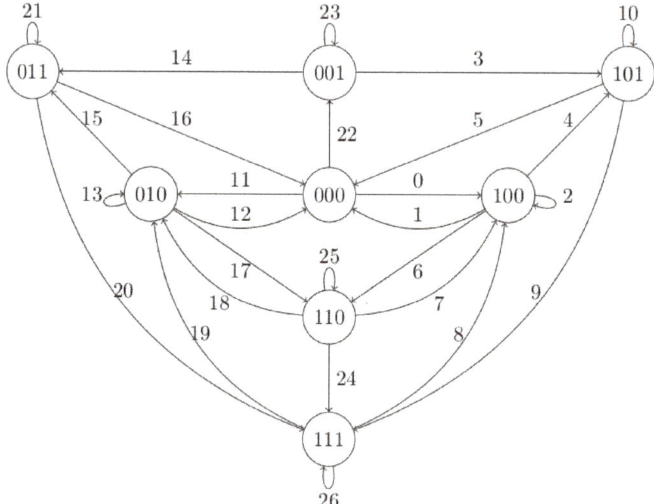

Figure A1. The SHS Markov chain for the model with two parallel data queues and a battery.

$l = 0$ A data packet arrives to data queue 1 when the system is empty. Therefore, the data packet starts getting service and the value of x_1 is set to zero, that is, $x_1' = 0$. The rest of values of $\mathbf{x}$ are not modified.

$l = 1$ A data packet of data queue 1 ends service when data queue 2 is empty. The data packet is sent to the energy queue, which is empty. Hence, the data packet is lost and the system gets empty. The vector $\mathbf{x}$ is not modified.

$l = 2$ A data packet arrives to data queue 1 when it is full. For this case, the last arrived packet replaces the packet in service and the age of x_1 changes to zero since the last arrived packet is fresh, that is, $x_1' = 0$. The rest of values of $\mathbf{x}$ do not change.

$l = 3$ A data packet arrives to data queue 1 when the energy queue is full. Therefore, the data packet starts getting service and the value of x_1 is set to zero, that is, $x_1' = 0$. The rest of values of $\mathbf{x}$ are not modified.

$l = 4$ An energy packet arrives to the system when data queue 1 is full and data queue 2 is empty. The arrival of an energy packet does not modify the age of the packets in the system and of the monitor, therefore the vector $\mathbf{x}$ is not modified.

$l = 5$ A data packet is data queue 1 ends service and it is sent to the energy queue, which contains one energy packet. Therefore, the packet is successfully delivered and the age of information is modified by the age of the served packet, that is, $x_0' = x_1$.

$l = 6$ A data packet arrives to data queue 2 when the data queue 1 is full and the data queue 2 and the energy queue were empty. For this case, the value of x_2 is set to zero, that is, $x_2' = 0$ and the rest of the values of $\mathbf{x}$ are not modified.

$l = 7$ A data packet in data queue 2 ends service when the energy queue is empty and data queue 1 is full. Therefore, the served packet is sent to the energy queue and it is lost. The vector $\mathbf{x}$ is not modified.

$l = 8$ A data packet in data queue 2 ends service and the system is full. Therefore, the served packet is delivered and thus, the value of x_0 is replaced by that of x_2, that is, $x_0' = x_2$. The rest of the values of $\mathbf{x}$ are not modified.

$l = 9$ A data packet arrives to data queue 2 when the energy queue and data queue 1 are full. Therefore, all the queues are full and the value of x_2 is set to zero since a fresh packet arrived to that queue, that is, $x_2' = 0$. The rest of the values of $\mathbf{x}$ are not modified.

$l = 10$ A data packet arrives to data queue 1 or there is an energy packet arrival when the energy queue and data queue 1 are full and data queue 2 is empty. In the former case, the value of x_1 changes to zero since a fresh packet arrived to that queue, that is, $x_1' = 0$ and the rest of the values of $\mathbf{x}$ are not modified, whereas in the later case the vector $\mathbf{x}$ does not change.

With the roles of data queue 1 and data queue 2 reversed, transitions 11–21 coincide with 0–10, respectively. Besides, transitions 21–24 represent an arrival of an energy queue, which increases by one the number of energy packets if the energy queue is empty, but the vector $\mathbf{x}$ is never modified. Finally, we focus on transitions 25 and 26:

$l = 25$ The data queues are full the energy queue is empty. If a data packet arrives to data queue 1, the value of x_1 changes to zero since a fresh packet arrived to that queue, that is, $x_1' = 0$ and the rest of the values of $\mathbf{x}$ are not modified. Likewise, if a data packet arrives to data queue 2, the value of x_2 changes to zero since a fresh packet arrived to that queue, that is, $x_2' = 0$ and the rest of the values of $\mathbf{x}$ are not modified.

$l = 26$ The system is full and one of the following events occur: (i) a data packet arrives to data queue 1, (ii) a data packet arrives to data queue 2 and (iii) an energy packet arrives to the energy queue. For the case (i), (resp. the case (ii)) the value of x_1 changes to zero (resp. x_2 changes to zero) since a fresh packet arrived to that queue, that is, $x_1' = 0$ (resp. $x_2' = 0$) and the rest of the values of $\mathbf{x}$ are not modified. Finally, for the case (iii), the vector $\mathbf{x}$ is not modified since an arrival of an energy queue does not change the age of the packets in the system.

Table A1. Table of SHS transitions of Figure A1.

l	$q_l \rightarrow q_{l'}$	λ^l	$x' = xA_l$	$\bar{v}_{q_l} A_l$
0	$000 \rightarrow 100$	λp	$[x_0\ 0\ 0]$	$[v_{000}(0)\ 0\ 0]$
1	$100 \rightarrow 000$	μ	$[x_0\ 0\ 0]$	$[v_{100}(0)\ 0\ 0]$
2	$100 \rightarrow 100$	λp	$[x_0\ 0\ 0]$	$[v_{100}(0)\ 0\ 0]$
3	$001 \rightarrow 101$	λp	$[x_0\ 0\ 0]$	$[v_{001}(0)\ 0\ 0]$
4	$100 \rightarrow 101$	α	$[x_0\ 0\ 0]$	$[v_{100}(0)\ 0\ 0]$
5	$101 \rightarrow 000$	μ	$[x_1\ 0\ 0]$	$[v_{101}(1)\ 0\ 0]$
6	$100 \rightarrow 110$	$\lambda(1-p)$	$[x_0\ x_1\ 0]$	$[v_{100}(0)\ v_{100}(1)\ 0]$
7	$110 \rightarrow 100$	μ	$[x_0\ x_1\ 0]$	$[v_{110}(0)\ v_{110}(1)\ 0]$
8	$111 \rightarrow 100$	μ	$[x_2\ x_1\ 0]$	$[v_{111}(2)\ v_{111}(1)\ 0]$
9	$101 \rightarrow 111$	$\lambda(1-p)$	$[x_0\ x_1\ 0]$	$[v_{101}(0)\ v_{101}(1)\ 0]$
10	$101 \rightarrow 101$	λp	$[x_0\ 0\ 0]$	$[v_{101}(0)\ 0\ 0]$
		α	$[x_0\ x_1\ 0]$	$[v_{101}(0)\ v_{101}(1)\ 0]$
11	$000 \rightarrow 010$	$\lambda(1-p)$	$[x_0\ 0\ 0]$	$[v_{000}(0)\ 0\ 0]$
12	$010 \rightarrow 000$	μ	$[x_0\ 0\ 0]$	$[v_{010}(0)\ 0\ 0]$
13	$010 \rightarrow 010$	$\lambda(1-p)$	$[x_0\ 0\ 0]$	$[v_{010}(0)\ 0\ 0]$
14	$001 \rightarrow 011$	$\lambda(1-p)$	$[x_0\ 0\ 0]$	$[v_{001}(0)\ 0\ 0]$
15	$010 \rightarrow 011$	α	$[x_0\ 0\ x_2]$	$[v_{010}(0)\ 0\ v_{010}(2)]$
16	$011 \rightarrow 000$	μ	$[x_2\ 0\ 0]$	$[v_{011}(2)\ 0\ 0]$
17	$010 \rightarrow 110$	λp	$[x_0\ 0\ x_2]$	$[v_{010}(0)\ 0\ v_{010}(2)]$
18	$110 \rightarrow 010$	μ	$[x_0\ 0\ x_2]$	$[v_{110}(0)\ 0\ v_{110}(2)]$
19	$111 \rightarrow 010$	μ	$[x_1\ 0\ x_2]$	$[v_{111}(1)\ 0\ v_{111}(2)]$
20	$011 \rightarrow 111$	λp	$[x_0\ 0\ x_2]$	$[v_{011}(0)\ 0\ v_{011}(2)]$
21	$011 \rightarrow 011$	$\lambda(1-p)$	$[x_0\ 0\ 0]$	$[v_{011}(0)\ 0\ 0]$
		α	$[x_0\ 0\ x_2]$	$[v_{011}(0)\ 0\ v_{011}(2)]$
22	$000 \rightarrow 001$	α	$[x_0\ 0\ 0]$	$[v_{000}(0)\ 0\ 0]$
23	$001 \rightarrow 001$	α	$[x_0\ 0\ 0]$	$[v_{001}(0)\ 0\ 0]$
24	$110 \rightarrow 111$	α	$[x_0\ x_1\ x_2]$	$[v_{110}(0)\ v_{110}(1)\ v_{110}(2)]$
25	$110 \rightarrow 110$	λp	$[x_0\ 0\ x_2]$	$[v_{110}(0)\ 0\ v_{110}(2)]$
		$\lambda(1-p)$	$[x_0\ x_1\ 0]$	$[v_{110}(0)\ v_{110}(1)\ 0]$
26	$111 \rightarrow 111$	λp	$[x_0\ 0\ x_2]$	$[v_{111}(0)\ 0\ v_{111}(2)]$
		$\lambda(1-p)$	$[x_0\ x_1\ 0]$	$[v_{111}(0)\ v_{111}(1)\ 0]$
		α	$[x_0\ x_1\ x_2]$	$[v_{111}(0)\ v_{111}(1)\ v_{111}(2)]$

We represent the evolution of **x** for all the above transitions in Table A1. We now that the information is presented in the same way as in Table A1, that is, in each row, each transition is represented in a different row, in the second column of the table, the origin and the destination state of the Markov Chain, in the third column the rate of each transition, whereas in the last two columns we show the evolution of **x** and $\bar{v}_{q_l} A_l$, respectively.

Let $\mathcal{Q}^p$ be the set of state of the Markov Chain of Figure A1. The stationary distribution of state $q \in \mathcal{Q}^p$ is denoted by π_q. We now focus on the stationary distribution of the Markov Chain of Figure A1. The balance equations are provided next:

$$\pi_{000}(\lambda + \alpha) = \pi_{101}\mu + \pi_{100}\mu + \pi_{011}\mu + \pi_{010}\mu$$

$$\pi_{001}\lambda = \pi_{000}\alpha$$

$$\pi_{010}(\lambda p + \mu + \alpha) = \pi_{000}\lambda(1-p) + \pi_{110}\mu + \pi_{111}\mu$$

$$\pi_{011}(\lambda p + \mu) = \pi_{010}\alpha + \pi_{001}\lambda(1-p)$$

$$\pi_{100}(\lambda(1-p) + \alpha + \mu) = \pi_{000}\lambda p + \pi_{110}\mu + \pi_{111}\mu$$

$$\pi_{101}(\lambda(1-p) + \mu) = \pi_{001}\lambda p + \pi_{100}\alpha$$

$$\pi_{110}(\alpha + 2\mu) = \pi_{010}\lambda p + \pi_{100}\lambda(1-p)$$

$$\pi_{111}2\mu = \pi_{011}\lambda p + \pi_{101}\lambda(1-p) + \pi_{110}\alpha$$

The Markov chain under study is clearly ergodic. Therefore, there always exists a unique solution of the above system of equations that satisfies $\sum_{q \in \mathcal{Q}^p} \pi_q = 1$.

We remark that the above system of equations coincides with Equation (1a–h).

We now define the vector $\mathbf{b}_q$ for any $q \in \mathcal{Q}^p$. For $q \in \{000, 001\}$, we have that $\mathbf{b}_q = [1\,0\,0]$, whereas for $q \in \{100, 101\}$, $\mathbf{b}_q = [1\,1\,0]$, for $q \in \{010, 011\}$, $\mathbf{b}_q = [1\,0\,1]$ and for $q \in \{110, 111\}$, $\mathbf{b}_q = [1\,1\,1]$. Besides, for all $q \in \mathcal{Q}^p$, we denote by $v_q(i)$ the i-th component of vector $\mathbf{v}_q$, with $i = 0, 1, 2$.

We now aim to apply Theorem A1 to this model and, from (A1), it follows that

$$\mathbf{v}_{000}(\lambda + \alpha) = [\pi_{000}\,0\,0] + \mu[v_{100}(0)\,0\,0]$$
$$+ \mu[v_{101}(1)\,0\,0] + \mu[v_{010}(0)\,0\,0]$$
$$+ \mu[v_{011}(2)\,0\,0] \tag{A2a}$$

$$\mathbf{v}_{001}(\lambda + \alpha) = [\pi_{001}\,0\,0] + \alpha[v_{000}(0)\,0\,0]$$
$$+ \alpha[v_{001}(0)\,0\,0] \tag{A2b}$$

$$\mathbf{v}_{010}(\lambda + \alpha + \mu) = [\pi_{010}\,0\,\pi_{010}] + \lambda(1-p)[v_{000}(0)\,0\,0]$$
$$+ \lambda(1-p)[v_{010}(0)\,0\,0]$$
$$+ \mu[v_{111}(1)\,0\,v_{111}(2)] \tag{A2c}$$

$$\mathbf{v}_{011}(\lambda + \alpha + \mu) = [\pi_{011}\,0\,\pi_{011}] + \alpha[v_{010}(0)\,0\,v_{010}(2)]$$
$$+ \alpha[v_{011}(0)\,0\,v_{011}(2)]$$
$$+ \lambda(1-p)[v_{001}(0)\,0\,0]$$
$$+ \lambda(1-p)[v_{011}(0)\,0\,0] \tag{A2d}$$

$$\mathbf{v}_{100}(\lambda + \alpha + \mu) = [\pi_{100}\,\pi_{100}\,0] + \lambda p[v_{000}(0)\,0\,0]$$
$$+ \lambda p[v_{100}(0)\,0\,0]$$
$$+ \mu[v_{111}(2)\,v_{111}(1)\,0] \tag{A2e}$$

$$\mathbf{v}_{101}(\lambda + \alpha + \mu) = [\pi_{101}\,\pi_{101}\,0] + \lambda p[v_{001}(0)\,0\,0]$$
$$+ \lambda p[v_{101}(0)\,0\,0]$$
$$+ \alpha[v_{101}(0)\,v_{101}(1)\,0]$$
$$+ \alpha[v_{100}(0)\,v_{100}(1)\,0] \tag{A2f}$$

$$\mathbf{v}_{110}(\lambda + 2\mu + \alpha) = [\pi_{110}\,\pi_{110}\,\pi_{110}]$$
$$+ \lambda p[v_{010}(0)\,0\,v_{010}(2)]$$
$$+ \lambda p[v_{110}(0)\,0\,v_{110}(2)]$$
$$+ \lambda(1-p)[v_{100}(0)\,v_{100}(1)\,0]$$
$$+ \lambda(1-p)[v_{110}(0)\,v_{110}(1)\,0] \tag{A2g}$$

$$\mathbf{v}_{111}(\lambda + 2\mu\alpha) = [\pi_{111}\,\pi_{111}\,\pi_{111}]$$
$$+ \lambda p[v_{011}(0)\,0\,v_{011}(2)]$$
$$+ \lambda p[v_{111}(0)\,0\,v_{111}(2)]$$
$$+ \lambda(1-p)[v_{101}(0)\,v_{101}(1)\,0]$$
$$+ \lambda(1-p)[v_{111}(0)\,v_{111}(1)\,0]$$
$$+ \alpha[v_{110}(0)\,v_{110}(1)\,v_{110}(2)]$$
$$+ \alpha[v_{111}(0)\,v_{111}(1)\,v_{111}(2)]. \tag{A2h}$$

We note that the above expression is the same as Equation (2a–p) with the following change of variable: $v_{000}(0) = x_1, v_{001}(0) = x_2, \ldots, v_{111}(1) = x_{15}, v_{111}(2) = x_{16}$ and $s_0 = \pi_{000}, s_1 = \pi_{001}, \ldots, s_7 = \pi_{111}$. Using Theorem A1, the desired result follows.

Appendix C. Proof of Lemma 1

We first write the Equation (2d,f,m,p), :

$$-s_2 = -x_4(\lambda + \alpha + \mu) + \mu x_{16} \tag{A3a}$$

$$-s_3 = -x_6(\lambda + \mu) + \alpha x_4 \tag{A3b}$$

$$-s_6 = -x_{13}(\lambda(1-p) + 2\mu + \alpha) + \lambda p x_4 \tag{A3c}$$

$$-s_7 = -x_{16}(\lambda(1-p) + 2\mu) + \lambda p x_6 + \alpha x_{13}, \tag{A3d}$$

And we observe that it is a system with 4 linear equations with 4 variables. Moreover, from (A3b), we get

$$x_6 = \frac{\alpha x_4 + s_3}{\lambda + \mu},$$

which gives (6), whereas from (A3c) we get

$$x_{13} = \frac{\lambda p x_4 + s_6}{\lambda(1-p) + 2\mu + \alpha}.$$

Substituting these expression in Equation (A3a–d), we obtain the following system of equations:

$$-s_2 = -x_4(\lambda + \alpha + \mu) + \mu x_{16} \tag{A4a}$$

$$-s_7 = -x_{16}(\lambda(1-p) + 2\mu) + \lambda p \frac{\alpha x_4 + s_3}{\lambda + \mu}$$

$$+ \alpha \frac{\lambda p x_4 + s_6}{\lambda(1-p) + 2\mu + \alpha}, \tag{A4b}$$

From (A4a), it results that

$$x_4(\lambda + \alpha + \mu) = \mu x_{16} + s_2 \iff x_4 = \frac{\mu x_{16} + s_2}{\lambda + \alpha + \mu},$$

which is equal to (5) and substituting the obtained expression in (A4b) and simplifying, we get (4). And the desired result follows.

Appendix D. Proof of Lemma 3

We first note that, from (1a), we have that

$$s_0 = \frac{s_2 \mu + s_3 \mu + s_4 \mu + s_5 \mu}{\lambda + \alpha},$$

which tends to zero when $\lambda \to \infty$ because $s_i < 1$ for all i and $\mu < \infty$. From (1b), we get that

$$s_1 = \frac{s_0 \alpha}{\lambda},$$

which, using the previous result and $\alpha < \infty$, also proves that s_1 tends to zero when $\lambda \to \infty$ because $s_0 < 1$. We now focus on (1c),:

$$s_2 = \frac{s_0 \lambda(1-p) + s_6 \mu + s_7 \mu}{\lambda p + \mu + \alpha},$$

and when $\lambda \to \infty$, we have that s_2 tends to $s_0(1-p)/p$ and this tends to zero because s_0 tends to zero. We now study (1d) and, using that $\alpha s_0 = \lambda s_1$ (see (1b)), we get

$$s_3 = \frac{s_0 \alpha(1-p) + s_2 \alpha}{\lambda p + \mu},$$

F which also tends to zero when $\lambda \to \infty$ because $s_0 < 1$ and $\alpha < \infty$. From (1e), we obtain

$$s_4 = \frac{s_0 \lambda p + s_6 \mu + s_7 \mu}{\lambda(1-p) + \alpha + \mu},$$

and we observe that s_4 tends to $s_0 p / (1-p)$ when $\lambda \to \infty$ and, since s_0 tends to zero, so does s_4. Finally, we have from (1f) and using that $\alpha s_0 = \lambda s_1$ (see (1b)),

$$s_5 = \frac{s_0 \alpha p + s_4 \alpha}{\lambda(1-p) + \mu},$$

which also tends to zero when $\lambda \to \infty$. And the desired result follows.

References

1. Kaul, S.; Gruteser, M.; Rai, V.; Kenney, J. Minimizing age of information in vehicular networks. In Proceedings of the 2011 8th Annual IEEE Communications Society Conference on Sensor, Mesh and Ad Hoc Communications and Networks, Salt Lake City, UT, USA, 27–30 June 2011; IEEE: Piscataway, NJ, USA, 2011; pp. 350–358.
2. Kaul, S.; Yates, R.; Gruteser, M. Real-time status: How often should one update? In Proceedings of the 2012 Proceedings IEEE INFOCOM, Orlando, FL, USA, 25–30 March 2012; IEEE: Piscataway, NJ, USA, 2012; pp. 2731–2735.
3. Costa, M.; Codreanu, M.; Ephremides, A. On the age of information in status update systems with packet management. *IEEE Trans. Inf. Theory* **2016**, *62*, 1897–1910. [CrossRef]
4. Bedewy, A.M.; Sun, Y.; Shroff, N.B. Minimizing the age of information through queues. *IEEE Trans. Inf. Theory* **2019**, *65*, 5215–5232. [CrossRef]
5. Costa, M.; Codreanu, M.; Ephremides, A. Age of information with packet management. In Proceedings of the 2014 IEEE International Symposium on Information Theory, Honolulu, HI, USA, 29 June–4 July 2014; IEEE: Piscataway, NJ, USA, 2014; pp. 1583–1587.
6. Maatouk, A.; Kriouile, S.; Assaad, M.; Ephremides, A. The Age of Incorrect Information: A New Performance Metric for Status Updates. *arXiv* **2019**, arXiv:1907.06604.
7. Sun, Y.; Kadota, I.; Talak, R.; Modiano, E. Age of Information: A New Metric for Information Freshness. *Synth. Lect. Commun. Netw.* **2019**, *12*, 1–224. [CrossRef]
8. Kosta, A.; Pappas, N.; Angelakis, V. Age of information: A new concept, metric, and tool. *Found. Trends Netw.* **2017**, *12*, 162–259. [CrossRef]
9. Yates, R.D.; Sun, Y.; Brown, D.R.; Kaul, S.K.; Modiano, E.; Ulukus, S. Age of information: An introduction and survey. *IEEE J. Sel. Areas Commun.* **2021**, *39*, 1183–1210. [CrossRef]
10. Wu, X.; Yang, J.; Wu, J. Optimal status update for age of information minimization with an energy harvesting source. *IEEE Trans. Green Commun. Netw.* **2017**, *2*, 193–204. [CrossRef]
11. Arafa, A.; Yang, J.; Ulukus, S.; Poor, H.V. Age-minimal transmission for energy harvesting sensors with finite batteries: Online policies. *IEEE Trans. Inf. Theory* **2019**, *66*, 534–556. [CrossRef]
12. Feng, S.; Yang, J. Age of information minimization for an energy harvesting source with updating erasures: Without and with feedback. *IEEE Trans. Commun.* **2021**, *69*, 5091–5105. [CrossRef]
13. Arafa, A.; Yang, J.; Ulukus, S.; Poor, H.V. Online timely status updates with erasures for energy harvesting sensors. In Proceedings of the 2018 56th Annual Allerton Conference on Communication, Control, and Computing (Allerton), Monticello, IL, USA, 2–5 October 2018; IEEE: Piscataway, NJ, USA, 2018; pp. 966–972.
14. Arafa, A.; Yang, J.; Ulukus, S.; Poor, H.V. Using erasure feedback for online timely updating with an energy harvesting sensor. In Proceedings of the 2019 IEEE International Symposium on Information Theory (ISIT), Paris, France, 7–12 July 2019; IEEE: Piscataway, NJ, USA, 2019; pp. 607–611.
15. Gelenbe, E. Energy packet networks: ICT based energy allocation and storage. In *International Conference on Green Communications and Networking*; Springer: Berlin/Heidelberg, Germany, 2011; pp. 186–195.
16. Gelenbe, E. Energy packet networks: Smart electricity storage to meet surges in demand. In Proceedings of the International ICST Conference on Simulation Tools and Techniques, SIMUTOOLS '12, Sirmione-Desenzano, Italy, 19–23 March 2012; ICST/ACM: New York, NY, USA, 2012; pp. 1–7.
17. Yates, R.D.; Kaul, S.K. The Age of Information: Real-Time Status Updating by Multiple Sources. *IEEE Trans. Inf. Theory* **2019**, *65*, 1807–1827. [CrossRef]
18. Doncel, J.; Assaad, M. Age of Information in a Decentralized Network of Parallel Queues with Routing and Packets Losses. *arXiv* **2020**, arXiv:2002.01696.
19. Doncel, J. Age of information of a server with energy requirements. *PeerJ Comput. Sci.* **2021**, *7*, e354. [CrossRef] [PubMed]

Article

Age-Aware Utility Maximization in Relay-Assisted Wireless Powered Communication Networks

Ning Luan [1,2], Ke Xiong [1,2,3,*], Zhifei Zhang [1,*], Haina Zheng [1,2], Yu Zhang [4], Pingyi Fan [5,6] and Gang Qu [7]

1   School of Computer and Information Technology, Beijing Jiaotong University, Beijing 100044, China; 19120388@bjtu.edu.cn (N.L.); 18112041@bjtu.edu.cn (H.Z.)
2   Beijing Key Laboratory of Traffic Data Analysis and Mining, Beijing Jiaotong University, Beijing 100044, China
3   Frontiers Science Center for Smart High-Speed Railway System, Beijing Jiaotong University, Beijing 100044, China
4   State Grid Energy Research Institute Co., Ltd., Beijing 102209, China; zhangyu2@sgeri.sgcc.com.cn
5   Department of Electronic Engineering, Tsinghua University, Beijing 100084, China; fpy@tsinghua.edu.cn
6   Beijing National Research Center for Information Science and Technology, Tsinghua University, Beijing 100084, China
7   Department of Electrical and Computer Engineering, University of Maryland at College Park, College Park, MD 20742, USA; gangqu@umd.edu
*   Correspondence: kxiong@bjtu.edu.cn (K.X.); zhfzhang@bjtu.edu.cn (Z.Z.)

Abstract: This article investigates a relay-assisted wireless powered communication network (WPCN), where the access point (AP) inspires the auxiliary nodes to participate together in charging the sensor, and then the sensor uses its harvested energy to send status update packets to the AP. An incentive mechanism is designed to overcome the selfishness of the auxiliary node. In order to further improve the system performance, we establish a Stackelberg game to model the efficient cooperation between the AP–sensor pair and auxiliary node. Specifically, we formulate two utility functions for the AP–sensor pair and the auxiliary node, and then formulate two maximization problems respectively. As the former problem is non-convex, we transform it into a convex problem by introducing an extra slack variable, and then by using the Lagrangian method, we obtain the optimal solution with closed-form expressions. Numerical experiments show that the larger the transmit power of the AP, the smaller the age of information (AoI) of the AP–sensor pair and the less the influence of the location of the auxiliary node on AoI. In addition, when the distance between the AP and the sensor node exceeds a certain threshold, employing the relay can achieve better AoI performance than non-relaying systems.

Keywords: wireless powered communication networks; real-time state update; age of information; utility maximization

Citation: Luan, N.; Xiong, K.; Zhang, Z.; Zheng, H.; Zhang, Y.; Fan, P.; Qu, G. Age-Aware Utility Maximization in Relay-Assisted Wireless Powered Communication Networks. *Entropy* **2021**, *23*, 1177. https://doi.org/10.3390/e23091177

Academic Editors: Anthony Ephremides and Yin Sun

Received: 13 July 2021
Accepted: 9 August 2021
Published: 7 September 2021

Publisher's Note: MDPI stays neutral with regard to jurisdictional claims in published maps and institutional affiliations.

1. Introduction

With the large-scale deployment of the Internet of things (IoT) devices in applications such as environment surveillance and industrial control [1,2], status update systems that report real-time system information become increasingly more important. In such systems, it is required to make accurate decisions based on fresh information update and measurement of information freshness becomes necessary. However, traditional network performance metrics like delay and throughput cannot characterize the information freshness. Therefore, the concept of age of information (AoI) was introduced as the time duration from the generation time of the latest received status update packet to the current time moment [3]. For real-time update systems, the goal is to get status update information as fresh as possible, which can be considered as the minimization of AoI.

In IoT systems, information is normally collected by the edge devices or sensors. These sensors are generally powered by batteries of limited capacity, which require to be replaced or recharged periodically. However, battery replacement or frequent recharging is labor-intensive and could be impossible, in particular in large-scale network scenarios and harsh environments. For this, energy harvesting (EH) technology has emerged as a promising alternative to collect energy from the external environment to power the low-power IoT devices. It is believed that EH has great potential in the future sixth-generation (6G) communication networks. Therefore, it is also expected to be used in future industrial control, environment monitoring, and other real-time IoT applications.

Existing EH technologies are of two types: traditional natural energy source and radio-frequency (RF) signal-based energy source. Compared with traditional natural energy sources [4,5], RF signals are easy to control, can provide steady power, and have relatively low requirements on the deployment environment [6]. Evidently, we have seen many works on AoI-based wireless powered communication networks (WPCN) powered by RF EH [7–13]. In [7], an optimal online state update strategy to minimize the AoI over long-term time scale with energy constraints was studied. In [8], the performance of AoI in WPCN was analyzed, and it proved that the smaller the probability of packet generation, the smaller the average AoI. In [9], the emergency AoI (U-AoI) in WPCN was minimized. In [10], the uplink AoI in two-way wireless powered networks was discussed, where the nonlinear AoI expression was adopted. In [11], the AoI performance limit for the actual wireless power transfer (WPT) network was explored. In [12], the trade-off between storable energy and system AoI in the WPT network was investigated, and in [13], the optimal design of AoI-based fog computing WPCN was studied.

The above works only considered the basic three-node network model. Recently, researchers have extended this to the study of AoI performance in multiple-node networks. For example, the authors of [14] minimized the AoI in large-scale WPCN with multi-sensor nodes, and derived the solution of the average charging time of nodes in the network. The authors of [15] studied the WPT-powered networks and explored when to terminate energy collection and how to properly assign resources for uploading data in order to minimize AoI. In [16], an optimal online sampling strategy for joint optimization of update packet transmit was presented to minimize the AoI, and a deep reinforcement learning (DRL) approach was proposed to effectively learn the optimal AoI strategy. In [17], the AoI in WPCN with multi-user scheduling based on non-orthogonal multiple-access (NOMA) was discussed, and the closed-form expression of AoI was derived. However, the above work only considered the single-hop network. Due to factors such as too far or obstructions between the sensor and the AP, the connection between them cannot always be established. Therefore, relaying technology can be employed to help sensors transmit information to their sink node [18–21]. Particularly, in [18], a cooperative WPCN with flat fading was studied, in which a source and a relay collected energy from a remote power station. In [19], the AoI in a cooperative wireless communication system with synchronous wireless information and power transmission (SWIPT) was studied, where two protocols were discussed.

We observe that in most of these works on AoI-based relay-assisted WPCNs, it was assumed that the relay node (or auxiliary node) could directly participate in charging or information transmission. As a matter of fact, this may not be realistic in real-life WPCN because relay nodes may also have limited energy and thus are reluctant or refuse to collaborate in the transmission of energy and/or information. We consider this selfish nature of the relay nodes in this article and propose to design an effective incentive mechanism to improve the system AoI performance. In our proposed mechanism, the access point (AP) will coordinate auxiliary nodes to charge the sensor node based on RF until a sufficient amount of energy is harvested by the sensor node; then, the sensor node will send real-time status update information to the AP. Unlike a similar work [22] that uses a vague incentive mechanism and assumes the sensor node transmit packets directly to the AP, we use the spectrum priority as the incentive for the auxiliary nodes to participate in the charging process, and use the auxiliary node as the relay on the rout of the status update packets from the sensor node to the AP. For such a system, utility functions are defined based on AoI for AP and auxiliary nodes respectively, and the utility maximization problems are formulated. In order to achieve a win–win benefit, we apply the Stackelberg game model to design the effective cooperation between the AP-sensor pair and auxiliary node.

We make the following key contributions in this article.

First, we extend the previous work on WPCN to the more realistic scenarios of relay-assisted WPCN with selfish auxiliary nodes. We describe the system model and propose a protocol to encourage the cooperation between AP and auxiliary node to keep information fresh.

Second, we introduce utility functions for the AP-sensor pair and the auxiliary node and then establish a Stackelberg game in order to improve the system's performance. To solve the non-convex utility maximization problem, we use the Lagrangian method based on a new slack variable and are able to obtain the optimal solution in the closed form.

Third, we conduct numerical simulations to show that larger transmit power of the AP results in smaller AoI and less dependency on the location of the auxiliary nodes. In addition, when the distance between the AP and the sensor exceeds a certain threshold, employing the relay can achieve better AoI performance than the current non-relaying systems.

The rest of this article is organized as follows. We elaborate the system model of relay-assisted WPCN in the next section and then formulate the AoI-aware utility maximization problem in Section 3. Our proposed solution to this problem is explained in detail in Section 4. The simulation results are reported in Section 5 before we conclude the article.

2. System Model

Consider a WPT-driven network, as illustrated in Figure 1, which includes an AP, multiple auxiliary nodes, and a sensor node with limited energy. In this system, the AP needs to collect status update information from the sensor and this is performed in two stages: first, the AP broadcasts the accessible bandwidth resources to the auxiliary node as an incentive to participate in the cooperation to charge the sensor, and the auxiliary node judges whether to participate in the power supply to the sensor according to the received incentive. When the sensor harvests enough energy, it utilizes the accumulated energy to send the status update to the AP. In the network, the frequency band authorized by the AP is fixed, so multiple auxiliary nodes compete to act as the helping node and only one auxiliary node will be selected to help charge the sensor node. The selected auxiliary node is allowed to use the bandwidth resource and can be used as a relay.

We use k to denote the k-th auxiliary node, where $k \in \{1, \ldots, K\}$. We use i to denote the index of the data package, where $i \in \{1, 2, \ldots, I\}$. We assume that the AP has steady power supply and the energy and information is transmitted through orthogonal channels to avoid any interference. We further assume that all wireless links expose in additive white Gaussian noise.

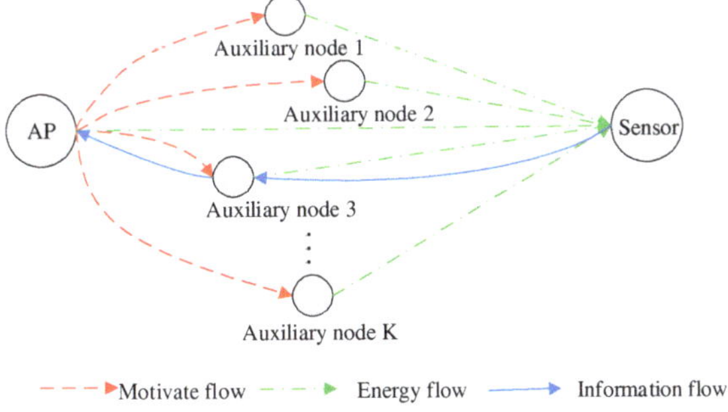

Figure 1. Illustration of the relay-assisted WPCN model.

2.1. Energy Harvesting Model

Let P_{AP} and P_k denote the transmit power of AP and the k-th auxiliary node, respectively. Let $h_{a,d}(t)$ and $h_{k,d}(t)$ be the wireless channel gain between the AP and the sensor and between the k-th auxiliary node and the sensor at time t, respectively. Let $b^{[i]}$ be the time when the i-th packet starts transmission. We assume that the sensor's battery capacity is B_s Joule. Once the battery is fully charged, the sensor node is triggered to send a newly generated packet with the harvested energy. It is well known that energy is the integral of power over time. Considering the energy harvesting efficiency and our system model, the accumulated energy by RF-based EH for data packet i can be expressed as

$$E_h^{[i]} = \int_{b^{[i-1]}}^{b^{[i]}} \eta \left(P_{AP} |h_{a,d}(t)|^2 + P_k |h_{k,d}(t)|^2 \right) dt, \tag{1}$$

where $\eta \in (0,1)$ denotes the energy harvesting efficiency. Note that $b^{[i]}$ is the time when the battery is fully charged, therefore, we have

$$B_s = E_h^{[i]}. \tag{2}$$

We partition the time into I subintervals at time instants, i.e., $0 = b^{[0]} < b^{[1]} < ... < b^{[i-1]} < b^{[i]} < ... < b^{[I]}$, where the i-th subinterval $[b^{[i-1]}, b^{[i]}]$ is the time that the sensor harvests energy for information packet i. The length of this subinterval is

$$T_s^{[i]} = b^{[i]} - b^{[i-1]}. \tag{3}$$

Let $d^{[i]}$ denote the arriving time of the i-th information packet at the AP, i.e., the time when the packet i transmission is completed. Thus, the transmitting time of the i-th information packet is expressed by

$$T_t^{[i]} = d^{[i]} - b^{[i]}. \tag{4}$$

Once the i-th information packet arrives at the AP, the sensor node can start generating the $(i+1)$-th packet if the status information becomes available. Therefore, we have the following:

$$T_t^{[i]} \leq T_s^{[i+1]}. \tag{5}$$

The auxiliary relay node has a single antenna and adopts the decoding and forwarding relaying strategy. That is, decoding the received information before encoding and forwarding. The relay node is installed with an information decoder, an encoder, and energy storage. The length of the information packet is denoted by l.

Let $T_{t1}^{[i]}$ and $T_{t2}^{[i]}$ be the transmitting time of the i-th packet from the sensor to the k-th auxiliary node and from the k-th auxiliary node to the AP, respectively. According to Shannon theory, the rate can be calculated in terms of the signal-to-noise ratio (SNR) and bandwidth, so we have the expression of transmitting time, i.e.,

$$T_{t1}^{[i]} = \frac{L}{W \log(1 + \gamma_1)} \text{ and } T_{t2}^{[i]} = \frac{L}{W \log(1 + \gamma_2)}, \tag{6}$$

where W represents the bandwidth, γ_1 and γ_2 are the received SNR of the two communication links, that between sensor and the k-th auxiliary node and that between the k-th auxiliary node and the AP, respectively.

Considering the situation that the AP can collect the update packet directly from the sensor node, we can see that the total transmitting time of the i-th update packet from the sensor to the AP will be

$$T_t^{[i]} = \min \left(\frac{L}{W \log(1 + \gamma_1)} + \frac{L}{W \log(1 + \gamma_2)}, \frac{L}{W \log(1 + \gamma)} \right). \tag{7}$$

Finally, the SNRs are given by

$$\gamma = \frac{P_d|h_{d,a}(t)|^2}{N_0 W}, \gamma_1 = \frac{P_d|h_{d,k}(t)|^2}{N_0 W} \text{ and } \gamma_2 = \frac{P_k'|h_{k,a}(t)|^2}{N_0 W}, \tag{8}$$

where $|h_{d,a}(t)|^2$, $|h_{d,k}(t)|^2$ and $|h_{k,a}(t)|^2$ denote the wireless channel power gain between the sensor and the AP, between the sensor and the auxiliary node k, and between the auxiliary node k and the AP, respectively. N_0 represents the noise spectral density. P_d and P_k' are the transmit power of the sensor and the k-th auxiliary node, respectively.

2.2. AoI Modeling

The current information packet's AoI, i.e., packet i at time t, is described by

$$\Delta^{[i]}(t) = t - U^{[i]}(t), \tag{9}$$

where $U^{[i]}(t)$ represents the time of generating the latest update packet received by AP, i.e.,

$$U^{[i]}(t) = \max\left\{b^{[j]} \Big| d^{[j]} \leq t\right\}. \tag{10}$$

Figure 2 illustrates the evolution AoI versus time. It is seen from the figure that when the destination node does not receive the state update packet, AoI increases linearly with time, which shows a sawtooth shape. When the destination node receives a new data packet, the AoI is reset to the delay that the status update experiences. In the time interval $[d^{[i]}, d^{[i+1]}]$, the integral of AoI is the area under the $\Delta^{[i]}(t)$ curve. Therefore, its average AoI is expressed by

$$\bar{\Delta}^{[i]} = \frac{1}{d^{[i+1]} - d^{[i]}} \int_{d^{[i]}}^{d^{[i+1]}} \Delta^{[i]}(t) dt. \tag{11}$$

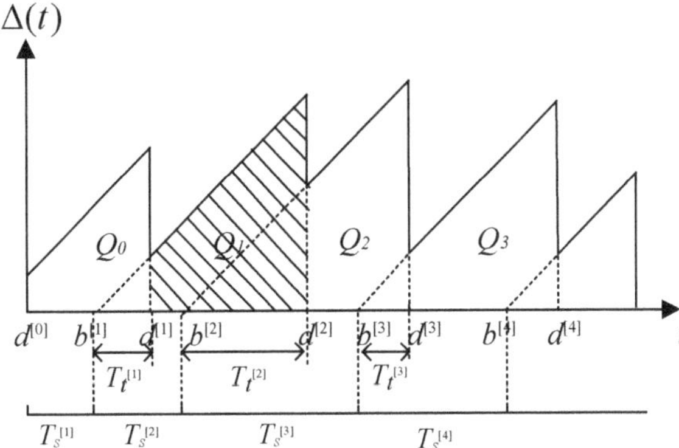

Figure 2. Evolution of the AoI.

The integral term in the above formula is obtained by summing the area of $Q_{[i]}$,

$$\int_{d^{[i]}}^{d^{[i+1]}} \Delta^{[i]}(t) dt = Q_{[i]}, \tag{12}$$

which is expressed by

$$Q_{[i]} = \frac{\left(T_t^{[i]} + T_s^{[i+1]} + T_t^{[i+1]}\right)\left(T_s^{[i+1]} - T_t^{[i]} + T_t^{[i+1]}\right)}{2}. \tag{13}$$

Therefore, the average AoI of the i-th data packet is expressed by

$$\bar{\Delta}^{[i]} = \frac{Q_{[i]}}{d^{[i+1]} - d^{[i]}} = \frac{1}{2}\left(T_t^{[i]} + T_s^{[i+1]} + T_t^{[i+1]}\right). \tag{14}$$

The harvesting energy time and transmitting update packet time are independent, so they could be regarded as independent and identically distributed variables. Therefore, for the long-term running, the system will be in a quasi-stationary state, which guarantees $T_s^{[i+1]} = T_s^{[i]}$ and $T_t^{[i+1]} = T_t^{[i]}$. In consequence, the average AoI is given by

$$\bar{\Delta} = T_t + \frac{1}{2}T_s. \tag{15}$$

3. Problem Formulation

In the system described above, the auxiliary node may become selfish and not willing to charge the sensor due to their own limited energy. In order to motivate them to participate in charging the sensor node, the AP may use spectrum priority access as the incentive to encourage the auxiliary nodes. In this section, we will define the utility functions for the AP and the auxiliary node. We design a Stackelberg game method to build an effective cooperation between auxiliary node and AP–sensor pair. The basic idea of Stackelberg game is that one side is the leader and the other side is the follower. The leader acts first and the follower chooses his own action according to the leader's strategy [23]. Their goal is to maximize their own interests.

Specifically, first, the AP issues a certain bandwidth as the incentive for the auxiliary nodes to participate in charging. Second, the auxiliary nodes optimize their transmit power based on the incentive. Third, the AP optimizes its transmit power and then transfers energy to the sensor together with the selected auxiliary node. At last, the sensor utilizes the harvested energy to send update packets to the AP. The flow chart of the system is illustrated in Figure 3. Energy flow is shown in green on the right; update packet transmission flow is shown in red on the left.

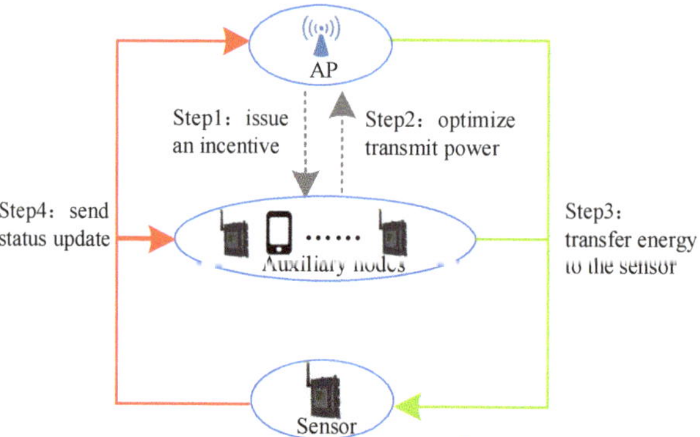

Figure 3. Overview of the incentive-based update packet collection system.

3.1. Utilities of the Auxiliary Node

Let $\Gamma_k(x)$ denote the cost of the k-th auxiliary node to transmit energy to the sensor and information to the AP at power level x. Similar to the work in [24], we can model this cost as

$$\Gamma_k(x) = a_k x^2 - b_k x + c_k, \tag{16}$$

where a_k, b_k, and c_k are predetermined parameters related to auxiliary node k. If node k uses power level P_k and P_k' to transmit to the sensor and to the AP, respectively, the utility of the auxiliary node k can be expressed by

$$U_k\left(P_k, P_k', T_s, T_s'\right) = \alpha B - T_s \Gamma_k(P_k) - T_s' \Gamma_k\left(P_k'\right), \tag{17}$$

where αB is the incentive from the AP with α as the uniform factor of bandwidth and revenue, T_s is the energy harvesting time and T_s' is the transmitting time from node k to the AP.

For the auxiliary node, it expects to maximize its utility. Therefore, the maximization problem P_k is expressed by

$$\begin{aligned} \mathrm{P}_k : \ &\max_{P_k, P_k'} U_k\left(P_k, P_k', T_s, T_s'\right) \\ &s.t.\ 0 < P_k, P_k' \leq P_k^{\max}, \end{aligned} \tag{18}$$

where $P_k^{\max}$ is the power threshold of the k-th auxiliary node.

3.2. Utilities of the AP

As mentioned earlier, the ratio factor of revenue to bandwidth is defined as α, and the overhead of AP is given by

$$\Xi = \alpha B. \tag{19}$$

Therefore, the utility associated with the sensor-AP pair is given by

$$U_{AP}^{(0)} = \tilde{U}_{AP} - \mu\bar{\Delta} - \omega T_s P_{AP}|h_{a,d}|^2 - \Xi, \tag{20}$$

where $\tilde{U}_{AP}$ is a constant, which is pre-defined. $\mu > 0$, $\omega > 0$ are the cost coefficient based on AoI and the cost coefficient based on energy, respectively.

Then, the AoI-based utility maximization problem is formulated as

$$\begin{aligned} \mathrm{P}_{AP}^{(1)} : \ &\max_{P_{AP}, T_s} U_{AP}^{(0)}(B, T_s, P_{AP}, P_k) \\ &s.t.\ B_s \leq E_h;\ T_t \leq T_s;\ B \geq 0; 0 \leq P_{AP} \leq P_{AP}^{\max}, \end{aligned} \tag{21}$$

where $P_{AP}^{\max}$ is the maximum available power of the AP. As $\tilde{U}_{AP}$ is a constant, $\mathrm{P}_{AP}^{(1)}$ can be transformed to $\mathrm{P}_{AP}^{(2)}$, i.e.,

$$\begin{aligned} \mathrm{P}_{AP}^{(2)} : \ &\min_{P_{AP}, T_s} U_{AP}(B, T_s, P_{AP}, P_k) \\ &s.t.\ B_s \leq E_h;\ T_t \leq T_s;\ B \geq 0; 0 \leq P_{AP} \leq P_{AP}^{\max}, \end{aligned} \tag{22}$$

where

$$\begin{aligned} U_{AP}(B, T_s, P_{AP}, P_k) &= \mu\bar{\Delta} + \omega T_s P_{AP}|h_{a,d}|^2 + \Xi \\ &= \mu T_t + \tfrac{1}{2}\mu T_s + \omega T_s P_{AP}|h_{a,d}|^2 + \alpha B. \end{aligned} \tag{23}$$

As μ, T_t, α, and B are fixed values, the problem $\mathrm{P}_{AP}^{(2)}$ is transformed to $\mathrm{P}_{AP}^{(3)}$, i.e.,

$$\begin{aligned} \mathrm{P}_{AP}^{(3)} : \ &\min_{P_{AP}, T_s} \tfrac{1}{2}\mu T_s + \omega T_s P_{AP}|h_{a,d}|^2 \\ &s.t.\ B_s \leq E_h;\ T_t \leq T_s; 0 \leq P_{AP} \leq P_{AP}^{\max}. \end{aligned} \tag{24}$$

4. Solution Method

We elaborate our proposed method and the derived solution for the above utility optimization problem.

4.1. Optimization of P_k and P_k' with a Given $\{P_{AP}, T_s\}$

As mentioned above, the problem P_k is expressed by

$$P_k : \max_{P_k, P_k'} U_k\left(P_k, P_k', T_s, T_s'\right)$$
$$s.t. \ \ 0 < P_k, P_k' \le P_k^{\max},$$

$$(25)$$

where

$$U_k\left(P_k, P_k', T_s, T_s'\right) = \alpha B - T_s \Gamma_k(P_k) - T_s' \Gamma_k\left(P_k'\right). \tag{26}$$

By substituting the cost function expression into the above problem, problem P_k can be expressed as follows,

$$P_k{}^{(1)} : \max_{P_k, P_k'}\left\{-a_k T_s P_k{}^2 + b_k T_s P_k - a_k T_s' P_k'^2 + b_k T_s' P_k' + \alpha B - c_k T_s - c_k T_s'\right\}$$
$$s.t. \ \ 0 < P_k, P_k' \le P_k^{\max}.$$

$$(27)$$

Lemma 1. *For a given* $\{P_{AP}, T_s\}$*, the optimal solution to Problem* P_k *is*

$$\begin{cases} P_k^* = \frac{b_k}{2a_k}, \\ P_k'^* = \frac{b_k}{2a_k}. \end{cases} \tag{28}$$

Proof. The objective function is about the quadratic function of two independent variables, and the second derivative is negative, so it is a concave function. The maximum value can be solved based on the quadratic formula directly. $\square$

4.2. Optimization of P_{AP} and T_s with a Given $\{P_k, P_k'\}$

In problem $P_{AP}^{(3)}$, two variables are multiple coupled, so the problem is non-convex. To tackle it, we introduce a new slack variable, i.e., $\pi = T_s \cdot P_{AP}$. As a result, the problem can be transformed into the following problem by variable substitution method:

$$P_{AP}^{(4)} : \min_{T_s, \pi} \tfrac{1}{2} \mu T_s + \omega |h_{a,d}|^2 \pi$$
$$s.t. \ -T_s + T_t \le 0$$
$$-T_s + \frac{1}{P_{AP}^{\max}} \pi \le 0$$
$$-T_s - \frac{|h_{a,d}|^2}{P_k |h_{k,d}|^2} \pi + \frac{B_s}{\eta P_k |h_{k,d}|^2} \le 0.$$

$$(29)$$

As both of the objective function and the constraints of $P_{AP}^{(4)}$ are linear, it is a convex problem, and the corresponding Lagrange function can be given by

$$L(T_s, \pi, \lambda) = \tfrac{1}{2} \mu T_s + \omega |h_{a,d}|^2 \pi + \lambda_1(-T_s + T_t) + \lambda_2\left(-T_s + \frac{1}{P_{AP}^{\max}} \pi\right)$$
$$+ \lambda_3\left(-T_s - \frac{|h_{a,d}|^2}{P_k |h_{k,d}|^2} \pi + \frac{B_s}{\eta P_k |h_{k,d}|^2}\right). \tag{30}$$

The Karush–Kuhn–Tucker (KKT) condition of the problem is

$$\frac{1}{2}\mu - \lambda_1 - \lambda_2 - \lambda_3 = 0, \tag{31a}$$

$$\omega |h_{a,d}|^2 + \frac{1}{P_{AP}^{\max}} \lambda_2 - \frac{|h_{a,d}|^2}{P_k |h_{k,d}|^2} \lambda_3 = 0, \tag{31b}$$

$$\lambda_1(-T_s + T_t) = 0, \tag{31c}$$

$$\lambda_2\left(-T_s + \frac{1}{P_{AP}{}^{\max}}\pi\right) = 0,$$ (31d)

$$\lambda_3\left(-T_s - \frac{|h_{a,d}|^2}{P_k|h_{k,d}|^2}\pi + \frac{B_s}{\eta P_k|h_{k,d}|^2}\right) = 0,$$ (31e)

$$\lambda_1 \geq 0,$$ (31f)

$$\lambda_2 \geq 0,$$ (31g)

$$\lambda_3 \geq 0,$$ (31h)

$$-T_s + T_t \leq 0,$$ (31i)

$$-T_s + \frac{1}{P_{AP}{}^{\max}}\pi \leq 0,$$ (31j)

$$-T_s - \frac{|h_{a,d}|^2}{P_k|h_{k,d}|^2}\pi + \frac{B_s}{\eta P_k|h_{k,d}|^2} \leq 0.$$ (31k)

According to (31b), the sum of the first two terms must be greater than 0, so $\lambda_3 \neq 0$. Otherwise, (31b) will not hold.

It can be seen from (31a) and (31b) that λ_1 and λ_2 must not be 0 at the same time. Otherwise, λ_3 will be equal to two different values.

From (31c) and (31d), one can see that λ_1 and λ_2 cannot not be equal 0 at the same time. Otherwise, T_s will be equal to two different values.

With above observations, there are two cases to analyze the solutions to the optimization problem.

Case 1: $\lambda_1 = 0$, $\lambda_2 \neq 0$ and $\lambda_3 \neq 0$, the optimal solution to $P_{AP}^{(4)}$ is

$$\begin{cases} T_s{}^* = \dfrac{B_s}{\eta\left(P_k|h_{k,d}|^2 + P_{AP}{}^{\max}|h_{a,d}|^2\right)}, \\ \pi^* = \dfrac{B_s P_{AP}{}^{\max}}{\eta\left(P_k|h_{k,d}|^2 + P_{AP}{}^{\max}|h_{a,d}|^2\right)}, \\ P_{AP}{}^* = P_{AP}{}^{\max}. \end{cases}$$ (32)

Case 2: $\lambda_1 \neq 0$, $\lambda_2 = 0$ and $\lambda_3 \neq 0$, the optimal solution to $P_{AP}^{(4)}$ is

$$\begin{cases} T_s{}^* = T_t, \\ \pi^* = \left(\dfrac{B_s}{\eta P_k|h_{k,d}|^2} - T_t\right)\dfrac{P_k|h_{k,d}|^2}{|h_{a,d}|^2}, \\ P_{AP}{}^* = \dfrac{\pi^*}{T_t}. \end{cases}$$ (33)

For the solutions of the two cases, it can be seen from the restrictive conditions that when (31i) is true, the solution is the case 1. Otherwise, it is the case 2. That is to say, the following formula is the judgment condition.

$$\frac{B_s}{\eta\left(P_k|h_{k,d}|^2 + P_{AP}{}^{\max}|h_{a,d}|^2\right)} \geq T_t,$$ (34)

where the expression of T_t is given by (7).

Theorem 1. *The optimal solution to* $P_{AP}^{(4)}$ *is expressed by*

$$
\begin{cases}
T_s{}^* = \begin{cases} \dfrac{B_s}{\eta\left(P_k|h_{k,d}|^2 + P_{AP}{}^{max}|h_{a,d}|^2\right)}, & if \dfrac{B_s}{\eta\left(P_k|h_{k,d}|^2 + P_{AP}{}^{max}|h_{a,d}|^2\right)} \geq T_t \\ T_t, & otherwise \end{cases} \\[4mm]
\pi^* = \begin{cases} \dfrac{B_s P_{AP}{}^{max}}{\eta\left(P_k|h_{k,d}|^2 + P_{AP}{}^{max}|h_{a,d}|^2\right)}, & if \dfrac{B_s}{\eta\left(P_k|h_{k,d}|^2 + P_{AP}{}^{max}|h_{a,d}|^2\right)} \geq T_t \\ \left(\dfrac{B_s}{\eta P_k|h_{k,d}|^2} - T_t\right)\dfrac{P_k|h_{k,d}|^2}{|h_{a,d}|^2}, & otherwise \end{cases} \\[4mm]
P_{AP}{}^* = \begin{cases} P_{AP}{}^{max}, & if \dfrac{B_s}{\eta\left(P_k|h_{k,d}|^2 + P_{AP}{}^{max}|h_{a,d}|^2\right)} \geq T_t \\ \dfrac{\pi^*}{T_t}. & otherwise \end{cases}
\end{cases}
\tag{35}
$$

Proof. As problem $P_{AP}^{(4)}$ is convex, Theorem 1 can be proved with KKT conditions. $\square$

5. Simulation Results

Some numerical results are shown to discuss the system performance in terms of achievable AoI, where the simulational parameters are set as follows. For clarity, the distance between the AP and the sensor is set to 6 m. The auxiliary nodes are randomly placed between AP and sensor. Assume that the distance between all auxiliary nodes and sensors is the same, that is, 3 m. Note that the location of the auxiliary nodes will be varied according to the purpose of the simulation. The channel is generated according to Rayleigh distribution. The path loss factor is set to 2. Other simulational parameters are summarized in Table 1.

Table 1. Parameter list.

Meaning	Parameter	Value
noise spectral density	N_0	10^{-9} mW/Hz
bandwidth	W	1 kHz
maximum transmit power of k-th node	P_k^{max}	100 mW
maximum transmit power of sensor	P_d^{max}	10 mW
maximum transmit power of AP	P_{AP}^{max}	5 W
battery capacity	B_s	0.1 mj
energy harvesting efficiency	η	0.8
packet length	L	100 bytes
the cost coefficient based on AoI	μ	1
the cost coefficient based on energy	ω	10^4
the cost coefficient based on bandwidth	α	1
predetermined parameters	a_k	1×10^4
predetermined parameters	b_k	2×10^3
bandwidth incentive	B	100 Hz

Figure 4 plots the minimized average AoI versus the P_{AP}. One can observe that the AoI decreases gradually with the increase of P_{AP}. The reason is that increasing the AP's transmit power will shorten the time for the energy harvested by the sensor to reach the battery capacity. Thus, it improves the freshness of the information. However, beyond a certain range, the change of AoI with transmit power is no longer significant because the average AoI is dominated by the information packet transmission time. This figure also shows the influence of different locations of auxiliary nodes on AoI is also different. The closer the auxiliary node to the sensor, the shorter time it will take to charge the sensor and the smaller the AoI.

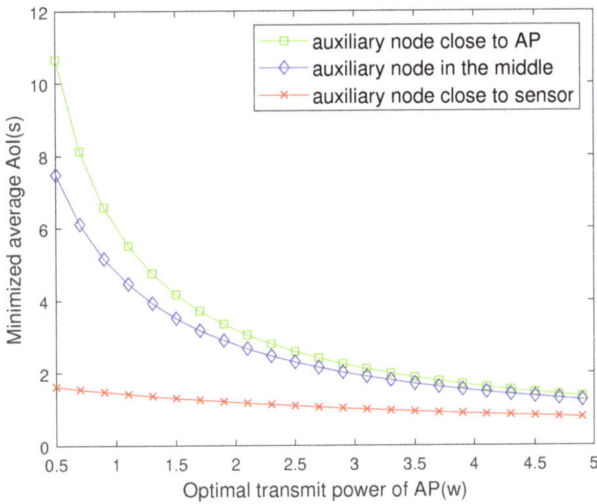

Figure 4. The minimized AoI versus P_{AP}.

Figure 5 plots the minimized AoI versus the distance between the auxiliary nodes and the sensor. It is seen that the closer the auxiliary node deployed to the sensor, the lower the AoI. This observation is consistent with the real situation as the closer the auxiliary node transmitting energy is to the sensor, the smaller the attenuation of electromagnetic wave is. Therefore, the time for sensor to collect energy will be reduced, so as to improve AoI. In addition, it is seen that the influence of the location on the AoI is different under different P_{AP}. The smaller the P_{AP}, the more obvious the influence of the location of the auxiliary node on AoI. When the P_{AP} is small, the charging mainly depends on the auxiliary node, and the position of the auxiliary node is more critical. When the P_{AP} is normal or larger, the P_{AP} is the main factor affecting the AoI, so that the position of the auxiliary node has no obvious effect on AoI.

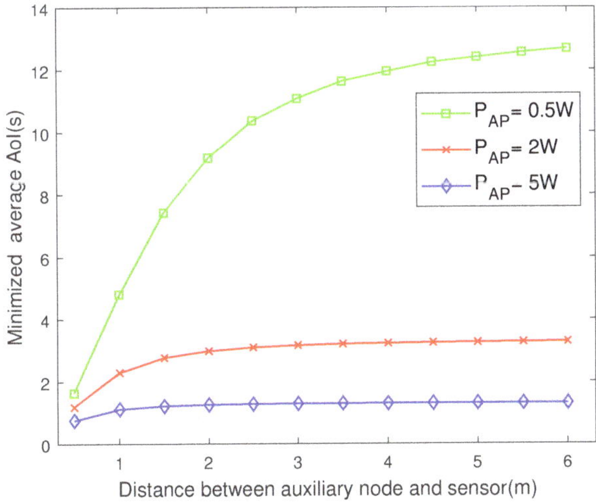

Figure 5. The minimized AoI versus D_{kd}.

Figure 6 plots the AP's maximum utility value versus the transmit power of AP. From this figure, it is seen that the larger the P_{AP}, the greater the utility value, and finally tends to remain unchanged. The reason may be that in a certain range, by increasing the P_{AP}, although the cost of transmitting energy increases, it makes the information fresher. The benefit of AoI is greater than the cost of transmitting energy. Therefore, the utility value of the AP based on AoI becomes larger, and beyond a certain range, with the increase of P_{AP}, the impact on AoI is negligible and the increase of utility value is not obvious. Besides, it is also seen that the auxiliary nodes in different positions also have an impact on the utility value. The specific relationship is shown in the figure below.

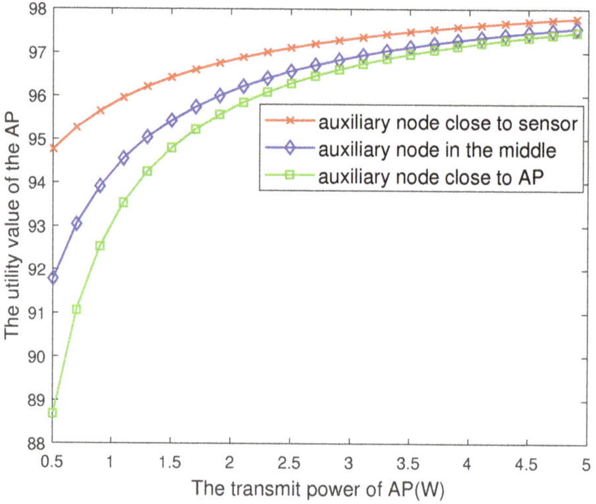

Figure 6. The AP's utility value versus the P_{AP}.

Figure 7 plots the utility of AP versus the distance between the auxiliary nodes and the sensor. The shorter the distance between the auxiliary node and the sensor, the greater the AP's utility value, and the smaller the AP transmit power, the more obvious that effect. The reason is similar to that associated with Figure 2.

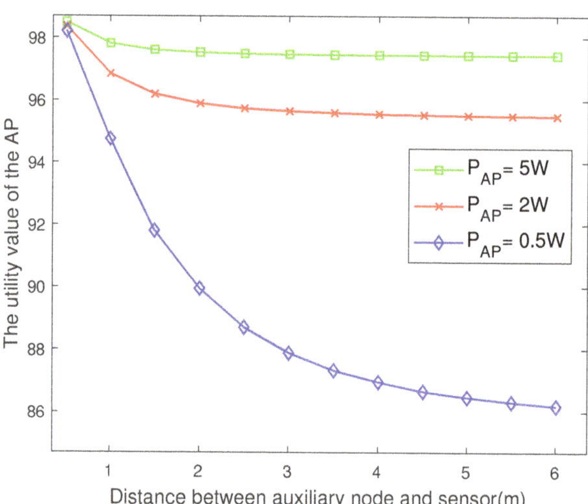

Figure 7. The AP's utility value versus the D_{ak}.

Figure 8 plots the change of AoI versus packet length. It is seen that the larger the length of the information packet, the larger the average AoI of the system. This may be because the length of the packet directly affects the transmit time of the packet. The larger the packet, the longer the transmitting time, which affects the freshness of the information. In addition, the influence of packet length on AoI varies with the location of auxiliary nodes. The closer the auxiliary node to the sensor, the less influence of packet length on AoI, because in the process of the auxiliary node serving as the relay and the sensor sending packets to the AP, the transmit power of the auxiliary node is larger than that of the sensor node, and the same packet length change has less impact on the transmit time.

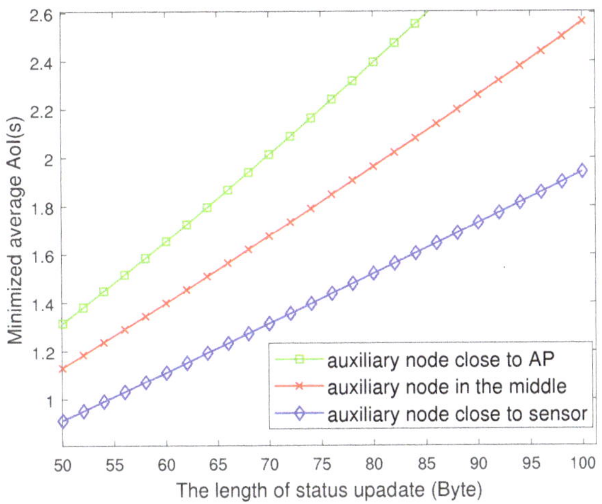

Figure 8. The minimized AoI versus L.

Figure 9 depicts the change of AoI with the distance between the AP and the sensor. Obviously, the farther the distance, the larger the AoI. The figure also compares the AoI in the system with and without auxiliary node as the relay. When the distance is greater than 13 m, the information with auxiliary node as the relay is fresher when the sensor transmits update packets to the AP. Thus, for the actual system, if the relay within 13 m can choose not to activate them, it has important guiding significance.

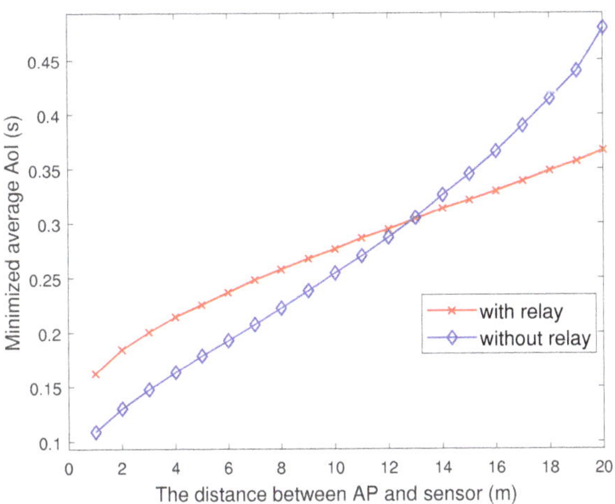

Figure 9. The minimized AoI versus D_{ad}.

Note that in this paper, the AoI is calculated by using the current packet's transmit time to approximate the transmit time of the next packet, that is, using the current channel to approximate the channel of the next slot. Figure 10 shows the relationship between the approximate value and the exact value of AoI. It is seen that the change of the approximate value lags behind the exact value by one time slot. Due to the randomness of the channel, the approximate value may be larger or smaller than the true value. If the channel of the current time slot is better than the next time slot, the approximate value of the current AoI is a little smaller than the true value. If the channel of the current time slot is worse than the next time slot, the approximate value of the current AoI is larger than the real value. However, on average, the difference between our modeling method and the real value is very small, which indirectly proves the effectiveness of the modeling method.

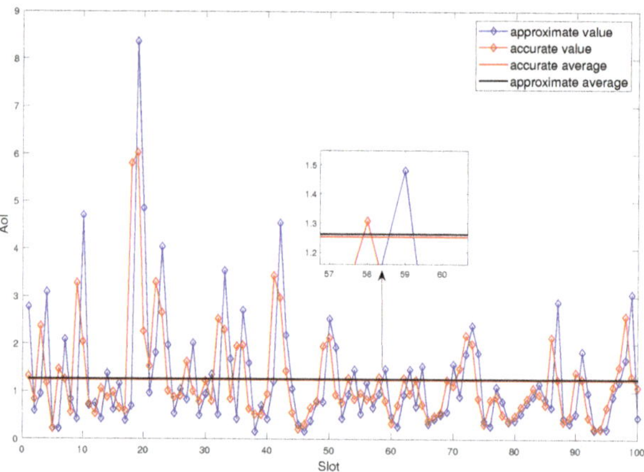

Figure 10. Accurate AoI versus approximate AoI.

6. Conclusions

In this article, we study a relay-assisted WPCN based on AoI with a focus on the case when the auxiliary nodes are selfish. The main idea behind our proposed solution is an incentive scheme that encourages the auxiliary nodes to collaborate. We formulate the problem and use the Stackelberg game theory to design an effective collaboration between AP–sensor pair and auxiliary node. More specifically, two utility functions for the AP–sensor pair and the auxiliary node were formulated. As maximizing the utility of the AP–sensor pair was non-convex, we transformed it into a convex problem by introducing a new slack variable, and then solved it by the Lagrangian method to obtain optimal solutions in the closed form. Simulation results showed that the larger the transmit power of the AP, the smaller the AoI and the less the influence of the location of the auxiliary node on AoI. In addition, when the distance from the AP to the sensor node exceeds a certain threshold, employing the relay can achieve better AoI performance than non-relaying systems. These results provide insightful and practical guidance for the design of relay-assisted WPCN in real life.

Author Contributions: N.L., K.X., H.Z. and Z.Z. equally contributed to this work on system modeling, methodology and simulation; N.L. also contributed to writing and editing; K.X. also contributed to project administration; G.Q. and P.F. contributed to the review; K.X. and Y.Z. contributed to funding acquisition. All authors have read and agreed to the published version of the manuscript.

Funding: This work was supported in part by the General Program of the National Natural Science Foundation of China (NSFC) (No. 62071033 and U1834210), in part by Frontiers Science Center for Smart High-speed Railway System with the Fundamental Research Funds for the Central Universities (No. 2020JBZD010), and also in part by the Self-developed project of State Grid Energy

Research Institute Co., Ltd. (Electrical internet of Things Edge Computing Performance Analysis and Simulation Based on Typical Scenarios, No. SGNY202009014).

Institutional Review Board Statement: Not applicable.

Informed Consent Statement: Not applicable.

Data Availability Statement: Not applicable.

Acknowledgments: We would like to thank all the reviewers for their constructive comments and helpful suggestions. In addition, we would like to give special thanks to Haina Zheng, who discussed the relative methods and also gave some helpful suggestions.

Conflicts of Interest: The authors declare no conflict of interest.

References

1. Feltrin, L.; Tsoukaneri, G.; Condoluci, M.; Buratti, C.; Mahmoodi, T.; Dohler, M.; Verdone, R. Narrowband IoT: A survey on downlink and uplink perspectives. *IEEE Wirel. Commun* **2019**, *26*, 78–86. [CrossRef]
2. Zhang, S.; Zhang, N.; Kang, G. Energy efficiency for NPUSCH in NB-IoT with guard band. *ZTE Commun.* **2018**, *16*, 46–51.
3. Kaul, S.K.; Yates, R.D.; Gruteser, M. *Real-Time Status: How Often Should One Update*; IEEE INFOCOM: Orlando, FL, USA, 2012.
4. Arafa, A.; Ulukus, S. Age Minimization in Energy Harvesting Communications: Energy-Controlled Delays. In Proceedings of the 51st Asilomar Conference on Signals, Systems, and Computers, Pacific Grove, CA, USA, 29 October–1 November 2017.
5. Arafa, A.; Ulukus, S. Age-Minimal Transmission in Energy Harvesting Two-Hop Networks. In Proceedings of the IEEE Global Communications Conference, Singapore, 4–8 December 2017.
6. Liu, J.; Xiong, k.; Fan, P.; Zhong, Z. RF energy harvesting wireless powered sensor networks for smart cities. *IEEE Access* **2017**, *5*, 9348–9358. [CrossRef]
7. Wu, X.; Yang, J.; Wu, J. Optimal status update for age of information minimization with an energy harvesting source. *IEEE Trans. Green Commun. Netw.* **2018**, *2*, 193–204. [CrossRef]
8. Hu, H.; Xiong, K.; Zhang, Y.; Fan, P.; Liu, T.; Kang, S. Age of information in wireless powered networks in low SNR region for future 5G. *Entropy* **2018**, *20*, 948. [CrossRef] [PubMed]
9. Lu, Y.; Xiong, K.; Fan, P.Y.; Zhong, Z.; Letaief, K.B. *Optimal Online Transmission Policy in Wireless Powered Networks with Urgency-Aware Age of Information*; IEEE IWCMC: Tangier, Morocco, 2019.
10. Dong, Y.; Chen, Z.; Fan, P. *Uplink Age of Information of Unilaterally Powered Two-Way Data Exchanging Systems*; INFOCOM WKSHPS: Honolulu, HI, USA, 2018.
11. Krikidis, I. Average age of information in wireless powered sensor networks. *IEEE Wirel. Commun. Lett.* **2019**, *8*, 628–631. [CrossRef]
12. Zheng, H.; Xiong, K.; Fan, P.; Zhong, Z.; Letaief, K.B. *Age-Energy Region in Wireless Powered Communication Networks*; IEEE INFOCOM WKSHPS: Toronto, ON, Canada, 2020.
13. Zheng, H.; Xiong, K.; Fan, P.; Zhong, Z.; Letaief, K.B. Minimum Age-Energy Aware Cost in Wireless Powered Fog Computing Networks. In Proceedings of the ICC 2020 IEEE International Conference on Communications (ICC), Dublin, Ireland, 7–11 June 2020.
14. Sleem, O.M.; Leng, S.; Yener, A. Age of Information Minimization in Wireless Powered Stochastic Energy Harvesting Networks. In Proceedings of the 54th Annual Conference on Information Sciences and Systems(CISS), Princeton, NJ, USA, 18–20 March 2020.
15. Gu, Q.; Wang, G.; Fan, R.; Li, F.; Jiang, H.; Zhong, Z. Optimal Resource Allocation for Wireless Powered Sensors: A Perspective from Age of Information. *IEEE Commun. Lett.* **2020**, *24*, 2559–2563.
16. Abd-Elmagid, M.A.; Dhillon, H.S.; Pappas, N. A Reinforcement Learning Framework for Optimizing Age of Information in RF-Powered Communication Systems. *IEEE Trans. Commun.* **2020**, *68*, 4747–4760. [CrossRef]
17. Azarhava, H.; Abdollahi, M.P.; Niya, J.M. Age of information in wireless powered IoT networks: NOMA vs. TDMA. *Ad Hoc Netw.* **2020**, *104*, 102179. [CrossRef]
18. Khorsandmanesh, Y.; Emadi, M.J.; Krikidis, I. Average Peak Age of Information Analysis for Wireless Powered Cooperative Networks. *arXiv* **2020**, arXiv:2007.06598.
19. Perera, T.D.P.; Jayakody, D.N.K.; Pitas, I.; Garg, S. Age of Information in SWIPT-Enabled Wireless Communication System for 5GB. *IEEE Wirel. Commun.* **2020**, *27*, 162–167. [CrossRef]
20. Moradian, M.; Dadlani, A. Age Of Information In Scheduled Wireless Relay Networks. In Proceedings of the IEEE Wireless Communications and Networking Conference (WCNC), Seoul, Korea, 25–28 May 2020.
21. Hu, H.; Xiong, K.; Qu, G.; Ni, Q.; Fan, P.; Letaief, K.B. AoI-Minimal Trajectory Planning and Data Collection in UAV-Assisted Wireless Powered IoT Networks. *IEEE Internet Things J.* **2021**, *8*, 1211–1223. [CrossRef]
22. Zheng, H.; Xiong, K.; Fan, P.; Zhong, Z.; Letaief, K.B. Age-Based Utility Maximization for Wireless Powered Networks: A Stackelberg Game Approach. In Proceedings of the IEEE Global Communications Conference, Waikoloa, HI, USA, 9–13 December 2019.

23. Xin, K.; Rui, Z.; Motani, M. Price-Based Resource Allocation for Spectrum-Sharing Femtocell Networks: A Stackelberg Game Approach. In Proceedings of the IEEE Global Telecommunications Conference, Houston, TX, USA, 3–7 December 2012.
24. Mohsenian-Rad, A.; Wong, V.W.; Jatskevich, J.; Schober, R.; Leon-Garcia, A. Autonomous demand-side management based on game-theoretic energy consumption scheduling for the future smart grid. *IEEE Trans.* **2010**, *1*, 320–331. [CrossRef]

Article

A UoI-Optimal Policy for Timely Status Updates with Resource Constraint

Lehan Wang, Jingzhou Sun, Yuxuan Sun, Sheng Zhou * and Zhisheng Niu

Beijing National Research Center for Information Science and Technology, Department of Electronic Engineering, Tsinghua University, Beijing 100084, China; wang-lh19@mails.tsinghua.edu.cn (L.W.); sunjz18@mails.tsinghua.edu.cn (J.S.); sunyuxuan@tsinghua.edu.cn (Y.S.); niuzhs@tsinghua.edu.cn (Z.N.)
* Correspondence: sheng.zhou@tsinghua.edu.cn

Abstract: Timely status updates are critical in remote control systems such as autonomous driving and the industrial Internet of Things, where timeliness requirements are usually context dependent. Accordingly, the Urgency of Information (UoI) has been proposed beyond the well-known Age of Information (AoI) by further including context-aware weights which indicate whether the monitored process is in an emergency. However, the optimal updating and scheduling strategies in terms of UoI remain open. In this paper, we propose a UoI-optimal updating policy for timely status information with resource constraint. We first formulate the problem in a constrained Markov decision process and prove that the UoI-optimal policy has a threshold structure. When the context-aware weights are known, we propose a numerical method based on linear programming. When the weights are unknown, we further design a reinforcement learning (RL)-based scheduling policy. The simulation reveals that the threshold of the UoI-optimal policy increases as the resource constraint tightens. In addition, the UoI-optimal policy outperforms the AoI-optimal policy in terms of average squared estimation error, and the proposed RL-based updating policy achieves a near-optimal performance without the advanced knowledge of the system model.

Keywords: age of information; constrained Markov decision process; reinforcement learning; context-awareness; timely status updates

Citation: Wang, L.; Sun, J.; Sun, Y.; Zhou, S.; Niu, Z. A UoI-Optimal Updating Policy for Timely Status Information with Resource Constraint. *Entropy* **2021**, *23*, 1084. https://doi.org/10.3390/e23081084

Academic Editors: Yin Sun and Anthony Ephremides

Received: 20 June 2021
Accepted: 16 August 2021
Published: 20 August 2021

Publisher's Note: MDPI stays neutral with regard to jurisdictional claims in published maps and institutional affiliations.

1. Introduction

With the development of 5G and the Internet of Things (IoT), requirements for wireless communication have shifted from merely providing communication channels to covering the entire process of various IoT applications, e.g., autonomous vehicle [1] and virtual reality (VR) [2], where sensing, communication, computation, and control form a closed loop. Therefore, in addition to the communication delay, it is necessary to consider the information delay counted from the generation of the state information to the execution, namely the timeliness of information. For this purpose, Age of Information (AoI) has been proposed, which is defined as the time elapsed since the generation time of the latest received packets [3]. Due to its concise definition and clear physical meaning, AoI has been widely used for the design of scheduling and updating policies in remote estimation [4–6] and wireless communication networks [7–12]. Most existing works focus on optimizing average AoI or peak age. In [13], the authors claim that minimizing average age cannot satisfy the requirements for ultra-reliable low-latency communication (URLLC) and study the tail distribution of AoI. The violation probability for peak age is derived in [14] and the stationary distribution of AoI is studied in [15].

Nevertheless, the AoI still has some limitations. First, it fails to measure the nonlinear performance degradation caused by information staleness. In [16–19], nonlinear age penalty functions were introduced to solve this problem. Meanwhile, the Age of Synchronization (AoS) [20] and Age of Incorrect Information (AoII) [21] are defined to associate information freshness with the content of information. AoS is the time elapsed since the information at the receiver becomes desynchronized with the actual status of the monitored process. AoII is defined as the product of an increasing time penalty function and a penalty function of the estimation error. In addition, the status of heterogeneous data sources may change at different rates. A fast-changing process may require information with a lower age. However, age is independent of the changing rate and thus is not proper in the cases when heterogeneous data sources are jointly considered. To solve this problem, weighted age was introduced in [22,23] to distinguish important monitored processes. In [24], the metric based on information theory is proposed as a replacement of the time-based metric, AoI, to characterize the changing rate. In [5], the authors claim that minimizing age is not equivalent to minimizing the estimation error in a remote estimation problem and propose an effective age to solve this problem [25].

Practical systems (e.g., V2X-communication systems) may have different requirements for information freshness with different contexts. The context refers to all environmental factors that affect the requirement for information freshness. Therefore, resources should be reserved for frequent status updates in emergency to ensure safety.

However, the timeliness metrics mentioned above pay no attention to the significance of context information. To solve this problem, Urgency of Information (UoI) has been proposed in [26–28] to measure the influence of inaccurate information on performance under different contexts. To be specific, UoI uses a time-variant context-aware weight $\omega(t)$ to distinguish different contexts. A higher $\omega(t)$ indicates that the system is in more urgent situations (e.g., when a vehicle is approaching an intersection or overtaking) and therefore requires frequent updates. For example, when a vehicle passes through an intersection, the context-aware weight increases as the distance between the vehicle and the center of the intersection decreases. Meanwhile, the estimation error $Q(t)$ is introduced to measure the information inaccuracy, which is defined as the difference between the actual status and the estimated status at the receiver. The larger the absolute value of $Q(t)$ is, the less accurate the estimated status is. Therefore, UoI is defined as the product of context-aware weight and a cost function of the estimation error $Q(t)$:

$$F(t) = \omega(t)\delta(Q(t)). \tag{1}$$

In discrete-time systems, the estimation error $Q(t)$ is:

$$Q(t) = \sum_{\tau=g(t)}^{t-1} A(\tau), \tag{2}$$

where $g(t)$ is the generation time of the latest status update at the receiver and $A(t)$ is the increment in estimation error in time slot t. Specifically, if the context-aware weight is time-invariant (i.e., $\omega(t) = 1$), and $A(t) = 1$ as well as $\delta(Q(t)) = Q(t)$, UoI is the same as AoI. If the context-aware weight is process-dependent, UoI can represent weighted age. If the cost function $\delta(Q(t))$ is nonlinear, UoI can represent the nonlinear age penalty function. For example, when the outdated information is worthless, e.g., the information is about sales that expire after some time [29], then the shifted unit step cost function $\delta(Q(t)) = u(Q(t) - \tau), \tau > 0$ is recommended. For the unit step function, $u(x) = 1$ when $x \geq 0$ and otherwise $u(x) = 0$.

In this work, we considered a single-user remote monitoring system, and the objective was to find an updating policy minimizing the average UoI over time under the constraint on average update frequency. To solve this problem, Refs. [27,30] proposed update-index-based adaptive schemes with Lyapunov optimization but did not conduct a theoretical analysis of their optimality. In addition, the constrained Markov decision process (CMDP) formulation was only used in the simulation for a numerically solved benchmark. Based on the existing works, in this paper, we theoretically analyzed the structure of the UoI-optimal policy and focused on how to derive an updating policy in an unknown environment.

The main contributions of this paper are summarized as follows.

- In contrast to [27,30], we assumed that the context-aware weight is a first-order irreducible positive recurrent Markov process or independent and identically distributed (i.i.d.) over time. We formulated the updating problem as a CMDP problem and proved the single threshold structure of the UoI-optimal policy. We then derived the policy through LP with the threshold structure and discussed the conditions that the monitored process needs to satisfy for the threshold structure.
- When the distributions of the context-aware weight and the increment in estimation error were unknown, we used model-based RL method to learn the state transitions of the whole system and derive a near-optimal RL-based updating policy.
- Simulations were conducted to verify the theoretical analysis of the threshold structure and show the near-optimal performance of the RL-based updating policy. The results indicate that: (i) the update thresholds decrease when the maximum average update frequency becomes large; (ii) the update threshold for emergency can actually be larger than that for ordinary states when the probability of transferring from emergency to ordinary states tends to 1.

The rest of this paper is organized as follows. The system model and the problem formulation are described in Section 2. In Section 3, we obtain the CMDP formulation of the problem with the given distribution of context-aware weight and prove the threshold structure of the UoI-optimal policy. The proposed model-based RL updating policy is obtained in Section 4. In Section 5, the simulation results are shown and discussed while the conclusions are drawn in Section 6.

2. System Model and Problem Formulation

In this paper, we considered a remote monitoring system, in which a fusion center collects the status information (e.g., current location, velocity, information of surrounding) from a vehicle of interest via a wireless channel with limited resources, as shown in Figure 1. The whole system is considered as a discrete-time system and the status can be generated at will. Due to the limitations on the wireless resources and energy supply, there is a constraint on the average update frequency of the vehicle. The update decision in time slot t is denoted by $U(t) \in \{0,1\}$, where $U(t) = 1$ means that the vehicle decides to transmit the current status to the center, and $U(t) = 0$ denotes that the vehicle decides to stay idle.

The wireless channel is assumed as a block fading channel with successful transmission probability p_s. Let $S(t) \in \{0,1\}$ be the state of the channel. $S(t) = 0$ represents that the channel is in deep fading, and no packet can be successfully transmitted. $S(t) = 1$ means the packets can be successfully transmitted to the center through the channel. If the center receives an update, then $U(t)S(t) = 1$ and an ACK will be sent to the vehicle.

Let $x(t)$ and $\hat{x}(t)$ denote the current status of the monitored vehicle and the estimated status of the vehicle at the center, and $Q(t) = x(t) - \hat{x}(t)$ denotes the estimation error. Similar to [26], we further assume that the time period of a packet transmission is less than a time slot and the estimation at the center equals the latest status information received by the center. This estimation scheme is easy to implement, theoretically tractable and has been proven to be an optimal policy that can minimize the average squared error of status estimation in a remote estimation system under energy constraints when the monitored process is a Wiener process [31]. Then, the recurrence relation of the estimation error $Q(t)$ is:

$$Q(t+1) = (1 - U(t)S(t))Q(t) + A(t). \tag{3}$$

Equation (3) indicates that the estimation error will be the amount of variation of the monitored process from the generation time of the latest received status to the current time. The increment $A(t)$ represents the variation of the monitored process. For example, when $A(t)$ follows a Gaussian distribution with a mean of zero and variance of σ^2, represented by $N(0,\sigma^2)$, the monitored status follows a Wiener process. When $A(t)$ takes values from $\{0, 1, -1\}$ with a probability of $\{1 - 2p_{rw}, p_{rw}, p_{rw}\}$, where $0 < p_{rw} < \frac{1}{2}$, then the status of the monitored source will be a one-dimensional random walk. In this paper, we assumed that the monitored status of the vehicle is a Wiener process and $A(t)$ is i.i.d. over time. However, the increment in estimation error during a single slot cannot be infinite in practical systems. Therefore, in contrast to [27,30], we assumed that increment $A(t)$ obeys a truncated Gaussian distribution, i.e., the probability density function (PDF) of $A(t)$ is:

$$f_{A(t)}(a) = \frac{\frac{1}{\sigma}\phi\left(\frac{a-\mu}{\sigma}\right)}{\Phi\left(\frac{A_{max}-\mu}{\sigma}\right) - \Phi\left(\frac{-A_{min}-\mu}{\sigma}\right)}, \tag{4}$$

where μ and σ are the expectation and standard deviation of increment $A(t)$. ϕ and Φ are the PDF and the cumulative distribution function (CDF) of standard normal distribution. We also assumed $A(t) \in [-A_{min}, A_{max}]$, $A_{max} = A_{min} > 0$ and $\mu = 0$.

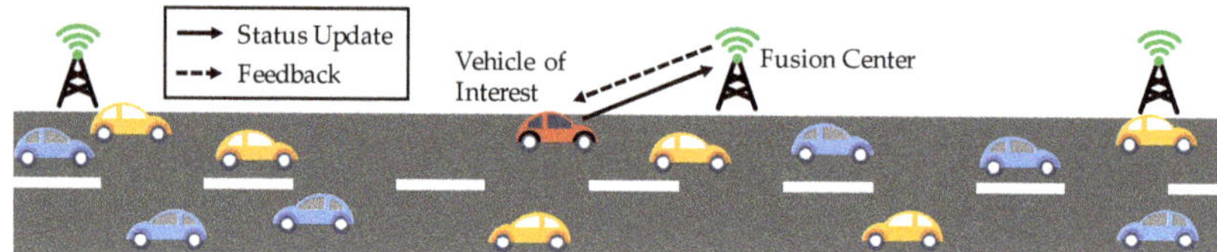

Figure 1. Remote control and monitoring model. The vehicle of interest is shown in red.

Meanwhile, the scheduling policy of information updates should also be related to the situation and environment of the system. For example, when the system is in an emergency, it should be very sensitive to the accuracy and the delay of the status information, thus the status should be updated more frequently. Therefore, our objective is to find a policy telling the vehicle whether to transmit status information or not in each slot for a minimum average UoI over time under the constraint:

$$\min_{U(t)} \limsup_{T \to \infty} \frac{1}{T} E\left[\sum_{t=0}^{T-1} w(t)Q(t)^2\right]$$
$$s.t. \limsup_{T \to \infty} \frac{1}{T} \sum_{t=0}^{T-1} E[U(t)] \le \rho, \tag{5}$$

where $\omega(t) > 0$ is the context-aware weight, which is independent with $Q(t)$. $\rho \in (0,1]$ is the maximum average update frequency. The cost function of the estimation error used here is $\delta(Q(t)) = (Q(t))^2$, which is inspired by the squared error of status estimation.

3. Scheduling with CMDP-Based Approach

In this section, we start by formulating problem (5) into a constrained Markov decision process (CMDP) with assumptions on the distribution of the context-aware weight. We will prove the threshold structure of the UoI-optimal updating policy and derive the optimal policy through a linear programming (LP) formulation.

3.1. Constrained Markov Decision Process Formulation

In the remote monitoring system, the context may be related to the distance between adjacent vehicles/mobile devices, the unexpected maneuver of the neighboring vehicles, etc. In [32], the authors prove that whether the distance between two mobile wireless devices with Ornstein–Uhlenbeck mobility is less than a certain threshold follows a first-order Markov process. When the two devices are closer, they are more interested in each other's status information, communication and computing resources to facilitate cooperation, share resources, and avoid collisions. At this time, the transmission of status information is more urgent than when the two devices are far apart. As for the unexpected maneuver of the neighboring vehicles, it is very challenging to find a proper formulation. Instead, we assumed that such emergencies occur independently in each slot according to a certain probability. Therefore, in contrast to [27,30], we assumed that the context-aware weight $\omega(t)$ is i.i.d. over time or a first-order irreducible positive recurrent Markov process and formulated the problem (5) as a CMDP problem. The irreducible positive recurrent Markov formulation guarantees the existence of the UoI-optimal policy (see Appendix A). In this section, we will first focus on the situation where $\omega(t)$ is a first-order Markov process:

- State space: The state of the vehicle in slot t, denoted by $s(t) = (Q(t), \omega(t))$, includes the current estimation error and the context-aware weight. Then, we discretize $Q(t)$ with the step size $\Delta_Q > 0$, i.e., the estimation error $Q(t) \in \mathbb{Q} = \{0, \pm\Delta_Q, \pm 2\Delta_Q, \cdots, \pm n\Delta_Q, \cdots\}$. For example, when $Q(t) \in [n\Delta_Q - \frac{1}{2}\Delta_Q, n\Delta_Q + \frac{1}{2}\Delta_Q)$, its value will be taken as $n\Delta_Q$. The smaller the step size Δ_Q, the smaller the performance degradation caused by discretization. In addition, the value set of the context-aware weight is denoted by $\mathbb{W}$. Then, the state space $\mathbb{S} = \{\mathbb{Q} \times \mathbb{W}\}$ is thus countable but infinite.
- Action space: At each slot, the vehicle can take two actions, namely $U(t) \in \mathbb{U} = \{0, 1\}$, where $U(t) = 1$ denotes the vehicle deciding to transmit updates in slot t and $U(t) = 0$ denotes the vehicle deciding to wait.
- Probability transfer function: After taking action U at state $s = (Q, \omega)$, the next state is denoted by $s' = (Q', \omega')$. When the vehicle decides not to transmit or the transmission fails, the probability of the estimation error transferring from Q to Q' is written as $\Pr\{Q' - Q = a\} = p_a$. Due to the discretization of the estimation error, the increment $a \in \mathbb{A} = \{0, \pm\Delta_Q, \pm 2\Delta_Q, \cdots, \pm A_m\}$, where $A_m = \lfloor \frac{A_{max}}{\Delta_Q} \rfloor \Delta_Q > 0$. In addition, $p_a = F_A(a + \frac{1}{2}\Delta_Q) - F_A(a - \frac{1}{2}\Delta_Q)$, where $F_A(a)$ is the CDF of increment $A(t)$. In addition, the probability of the context-aware weight transferring from ω to ω' is written as $\Pr\{\omega \to \omega'\} = p_{\omega\omega'}$. Based on the assumption that the context-aware weight $\omega(t)$ is independent with the estimation error $Q(t)$, then the probability of the state transferring from $s = (Q, \omega)$ to $s' = (Q', \omega')$ given action U is:

$$\Pr\{s \to s'|U\} = \Pr\{(Q, \omega) \to (Q', \omega')|U\}$$

$$= \begin{cases} p_{\omega\omega'} p_{Q'-Q} & , U = 0, \\ p_{\omega\omega'}((1 - p_s)p_{Q'-Q} + p_s p_{Q'-0}) & , U = 1. \end{cases} \tag{6}$$

- One-step cost: The cost caused by taking action U in state (Q, ω) is:

$$C(Q, \omega, U) = \omega Q^2, \tag{7}$$

while the one-step updating penalty only depends on the chosen action:

$$D(Q, \omega, U) = U. \tag{8}$$

The average cost caused under a certain policy π is the average UoI, which is defined as $\bar{C}^\pi$ and the average updating penalty under π is defined as $\bar{D}^\pi$. We aimed to find the UoI-optimal policy which minimizes the average cost under the resources constraint. Therefore, problem (5) can be formulated into the following CMDP problem:

$$\min_\pi \bar{C}^\pi = \lim_{T \to \infty} \frac{1}{T} \mathbb{E}_\pi \left[\sum_{t=1}^T C(Q(t), \omega(t), U(t)) \right]$$

$$s.t. \ \bar{D}^\pi = \lim_{T \to \infty} \frac{1}{T} \mathbb{E}_\pi \left[\sum_{t=1}^T D(Q(t), \omega(t), U(t)) \right] \le \rho. \tag{9}$$

3.2. Threshold Structure of the Optimal Policy

We start from some basic definitions in [33] and show the properties of problem (9).

Definition 1. *A stationary deterministic policy is a policy that takes the same action whenever in a given state $s = (Q, \omega)$, while a stationary randomized policy chooses to update or not in state s with a certain probability.*

Theorem 1. *There exists an optimal stationary randomized policy for problem (9). The optimal policy is a probabilistic combination of two stationary deterministic policies. The two deterministic policies only differ on at most one state and each policy minimizes the unconstrained cost in (10) with a different Lagrange multiplier λ:*

$$L_\lambda^\pi = \lim_{T \to \infty} \frac{1}{T} \mathbb{E}_\pi \left[\sum_{t=1}^{T} [C(Q(t), \omega(t), U(t)) + \lambda D(Q(t), \omega(t), U(t))] \right]. \tag{10}$$

Proof of Theorem 1. The proof is shown in Appendix A. $\square$

We denote the optimal policy that minimizes the unconstrained cost in (10) with a given λ by π^* and the cost obtained under policy π^* by $L_\lambda^{\pi^*}$, namely $L_\lambda^{\pi^*} = \min_\pi L_\lambda^\pi$. Then, there exists a differential cost function $V(Q, \omega)$ that satisfies the Bellman Equation [34]:

$$V(Q, \omega) + L_\lambda^{\pi^*} = \min \Bigg\{ C(Q, \omega, 1) + \lambda D(Q, \omega, 1)$$

$$+ (1 - p_s) \sum_{\omega' \in W} p_{\omega\omega'} \sum_{a=-A_m}^{A_m} p_a V(Q + a, \omega') + p_s \sum_{\omega' \in W} p_{\omega\omega'} \sum_{a=-A_m}^{A_m} p_a V(a, \omega'),$$

$$C(Q, \omega, 0) + \sum_{\omega' \in W} p_{\omega\omega'} \sum_{a=-A_m}^{A_m} p_a V(Q + a, \omega') \Bigg\}. \tag{11}$$

To solve problem (5), we first prove that with a given λ, the optimal stationary deterministic policy π^* has a threshold structure. We then introduce a discounted problem with a discount factor α and the discounted cost starting from state (Q, ω) under a certain policy π is:

$$J_{\alpha, \pi}(Q, \omega) = \lim_{T \to \infty} \mathbb{E}_\pi \Bigg[\sum_{t=0}^{T} \alpha^t [C(Q(t), \omega(t), U(t))$$

$$+ \lambda D(Q(t), \omega(t), U(t))] \mid (Q(0) = Q, \omega(0) = \omega) \Bigg]. \tag{12}$$

Denote the minimum cost starting from state (Q, ω) by $V_\alpha(Q, \omega) = \min_\pi J_{\alpha, \pi}(Q, \omega)$. Then, we have:

$$V_\alpha(Q, \omega) = \min \Bigg\{ C(Q, \omega, 1) + \lambda D(Q, \omega, 1) + (1 - p_s)\alpha \sum_{\omega' \in W} p_{\omega\omega'} \sum_{a=-A_m}^{A_m} p_a V_\alpha(Q + a, \omega')$$

$$+ p_s \alpha \sum_{\omega' \in W} p_{\omega\omega'} \sum_{a=-A_m}^{A_m} p_a V_\alpha(a, \omega'), C(Q, \omega, 0)$$

$$+ \alpha \sum_{\omega' \in W} p_{\omega\omega'} \sum_{a=-A_m}^{A_m} p_a V_\alpha(Q + a, \omega') \Bigg\}. \tag{13}$$

Define $\Delta(Q, \omega)$ as the difference between the value functions by taking the two different actions $U = 0, 1$, meaning that:

$$\Delta(Q,\omega) = C(Q,\omega,0) + \alpha \sum_{\omega' \in W} p_{\omega\omega'} \sum_{a=-A_m}^{A_m} p_a V_\alpha(Q+a,\omega')$$

$$-C(Q,\omega,1) - \lambda D(Q,\omega,1) - p_s \alpha \sum_{\omega' \in W} p_{\omega\omega'} \sum_{a=-A_m}^{A_m} p_a V_\alpha(a,\omega')$$

$$-(1-p_s)\alpha \sum_{\omega' \in W} p_{\omega\omega'} \sum_{a=-A_m}^{A_m} p_a V_\alpha(Q+a,\omega')$$

$$= p_s \alpha \sum_{\omega' \in W} p_{\omega\omega'} \sum_{a=-A_m}^{A_m} p_a \{V_\alpha(Q+a,\omega') - V_\alpha(a,\omega')\} - \lambda. \tag{14}$$

Define $\sum_{a=-A_m}^{A_m} p_a V_\alpha(Q+a,\omega)$ as a function $f_\alpha(Q,\omega)$. Then we will prove that for $\forall |Q_1| < |Q_2|$, we have $f_\alpha(Q_1,\omega) < f_\alpha(Q_2,\omega)$. To this end, we first prove the following Lemma 1.

Lemma 1. *For a given discount factor α and a fixed context-aware weight ω, the value function for Q equals the value function for $-Q$, namely:*

$$V_\alpha(Q,\omega) = V_\alpha(-Q,\omega).$$

Proof of Lemma 1. The Lemma is proven by induction. Define $V_\alpha^{(k)}(Q,\omega)$ as the value function obtained after the kth iteration. Assume that for $\forall Q$, we have: $V_\alpha^{(k)}(Q,\omega) = V_\alpha^{(k)}(-Q,\omega)$. If action U is taken in the kth iteration, then the expected discounted cost is defined as $J_{\alpha,U}^{(k)}(Q,\omega)$. Therefore, $V_\alpha^{(k+1)}(Q,\omega) = \min_U J_{\alpha,U}^{(k)}(Q,\omega)$. We have:

$$J_{\alpha,0}^{(k)}(Q,\omega) = C(Q,\omega,0) + \alpha \sum_{\omega' \in W} p_{\omega\omega'} \sum_{a=-A_m}^{A_m} p_a V_\alpha^{(k)}(Q+a,\omega')$$

$$= \omega(-Q)^2 + \alpha \sum_{\omega' \in W} p_{\omega\omega'} \sum_{a=-A_m}^{A_m} p_a V_\alpha^{(k)}(-Q-a,\omega')$$

$$= C_X(-Q,\omega,0) + \alpha \sum_{\omega' \in W} p_{\omega\omega'} \sum_{a=-A_m}^{A_m} p_a V_\alpha^{(k)}(-Q+a,\omega') = J_{\alpha,0}^{(k)}(-Q,\omega). \tag{15}$$

Similarly, we can further prove that $J_{\alpha,1}^{(k)}(Q,\omega) = J_{\alpha,1}^{(k)}(-Q,\omega)$. Notice that the value function obtained in $(k+1)^{\text{th}}$ iteration is obtained by: $V_\alpha^{(k+1)}(Q,\omega) = \min_U J_{\alpha,U}^{(k)}(Q,\omega)$, and for any action U, $J_{\alpha,U}^{(k)}(Q,\omega) = J_{\alpha,U}^{(k)}(-Q,\omega)$. Thus, $V_\alpha^{(k+1)}(Q,\omega) = V_\alpha^{(k+1)}(-Q,\omega)$. By letting $k \to \infty$, $V_\alpha^{(k)}(Q,\omega) \to V_\alpha(Q,\omega)$. Hence, $V_\alpha(Q,\omega) = V_\alpha(-Q,\omega)$. $\square$

Lemma 2. *For a given discount factor α and a fixed context-aware weight ω, function $f_\alpha(Q,\omega)$ for Q increases monotonically with the absolute value of Q, namely: for $\forall |Q_1| < |Q_2|$, $f_\alpha(Q_1,\omega) < f_\alpha(Q_2,\omega)$.*

Proof of Lemma 2. Using the induction method, we first assume that for $\forall |Q_1| < |Q_2|$, we have $f_\alpha^{(k)}(Q_1,\omega) < f_\alpha^{(k)}(Q_2,\omega)$. Therefore:

$$J_{\alpha,0}^{(k)}(Q_1,\omega) = C(Q_1,\omega,0) + \alpha \sum_{\omega' \in \mathbb{W}} p_{\omega\omega'} \sum_{a=-A_m}^{A_m} p_a V_\alpha^{(k)}(Q_1 + a, \omega')$$

$$= \omega Q_1^2 + \alpha \sum_{\omega' \in \mathbb{W}} p_{\omega\omega'} f_\alpha^{(k)}(Q_1,\omega')$$

$$< C(Q_2,\omega,0) + \alpha \sum_{\omega' \in \mathbb{W}} p_{\omega\omega'} f_\alpha^{(k)}(Q_2,\omega')$$

$$= J_{\alpha,0}^{(k)}(Q_2,\omega). \tag{16}$$

Similarly, we can obtain $J_{\alpha,1}^{(k)}(Q_1,\omega) < J_{\alpha,1}^{(k)}(Q_2,\omega)$. Meanwhile, $V_\alpha^{(k+1)}(Q,\omega) = \min_U J_{\alpha,U}^{(k)}(Q,\omega)$, then we have $V_\alpha^{(k+1)}(Q_1,\omega) < V_\alpha^{(k+1)}(Q_2,\omega)$, for $\forall |Q_1| < |Q_2|$. Obviously, if we want to use induction to complete the proof of Lemma 2, we have to prove that: $f_\alpha^{(k+1)}(Q_1,\omega) < f_\alpha^{(k+1)}(Q_2,\omega)$, for $\forall |Q_1| < |Q_2|$. To simplify the proof, it is assumed that $Q_2 > Q_1 > 0$. The discussion will be divided into the following three situations.

- When $A_m \leq |Q_1|$, then $|Q_1 + a| < |Q_2 + a|$, for $\forall a \in [-A_m, A_m]$, we can derive that:

$$f_\alpha^{(k+1)}(Q_1,\omega) = \sum_{a=-A_m}^{A_m} p_a V_\alpha^{(k+1)}(Q_1 + a, \omega)$$

$$< \sum_{a=-A_m}^{A_m} p_a V_\alpha^{(k+1)}(Q_2 + a, \omega) = f_\alpha^{(k+1)}(Q_2,\omega). \tag{17}$$

- When $A_m > |Q_2|$, there exists an increment $a' \in \mathcal{A}' = \{a | a \in [-A_m, -\frac{1}{2}(Q_1 + Q_2)]\}$, such that $|Q_1 + a'| > |Q_2 + a'|$, and $V_\alpha^{(k+1)}(Q_1 + a', \omega') > V_\alpha^{(k+1)}(Q_2 + a', \omega')$. Notice that $-Q_1 - a' \in (\frac{1}{2}(Q_2 - Q_1), A_m - Q_1]$ and $Q_2 + a \in [-A_m + Q_2, A_m + Q_2]$, then $p_{-Q_1 - a' - Q_2} V_\alpha^{(k+1)}(-Q_1 - a', \omega)$ is a term in the summation $f_\alpha^{(k+1)}(Q_2,\omega)$, namely $\sum_{a=-A_m}^{A_m} p_a V_\alpha(Q + a, \omega)$. Similarly, $p_{-Q_2 - a' - Q_1} V_\alpha^{(k+1)}(-Q_2 - a', \omega)$ is a term in the summation $f_\alpha^{(k+1)}(Q_1,\omega)$. We further define $\mathcal{A}'' = \{a | a = -Q_1 - Q_2 - a'\}$, since $-Q_1 - Q_2 - a' \in \left(-\frac{1}{2}(Q_1 + Q_2), A_m - Q_1 - Q_2\right]$, then $\mathcal{A}' \cap \mathcal{A}'' = \varnothing$.
  Furthermore, the probability of the estimation error transferring from Q_1 to $-Q_2 - a'$, i.e., $p_{-Q_2 - a' - Q_1}$ equals $p_{-Q_1 - a' - Q_2}$, the probability of the estimation error transferring from Q_2 to $-Q_1 - a'$. Since $-a' \in (\frac{1}{2}(Q_1 + Q_2), A_m]$, then $|a'| > |-Q_1 - Q_2 - a'|$. According to our assumption of the increment, we can prove that for any $a' \in \mathcal{A}'$, $p_{a'} < p_{-Q_1 - Q_2 - a'}$. Then, we can derive:

$$f_\alpha^{(k+1)}(Q_1,\omega) - f_\alpha^{(k+1)}(Q_2,\omega)$$

$$= \sum_{a \in \mathcal{A}'} p_a V_\alpha^{(k+1)}(Q_1 + a, \omega) + \sum_{a \in \mathcal{A}''} p_a V_\alpha^{(k+1)}(Q_1 + a, \omega)$$

$$- \sum_{a \in \mathcal{A}'} p_a V_\alpha^{(k+1)}(Q_2 + a, \omega) - \sum_{a \in \mathcal{A}''} p_a V_\alpha^{(k+1)}(Q_2 + a, \omega) + M(Q_1, Q_2)$$

$$= \sum_{a \in \mathcal{A}'} p_a \{V_\alpha^{(k+1)}(Q_1 + a, \omega) - V_\alpha^{(k+1)}(Q_2 + a, \omega)\}$$

$$+ \sum_{a \in \mathcal{A}'} p_{-Q_1 - Q_2 - a} \{V_\alpha^{(k+1)}(Q_2 + a, \omega) - V_\alpha^{(k+1)}(Q_1 + a, \omega)\} + M(Q_1, Q_2)$$

$$= \sum_{a \in \mathcal{A}'} (p_a - p_{-Q_1 - Q_2 - a}) \{V_\alpha^{(k+1)}(Q_1 + a, \omega) - V_\alpha^{(k+1)}(Q_2 + a, \omega)\} + M(Q_1, Q_2) < 0, \tag{18}$$

where $M(Q_1, Q_2) = \sum_{a \notin \mathcal{A}' \cup \mathcal{A}''} p_a \left(V_\alpha^{(k+1)}(Q_1 + a, \omega) - V_\alpha^{(k+1)}(Q_2 + a, \omega)\right) < 0$.

- When $|Q_2| > A_m > |Q_1|$, since $a' \in [-A_m, -\frac{1}{2}(Q_1 + Q_2))$, we only need to consider the case when $A_m > \frac{1}{2}(Q_1 + Q_2)$, in this case $-Q_1 - a' > \frac{1}{2}(Q_2 - Q_1) > Q_2 - A_m$. Therefore, $p_{-Q_1 - a' - Q_2} V_\alpha^{(k+1)}(-Q_1 - a', \omega)$ is a term in the summation $f_\alpha^{(k+1)}(Q_2, \omega)$. Similarly, we can also prove that $f_\alpha^{(k+1)}(Q_1, \omega) < f_\alpha^{(k+1)}(Q_2, \omega)$ when $|Q_2| > A_m > |Q_1|$.

According to Lemma 1, the conclusions above can be easily generalized to the cases without the condition $Q_2 > Q_1 > 0$. Finally, by letting $k \to \infty$, $V_\alpha^{(k+1)}(Q, \omega) \to V_\alpha(Q, \omega)$, therefore: $f_\alpha^{(k+1)}(Q, \omega) \to f_\alpha(Q, \omega)$. Hence: $f_\alpha(Q_1, \omega) < f_\alpha(Q_2, \omega)$. $\square$

Remark 1. *Lemma 2 holds when $f_A(a)$, i.e., the PDF of increment $A(t)$ satisfies the following conditions:*

- $f_A(a) = f_A(-a), \mu = 0;$
- $f_A(a_2) \leq f_A(a_1), \forall a_2 \geq a_1 \geq 0.$

Then, with Lemmas 1 and 2, we can prove the threshold structure of the optimal stationary deterministic policy which minimizes L_λ^π in (10).

Theorem 2. *For a given λ, the optimal stationary deterministic policy which minimizes L_λ^π in (10) has a threshold structure when the context-aware weight is a first-order irreducible positive recurrent Markov process.*

Proof of Theorem 2. Let $s_\alpha^*(Q, \omega)$ denote the optimal action which minimizes the discounted cost $V_\alpha(Q, \omega)$ at state (Q, ω). If the optimal action $s_\alpha^*(Q, \omega) = 1$, then the vehicle will transmit its status update to the center at state (Q, ω) and $\Delta(Q, \omega) \geq 0$. Thus, we have:

$$\Delta(Q, \omega) = p_s \alpha \sum_{\omega' \in \mathbb{W}} p_{\omega\omega'} \sum_{a=-A_m}^{A_m} p_a \{V_\alpha(Q + a, \omega') - V_\alpha(a, \omega')\} - \lambda \geq 0. \tag{19}$$

According to Lemma 2, for any $|Q'| > |Q|$, $\Delta(Q', \omega)$ can be lower bounded by

$$\Delta(Q', \omega) = p_s \alpha \sum_{\omega' \in \mathbb{W}} p_{\omega\omega'} \sum_{a=-A_m}^{A_m} p_a \{V_\alpha(Q' + a, \omega') - V_\alpha(a, \omega')\} - \lambda$$

$$\geq p_s \alpha \sum_{\omega' \in \mathbb{W}} p_{\omega\omega'} \sum_{a=-A_m}^{A_m} p_a \{V_\alpha(Q + a, \omega') - V_\alpha(a, \omega')\} - \lambda \geq 0. \tag{20}$$

If $\Delta(Q, \omega) > 0$, then for any states with $|Q'| > |Q|$, the optimal policy is to transmit the status to the center. If $\Delta(Q, \omega) < 0$, then for any states with $|Q'| < |Q|$, the optimal action is not to transmit. In addition, the optimal policy will not be choosing to wait in all the slots. Therefore, for each context-aware weight ω, there must be a threshold $\tau_\omega \geq 0$. For any state (Q, ω) with $|Q| > \tau_\omega$, the optimal choice is to transmit the status update. We can then conclude that for a given weight ω, the optimal policy with a discount factor α has a threshold structure.

Let $\{\alpha_1, \alpha_2, \cdots, \alpha_k\}$ denote a sequence of discount factors and α_k converges to 1 when $k \to \infty$. Then, the optimal deterministic policy for $\alpha = 1$ will also converge to the optimal policy with a discount factor which is less than 1 [35]. Similar derivation is also applied in [12]. Therefore, we can prove the threshold structure of the optimal stationary deterministic policy which minimizes L_λ^π. $\square$

Similarly, when the context-aware weight is i.i.d. over time, we can obtain the following theorem:

Theorem 3. *For a given λ, the optimal stationary deterministic policy which minimizes L_λ^π in (10) has a threshold structure when the context-aware weight is i.i.d. over time. The thresholds are the same for each state of the context-aware weight.*

Proof of Theorem 3. If the context-aware weight is i.i.d. over time, then we have:

$$\Delta(Q,\omega) = p_s \alpha \sum_{\omega'\in W} p_{\omega'} \sum_{a=-A_m}^{A_m} p_a\{V_\alpha(Q+a,\omega') - V_\alpha(a,\omega')\} - \lambda = \Delta(Q), \qquad (21)$$

where p_ω is the probability of the value of the context-aware weight being in state ω. Therefore, in this case, the state will be reduced to one dimension and the thresholds will be the same for all the states of the context-aware weight. $\square$

According to Theorems 2 and 3, we proved the threshold structure of the two stationary deterministic policies that compose the UoI-optimal policy. Since the UoI-optimal policy for problem (9) is a probabilistic combination of two deterministic policies with threshold structures, we can finally draw the conclusion that the UoI-optimal policy also has a threshold structure.

3.3. Numerical Solution of Optimal Strategy

Based on Theorem 2, we only need to consider the policy that chooses to update with a probability of 1 in state (Q,ω), for $\forall |Q| \geq Q_{max} = \max_\omega \tau_\omega$. Let $\mu_{Q,\omega}$ denote the probability that the state of the vehicle is (Q,ω). $y_{Q,\omega}$ denotes the probability that the state is (Q,ω) and the vehicle chooses to transmit an update. Therefore, we have:

Theorem 4. *When the context-aware weight is a first-order irreducible positive recurrent Markov process, the UoI-optimal policy can be derived by solving the following LP problem:*

$$\{\mu^*_{Q,\omega}, y^*_{Q,\omega}\} = \arg \min_{\{\mu_{Q,\omega}, y_{Q,\omega}\}} \sum_{\omega\in W} \sum_{Q=-Q_{max}}^{Q_{max}} \omega Q^2 \mu_{Q,\omega}, \qquad (22a)$$

$$s.t. \ \sum_{\omega\in W} \sum_{Q=-Q_{max}}^{Q_{max}} \mu_{Q,\omega} = 1, \qquad (22b)$$

$$\sum_{\omega\in W} \sum_{Q=-Q_{max}}^{Q_{max}} y_{Q,\omega} \leq \rho, \qquad (22c)$$

$$y_{Q,\omega} \leq \mu_{Q,\omega}, \forall Q, \omega, \qquad (22d)$$

$$0 \leq y_{Q,\omega} \leq 1, 0 \leq \mu_{Q,\omega} \leq 1, \forall Q, \omega, \qquad (22e)$$

$$\mu_{Q,\omega} = \sum_{\omega'\in W} \sum_{Q'=-Q_{max}}^{Q_{max}} y_{Q',\omega'} p_s p_Q p_{\omega\omega'}$$

$$+ \sum_{\omega'\in W} \sum_{Q'=-Q_{max}}^{Q_{max}} (\mu_{Q',\omega'} - y_{Q',\omega'} p_s) p_{Q'-Q} p_{\omega\omega'}. \qquad (22f)$$

Proof of Theorem 4. We first derive the average UoI $\bar{C}^\pi$ as a function of $\mu_{Q,\omega}$ and $y_{Q,\omega}$. The vehicle is in state (Q,ω) and produces a cost of $C(Q,\omega,u) = \omega Q^2$ with a probability of $\mu_{Q,\omega}$. Therefore, the average UoI is:

$$\sum_{\omega\in W} \sum_{Q=-Q_{max}}^{Q_{max}} \omega Q^2 \mu_{Q,\omega}. \qquad (23)$$

As for the constraints, (22b) means that the sum of the probabilities of all the states should be 1. To explain (22c), we note that $y_{Q,\omega}$ is the probability of the vehicle being in state (Q,ω) and choosing to transmit the update, then the expectation of a one-step updating penalty for state (Q,ω) in (8) is $\mu_{Q,\omega}$. Therefore, the constraint on average update frequency $\bar{D}^\pi$ can be illustrated by

$$\sum_{\omega \in \mathbb{W}} \sum_{Q=-Q_{\max}}^{Q_{\max}} \mu_{Q,\omega} \leq \rho. \tag{24}$$

Then, we introduce $\xi_{Q,\omega} \in [0,1]$ to represent that the probability of the vehicle choosing to transmit updates in state (Q,ω) and (22d) can be obtained by the fact that $y_{Q,\omega} = \mu_{Q,\omega}\xi_{Q,\omega}$, while (22e) is derived by the nature of probability.

The right-hand side of (22f) can be viewed as two terms. The first term is the sum of transition probability from all the states to state (Q,ω) when the vehicle chooses to update and the transmission of status is successful. The second term is the sum of transition probability from all the states to state (Q,ω) when the transmission is failed or the vehicle chooses to wait. Therefore, we can prove that the optimal solution of problem (5) equals the solution of the LP problem. $\square$

When $\omega(t)$ is i.i.d. over time, we can also obtain the UoI-optimal policy through the LP problem proposed in Theorem 4 and only need to use $p_{\omega'}$ as a replacement of $p_{\omega\omega'}$.

4. Scheduling in Unknown Contexts

To make decisions, the UoI-optimal updating policy obtained in Section 3 still needs the distributions of the context-aware weight $\omega(t)$, the increment $A(t)$ and the successful transmission probability, which may not be available in advance or may change over time in most practical systems. To solve this problem, we will assume that the distribution of the context-aware weight is not pre-determined and the vehicle has to learn it. In this section, we use the reinforcement learning (RL) algorithm to learn the dynamic of the context and the characteristic of the wireless channel.

To solve this problem, we turn to the model-based RL framework proposed in [36]. We only consider the cases when the UoI-optimal policy has a threshold structure. This assumption makes the optimal policy based on the truncated state space equal the optimal policy of the original problem.

We use the 3-tuple (s, s', U) to formulate the proposed RL-based updating policy. The states in the current slot and next slot are denoted by s and s', respectively. U denotes the action chosen in the current slot. The settings of the discretized state space and the action space are the same as the settings proposed in Section 3.1. The smaller the step size used in the discretization is, the closer our results are to those in continuous state space. In addition, the selection of the step size only affects the accuracy of the update threshold. Therefore, the performance loss caused by discretization can be reduced by choosing a smaller step size.

We display details about the proposed RL-based updating policy in Algorithm 1. At the beginning of episode k, we randomly decide whether to explore or exploit. $l \in [0,1]$ represents the trade-off between exploration and exploitation during the following episode. A larger l means a higher frequency of exploration and vice versa. If the algorithm chooses to explore during this episode, a random policy $\pi_{rand}(s)$ will be used, i.e., we randomly choose to update or not in each state to find more valuable actions. If the algorithm chooses to exploit, then we have to obtain the probability transfer functions $\tilde{p}_k(s'|s, U)$ for each state transmission pair. In Algorithm 1, $N(s, U)$ and $N(s, U, s')$ represent the number of occurrences of state–action pair s, U and state transition from s to s' given action U, respectively. Based on the assumption that the optimal policy has a threshold structure, the policy $\pi(k)$ which can minimize the average UoI with the estimated probability transfer functions, can be directly solved through the LP problem proposed in Theorem 4. Then, the vehicle will use policy π_k to derive state–action pairs and the state transitions in the following $\lceil L_k \rceil$ slots. Here, $L > 0$ is defined to control the number of state transitions observed in each episode. At the end of each episode, the model will be updated according to the state–action pairs and the state transitions observed during the episode. Finally, after K episodes, the algorithm will output the RL-based updating policy $\pi^\star(s)$, which is derived based on $\tilde{p}_K(s'|s, U)$.

Algorithm 1 RL-based Updating Policy

Input: $l \in [0,1], L > 0, K > 0$

1: **for** episodes $k = 1, 2, \ldots, K$ **do**
2:      Set $L_k = L\sqrt{k}, \epsilon_k = 1/\sqrt{k}$, uniformly draw $\alpha \in [0,1]$.
3:      **if** $\alpha < \epsilon_k$ **then**
4:          Set $\pi_k(s) = \pi_{rand}(s)$,
5:      **else**
6:          **for** each state $s, s' \in \mathbb{S}$ and $U \in \mathbb{U}$ **do**
7:              **if** $N(s, U) > 0$ **then**
8:                  Let $\tilde{p}_k(s'|s, U) = N(s, U, s')/N(s, u)$,
9:              **else**
10:                  $\tilde{p}_k(s'|s, U) = 1/|\mathbb{S}|$.
11:              **end if**
12:          **end for**
13:          obtain policy $\pi_k(s)$ by solving the estimated CMDP
14:      **end if**
15:      Randomly choose an initial state $s(1)$.
16:      **for** slots $t = 1, 2, \ldots, \lceil L_k \rceil - 1$ **do**
17:          Choose action $U(t)$ as $\pi_k(s(t))$.
18:          Observe the next state $s(t+1)$.
19:          $N(s(t), U(t), s(t+1)) = N(s(t), U(t), s(t+1)) + 1$.
20:          $N(s(t), U(t)) = N(s(t), U(t)) + 1$.
21:          $s(t) \leftarrow s(t+1)$.
22:      **end for**
23: **end for**
24: obtain policy $\pi^\star(s)$ by solving the estimated CMDP based on $\tilde{p}_k(s'|s, U), s, s' \in \mathbb{S}, U \in \mathbb{U}$.

Output: output the RL-based updating policy $\pi^\star(s)$

---

5. Simulation Results and Discussion

5.1. Simulation Setup

To facilitate the simulation, we consider the case where the context-aware weight of the vehicle only has two different states: the 'normal' state and 'urgent' state. The 'normal' state means that the vehicle is in ordinary situations and the significance of accuracy of status information is relatively low. We set $w(t)$ as 1 in 'normal' state while $w(t)$ is set as a constant much larger than 1, w_e, in 'urgent' state to show that the vehicle is in emergencies. Two different distributions of the context-aware weight are taken into consideration to conform to the assumptions about $w(t)$ used in Section 3.1:

1. The context-aware weight $w(t)$ has the first-order Markov property. The state transition diagram of $w(t)$ is shown in Figure 2 and $w(t)$ is irreducible and positive recurrent. p_1 is the probability of the context-aware weight transferring from the normal state to the urgent state, while p_2 is the probability of the weight transferring from the urgent state to the normal state;

2. The context-aware weight $w(t)$ is i.i.d. over time. The probability of the weight being in the urgent state and the normal state are denoted by p_h and p_l, respectively.

As for the increment $A(t)$, A_{max} is set to a large enough positive number to simplify the simulations.

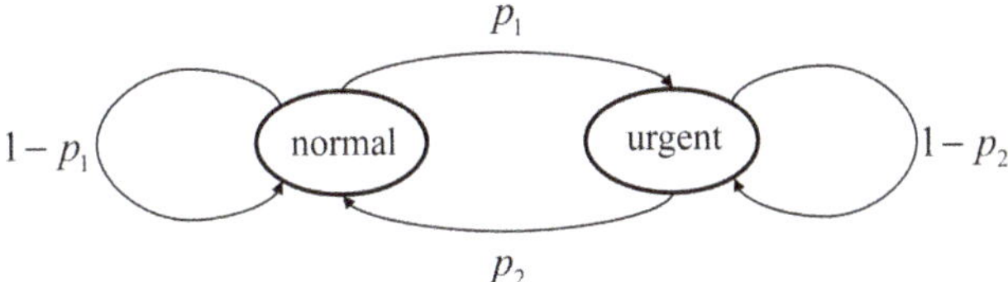

Figure 2. The state transition diagram of $\omega(t)$.

5.2. Numerical Results

Figure 3 shows the structure of the UoI-optimal updating policy. For the discretization of the estimation error, the step size used is 1. It can be seen that under the two different distributions of the context-aware weight mentioned above, the optimal updating policies all have threshold structures. Especially when the context-aware weight is i.i.d. over time, Figure 3b shows that thresholds for all the states of the context-aware weight are the same, which matches well with theoretical analysis. From Figure 3c, we can find that the UoI-optimal policy also has threshold structure when increment $A(t)$ obeys a uniform distribution $Unif(-3,3)$, which verifies Remark 1. We then simulate the UoI-optimal policy under the contexts with more states to show the policy is generic. We consider a three-state context-aware weight which takes value from $\omega_1 = 1, \omega_2 = 50, \omega_3 = 100$. The state transition matrix P_3 of the three-state context-aware weight is:

$$P_3 = \begin{bmatrix} 0.997 & 0.002 & 0.001 \\ 0.02 & 0.97 & 0.01 \\ 0.2 & 0.1 & 0.7 \end{bmatrix}, \tag{25}$$

where the j-th element on the i-th row indicates the probability that the context transfers from state ω_i to state ω_j. The numerical results (Figure 3d) show that when the context-aware weight has more states, the UoI-optimal policy still has a threshold structure, which verifies our theoretical results.

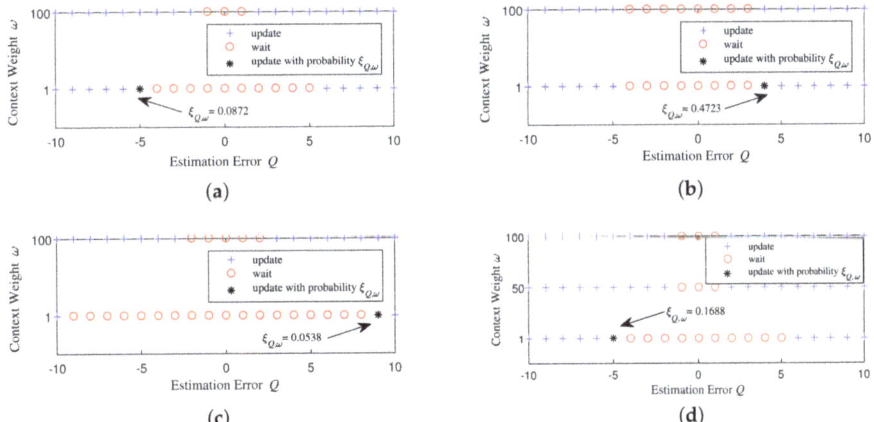

Figure 3. Threshold structure of the UoI-optimal updating policy when: (**a**) the context-aware weight is a first-order Markov process, $\rho = 0.05, p_1 = 0.001, p_2 = 0.01, p_s = 0.9, \sigma^2 = 1, \omega_e = 100$; (**b**) the context-aware weight is i.i.d. over time, $\rho = 0.05, p_l = 0.999, p_h = 0.001, p_s = 0.9, \sigma^2 = 1, \omega_e = 100$; (**c**) the context-aware weight is a first order Markov process, $\rho = 0.05, p_1 = 0.001, p_2 = 0.01, p_s = 0.9, \omega_e = 100$, increment in the estimation error during one slot, i.e., $A(t) \sim Unif(-3,3)$, for $\forall t$; and (**d**) the context-aware weight is a three-state first-order Markov process, which takes value from $\omega_1 = 1, \omega_2 = 50, \omega_3 = 100$ and evolves according to the state transition matrix $P_3, \rho = 0.05, p_s = 0.9, \sigma^2 = 1$.

Then, we will focus on the results obtained when the context-aware weight is a first-order irreducible positive recurrent Markov process, as shown in Figure 2. Figure 4 shows the average UoI of the UoI-optimal policy, the AoI-optimal policy derived by CMDP, the RL-based updating policy, and the update-index-based adaptive scheme [27]. In the RL-based updating policy, $L = 8000$, $l = 1$ and $K = 50$. All the numerical results of the RL-based policy are averaged over 100 runs.

First of all, the UoI-optimal policy can only be obtained based on advanced information about the system dynamics. However, the RL-based policy achieves near-optimal without knowing the system dynamics, which indicates that Algorithm 1 learns relatively accurate probability transfer functions from the observed state–action pairs and state transitions during the training.

Secondly, according to Figure 4, the AoI-optimal policy yields a much higher UoI than the three UoI-based policies, namely the UoI-optimal policy, the RL-based updating policy, and the update-index-based adaptive scheme. On the one hand, AoI is one special case of UoI. When the context-aware weight $\omega(t) = 1$, the increment $A(t) = 1$, and the cost function $\delta(Q(t)) = Q(t)$, then UoI equals AoI. Therefore, the AoI-optimal policy ignores the fact that different contexts have different requirements for information freshness. In the proposed UoI-based updating policies, different contexts have different policies and update thresholds, while the AoI-optimal updating policies for different contexts are the same. On the other hand, Figure 5 reveals that the AoI-optimal policy leads to a much higher estimation error, which results in worse performance in terms of UoI. The AoI-optimal policy is an oblivious policy, which is independent of the monitored process. Since AoI increases linearly with time, the AoI-optimal policy can only minimize the linear performance degradation in terms of time. However, the UoI-based policies (the cost function $\delta(Q(t)) = (Q(t))^2$) considered in this paper are process-dependent, which are called non-oblivious policies, and can benefit from both age and process realization [37]. These policies can directly minimize the nonlinear impact exerted by information staleness and the gap between the actual status and the estimated status.

Thirdly, our updating policies outperform the update-index-based adaptive scheme [27] in terms of UoI. Under the adaptive scheme, the vehicle will derive an update index as a function of the current estimation error and the context-aware weight for the next slot. If the index is larger than the adaptive update threshold, then the vehicle is supposed to transmit its status information to the center. If the vehicle transmits an update in slot t, then the adaptive threshold will increase in the next slot; otherwise, the adaptive threshold will decrease. The adaptive scheme will cause an overuse of the resource in 'urgent' states and lead to the fact that the vehicles cannot receive resources in 'normal' states. However, the UoI-optimal policy and the trained RL-updating policy are fixed schemes, which can avoid the extremely unbalanced resource allocation between the two contexts and achieve better performance.

Figure 6 shows the influence of the maximum average update frequency ρ and the context weight for emergency, ω_e, on update threshold of UoI-optimal policy. In order to obtain more accurate results, the step size used here is 0.25. The solid curves show update thresholds for the normal state while the dashed curves show update thresholds for the urgent state. When the constraint on update resources is strict, the update thresholds fall faster. Furthermore, a larger ω_e results in a lower update threshold for the urgent state and a higher threshold for the normal state. This phenomenon indicates that the value of ω_e means the tolerance of estimation error in the emergency. When $\rho < 0.1$, the influence of ω_e on the update threshold for the normal state is larger than the urgent state. For the cases where the maximum average update frequency is relatively large, ω_e has little effect on update thresholds for both normal state and urgent state.

Figure 7 shows that the update thresholds also depend on the dynamic of context-aware weight when the weight has first-order Markov property. When p_2 is approaching $1 - p_1$, the gap between update thresholds for the urgent state and the normal state becomes smaller for the context-aware weight which tends to be i.i.d. over time. When $p_2 = 1$, the update threshold for the urgent state exceeds the threshold for the normal state. Therefore, the update threshold for the urgent state is not necessarily lower than the update threshold for the normal state.

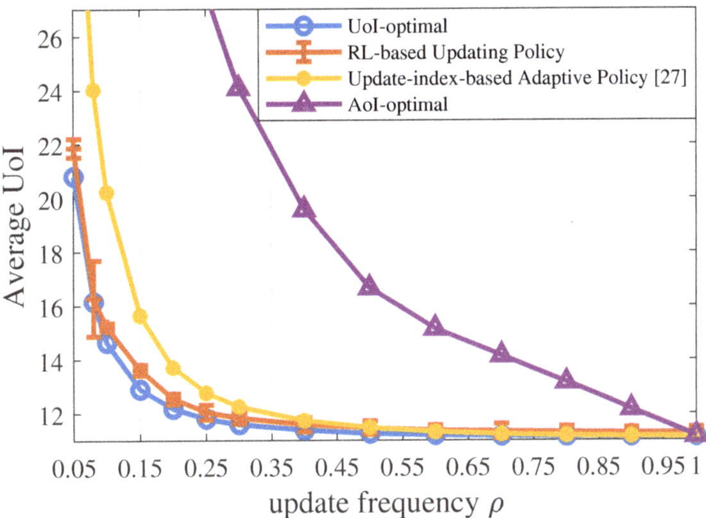

Figure 4. Average UoI of the UoI-optimal updating policy, the RL-based updating policy, the update-index-based adaptive scheme [27], and the AoI-optimal updating policy when $p_1 = 0.001$, $p_2 = 0.01, p_s = 0.9, \sigma^2 = 1, \omega_e = 100$.

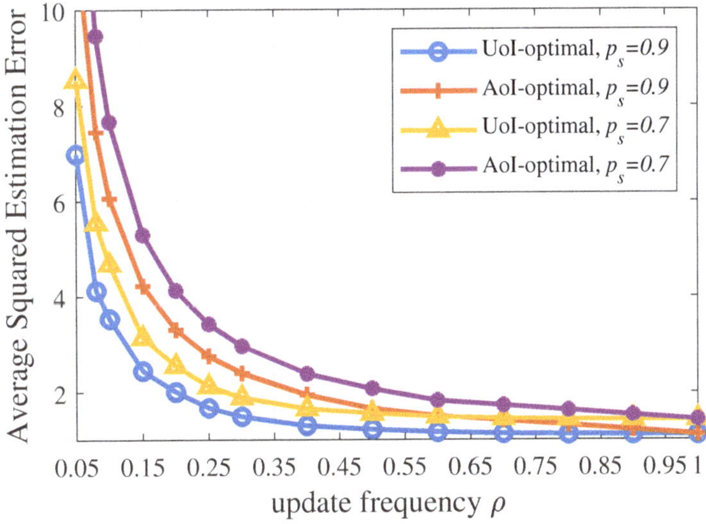

Figure 5. Average squared estimation error of the UoI-optimal updating policy and the AoI-optimal updating policy when $p_1 = 0.001, p_2 = 0.01, \sigma^2 = 1, \omega_e = 100$.

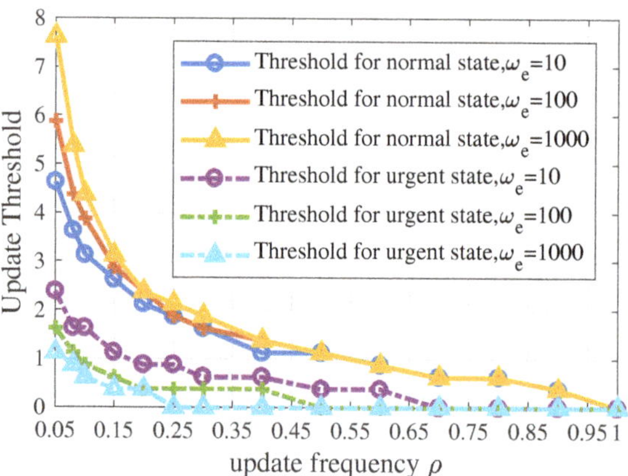

Figure 6. Update thresholds of the UoI-optimal updating policy with different values of ω_e when $p_1 = 0.001, p_2 = 0.01, p_s = 0.9, \sigma^2 = 1$.

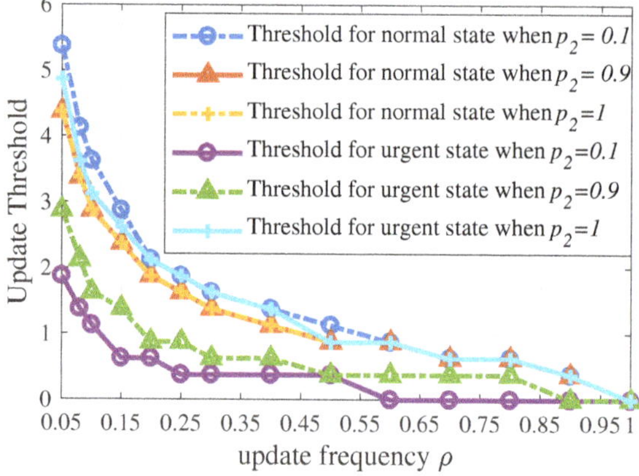

Figure 7. Update thresholds of the UoI-optimal updating policy with different values of p_2 when $p_1 = 0.01, p_s = 0.9, \sigma^2 = 1, \omega_e = 100$.

Figure 8 shows the performance of the RL-based updating policy with different values of L. According to Algorithm 1, the number of state transitions observed in episode k is $\lceil L\sqrt{k} \rceil$. Therefore, L denotes the number of state transitions observed during the whole learning process. Generally speaking, a larger L reduces the randomness of the performance and achieves a better UoI. The performance of the RL-based updating policy depends on the accuracy of the model obtained through training, namely whether the estimated probability transfer function of the system is accurate. A larger L means that the algorithm can collect more data or state transitions and obtain a more accurate model.

Figure 9 shows the influence of the number of episodes, i.e., K, on the performance of the RL-based updating policy. A larger K leads to a lower average UoI and smaller randomness over 100 runs. On the one hand, the more episodes and the more data the algorithm observes, the more accurate the model obtained will be and the better the performance of the updating policy will be. On the other hand, the value of K is the number of iterations for the policy obtained through the estimated CMDP. The policy $\pi_k(s)$ used in episode k is derived based on the state–action pairs and the state transitions observed in the previous $k - 1$ episodes. Therefore, more frequent iterations of the updating policy can obtain more valuable state–action pairs and better performance.

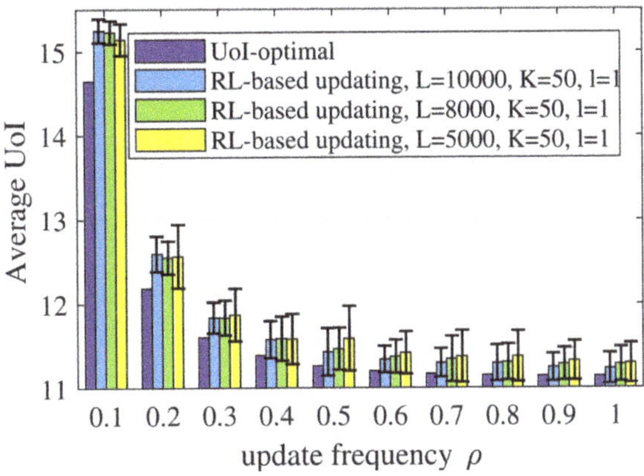

Figure 8. Average UoI of the RL-based updating policy with different values of L when $p_1 = 0.001, p_2 = 0.01, \sigma^2 = 1, \omega_e = 100$.

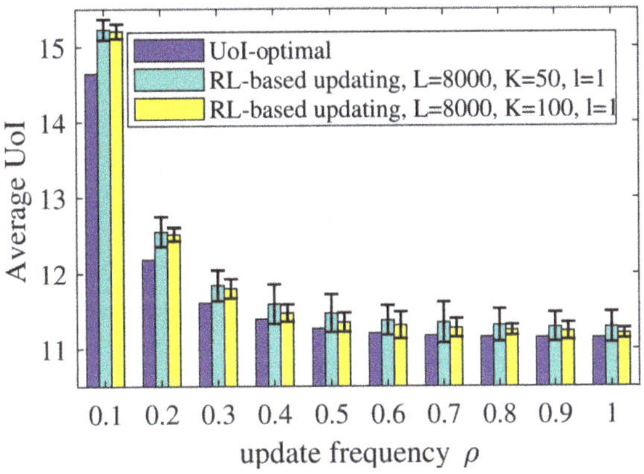

Figure 9. Average UoI of the RL-based updating policy with different values of K when $p_1 = 0.001, p_2 = 0.01, \sigma^2 = 1, \omega_e = 100$.

6. Conclusions

In this work, we studied how to minimize the performance degradation caused by outdated information in terms of UoI, which is a new metric jointly considering context and information freshness. We proved that the UoI-optimal updating policy for the considered single-user remote monitoring system has a single threshold structure. Then, the policy was obtained through linear programming by assuming that the state transition probability of the system is known in advance. In unknown contexts, we further used a reinforcement learning algorithm to learn the dynamics of the system. Simulations verified the threshold structure of the UoI-optimal policies and showed that the update thresholds decrease as the maximum average update frequency increases. In addition, a larger context-aware weight in emergencies resulted in a lower update threshold for urgent states. However, since the state transition probability also influenced the update thresholds, the update threshold for emergencies was not necessarily higher than the update threshold for normal states, especially when the probability of transferring from urgent states to normal states tended towards 1. Furthermore, the numerical results showed that the proposed RL-based updating policy achieved a near-optimal performance without advanced knowledge of the system model.

In fact, determining the context-aware weight in practical systems, where the models of the context are often very complicated and difficult to obtain in advance, remains open. As for future work, we plan to use deep RL algorithms to learn the models of the context variation. We believe that UoI can provide a new performance metric for information timeliness measurement in the future V2X scenario. In addition, we believe the proposed UoI metric and the context-aware scheduling policy can shed some light on low-latency and ultra-reliable wireless communication in the future 5G/6G systems.

Author Contributions: Conceptualization, L.W.; formal analysis, L.W.; methodology, L.W.; software, L.W.; supervision, S.Z. and Z.N.; writing—original draft, L.W.; writing—review and editing, J.S., Y.S., S.Z. and Z.N. All authors have read and agreed to the published version of the manuscript.

Funding: This work is sponsored in part by the National Key R&D Program of China No. 2020YFB1806605, by the Nature Science Foundation of China (No. 62022049, No. 61871254, No. 61861136003), by the China Postdoctoral Science Foundation No. 2020M680558, and Hitachi Ltd.

Institutional Review Board Statement: Not applicable.

Informed Consent Statement: Not applicable.

Data Availability Statement: Not applicable.

Conflicts of Interest: The authors declare no conflict of interest.

Abbreviations

The following abbreviations are used in this manuscript:

AoI	Age of Information
AoII	Age of Incorrect Information
AoS	Age of Synchronization
CDF	Cumulative Distribution Function
CMDP	Constrained Markov Decision Process
CoUD	Cost of Update Delay
i.i.d.	Independent and Identically Distributed
IoT	Internet of Things
LP	Linear Programming
PDF	Probability Density Function
RL	Reinforcement Learning
UoI	Urgency of Information
URLLC	Ultra-Reliable Low-Latency Communication
VR	Virtual Reality
V2X	Vehicle to Everything

Appendix A. Proof of Theorem 1

Given a state $s = (Q, \omega) \in \mathbb{S}$ and a nonempty subset of the state space, $\mathbb{G} \subset \mathbb{S}$, let $\mathcal{R}(s, \mathbb{G})$ denote the class of policies θ such that the probability $P^\theta(s(t) \in \mathbb{G}$ for some $t \geq 1|s(0) = s) = 1$ and the expected time $m_{s,\mathbb{G}}(\theta)$ of the first passage from s to $\mathbb{G}$ under policy θ is finite. Then, let $\mathcal{R}^\star(s, \mathbb{G})$ denote the class of policies θ such that the expected average UoI $c_{s,\mathbb{G}}(\theta)$ and the expected transmission cost $d_{s,\mathbb{G}}(\theta)$ of the first passage from s to $\mathbb{G}$ are finite and $\theta \in \mathcal{R}(s, \mathbb{G})$. To prove Theorem 1, we then introduce Assumptions A1–A5 in [33]:

Assumption A1. *For all $b > 0$, the set $\mathbb{G}(b) \triangleq \{s|$ there exists an action U such that $C(s, U) + D(s, U) \leq b\}$ is finite.*

Assumption A2. *There exists a stationary deterministic policy π that induces a Markov chain with the following properties: the state space consists of a single (nonempty) positive recurrent class $\mathbb{R}^\pi$ and a set $\mathbb{T}^\pi$ of transient states such that $\pi \in \mathcal{R}^\star(s, \mathbb{R}^\pi)$, for any $s \in \mathbb{T}^\pi$, and both the average UoI $\bar{C}^\pi$ and the average transmission cost $\bar{D}^\pi$ on $\mathbb{R}^\pi$ are finite.*

Assumption A3. *Given any two states $s, s' \in \mathbb{S}$ and $s \neq s'$, there exists a policy π (a function of s and s') such that $\pi \in \mathcal{R}^\star(s, \{s'\})$.*

Assumption A4. *If a stationary deterministic policy has at least one positive recurrent state, then it has a single positive recurrent class, and this class contains the state (Q, ω) with $Q = 0$.*

Assumption A5. *There exists a policy π such that the average UoI $\bar{C}^\pi < \infty$ and average transmission cost $\bar{D}^\pi < \rho$.*

Furthermore, the problem (9) has the following property:

Lemma A1. *Assumptions A1–A5 hold for problem (9).*

Proof of Lemma A1. First of all, we focus on the cases where the context-aware weight is assumed as a first-order irreducible positive recurrent Markov process:

- Assumption A1: In this problem, $C(s, U)$ is the UoI at state s, namely $C(Q, \omega, U) = \omega Q^2$. $D(s, U)$ is 1 if the vehicle chooses to transmit its status and $D(s, U)$ is 0 otherwise, namely $D(Q, \omega, U) = U$. Therefore, Assumption A1 holds, for any $b > 0$, the number of states (Q, ω) with $\omega Q^2 \leq b$ is finite.

- Assumption A2: Due to the current high-level wireless communication technology, we reasonably assumed that the successful transmission probability p_s is relatively close to 1. Based on the assumptions mentioned above, the Markov chain of context-aware weight obviously satisfies Assumption A2. Define the probability of the context-aware weight transferring from ω to ω' in k steps for the first time as $P_{\omega,\omega',k}$. Then, we consider the policy $\pi(Q, \omega) = 1$ for all $(Q, \omega) \in \mathbb{S}$, namely this policy chooses to transmit in all the states.

  Since the evolution of the context-aware weight is independent with the evolution of the estimation error and the updating policy. Therefore, we first focused on the estimation error, which can be formulated as a one-dimensional irreducible Markov chain with state space $\mathbb{Q} = \{0, \pm\Delta_Q, \pm 2\Delta_Q, \cdots, \pm n\Delta_Q, \cdots\}$. We denote the set of states which can transfer to state Q in a single step by $\mathcal{Z}_Q$. The probability of the estimation error transferring from state Q to state Q' at the k-th step without an arrival to state $Q = 0$ is defined as $P'_{Q,Q',k}$. Obviously, $\sum_{Q' \in \mathbb{Q}} P'_{Q,Q',k} < (1 - p_s)^k$. Then, the probability of the first passage from state $Q(Q \neq 0)$ to 0 taking $k + 1$ steps is $\sum_{Q' \notin \mathcal{Z}_0} P'_{Q,Q',k} p_s + \sum_{Q' \in \mathcal{Z}_0} P'_{Q,Q',k}(p_s + (1 - p_s)p_{0-Q'}) < (1 - p_s)^k$, where $p_{0-Q'}$ is the probability that the increment in estimation error is $-Q'$. Therefore, the expected time of the first passage from $Q(Q \neq 0)$ to 0 is finite.

  For state $Q = 0$, the estimation error will stay in this state in the next step with a probability of $p_s + p_{0-0}$ and will first return to state $Q = 0$ in the second transition with a probability smaller than $(1 - p_s - p_{0-0})$. Then, starting from state $Q = 0$, the estimation error will first return to state $Q = 0$ in the $k + 1$-th ($k > 2$) step will be smaller than $(1 - p_s - p_{0-0})(1 - p_s)^{k-1}$. Therefore, we can prove that state $Q = 0$ is a positive recurrent state, and $\mathbb{R}^\pi_Q = \{Q = 0\}$ is a positive recurrent class of the induced Markov chain of the estimation error. Furthermore, for any states in $\mathbb{T}^\pi_Q = \mathbb{Q} \backslash \mathbb{R}^\pi_Q$, the expected time of the first passage from the state in $\mathbb{T}^\pi_Q$ to state $Q = 0$ under π is finite and the probability of the states in $\mathbb{T}^\pi_Q$ not getting to state $Q = 0$ in k steps is smaller than $(1 - p_s)^k$.

  Define the probability of state Q transferring to state Q' in k steps for the first time as $P_{Q,Q',k}$. Then, the probability of state (Q, ω) transferring to state (Q', ω') in k steps for the first time is $P_{Q,Q',k}P_{\omega,\omega',k}$. Since $\sum_{k=1}^{\infty} P_{Q,Q',k}k < \infty$ and $\sum_{k=1}^{\infty} P_{\omega,\omega',k}k < \infty$, then $\sum_{k=1}^{\infty} P_{Q,Q',k}P_{\omega,\omega',k}k < \infty$. Therefore, the set of states $\mathbb{R}^\pi = \{(Q, \omega) | Q \in \mathbb{R}^\pi_Q, \omega \in \mathbb{W}\}$ is a positive recurrent class. Similarly, we can prove that $\mathbb{T}^\pi = \mathbb{S} \backslash \mathbb{R}^\pi$ satisfies Assumption A2. Finally, $\bar{D}^\pi = 1 < \infty$, $\bar{C}^\pi = E[\omega]\frac{1}{p_s}\sigma^2 < \infty$.

- Assumption A3: Define $P_{Q,min} = \min_{Q'} p_{Q-Q'}$, $P_{Q,max} = \max_{Q'} p_{Q-Q'}$. Consider the policy $\pi'(Q,\omega) = 0$ for all states $(Q,\omega) \in \mathbb{S}$, notably that this policy chooses not to transmit in any states. Similarly, we first focus on the Markov chain of estimation error. Starting from state Q, the probability of transferring to state Q' in $k+1$-th $(k \geq 2)$ steps for the first time is smaller than $(1 - p_{Q'-Q})P_{Q',max}(1 - P_{Q',min})^{k-1}$. Then, the expected time of the first passage from state Q to state Q' under policy π' is finite. Similarly, since the Markov chain of context-aware weight is irreducible positive recurrent and independent with the updating policy, we can therefore prove that the expected time of the first passage from state (Q,ω) to state (Q',ω') under policy π' is finite.
- Assumption A4: For the Markov chain of the estimation error, any state will return to state $Q = 0$ if a successful transmission occurs. For the policy without transmission, namely $\pi'(Q,\omega) = 0$, state $Q = 0$ still exists in only one positive recurrent class. For each positive recurrent class containing state $Q = 0$, we can prove that there is only one positive recurrent class. Since the Markov chain of the context-aware weight is irreducible positive recurrent, we can similarly prove Assumption A4 .
- Assumption A5: The policy π_ρ that updates the status with a probability of $\rho - \delta$ satisfies Assumption A5. Here, δ is a small positive number. Under this policy, $\bar{D}^\pi = \rho - \delta < \rho$ and $\bar{C}^\pi = E[\omega]\frac{1}{p_s(\rho-\delta)}\sigma^2 < \infty$.

Similarly, we can prove that Assumptions A1–A5 also holds for problem (9) when the context-aware weight is i.i.d. over time. $\quad \square$

Since Assumptions A1–A5 hold for problem (9), then according to Theorem 2.5 in [33], there exists an optimal stationary randomized policy for problem (9). Meanwhile, the optimal policy is a probabilistic combination of two stationary deterministic policies which only differ on at most one state.

Furthermore, according to Lemma 3.9 in [33], the two stationary deterministic policies each optimize the unconstrained cost in (10) with a different λ.

References

1. Talak, R.; Karaman, S.; Modiano, E. Speed limits in autonomous vehicular networks due to communication constraints. In Proceedings of the 2016 IEEE 55th Conference on Decision and Control (CDC), Las Vegas, NV, USA, 12–14 December 2016; pp. 4998–5003.
2. Hou, I.H.; Naghsh, N.Z.; Paul, S.; Hu, Y.C.; Eryilmaz, A. Predictive Scheduling for Virtual Reality. In Proceedings of the IEEE INFOCOM 2020-IEEE Conference on Computer Communications, Toronto, ON, Canada, 6–9 July 2020; pp. 1349–1358.
3. Kaul, S.; Yates, R.; Gruteser, M. Real-time status: How often should one update? In Proceedings of the 2012 Proceedings IEEE INFOCOM, Orlando, FL, USA, 25–30 March 2012; pp. 2731–2735.
4. Sun, Y.; Polyanskiy, Y.; Uysal-Biyikoglu, E. Remote estimation of the Wiener process over a channel with random delay. In Proceedings of the 2017 IEEE International Symposium on Information Theory (ISIT), Aachen, Germany, 25–30 June 2017; pp. 321–325.
5. Sun, Y.; Polyanskiy, Y.; Uysal, E. Sampling of the wiener process for remote estimation over a channel with random delay. *IEEE Trans. Inf. Theory* **2019**, *66*, 1118–1135. [CrossRef]
6. Jiang, Z.; Zhou, S. Status from a random field: How densely should one update? In Proceedings of the 2019 IEEE International Symposium on Information Theory (ISIT), Paris, France, 7–12 July 2019; pp. 1037–1041.
7. Bedewy, A.M.; Sun, Y.; Singh, R.; Shroff, N.B. Optimizing information freshness using low-power status updates via sleep-wake scheduling. In Proceedings of the Twenty-First International Symposium on Theory, Algorithmic Foundations, and Protocol Design for Mobile Networks and Mobile Computing, New York, NY, USA, 11–14 October 2020; pp. 51–60.
8. Ceran, E.T.; Gündüz, D.; György, A. Average age of information with hybrid ARQ under a resource constraint. *IEEE Trans. Wirel. Commun.* **2019**, *18*, 1900–1913. [CrossRef]
9. Sun, J.; Jiang, Z.; Krishnamachari, B.; Zhou, S.; Niu, Z. Closed-form Whittle's index-enabled random access for timely status update. *IEEE Trans. Commun.* **2019**, *68*, 1538–1551. [CrossRef]
10. Yates, R.D.; Kaul, S.K. Status updates over unreliable multiaccess channels. In Proceedings of the 2017 IEEE International Symposium on Information Theory (ISIT), Aachen, Germany, 25–30 June 2017; pp. 331–335.
11. Sun, J.; Wang, L.; Jiang, Z.; Zhou, S.; Niu, Z. Age-Optimal Scheduling for Heterogeneous Traffic with Timely Throughput Constraints. *IEEE J. Sel. Areas Commun.* **2021**, *39*, 1485–1498. [CrossRef]
12. Tang, H.; Wang, J.; Song, L.; Song, J. Minimizing age of information with power constraints: Multi-user opportunistic scheduling in multi-state time-varying channels. *IEEE J. Sel. Areas Commun.* **2020**, *38*, 854–868. [CrossRef]

13. Abdel-Aziz, M.K.; Samarakoon, S.; Liu, C.F.; Bennis, M.; Saad, W. Optimized age of information tail for ultra-reliable low-latency communications in vehicular networks. *IEEE Trans. Commun.* **2019**, *68*, 1911–1924. [CrossRef]
14. Devassy, R.; Durisi, G.; Ferrante, G.C.; Simeone, O.; Uysal-Biyikoglu, E. Delay and peak-age violation probability in short-packet transmissions. In Proceedings of the 2018 IEEE International Symposium on Information Theory (ISIT), Vail, CO, USA, 17–22 June 2018; pp. 2471–2475.
15. Inoue, Y.; Masuyama, H.; Takine, T.; Tanaka, T. A general formula for the stationary distribution of the age of information and its application to single-server queues. *IEEE Trans. Inf. Theory* **2019**, *65*, 8305–8324. [CrossRef]
16. Sun, Y.; Uysal-Biyikoglu, E.; Yates, R.D.; Koksal, C.E.; Shroff, N.B. Update or wait: How to keep your data fresh. *IEEE Trans. Inf. Theory* **2017**, *63*, 7492–7508. [CrossRef]
17. Zheng, X.; Zhou, S.; Jiang, Z.; Niu, Z. Closed-form analysis of non-linear age of information in status updates with an energy harvesting transmitter. *IEEE Trans. Wirel. Commun.* **2019**, *18*, 4129–4142. [CrossRef]
18. Kosta, A.; Pappas, N.; Ephremides, A.; Angelakis, V. Non-linear age of information in a discrete time queue: Stationary distribution and average performance analysis. In Proceedings of the ICC 2020—2020 IEEE International Conference on Communications (ICC), Dublin, Ireland, 7–11 June 2020; pp. 1–6.
19. Kosta, A.; Pappas, N.; Ephremides, A.; Angelakis, V. The cost of delay in status updates and their value: Non-linear ageing. *IEEE Trans. Commun.* **2020**, *68*, 4905–4918. [CrossRef]
20. Zhong, J.; Yates, R.D.; Soljanin, E. Two freshness metrics for local cache refresh. In Proceedings of the 2018 IEEE International Symposium on Information Theory (ISIT), Vail, CO, USA, 17–22 June 2018; pp. 1924–1928.
21. Maatouk, A.; Kriouile, S.; Assaad, M.; Ephremides, A. The age of incorrect information: A new performance metric for status updates. *IEEE/ACM Trans. Netw.* **2020**, *28*, 2215–2228. [CrossRef]
22. Kadota, I.; Sinha, A.; Uysal-Biyikoglu, E.; Singh, R.; Modiano, E. Scheduling policies for minimizing age of information in broadcast wireless networks. *IEEE/ACM Trans. Netw.* **2018**, *26*, 2637–2650. [CrossRef]
23. Song, J.; Gunduz, D.; Choi, W. Optimal scheduling policy for minimizing age of information with a relay. *arXiv* **2020**, arXiv:2009.02716.
24. Sun, Y.; Cyr, B. Information aging through queues: A mutual information perspective. In Proceedings of the 2018 IEEE 19th International Workshop on Signal Processing Advances in Wireless Communications (SPAWC), Kalamata, Greece, 25–28 June 2018; pp. 1–5.
25. Kam, C.; Kompella, S.; Nguyen, G.D.; Wieselthier, J.E.; Ephremides, A. Towards an effective age of information: Remote estimation of a markov source. In Proceedings of the IEEE INFOCOM 2018—IEEE Conference on Computer Communications Workshops (INFOCOM WKSHPS), Honolulu, HI, USA, 15–19 April 2018; pp. 367–372.
26. Zheng, X.; Zhou, S.; Niu, Z. Context-aware information lapse for timely status updates in remote control systems. In Proceedings of the 2019 IEEE Global Communications Conference (GLOBECOM), Waikoloa, HI, USA, 9–13 December 2019; pp. 1–6.
27. Zheng, X.; Zhou, S.; Niu, Z. Beyond age: Urgency of information for timeliness guarantee in status update systems. In Proceedings of the 2020 2nd IEEE 6G Wireless Summit (6G SUMMIT), Levi, Finland, 17–20 March 2020; pp. 1–5.
28. Zheng, X.; Zhou, S.; Niu, Z. Urgency of Information for Context-Aware Timely Status Updates in Remote Control Systems. *IEEE Trans. Wirel. Commun.* **2020**, *19*, 7237–7250. [CrossRef]
29. Ioannidis, S.; Chaintreau, A.; Massoulié, L. Optimal and scalable distribution of content updates over a mobile social network. In Proceedings of the IEEE INFOCOM 2009, Rio de Janeiro, Brazil, 19–25 April 2009; pp. 1422–1430.
30. Wang, L.; Sun, J.; Zhou, S.; Niu, Z. Timely Status Update Based on Urgency of Information with Statistical Context. In Proceedings of the 2020 32nd IEEE International Teletraffic Congress (ITC 32), Osaka, Japan, 22–24 September 2020; pp. 90–96.
31. Nayyar, A.; Başar, T.; Teneketzis, D.; Veeravalli, V.V. Optimal strategies for communication and remote estimation with an energy harvesting sensor. *IEEE Trans. Autom. Control* **2013**, *58*, 2246–2260. [CrossRef]
32. Cika, A.; Badiu, M.A.; Coon, J.P. Quantifying link stability in Ad Hoc wireless networks subject to Ornstein-Uhlenbeck mobility. In Proceedings of the ICC 2019—2019 IEEE International Conference on Communications (ICC), Shanghai, China, 20–24 May 2019; pp. 1–6.
33. Sennott, L.I. Constrained average cost Markov decision chains. *Probab. Eng. Inf. Sci.* **1993**, *7*, 69–83. [CrossRef]
34. Bertsekas, D.P. *Dynamic Programming and Optimal Control*; Athena Scientific: Belmont, MA, USA, 2000.
35. Sennott, L.I. Average cost optimal stationary policies in infinite state Markov decision processes with unbounded costs. *Oper. Res.* **1989**, *37*, 626–633. [CrossRef]
36. Liu, B.; Xie, Q.; Modiano, E. Rl-qn: A reinforcement learning framework for optimal control of queueing systems. *arXiv* **2020**, arXiv:2011.07401.
37. Chen, X.; Liao, X.; Bidokhti, S.S. Real-time Sampling and Estimation on Random Access Channels: Age of Information and Beyond. *arXiv* **2020**, arXiv:2007.03652.

Article

Relationship between Age and Value of Information for a Noisy Ornstein–Uhlenbeck Process

Zijing Wang, Mihai-Alin Badiu and Justin P. Coon *

Department of Engineering Science, University of Oxford, Oxford OX1 3PJ, UK;
zijing.wang@balliol.ox.ac.uk (Z.W.); mihai.badiu@eng.ox.ac.uk (M.-A.B.)
* Correspondence: justin.coon@eng.ox.ac.uk

Abstract: The age of information (AoI) has been widely used to quantify the information freshness in real-time status update systems. As the AoI is independent of the inherent property of the source data and the context, we introduce a mutual information-based value of information (VoI) framework for hidden Markov models. In this paper, we investigate the VoI and its relationship to the AoI for a noisy Ornstein–Uhlenbeck (OU) process. We explore the effects of correlation and noise on their relationship, and find logarithmic, exponential and linear dependencies between the two in three different regimes. This gives the formal justification for the selection of non-linear AoI functions previously reported in other works. Moreover, we study the statistical properties of the VoI in the example of a queue model, deriving its distribution functions and moments. The lower and upper bounds of the average VoI are also analysed, which can be used for the design and optimisation of freshness-aware networks. Numerical results are presented and further show that, compared with the traditional linear age and some basic non-linear age functions, the proposed VoI framework is more general and suitable for various contexts.

Keywords: value of information; age of information; noisy Ornstein–Uhlenbeck process

Citation: Wang, Z.; Badiu, M.-A.; Coon, J.P. Relationship between Age and Value of Information for a Noisy Ornstein-Uhlenbeck Process. *Entropy* **2021**, *23*, 940. https://doi.org/10.3390/e23080940

Academic Editors: Anthony Ephremides and Yin Sun

Received: 18 June 2021
Accepted: 20 July 2021
Published: 23 July 2021

Publisher's Note: MDPI stays neutral with regard to jurisdictional claims in published maps and institutional affiliations.

1. Introduction

Nowadays, there are more and more real-time monitoring and control applications, such as industrial control, Internet of Things, autonomous driving and so on. Such applications are modelled as status update systems in which sensors need to continuously monitor a targeted random process, and the sampled status updates are required to be transmitted through the communication network to a remote destination in a timely manner to enable precise control and management. Therefore, the freshness of data has emerged as an important part of network research.

The age of information (AoI) is proposed as a novel end-to-end metric in [1,2] to evaluate the timeliness of status updates from the receiver's perspective. The AoI is defined as the time difference between the current time and the generation time of the last received status update. The AoI and its variants (e.g., the average AoI and the peak AoI) are widely used as tools to improve the system-level data freshness by optimising the sampling and link scheduling in a variety of emerging networks [3–8]. Moreover, there are many works exploring the AoI in the context of different queue systems. General expressions of the average AoI were derived in [1], and the stationary distribution of the AoI was studied in [9,10] for first-come-first-serve (FCFS) M/M/1, M/D/1 and D/M/1 queue disciplines. The statistical characterisation and violation probability of the AoI were treated in [11,12] for last-come-first-serve (LCFS) queue disciplines. The influence of the queue's buffer size, packet management and service pre-emption on the AoI and its distribution was investigated in [13–15].

However, the basic notion of the AoI grows linearly with a unit slope as time goes by, and it is independent of the context and the inherent characterisation of the targeted random process (e.g., the correlation property of the underlying source data). In light of these issues, the concept of the value of information (VoI) has begun to be studied, which emphasises the idea that in some cases, old information may still have value while even fresh information may hold little value, as different sources require different update frequency.

The idea of a non-linear age has become a common approach to evaluate information value [16]. The concept of the "age penalty" was proposed in [17], where it was assumed to be a non-decreasing function of the AoI and provided a general way to measure the dissatisfaction of the staleness of information. Closed-form expressions of the general penalty functions were studied in energy harvesting networks in [18]. In [19–21], three specific penalty functions (exponential, linear and logarithmic functions) and their statistical characterisations were further investigated. Moreover, the connection of the AoI with signal processing and information theory has received much attention, as it can provide a theoretical basis for non-linear age functions. The mean square error (MSE) for remote estimation can add non-linearity, and it was used to evaluate the information value in [22–25]. The relationship between the AoI and the MSE was studied in the Wiener process [22] and the Ornstein–Uhlenbeck (OU) process [23]. It is interesting to note that the age-optimal sampling policy was not equivalent to the MSE-optimal sampling. The mutual information was utilised in [26] to quantify the timeliness of data, and the optimal sampling policy was explored for a Markov source. In [26], the samples were assumed to be directly observable when they were received. In practice, samples at the source can be corrupted by noise, errors or measurements, and thus, they may be latent at the receiver. However, properties of the information value in hidden Markov models have not been explicitly studied. Furthermore, the authors in [20] proposed that age penalty functions can be chosen and adjusted, according to the autocorrelation of the underlying random process, but theoretical interpretation or formal justification for how to choose non-linear functions and how they relate to the correlation of the underlying process were not provided.

In our previous work [27], we proposed a mutual information-based value of information framework for hidden Markov models and started to explore it in the context of a noisy OU process. We obtained the closed-form expression of the VoI, which relates to the correlation of the process under observation at the source and the noise in the transmission environment, but we did not investigate its relationship to the AoI and its statistical characterisations in more depth. In this paper, the connection of the proposed VoI with the AoI is studied for a noisy OU process. The OU process is considered, as it is an important continuous-time, stationary, Markov and Gaussian random process, which is practical to represent many real-world applications [28]. For example, it can be used to model the mobility of a drone that moves towards a target point but experiences positional fluctuations in unmanned aerial vehicle (UAV) networks. In this work, we give the formal justification for how the correlation and the noise in the context affect the VoI and its relationship to AoI, and obtain the functional dependency between them. We show that the proposed VoI framework is a general one that includes the special sample cases given in [20], and it is suitable to be applied in different network settings. Moreover, we study the VoI in a FCFS M/M/1 queue model, deriving the probability density function (PDF), cumulative distribution function (CDF), average VoI and moment-generating function (MGF). We also derive the upper and lower bounds of the average VoI, which are tractable and useful for the design and optimisation of freshness-aware applications. Through all of these results, we provide a clear statistical framework linking the VoI to the AoI and a formal justification for the selection of non-linear age functions.

The rest of this paper is organised as follows. The VoI formalism in the noisy OU process model is introduced in Section 2. Relationships between the VoI and the AoI for different network settings are investigated in Section 3. The statistical characterisation of the VoI in the FCFS M/M/1 queue model is given in Section 4. Numerical results are provided in Section 5. Conclusions are drawn in Section 6.

2. VoI with Application to OU Processes

Here, we provide a brief introduction to the VoI framework that is used in this paper, and we recount key results reported in [27] that will be used later in the paper.

2.1. VoI Definition

We consider a real-time status update system with a pair of transmitter and receiver nodes. The source samples the data of a targeted random process $\{X_t\}$ and sends status updates to the receiver node for further analysis. Denote X_{t_i} as the i-th status update of the underlying random process. Denote $Y_{t_i'}$ as the corresponding observation which is captured in the observed random process $\{Y_t\}$. Here, t_i represents the sampling time of the i-th sample, and t_i' represents its receiving time. We consider a latent variable model in which the observation $Y_{t_i'}$ may be different from the initial value, as the update X_{t_i} can be negatively affected by the transmission noise, error or measurement when it is received by the destination in the real world.

In this paper, the notion of the value of information is defined as the mutual information between the current status of the process under observation at the transmitter and a sequence of noisy measurements recorded by the receiver. Specifically, the VoI at the time t is given as the following:

$$v(t) = I(X_t; Y_{t_n'}, \ldots, Y_{t_{n-m+1}'}), \quad t > t_n'. \tag{1}$$

Here, n is denoted as the index of the last received update during the period $(0, t)$. We look back in time, and the most recent m of n noisy observations ($m \leq n$) are utilised to evaluate the information value. This definition gives the interpretation of the reduction in the uncertainty of the current hidden status, given that we have some past noisy measurements.

2.2. Noisy OU Process Model

We assume the random process $\{X_t\}$ under observation is an Ornstein–Uhlenbeck process, which can be used to represent the mean reversion behaviour in practice. The underlying OU process satisfies the following stochastic differential equation:

$$dX_t = \kappa(\theta - X_t)\,dt + \sigma\,dW_t. \tag{2}$$

Here, κ ($\kappa > 0$) is the rate of mean reversion, which can be used to represent the correlation property of status updates, θ is the long-term mean, σ is the volatility of the random fluctuation, and $\{W_t\}$ is the Wiener process. We assume that the initial value X_0 is a Gaussian variable with mean θ and variance $\frac{\sigma^2}{2\kappa}$.

We assume this OU process $\{X_t\}$ is observed through an additive noise channel, and the corresponding noisy observation is defined as the following:

$$Y_{t_i'} = X_{t_i} + N_{t_i'}, \tag{3}$$

where $N_{t_i'}$ is the sample of the noise process taken by the receiver at t_i'. Here, the samples $\{N_{t_i'}\}$ are assumed to be independent Gaussian variables with zero mean and constant variance σ_n^2. In reality, it can represent the measurement or error that undermines the status update X_{t_i} of the underlying OU process.

2.3. VoI for the Noisy OU Process

Based on the model we described, the samples of the underlying OU process are jointly Gaussian and the noise samples are also Gaussian variables, which allow us to calculate the VoI in our previous work [27]. The VoI for the noisy OU process is given as follows:

$$v(t) = -\frac{1}{2}\log\left(1 - e^{-2\kappa(t-t_n)} + e^{-2\kappa(t-t_n)}\frac{\det(\mathbf{A}_{mm})}{\gamma \det(\mathbf{A})}\right), \quad t > t_n'. \tag{4}$$

Here, $\mathbf{A} = \sigma_n^2 \mathbf{\Sigma_X}^{-1} + \mathbf{I}$ where $\mathbf{\Sigma_X}^{-1}$ represents the covariance matrix of the vector $\mathbf{X} = [X_{t_{n-m+1}}, \ldots, X_{t_n}]^{\mathrm{T}}$, and $\mathbf{I}$ represents the identity matrix of size m. $\mathbf{A}_{ij}$ represents the $(m-1) \times (m-1)$ matrix constructed by deleting the ith row and the jth column of the matrix $\mathbf{A}$, and γ is denoted as the ratio of the variance of the OU process and the variance of the noise, i.e., the following:

$$\gamma = \frac{\mathrm{Var}[X_{t_i}]}{\mathrm{Var}[N_{t_i'}]} = \frac{\sigma^2}{2\kappa\sigma_n^2}. \tag{5}$$

The parameter γ is similar to the concept of the signal-to-noise ratio (SNR) in a communication system. In the following, the concept "SNR" refers to this parameter, which is used to compare the randomness in the OU process and the noise in the communication channel.

3. Relationship between VoI and AoI

The result given in (4) shows the general expression of the VoI in the noisy OU process. In this section, we consider a special case with a single observation ($m = 1$) and explore the relationship between the proposed VoI and the AoI. In the definition of the AoI, we consider that the time instant t_n is fixed, i.e., we view the AoI as deterministic. What we do here is to create a relationship between the VoI and the conditional AoI (i.e., the AoI conditioned on the most recent sample time).

The concept of the AoI is given as follows [1]:

$$A(t) = t - t_n, \quad t > t_n'. \tag{6}$$

In the noisy OU process, when $m = 1$, the VoI in (4) can be simplified as follows:

$$v(t) = -\frac{1}{2} \log \left(1 - \frac{\gamma}{1+\gamma} e^{-2\kappa(t-t_n)} \right), \quad t > t_n', \tag{7}$$

which is supported by the following:

$$0 \le v \le \frac{1}{2} \log(1+\gamma). \tag{8}$$

Therefore, the VoI is further written as a function of the AoI. Let $a = A(t)$; then, the VoI can be written as follows:

$$V(a) = -\frac{1}{2} \log \left(1 - \frac{\gamma}{1+\gamma} e^{-2\kappa a} \right). \tag{9}$$

The VoI in (9) and its relationship to the AoI can be largely affected by system parameters. Fixing the random fluctuation parameter σ^2 of the OU process, the SNR γ relates to two parameters, κ and σ_n^2. κ can be used to represent the correlation property of the underlying OU process. If κ is small, the status updates are highly correlated; as κ increases, they become less correlated. σ_n^2 represents the noise level in the transmission environment. If σ_n^2 is small, the underlying hidden Markov process is dominant, and the VoI approaches its Markov counterpart in the OU model; otherwise, the noise process is dominant. In the following part, the relationship between the VoI and AoI in different SNR regimes is investigated, and we have the following corollaries.

Corollary 1. *In the low SNR regime, the VoI can be approximated as an exponential function of the AoI, which is given by the following:*

$$V(a) \approx \frac{\gamma}{2(1+\gamma)} e^{-2\kappa a}. \tag{10}$$

Proof. In the low SNR regime (small γ), large κ and $\sigma_n^2 > 0$ (or large σ_n^2 and $\kappa > 0$) can lead to small SNR in (5). When γ approaches 0, the term $\frac{\gamma}{1+\gamma}e^{-2\kappa a}$ in (9) is small. For small x, we have $\log(1 + x) \approx x$, thus the result in (10) is obtained. $\square$

In the low SNR regime, the dependency between the VoI and AoI is exponential. Less correlated samples or large noise can negatively affect the VoI at the receiver, thus the approximated VoI decreases faster as the AoI increases. For a less correlated data source, even fresh updates may contain little valuable information about the underlying OU process. For a high level of noise, status updates are corrupted, due to the indirect observation.

Corollary 2. *In the high SNR regime resulting from high correlation, the VoI can be approximated as a logarithmic function of the AoI, which is given by the following:*

$$V(a) \approx -\frac{1}{2}\log(2\kappa\gamma a + 1) + \frac{1}{2}\log(1 + \gamma). \tag{11}$$

Proof. For small x, we have $e^x \approx 1 + x$. Therefore, when $\kappa \to 0$ in (9), $e^{-2\kappa a} \approx 1 - 2\kappa a$. $\square$

For highly correlated status updates, the VoI is expressed as a logarithmic function, and this means that the VoI decreases slower as the AoI increases. In this case, correlated updates can be transmitted under good channel conditions, thus old samples may still hold enough valuable information.

Corollary 3. *In the intermediate SNR regime where $\kappa \to 0$, $\sigma_n^2 \to \infty$ with $\kappa\sigma_n^2$ being constant, the VoI can be approximated as a linear function of the AoI, which is given by the following:*

$$V(a) \approx -\kappa\gamma a + \frac{1}{2}\log(1 + \gamma). \tag{12}$$

In the intermediate SNR regime where $\kappa \to \infty$, $\sigma_n^2 \to 0$ with $\kappa\sigma_n^2$ being constant, the VoI can be approximated as an exponential function of the AoI, which is given by the following:

$$V(a) \approx \frac{\gamma}{2(1 + \gamma)}e^{-2\kappa a}. \tag{13}$$

Proof. The result in (12) can be derived from Corollary 2 directly. When $\sigma_n^2 \to \infty$, the term $2\kappa\gamma a$ in (11) is small. Therefore, we have $\log(2\kappa\gamma a + 1) \approx 2\kappa\gamma a$. The result in (13) matches Corollary 1. When $\kappa \to \infty$, the term $e^{-2\kappa a}$ in (9) is small. For small x, we have $\log(1 + x) \approx x$, thus the result in (13) is obtained. $\square$

The three corollaries stated above provide the compelling insight into the adoption of non-linear AoI functions. In some existing works, exponential and logarithmic non-linear age functions are widely utilised to measure the information value, but they do not give the formal justification for why these functions are selected. Corollaries 1 to 3 provide a theoretic interpretation and explain how the correlation, noise and SNR affect the VoI and its relationship to the AoI in the noisy OU process. Generally, low SNR and high SNR conditions yield exponential and logarithmic relationships. The intermediate SNR regime yields an exponential or linear relationship, which depends on the value of noise and correlation. Therefore, the proposed VoI framework is more complete, general and appropriate to measure the timeliness of information in different SNR regimes.

4. Statistical Properties of the VoI in the M/M/1 Queue Model

Equations (10)–(13) show general relationships between the VoI and the AoI in the noisy OU process. In this section, we relax the "fixed time instants" restriction given in Section 3 and view the AoI as a random variable to study the distribution of the VoI. We explore the VoI in a specific FCFS M/M/1 queue system and derive its statistical properties (including the PDF, CDF, expectation value and MGF).

4.1. Distribution of the VoI

We assume that status updates of the underlying OU process are transmitted through a FCFS M/M/1 queue in which they are sampled as a rate λ Poisson process, and the service time is a rate μ exponential process ($\lambda < \mu$). Let random variables $S_i = t_i - t_{i-1}$ ($2 \leq i \leq n$) be the sampling interval of two packets, which are independent and identically distributed (i.i.d.) exponential random variables with $E[S] = \frac{1}{\lambda}$. Similarly, service times of status updates are also i.i.d. exponential random variables with mean $\frac{1}{\mu}$. In the example of the M/M/1 queue, the stationary distribution of the AoI was studied in [11] and the PDF and CDF of the AoI are given as follows:

$$f_A(a) = \mu \left[\frac{\mu - \lambda}{\mu} e^{-(\mu - \lambda)a} - \left(\frac{\mu}{\mu - \lambda} + \lambda a - \frac{\lambda}{\mu} \right) e^{-\mu a} + \frac{\lambda}{\mu - \lambda} e^{-\lambda a} \right], \tag{14}$$

$$F_A(a) = 1 - e^{-(\mu - \lambda)a} + \left(\frac{\mu}{\mu - \lambda} + \lambda a \right) e^{-\mu a} - \frac{\mu}{\mu - \lambda} e^{-\lambda a}. \tag{15}$$

It can be seen that the distribution of the AoI only relates to the queue discipline, which means that it is independent of the inherent statistical characterisations of the underlying random process. As for the distribution of the VoI of a latent OU process with a single observation, we can state the following propositions.

Proposition 1. *In the M/M/1 queue model, the PDF of the VoI for the noisy OU process is given by the following:*

$$f_V(v) = \frac{\mu e^{-2v}}{\kappa(1 - e^{-2v})} \left[\frac{\mu - \lambda}{\mu} r(v)^{\frac{\mu - \lambda}{2\kappa}} - \left(\frac{\mu}{\mu - \lambda} - \frac{\lambda}{\mu} - \frac{\lambda}{2\kappa} \log r(v) \right) r(v)^{\frac{\mu}{2\kappa}} \right.$$
$$\left. + \frac{\lambda}{\mu - \lambda} r(v)^{\frac{\lambda}{2\kappa}} \right], \tag{16}$$

where $r(v)$ is denoted as follows:

$$r(v) = \frac{(1 + \gamma)(1 - e^{-2v})}{\gamma}. \tag{17}$$

Proof. Since (9) is a monotonically decreasing function, the PDF of the VoI can be calculated by the following:

$$f_V(v) = f_A(V^{-1}(v)) \left| \frac{\mathrm{d}}{\mathrm{d}v} \left(V^{-1}(v) \right) \right|. \tag{18}$$

Here, V^{-1} denotes the inverse function of the VoI given in (9), which can be written as follows:

$$V^{-1}(v) = -\frac{1}{2\kappa} \log \left(\frac{(1 + \gamma)(1 - e^{-2v})}{\gamma} \right), \tag{19}$$

and we have the following:

$$\frac{\mathrm{d}}{\mathrm{d}v} \left(V^{-1}(v) \right) = -\frac{e^{-2v}}{\kappa(1 - e^{-2v})}. \tag{20}$$

Therefore, the PDF of the VoI given in (16) is obtained by substituting (19), (14) and (20) into (18). $\square$

Proposition 2. *In the M/M/1 queue model, the CDF of the VoI for the noisy OU process is given as follows:*

$$F_V(v) = r(v)^{\frac{\mu - \lambda}{2\kappa}} - \left(\frac{\mu}{\mu - \lambda} - \frac{\lambda}{2\kappa} \log r(v) \right) r(v)^{\frac{\mu}{2\kappa}} + \frac{\mu}{\mu - \lambda} r(v)^{\frac{\lambda}{2\kappa}}. \tag{21}$$

Proof. The CDF is obtained directly by the integral of the PDF, i.e., $F_V(v) = \mathrm{P}(V \leq v) = \int_0^v f_V(x)\,dx.$ □

Propositions 1 and 2 show that the distribution of the VoI relates to the sampling rate λ, service rate μ, correlation parameter κ and noise parameter σ_n^2, while the AoI distribution only relates to parameters λ and μ for the M/M/1 queue system.

The CDF of the VoI given in Proposition 2 can be interpreted as the "VoI outage probability", i.e., the probability that the VoI is smaller than a given threshold. It is interesting to note that Proposition 2 implies that the VoI outage probability is a monotonically decreasing function of the service rate μ, and it converges to $r(v)^{\frac{\lambda}{2\kappa}}$ as μ goes to infinity. The proof of this is given in Appendix A.1. The reason for this decreasing nature of the VoI with μ is predictable because one would expect the information value to increase if the service time in the queue reduces.

Proposition 2 also implies that the VoI outage probability first decreases and then increases as the sampling rate λ increases. The optimal sampling rate λ^* satisfies $\frac{\partial \mathrm{P}(V \leq v)}{\partial \lambda}\big|_{\lambda = \lambda^*} = 0$. The proof of this is provided in Appendix A.2. It is not surprising that small sampling rate λ can lead to high outage, due to the lack of fresh updates at the source. It is interesting to find that large sampling rate can also lead to high outage probability, due to the traffic congestion in the queue.

4.2. Moments and Bounds

In this subsection, we derive the expectation and two bounds of the VoI with a single observation, and calculate the moment-generating function of the VoI. We can state the following two propositions.

Proposition 3. *In the M/M/1 queue model, the average VoI for the noisy OU process is given as the following:*

$$
\begin{aligned}
\mathrm{E}[V] = \frac{1}{2}\Bigg[&\log(1+\gamma) - g_1\left(\frac{\gamma}{1+\gamma}, \frac{\mu-\lambda}{2\kappa}\right) - \frac{\mu}{\mu-\lambda} g_1\left(\frac{\gamma}{1+\gamma}, \frac{\lambda}{2\kappa}\right) \\
&+ \left(\frac{\mu}{\mu-\lambda} + \frac{\lambda}{2\kappa}\log\frac{\gamma}{1+\gamma}\right) g_1\left(\frac{\gamma}{1+\gamma}, \frac{\mu}{2\kappa}\right) - \frac{\lambda}{2\kappa} g_2\left(\frac{\gamma}{1+\gamma}, \frac{\mu}{2\kappa}\right)\Bigg],
\end{aligned}
\tag{22}
$$

where two functions $g_1(x,y)$ and $g_2(x,y)$ are defined for $x > 0$ and $y > 0$ with the following:

$$
g_1(x,y) = \frac{1}{x^y}\int_0^x \frac{z^y}{1-z}\,dz,
\tag{23}
$$

$$
g_2(x,y) = \frac{1}{x^y}\int_0^x \frac{z^y \log z}{1-z}\,dz.
\tag{24}
$$

Moreover, the average VoI is lower bounded by the following:

$$
\mathrm{E}[V] \geq -\frac{1}{2}\log\left[1 - \frac{\gamma}{1+\gamma}\left(\frac{\frac{\mu-\lambda}{2\kappa}}{\frac{\mu-\lambda}{2\kappa}+1} - \frac{\frac{\mu-\lambda}{2\kappa}\left(\frac{\mu+\lambda}{2\kappa}+1\right)}{\left(\frac{\mu}{2\kappa}+1\right)^2\left(\frac{\lambda}{2\kappa}+1\right)}\right)\right],
\tag{25}
$$

and it is upper bounded by the following:

$$
\mathrm{E}[V] \leq \frac{1}{2}\left[H\left(\frac{\mu-\lambda}{2\kappa}\right) + \frac{\mu}{\mu-\lambda}H\left(\frac{\lambda}{2\kappa}\right) - \frac{\mu}{\mu-\lambda}H\left(\frac{\mu}{2\kappa}\right) + \frac{\lambda}{2\kappa}\psi^{(1)}\left(1+\frac{\mu}{2\kappa}\right)\right].
\tag{26}
$$

Here, $H(\cdot)$ represents the harmonic number and the integral representation is given by the following: $H(x) = \int_0^1 \frac{1-z^x}{1-z}\,dz$ [29]. $\psi^{(1)}(\cdot)$ represents the first order polygamma function which is given by $\psi^{(1)}(x) = -\int_0^1 \frac{z^{x-1}\log z}{1-z}\,dz$ [30].

Proof. See Appendix B. □

This proposition gives two bounds of the average VoI in the noisy OU process. Compared with the general average VoI, the bounds are more tractable and may be useful for network design and optimisation. The details of the bounds are given in Appendix B as stated above.

The lower bound is based on Jensen's inequality. The equality holds if the VoI is a linear function on the Laplace transform of the AoI ($\mathrm{E}[e^{-2\kappa a}]$). In Corollary 1, we show that the dependence between the VoI and $\mathrm{E}[e^{-2\kappa a}]$ is approximately linear under the low SNR condition. Therefore, the average VoI approaches this lower bound in the low SNR regime. Moreover, as stated in Appendix B, the upper bound is based on the average VoI in the Markov model. Hence, in the high SNR regime, the upper bound is tight.

Proposition 4. *In the M/M/1 queue, the MGF of the VoI for the noisy OU process is given as follows:*

$$M_v(t) = {}_2F_1\left(\frac{\mu-\lambda}{2\kappa},\frac{t}{2};\frac{\mu-\lambda}{2\kappa}+1;\frac{\gamma}{1+\gamma}\right) + \frac{\mu}{\mu-\lambda}{}_2F_1\left(\frac{\lambda}{2\kappa},\frac{t}{2};\frac{\lambda}{2\kappa}+1;\frac{\gamma}{1+\gamma}\right)$$
$$-\left(\frac{\mu}{\mu-\lambda}-\frac{\lambda}{\mu}\right){}_2F_1\left(\frac{\mu}{2\kappa},\frac{t}{2};\frac{\mu}{2\kappa}+1;\frac{\gamma}{1+\gamma}\right) - \frac{\lambda}{\mu}{}_3F_2\left(\frac{\mu}{2\kappa},\frac{\mu}{2\kappa},\frac{t}{2};\frac{\mu}{2\kappa}+1,\frac{\mu}{2\kappa}+1;\frac{\gamma}{1+\gamma}\right). \quad (27)$$

Here, ${}_pF_q(a_1,\ldots,a_p;b_1,\ldots,b_q;z)$ represents the generalised hypergeometric function which is given by the following series:

$${}_pF_q(a_1,\ldots,a_p;b_1,\ldots,b_q;z) = \sum_{n=0}^{\infty} \frac{(a_1)_n \cdots (a_p)_n}{(b_1)_n \cdots (b_q)_n} \frac{z^n}{n!}. \quad (28)$$

where $(\cdot)_n$ represents the Pochhammer symbol, which is given as follows:

$$(x)_n = \begin{cases} 1 & n = 0 \\ \prod\limits_{i=0}^{n-1}(x-i) & n \geq 1 \end{cases}. \quad (29)$$

Proof. See Appendix C. □

Moments of the VoI can be obtained by derivatives of the MGF at $t = 0$. The average VoI given in Proposition 3 is the first-order moment and can be derived from the MGF directly. Using this MGF, higher order moments can also be used for the system design and optimisation instead of just utilising the average value.

5. Numerical Results

In this section, the relationship between the VoI and AoI and the distribution of the VoI are investigated through Monte Carlo simulations. In the simulation, the volatility parameter σ^2 of the OU model is fixed and set as 1. The sampling times $\{t_i\}$ are randomly generated by the rate λ Poisson process. The service times of each sample are randomly generated by the rate μ exponential process. We set time $t = 100$. For each running round, we record the sampling time of the most recent received update as t_n, and the AoI and the VoI are calculated by (6) and (7), respectively.

Figures 1–3 show the non-linear relationships between the VoI and the AoI under low, high and intermediate SNR conditions, respectively. Figures 4–7 illustrate the distribution of the VoI, including the PDF, CDF and the outage probability. Figures 8–11 provide the numerical results about the VoI expectation and bounds.

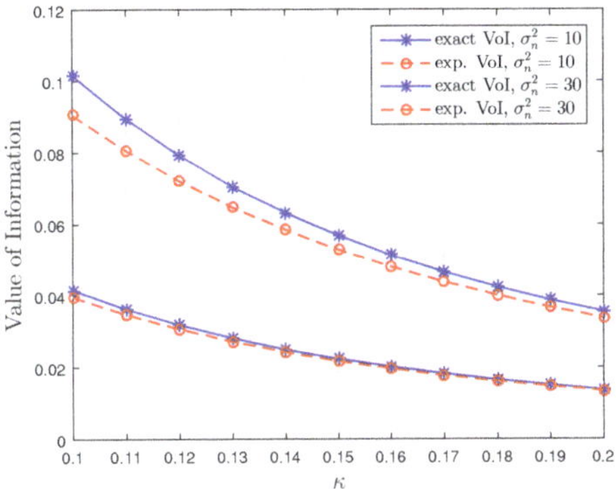

Figure 1. Low SNR regime: Comparison of the exact VoI and the exponential VoI versus κ for $\sigma_n^2 \in \{10, 30\}$ at $t = 100$, sampling rate $\lambda = 0.5$ and service rate $\mu = 1$.

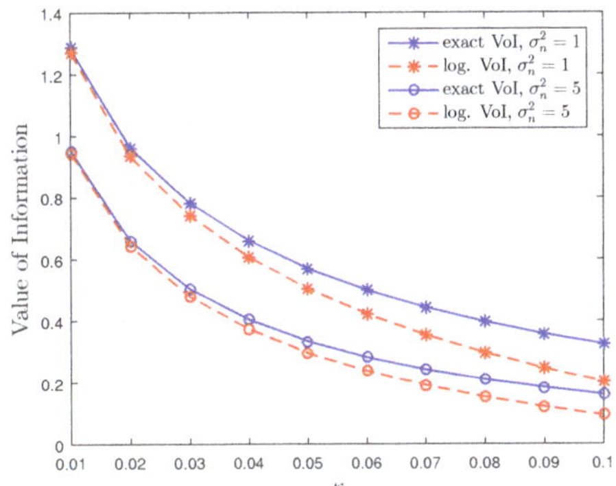

Figure 2. High SNR regime: Comparison of the exact VoI and the logarithmic VoI versus κ for $\sigma_n^2 \in \{1, 5\}$ at $t = 100$, sampling rate $\lambda = 0.5$ and service rate $\mu = 1$.

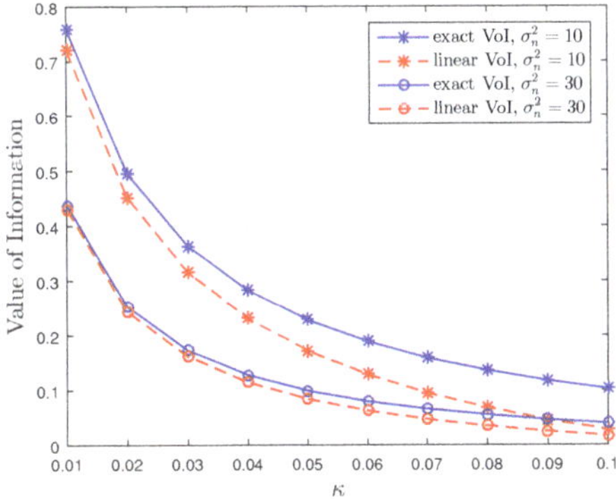

Figure 3. Intermediate SNR regime: Comparison of the exact VoI and the linear VoI versus κ for $\sigma_n^2 \in \{10, 30\}$ at $t = 100$, sampling rate $\lambda = 0.5$ and service rate $\mu = 1$.

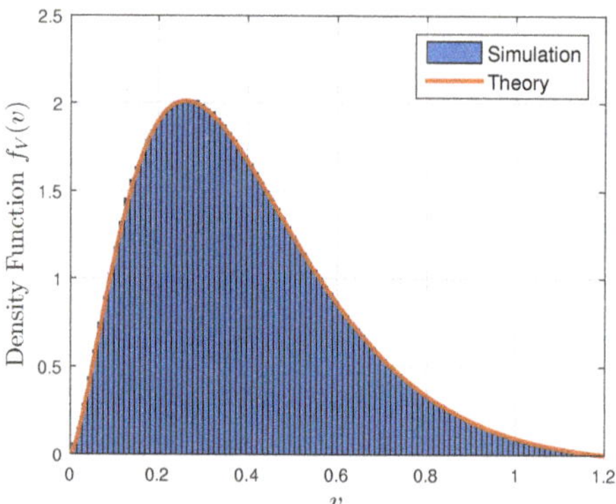

Figure 4. The density function of the VoI; correlation parameter $\kappa = 0.1$, noise parameter $\sigma_n^2 = 0.5$, sampling rate $\lambda = 0.5$ and service rate $\mu = 1$.

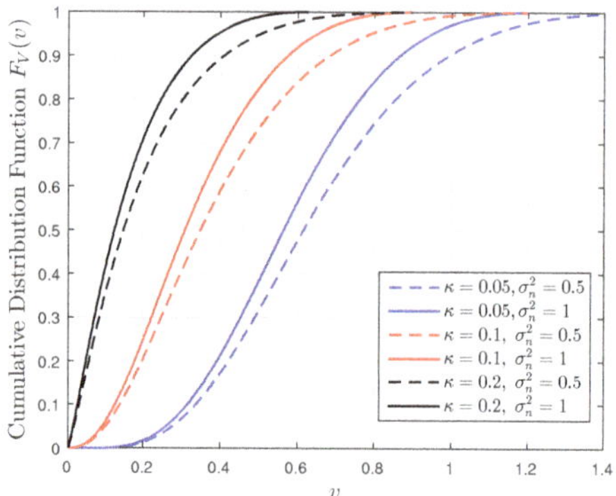

Figure 5. The cumulative distribution function of the VoI versus v for $\sigma_n^2 \in \{0.5, 1\}$ and $\kappa \in \{0.05, 0.1, 0.2\}$; sampling rate $\lambda = 0.5$ and service rate $\mu = 1$.

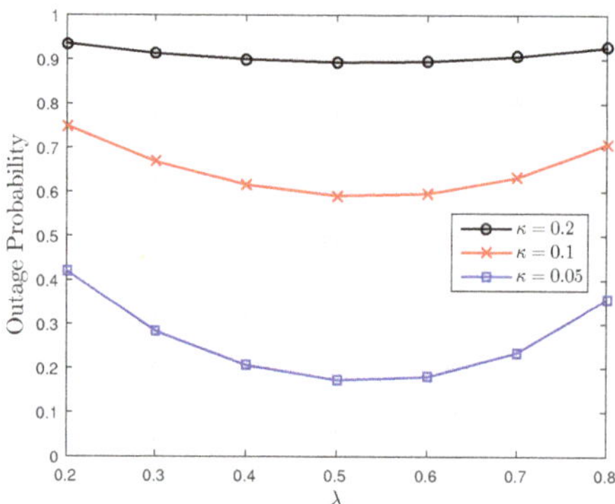

Figure 6. The VoI outage probability versus λ for $\kappa \in \{0.05, 0.1, 0.2\}$; threshold $v = 0.4$, noise parameter $\sigma_n^2 = 0.5$ and service rate $\mu = 1$.

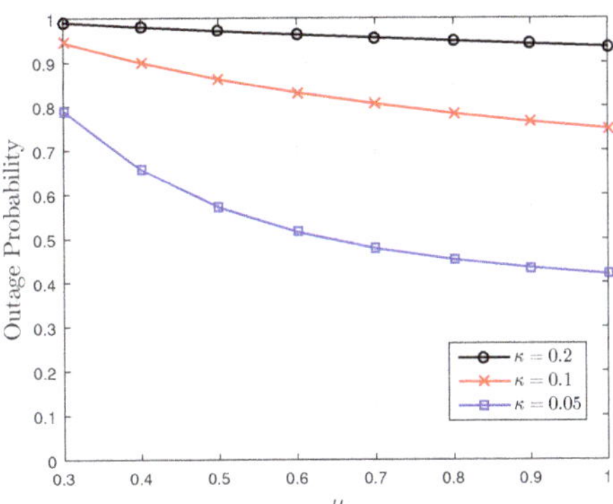

Figure 7. The VoI outage probability versus μ for $\kappa \in \{0.05, 0.1, 0.2\}$; threshold $v = 0.4$, noise parameter $\sigma_n^2 = 0.5$ and sampling rate $\lambda = 0.2$.

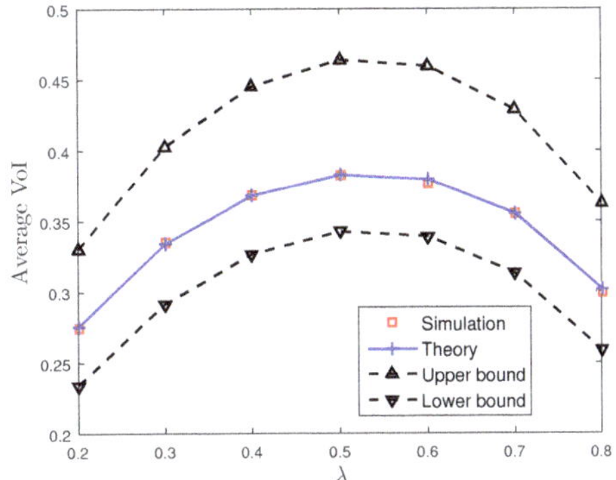

Figure 8. The average VoI and its bounds versus the sampling rate λ; correlation parameter $\kappa = 0.1$, noise parameter $\sigma_n^2 = 0.5$ and service rate $\mu = 1$.

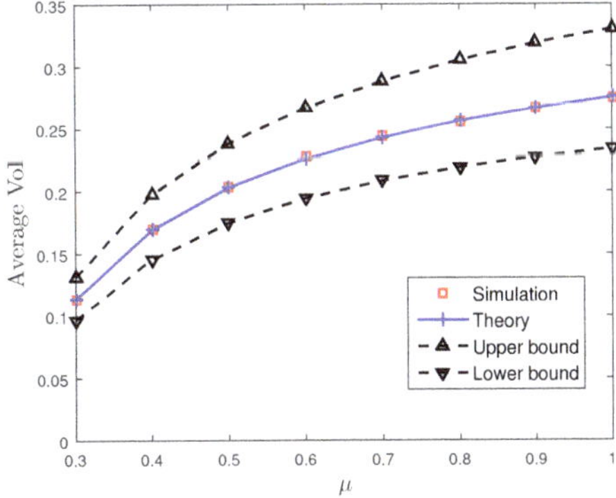

Figure 9. The average VoI and its bounds versus the service rate μ; correlation parameter $\kappa = 0.1$, noise parameter $\sigma_n^2 = 0.5$ and sampling rate $\lambda = 0.2$.

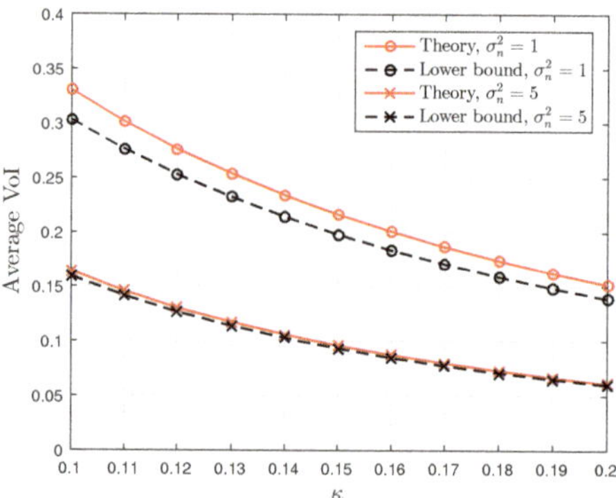

Figure 10. The average VoI and the lower bound versus κ for $\sigma_n^2 \in \{1, 5\}$; sampling rate $\lambda = 0.5$ and service rate $\mu = 1$.

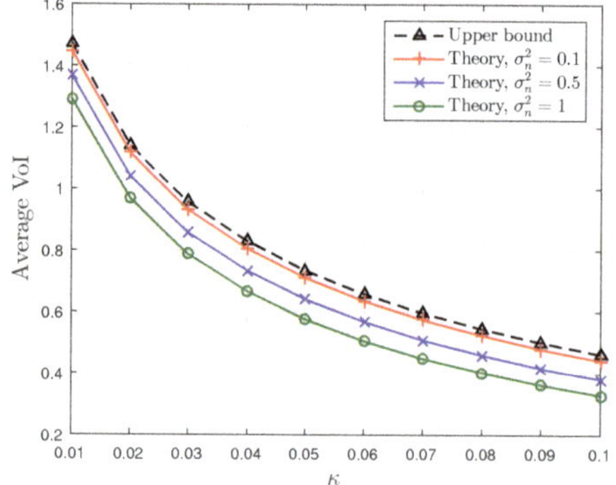

Figure 11. The average VoI and the upper bound versus κ for $\sigma_n^2 \in \{0.1, 0.5, 1\}$; sampling rate $\lambda = 0.5$ and service rate $\mu = 1$.

Figures 1–3 compare the exact VoI in (7) and the approximated VoI for different SNR regimes which are given in (10) to (12). Figure 1 shows that the exponential approximation is suitable when updates are less correlated and the noise is large. Figure 2 shows the opposite behaviour. The logarithmic approximation is more accurate when κ and σ_n^2 are small. Figure 3 shows that the linear approximation is accurate when κ is small but σ_n^2 is large. These results verify the functional dependencies between VoI and the AoI, which are discussed in Corollaries 1–3, illustrating that the low, high and intermediate SNR conditions yield exponential, logarithmic and linear relationships.

Figure 4 gives the numerical validation of the theoretical PDF given in Proposition 1 and the density of the discrete path of the VoI obtained from the Monte Carlo simulations. Figures 5–7 show the VoI outage probability given in Proposition 2 for different system parameters. In Figure 5, the VoI outage probability is high when the status updates are less correlated or when the system experiences large noise. For a particular threshold v, Figure 6 shows that either a too-small or too-large sampling rate can lead to a large VoI outage probability. Fixing μ, small λ means that we do not have sufficient newly generated status updates about the underlying OU process for prediction. Large λ means that enough newly generated updates have been sampled at the source, but they have to wait for a longer time, due to the packet congestion in the FCFS queue. Figure 7 shows that VoI outage probability decreases as the service rate μ increases. In the M/M/1 model, λ is smaller than μ. Fixing λ, large μ means that status updates can be served and transmitted more quickly, thus the receiver can hold more valuable information about the underlying process. These two figures verify the discussion given in Proposition 2.

Figures 8 and 9 show the effect of the sampling rate and service rate on the average VoI and its bounds given in Proposition 3. The average VoI and its bounds first increase and then decrease as λ increases, and they increase as μ increases. This behaviour is similar to the VoI outage and can be explained similar to Figures 6 and 7. Moreover, it can be seen that the theoretical average VoI is consistent with the result obtained from the Monte Carlo simulations.

Figures 10 and 11 plot the theoretical average VoI in (22) and the lower and upper bounds in (25) and (26) for different κ and σ_n^2. Small noise σ_n^2 and small κ can lead to large average VoI. In Figure 10, the gap between the exact value and the lower bound is small for large σ_n^2, and it decreases as κ increases. The gap between the exact value and the upper bound in Figure 11 shows the opposite behaviour; the gap narrows as σ_n^2 decreases. These two figures verify the discussion given in Proposition 3, illustrating that the average VoI approaches lower and upper bounds in low and high SNR regimes, respectively.

6. Conclusions

In this paper, we investigated the dependency between the proposed VoI and the AoI in a noisy OU process. The VoI is defined as the mutual information between the current status of the underlying random process and noisy observations captured by the receiver. Functional relationships between the VoI and the AoI were obtained in low, intermediate and high SNR regimes. Moreover, the distribution and moments of the VoI were investigated in the example of the M/M/1 queue model. Finally, we performed Monte Carlo simulations to obtain numerical validation of the theoretical analysis. The results presented in this paper provide insight into how the correlation and noise in a latent OU process influence the VoI of the observations of that process. We also elucidated the relationship between the VoI and the AoI. Our work has given a mathematical justification for selecting certain non-linear age functions. Future work can be focused on exploring the effect of multiple observations on the VoI and AoI relationship and on estimating the value of the status of the underlying process with multiple observations.

Author Contributions: Conceptualization, Z.W., M.-A.B. and J.P.C.; methodology, Z.W., M.-A.B. and J.P.C.; formal analysis, Z.W., M.-A.B. and J.P.C.; investigation, Z.W., M.-A.B. and J.P.C.; writing—original draft preparation, Z.W., M.-A.B. and J.P.C.; writing—review and editing, Z.W., M.-A.B. and J.P.C.; supervision, M.-A.B. and J.P.C. All authors have read and agreed to the published version of the manuscript.

Funding: This work was supported by EPSRC, grant number EP/T02612X/1.

Data Availability Statement: Not applicable.

Acknowledgments: The authors gratefully acknowledge the support of the Clarendon Fund Scholarships at the University of Oxford.

Conflicts of Interest: The authors declare no conflict of interest.

Appendix A

Appendix A.1. Proof of Monotonicity in μ

First, we prove the monotonicity in μ. For any particular VoI threshold v, the derivative of the VoI outage is given as follows:

$$\frac{\partial P(V \le v)}{\partial \mu} = \frac{\log r(v)}{2\kappa} \left[r(v)^{\frac{\mu-\lambda}{2\kappa}} - \left(\frac{\mu}{\mu-\lambda} - \frac{\lambda}{2\kappa} \log r(v) \right) r(v)^{\frac{\mu}{2\kappa}} \right]$$
$$+ \frac{\lambda}{(\mu-\lambda)^2} \left(r(v)^{\frac{\mu}{2\kappa}} - r(v)^{\frac{\lambda}{2\kappa}} \right). \quad (A1)$$

For simplicity, let $x_1 = \frac{\lambda}{2\kappa} \log r(v)$ and $x_2 = \frac{\mu}{2\kappa} \log r(v)$. Then, (A1) can be written as follows:

$$
\begin{aligned}
\frac{\partial P(V \leq v)}{\partial \mu} &= \frac{\log r(v)}{2\kappa} \left[e^{x_2 - x_1} + \left(x_1 - \frac{x_2}{x_2 - x_1} \right) e^{x_2} \right] + \frac{\log r(v)}{2\kappa} \frac{x_1}{(x_2 - x_1)^2} \left(e^{x_2} - e^{x_1} \right) \\
&= \frac{\log r(v)}{2\kappa} e^{x_2} \left[e^{-x_1} + x_1 - \frac{x_2}{x_2 - x_1} + \frac{x_1 (1 - e^{x_1 - x_2})}{(x_2 - x_1)^2} \right]. \quad \text{(A2)}
\end{aligned}
$$

Since $\lambda < \mu$ and $0 < r(v) < 1$, thus $x_2 < x_1 < 0$ and $\log r(v) < 0$. Moreover, for any x, we have $e^x \geq 1 + x$. Therefore, (A2) can be further given as follows:

$$
\begin{aligned}
\frac{\partial P(V \leq v)}{\partial \mu} &\leq \frac{\log r(v)}{2\kappa} e^{x_2} \left[1 - \frac{x_2}{x_2 - x_1} + \frac{x_1 (x_2 - x_1)}{(x_2 - x_1)^2} \right] \\
&= 0.
\end{aligned}
\quad \text{(A3)}
$$

As the derivative is non-positive, the VoI outage is a monotonic function of μ.

Appendix A.2. Proof of Optimal λ Exists

Next, we prove that the optimal sampling rate exists. The derivative of the VoI outage with respect to λ is given as follows:

$$
\begin{aligned}
\frac{\partial P(V \leq v)}{\partial \lambda} &= -\frac{\log r(v)}{2\kappa} \left(r(v)^{\frac{\mu - \lambda}{2\kappa}} - r(v)^{\frac{\mu}{2\kappa}} - \frac{\mu}{\mu - \lambda} r(v)^{\frac{\lambda}{2\kappa}} \right) \\
&\quad - \frac{\mu}{(\mu - \lambda)^2} \left(r(v)^{\frac{\mu}{2\kappa}} - r(v)^{\frac{\lambda}{2\kappa}} \right) \\
&= -\frac{\log r(v)}{2\kappa} \left[e^{x_2 - x_1} - e^{x_2} - \frac{x_2 e^{x_1}}{x_2 - x_1} + \frac{x_2 (e^{x_2} - e^{x_1})}{(x_2 - x_1)^2} \right]. \quad \text{(A4)}
\end{aligned}
$$

When λ approaches 0, we can write the following:

$$
\lim_{x_1 \to 0} \frac{\partial P(V \leq v)}{\partial \lambda} = -\frac{\log r(v)}{2\kappa} \left(\frac{e^{x_2} - 1}{x_2} - 1 \right) \leq 0. \quad \text{(A5)}
$$

When λ approaches μ, we have the following:

$$
\lim_{x_1 \to x_2} \frac{\partial P(V \leq v)}{\partial \lambda} = -\frac{\log r(v)}{2\kappa} \left[1 - e^{x_2} \left(1 - \frac{x_2}{2} \right) \right] \geq -\frac{\log r(v)}{2\kappa} \left(1 - e^{\frac{x_2}{2}} \right) \geq 0. \quad \text{(A6)}
$$

We show that the VoI outage probability decreases with λ when λ is small, and increases when λ is large. Therefore, there exists the optimal sampling rate λ^*, and the minimum outage probability is achieved when the derivative in (A4) is 0.

Appendix B. Proof of Proposition 3

The average VoI can be obtained directly by the following:

$$
\begin{aligned}
\mathrm{E}[V] = \int_0^{\frac{1}{2} \log(1 + \gamma)} v f_V(v) \, dv = -\frac{\mu}{4\kappa} \int_0^1 \log \left(1 - \frac{\gamma}{1 + \gamma} r \right) &\left[\frac{\mu - \lambda}{\mu} r^{\frac{\mu - \lambda}{2\kappa} - 1} \right. \\
&\left. + \frac{\lambda}{\mu - \lambda} r^{\frac{\lambda}{2\kappa} - 1} - \left(\frac{\mu}{\mu - \lambda} - \frac{\lambda}{\mu} - \frac{\lambda}{2\kappa} \log r \right) r^{\frac{\mu}{2\kappa} - 1} \right] dr. \quad \text{(A7)}
\end{aligned}
$$

Here, for the given x and y, we have the following:

$$\int_0^1 \log(1 - xr) \cdot r^{y-1}\, dr = \frac{1}{x^y} \int_0^x \log(1 - z) \cdot z^{y-1}\, dz$$

$$= \frac{1}{x^y} \frac{z^y}{y} \log(1 - z)\Big|_{z=0}^{z=x} + \frac{1}{yx^y} \int_0^x \frac{z^y}{1 - z}\, dz \tag{A8}$$

$$= \frac{1}{y}\left[\log(1 - x) + g_1(x, y)\right],$$

and

$$\int_0^1 \log r \log(1 - xr) \cdot r^{y-1}\, dr$$

$$= \frac{1}{x^y}\left(\int_0^x \log z \cdot \log(1 - z) \cdot z^{y-1}\, dz - \log x \int_0^x \log(1 - z) \cdot z^{y-1}\, dz\right)$$

$$= \frac{1}{x^y} \frac{z^y}{y} \log(1 - z) \cdot \log z\Big|_{z=0}^{z=x} - \frac{\log x}{x^y} \int_0^x \log(1 - z) \cdot z^{y-1} d\, dz$$

$$- \frac{1}{yx^y} \int_0^x z^y \left(\frac{\log(1 - z)}{z} - \frac{\log z}{1 - z}\right) dz$$

$$= \frac{\log(1 - x) \cdot \log x}{y} - \left(\frac{1}{yx^y} + \frac{\log x}{x^y}\right) \int_0^x \log(1 - z) \cdot z^{y-1}\, dz + \frac{1}{yx^y} \int_0^x \frac{z^y \log z}{1 - z}\, dz$$

$$= -\frac{\log(1 - x)}{y^2} - \left(\frac{1}{y^2} + \frac{\log x}{y}\right) g_1(x, y) + \frac{g_2(x, y)}{y}. \tag{A9}$$

Therefore, the average VoI is derived by substituting (A8) and (A9) into (A7).

The lower bound in (25) is obtained by applying Jensen's inequality, i.e., the following:

$$E[V] = E\left[-\frac{1}{2} \log\left(1 - \frac{\gamma}{1 + \gamma} e^{-2\kappa a}\right)\right]$$

$$\geq -\frac{1}{2} \log\left(1 - \frac{\gamma}{1 + \gamma} E[e^{-2\kappa a}]\right), \tag{A10}$$

where

$$E[e^{-2\kappa a}] = \int_0^{+\infty} e^{-2\kappa a} f_A(a)\, da$$

$$= \frac{\frac{\mu - \lambda}{2\kappa}}{\frac{\mu - \lambda}{2\kappa} + 1} - \frac{\frac{\mu - \lambda}{2\kappa}\left(\frac{\mu + \lambda}{2\kappa} + 1\right)}{\left(\frac{\mu}{2\kappa} + 1\right)^2 \left(\frac{\lambda}{2\kappa} + 1\right)}. \tag{A11}$$

The upper bound in (26) is the average VoI in the Markov OU process. In the hidden Markov model, we can write the following [31]:

$$v(t) = h(X_t) - h(X_t | Y_{t'_n}, \ldots, Y_{t'_{n-m+1}})$$

$$\leq h(X_t) - h(X_t | Y_{t'_n}, \ldots, Y_{t'_{n-m+1}}, X_{t_n})$$

$$= h(X_t) - h(X_t | X_{t_n}) \tag{A12}$$

$$= I(X_t; X_{t_n}).$$

Therefore, the VoI in the Markov model can be regarded as the upper bound of the VoI in the hidden Markov model. Denote $v_{OU}(t) = I(X_t; X_{t_n})$ as the VoI in the underlying OU process. Then, the result in (26) follows from the following calculation:

$$E[V] \leq E[V_{OU}] = E\left[-\frac{1}{2} \log\left(1 - e^{-2\kappa a}\right)\right]$$

$$= -\frac{\mu}{4\kappa} \int_0^1 \log(1 - r)\left[\frac{\mu - \lambda}{\mu} r^{\frac{\mu-\lambda}{2\kappa} - 1} + \frac{\lambda}{\mu - \lambda} r^{\frac{\lambda}{2\kappa} - 1} - \left(\frac{\mu}{\mu - \lambda} - \frac{\lambda}{\mu} - \frac{\lambda}{2\kappa} \log r\right) r^{\frac{\mu}{2\kappa} - 1}\right] dr. \tag{A13}$$

Similar to the calculation given in (A8) and (A9), for the given y, we have the following:

$$\int_0^1 \log(1-r) \cdot r^{y-1}\, \mathrm{d}r = -\frac{1}{y}H(y),$$

$$\int_0^1 \log r \cdot \log(1-r) \cdot r^{y-1}\, \mathrm{d}r = \frac{1}{y^2}H(y) - \frac{1}{y}\varphi^{(1)}(y). \tag{A14}$$

Therefore, the upper bound of the average VoI is derived by substituting (A14) into (A13).

Appendix C. Proof of Proposition 4

The MGF of the VoI is obtained directly by the following:

$$M_v(t) = \int_0^{\frac{1}{2}\log(1+\gamma)} e^{tv} f_V(v)\, \mathrm{d}v = \frac{\mu}{2\kappa} \int_0^1 \left(1 - \frac{\gamma}{1+\gamma}r\right)^{-\frac{t}{2}} \left[\frac{\mu-\lambda}{\mu} r^{\frac{\mu-\lambda}{2\kappa}-1}\right.$$
$$\left. + \frac{\lambda}{\mu-\lambda} r^{\frac{\lambda}{2\kappa}-1} - \left(\frac{\mu}{\mu-\lambda} - \frac{\lambda}{\mu} - \frac{\lambda}{2\kappa}\log r\right) r^{\frac{\mu}{2\kappa}-1}\right] \mathrm{d}r. \tag{A15}$$

Here, for the given x, y and t, we have the following [30]:

$$\int_0^1 (1-xr)^{-\frac{t}{2}} \cdot r^{y-1}\, \mathrm{d}r = \frac{1}{y} {}_2F_1\left(y, \frac{t}{2}; y+1; x\right), \tag{A16}$$

and

$$\int_0^1 \log r \cdot (1-xr)^{-\frac{t}{2}} \cdot r^{y-1}\, \mathrm{d}r = -\frac{1}{y^2} {}_3F_2\left(y, y, \frac{t}{2}; y+1, y+1; x\right). \tag{A17}$$

Therefore, the MGF of VoI is derived by substituting (A16) and (A17) into (A15).

References

1. Kaul, S.; Yates, R.; Gruteser, M. Real-time status: How often should one update? In Proceedings of the 2012 Proceedings IEEE INFOCOM, Orlando, FL, USA, 25–30 March 2012; pp. 2731–2735.
2. Yates, R.D.; Kaul, S. Real-time status updating: Multiple sources. In Proceedings of the 2012 IEEE International Symposium on Information Theory Proceedings, Cambridge, MA, USA, 1–6 July 2012; pp. 2666–2670.
3. Kosta, A.; Pappas, N.; Angelakis, V. Age of Information: A New Concept, Metric, and Tool. *Found. Trends Netw.* **2017**, *12*, 162–259. [CrossRef]
4. Wu, X.; Yang, J.; Wu, J. Optimal Status Update for Age of Information Minimization with an Energy Harvesting Source. *IEEE Trans. Green Commun. Netw.* **2018**, *2*, 193–204. [CrossRef]
5. Wang, Z.; Qin, X.; Liu, B.; Zhang, P. Joint Data Sampling and Link Scheduling for Age Minimization in Multihop Cyber-Physical Systems. *IEEE Commun. Lett.* **2019**, *8*, 765–768. [CrossRef]
6. He, Q.; Dán, G.; Fodor, V. Joint Assignment and Scheduling for Minimizing Age of Correlated Information. *IEEE/ACM Trans. Netw.* **2019**, *27*, 1887–1900. [CrossRef]
7. Abd-Elmagid, M.A.; Dhillon, H.S. Average Peak Age-of-Information Minimization in UAV-Assisted IoT Networks. *IEEE Trans. Veh. Technol.* **2019**, *68*, 2003–2008. [CrossRef]
8. Abd-Elmagid, M.A.; Pappas, N.; Dhillon, H.S. On the Role of Age of Information in the Internet of Things. *IEEE Commun. Mag.* **2019**, *57*, 72–77. [CrossRef]
9. Inoue, Y.; Masuyama, H.; Takine, T.; Tanaka, T. The stationary distribution of the age of information in FCFS single-server queues. In Proceedings of the 2017 IEEE International Symposium on Information Theory (ISIT), Aachen, Germany, 25–30 June 2017; pp. 571–575.
10. Hu, L.; Chen, Z.; Dong, Y.; Jia, Y.; Liang, L.; Wang, M. Status Update in IoT Networks: Age of Information Violation Probability and Optimal Update Rate. *IEEE Internet Things J.* **2021**, 11329–11344. [CrossRef]
11. Inoue, Y.; Masuyama, H.; Takine, T.; Tanaka, T. A General Formula for the Stationary Distribution of the Age of Information and Its Application to Single-Server Queues. *IEEE Trans. Inf. Theory* **2019**, *65*, 8305–8324. [CrossRef]
12. Bedewy, A.M.; Sun, Y.; Shroff, N.B. Minimizing the Age of Information Through Queues. *IEEE Trans. Inf. Theory* **2019**, *65*, 5215–5232. [CrossRef]
13. Kaul, S.K.; Yates, R.D. Age of Information: Updates with Priority. In Proceedings of the 2018 IEEE International Symposium on Information Theory (ISIT), Vail, CO, USA, 17–22 June 2018; pp. 2644–2648.

14. Kam, C.; Kompella, S.; Nguyen, G.D.; Wieselthier, J.E.; Ephremides, A. On the Age of Information with Packet Deadlines. *IEEE Trans. Inf. Theory* **2018**, *64*, 6419–6428. [CrossRef]
15. Champati, J.P.; Al-Zubaidy, H.; Gross, J. On the Distribution of AoI for the GI/GI/1/1 and GI/GI/1/2* Systems: Exact Expressions and Bounds. In Proceedings of the IEEE INFOCOM 2019—IEEE Conference on Computer Communications, Paris, France, 29 April–2 May 2019; pp. 37–45.
16. Sun, Y.; Cyr, B. Sampling for data freshness optimization: Non-linear age functions. *J. Commun. Netw.* **2019**, *21*, 204–219. [CrossRef]
17. Sun, Y.; Uysal-Biyikoglu, E.; Yates, R.D.; Koksal, C.E.; Shroff, N.B. Update or Wait: How to Keep Your Data Fresh. *IEEE Trans. Inf. Theory* **2017**, *63*, 7492–7508. [CrossRef]
18. Zheng, X.; Zhou, S.; Jiang, Z.; Niu, Z. Closed-Form Analysis of Non-Linear Age of Information in Status Updates with an Energy Harvesting Transmitter. *IEEE Trans. Wirel. Commun.* **2019**, *18*, 4129–4142. [CrossRef]
19. Kosta, A.; Pappas, N.; Ephremides, A.; Angelakis, V. Age and value of information: Non-linear age case. In Proceedings of the 2017 IEEE International Symposium on Information Theory (ISIT), Aachen, Germany, 25–30 June 2017; pp. 326–330.
20. Kosta, A.; Pappas, N.; Ephremides, A.; Angelakis, V. The Cost of Delay in Status Updates and Their Value: Non-Linear Ageing. *IEEE Trans. Commun.* **2020**, *68*, 4905–4918. [CrossRef]
21. Kosta, A.; Pappas, N.; Ephremides, A.; Angelakis, V. The Age of Information in a Discrete Time Queue: Stationary Distribution and Non-Linear Age Mean Analysis. *IEEE J. Sel. Areas Commun.* **2021**, *39*, 1352–1364. [CrossRef]
22. Sun, Y.; Polyanskiy, Y.; Uysal, E. Sampling of the Wiener Process for Remote Estimation over a Channel with Random Delay. *IEEE Trans. Inf. Theory* **2020**, *66*, 1118–1135. [CrossRef]
23. Ornee, T.Z.; Sun, Y. Sampling and Remote Estimation for the Ornstein-Uhlenbeck Process through Queues: Age of Information and Beyond. *IEEE/ACM Trans. Netw.* **2021**, 1–14. [CrossRef]
24. Zheng, X.; Zhou, S.; Niu, Z. Urgency of Information for Context-Aware Timely Status Updates in Remote Control Systems. *IEEE Trans. Wirel. Commun.* **2020**, *19*, 7237–7250. [CrossRef]
25. Kam, C.; Kompella, S.; Nguyen, G.D.; Wieselthier, J.E.; Ephremides, A. Towards an effective age of information: Remote estimation of a Markov source. In Proceedings of the IEEE INFOCOM 2018—IEEE Conference on Computer Communications Workshops (INFOCOM WKSHPS), Honolulu, HI, USA, 15–19 April 2018; pp. 367–372.
26. Sun, Y.; Cyr, B. Information Aging Through Queues: A Mutual Information Perspective. In Proceedings of the 2018 IEEE 19th International Workshop on Signal Processing Advances in Wireless Communications (SPAWC), Kalamata, Greece, 25–28 June 2018; pp. 1–5.
27. Wang, Z.; Badiu, M.A.; Coon, J.P. A Value of Information Framework for Latent Variable Models. In Proceedings of the GLOBECOM 2020—2020 IEEE Global Communications Conference, Taipei, Taiwan, 7–11 December 2020; pp. 1–6.
28. Doob, J.L. The Brownian Movement and Stochastic Equations. *Ann. Math.* **1942**, *43*, 351–369. [CrossRef]
29. Conway, J.; Guy, R. *The Book of Numbers*; Springer New York: New York, NY, USA, 1998.
30. Andrews, L. *Special Functions of Mathematics for Engineers*, 2nd ed.; Oxford University Press: Oxford, UK; SPIE Optical Engineering Press: Bellingham, WA, USA, 1998.
31. Cover, T.M.; Thomas, J.A. *Elements of Information Theory*; Wiley-Interscience: Hoboken, NJ, USA, 2006.

Article

Channel Quality-Based Optimal Status Update for Information Freshness in Internet of Things

Fuzhou Peng, Xiang Chen * and Xijun Wang

School of Electronics and Information Technology, Sun Yat-sen University, Guangzhou 510006, China;
pengfzh@mail2.sysu.edu.cn (F.P.); wangxijun@mail.sysu.edu.cn (X.W.)
* Correspondence: chenxiang@mail.sysu.edu.cn

Abstract: This paper investigates the status updating policy for information freshness in Internet of things (IoT) systems, where the channel quality is fed back to the sensor at the beginning of each time slot. Based on the channel quality, we aim to strike a balance between the information freshness and the update cost by minimizing the weighted sum of the age of information (AoI) and the energy consumption. The optimal status updating problem is formulated as a Markov decision process (MDP), and the structure of the optimal updating policy is investigated. We prove that, given the channel quality, the optimal policy is of a threshold type with respect to the AoI. In particular, the sensor remains idle when the AoI is smaller than the threshold, while the sensor transmits the update packet when the AoI is greater than the threshold. Moreover, the threshold is proven to be a non-increasing function of channel state. A numerical-based algorithm for efficiently computing the optimal thresholds is proposed for a special case where the channel is quantized into two states. Simulation results show that our proposed policy performs better than two baseline policies.

Keywords: age of information; status update; channel quality

Citation: Peng, F.; Chen, X.; Wang, X. Channel Quality-Based Optimal Status Update for Information Freshness in Internet of Things. *Entropy* **2021**, *23*, 912. https://doi.org/10.3390/e23070912

Academic Editors: Anthony Ephremides and Yin Sun

Received: 18 June 2021
Accepted: 16 July 2021
Published: 18 July 2021

Publisher's Note: MDPI stays neutral with regard to jurisdictional claims in published maps and institutional affiliations.

1. Introduction

Recently, the Internet of things (IoT) has been widely used in the field of industrial manufacturing, environment monitoring, and home automation. In these applications, the sensors generate and transmit new status updates to the destination, where the freshness of the status updates is crucial for the destination to track the state of the environment and to make decisions. Thus, a new information freshness metric, namely age of information (AoI), was proposed in [1] to measure the freshness of updates from the receiver's perspective. There are two widely used metrics, i.e., the average peak AoI [2] and the average AoI [3]. In general, the smaller the AoI is, the fresher the received updates are.

AoI was originally investigated in [1] for updating the status in vehicular networks. Considering the impact of the queueing system, the authors in [4] investigated the system performance under the M/M/1 and M/M/1/2 queueing systems with a first-come-first-served (FCFS) policy. Furthermore, the work of [5] studied how to keep the updates fresh by analyzing some general update policies, such as the zero-wait policy. The authors of [6] considered the optimal schedule problem for a more general cost that is the weighted sum of the transmission cost and the tracking inaccuracy for the information source. However, these works assumed that the communication channel is not error-prone. In practice, status updates are delivered through an erroneous wireless channel, which suffers from fading, interference, and noises. Therefore, the received updates may not be decoded correctly, which induces information aging and energy consumption.

There are several works that considered the erroneous channel [7,8]. The authors in [9] considered multiple communication channels and investigated the optimal coding and decoding schemes. The channel with an independent and identical packet error rate over time was considered in [10,11]. The work of [12] considered the impact of fading channels in packet transmission. A Markov channel was investigated in [13], where threshold policy was proven to be optimal, and a simulation-based approach was proposed to compute the corresponding threshold. However, how the information of channel quality should be exploited to improve system performance in information freshness remains to be investigated.

Channel quality indicator (CQI) feedback is commonly used in wireless communication systems [14]. In block fading channels, the channel quality, generally reported by the terminal, is highly relevant to the packet error rate (PER) [15] or, namely, the block error rate (BLER). It is probable that a received packet fails to be decoded when the channel suffers from a poor condition. However, a transmitter with the channel quality information is able to keep idle when there is deep fading, thereby saving energy. The channel quantization was also considered in [12,13], where the channel was quantized into multiple states. However, the decision making was not dependent on the channel state in [12], while [13] did not consider the freshness of information. These motivate us to introduce the information of channel quality into the design of the updating policy.

In this paper, a status update system with channel quality feedback is considered. In particular, the channel condition is quantized into multiple states, and the destination feeds the channel quality back to the sensor before the sensor updates the status. Our problem is to investigate the channel quality-based optimal status update policy, which minimizes the weighted sum of the AoI and the energy consumption. Our key contributions are summarized as follows:

- An average cost Markov decision process (MDP) is formulated to model this problem. Due to the infinite countable states and unbounded cost of the MDP, which makes analysis difficult, the discounted version of the original problem is first investigated, and the existence of the stationary and deterministic policy to the original problem is then proven. Furthermore, it is proven that the optimal policy is a threshold structure policy with respect to the AoI for each channel state by showing the monotonic property of the value function. We also prove that the threshold is a non-increasing function of channel state.

- By utilizing the threshold structure, a structure-aware policy iteration algorithm is proposed to efficiently obtain the optimal updating policy. Nevertheless, a numerical-based algorithm which directly computes the thresholds by non-linear fractional programming is also derived. Simulation results reveal the effects of system parameters and show that our proposed policy performs better than the zero-wait policy and periodic policy.

The rest of this paper is organized as follows. In Section 2, the system model is presented and the optimal updating problem is formulated. In Section 3, the optimal updating policy is proven to be of a threshold structure, and a threshold-based policy iteration algorithm is proposed to find the optimal policy. Section 4 presents the simulation results. Finally, we summarize our conclusions in Section 5.

2. System Model and Problem Formulation

2.1. System Description

In this paper, we consider a status update system that consists of a sensor and a destination, as shown in Figure 1. Time is divided into slots. Without loss of generality, we assume that each time slot has an equal length, which is normalized to unity. At the beginning of each slot, the destination feeds the CQI back to the sensor. It is worth noting that the PER is different for different CQIs. Based on the CQI, the sensor decides in each time slot whether it should generate and transmit a new update to the destination via a wireless channel or keep idle for saving energy. These updates are crucial for the destination to estimate the states of the surrounding environment of the sensor and to make in-time decisions. Let a_t, which takes value from the action set $\mathcal{A} = \{0, 1\}$, denote the action that the sensor performs in slot t, where $a_t = 1$ means that the sensor generates and transmits a new update to the destination, and $a_t = 0$ represents that the sensor is idle. If the sensor transmits an update packet in slot t, an acknowledgment will be fed back at the end of this time slot. In particular, an ACK is fed back when the destination successfully receives the update packet, and a NACK otherwise.

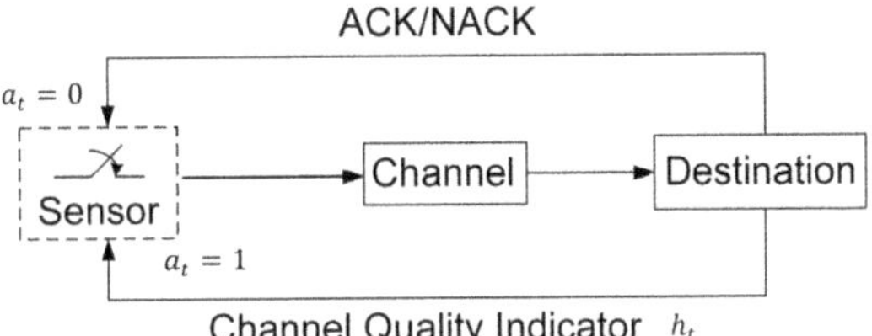

Figure 1. System model.

2.2. Channel Model

Suppose that the wireless channel is a block fading channel where the channel gain remains constant in each slot and varies independently over different slots. Let z_t denote the channel gain in slot t which takes value from $[0, +\infty)$. We quantize the channel gain into $N + 1$ levels which are denoted as $(z_0, z_1, ..., z_i, ..., z_N)$. The quantization levels are arranged in an increasing order where $z_0 = 0$ and $z_N = \infty$. Hence, the channel is said to be in state i if the channel gain z_t belongs to the interval $[z_i, z_{i+1})$. We denote by h_t the state of the channel in slot t, where $h_t \in \mathbb{H} \triangleq \{0, 1, 2..., N - 1\}$. With the aid of CQI fed back from the destination, the sensor has knowledge of the channel state at the beginning of each time slot.

Let $p_z(z)$ denote the distribution of the channel gain. Then, the probability of the channel being in state i is

$$p_i = \int_{z_i}^{z_{i+1}} p_z(z)dz. \tag{1}$$

We assume that the signal-to-noise ratio (SNR) per information bit during the transmission remains constant. Then, the PER depends only on the channel gain. In particular, the PER for channel state i is given by

$$g_i = \int_{z_i}^{z_{i+1}} P_{\text{PER}}(z)p_z(z|i)dz, \tag{2}$$

where $P_{\text{PER}}(z)$ is the PER of a packet with respect to the channel gain. The success probability q_i of a packet transmitted over channel state i is $q_i = 1 - g_i$. According to [15], the success probability is a non-decreasing function of the channel state.

2.3. Age of Information

This paper uses the AoI as the freshness metric, which is defined as the time elapsed since the generation time of the latest update packet that is successfully received by the destination [1]. Let G_i be the generation time of the ith successfully received update packet. Then, the AoI in time slot t, Δ_t, is defined as

$$\Delta_t = t - \max\{G_i : G_i \leq t\}. \tag{3}$$

In particular, if an update packet is successfully received, the AoI decreases to one. Otherwise, the AoI increases by one. Altogether, the evolution of the AoI is expressed by

$$\Delta_{t+1} = \begin{cases} 1, & \text{if the transmission is successful,} \\ \Delta_t + 1, & \text{otherwise.} \end{cases} \tag{4}$$

An example of the AoI evolution is shown in Figure 2, where the gray rectangle represents a successful reception of an update packet, and the mesh rectangle represents a transmission failure.

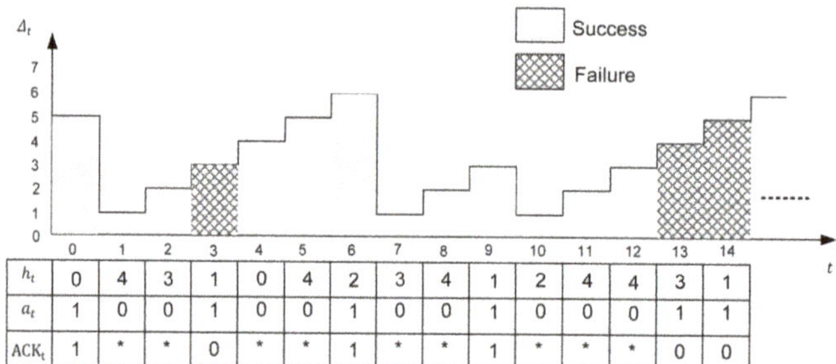

	0	1	2	3	4	5	6	7	8	9	10	11	12	13	14
h_t	0	4	3	1	0	4	2	3	4	1	2	4	4	3	1
a_t	1	0	0	1	0	0	1	0	0	1	0	0	0	1	1
ACK$_t$	1	*	*	0	*	*	1	*	*	1	*	*	*	0	0

Figure 2. An example of the AoI evolution with the channel state h_t, the action a_t, and the acknowledgment ACK$_t$. The asterisk stands for no acknowledgment from destination when the sensor keeps idle.

2.4. Problem Formulation

The objective of this paper is to find an optimal updating policy that minimizes the long-term average of the weighted sum of AoI and energy consumption. A policy π can be represented by the sequence of actions, i.e., $\pi = (a_0, a_1, \ldots, a_t, \ldots)$. Let Π be a set of stationary and deterministic policies. Then, the optimal updating problem is given by

$$\min_{\pi \in \Pi} \limsup_{T \to \infty} \frac{1}{T} \sum_{t=0}^{T} \mathbb{E}[\Delta_t + \omega a_t C_e], \tag{5}$$

where C_e is the energy consumption, and ω is the weighting factor.

3. Optimal Updating Policy

This section aims to investigate the optimal updating policy for the problem formulated in above section. In this section, our investigating problem is first formulated into an infinite horizon average cost MDP, and the existence of a stationary and deterministic policy that minimizes the average cost is proven. Then, the non-decreasing property of the value function is derived. Based on this property, we prove that the optimal update policy is of a threshold structure with respect to AoI, and the optimal threshold is a non-increasing function of the channel state. Aiming to reduce the computational complexity, a structure-aware policy iteration algorithm is proposed to find the optimal policy. Moreover, non-linear fractional programming is employed to directly compute the optimal thresholds in a special case where the channel is quantized into two states.

3.1. MDP Formulation

The Markov decision process (MDP) is typically applied to address the optimal decision problem when the investigation problem can be characterized by the evolution of the system state and the cost is per-stage. The optimization problem in (5) can be formulated as an infinite horizon average cost MDP, which is elaborated in the following.

- States: The state of the MDP in slot t is defined as $\mathbf{x}_t = (\Delta_t, h_t)$, which takes values in $\mathbb{Z}^+ \times \mathbb{H}$. Hence, the state space $\mathcal{S}$ is countable and infinite.
- Actions: The set of actions a_t chosen in slot t is $\mathcal{A} = \{0, 1\}$.
- Transition Probability: Let $\Pr(\mathbf{x}_{t+1}|\mathbf{x}_t, a_t)$ be the transition probability that the state $\mathbf{x}_t$ in slot t transits to $\mathbf{x}_{t+1}$ in slot $t+1$ after taking action a_t. According to the evolution of AoI in (4), the transition probability is given by

$$\begin{cases} \Pr(\mathbf{x}_{t+1} = (\Delta + 1, j)|\mathbf{x}_t = (\Delta, i), a_t = 0) = p_j, \\ \Pr(\mathbf{x}_{t+1} = (1, j)|\mathbf{x}_t = (\Delta, i), a_t = 1) = q_i p_j, \\ \Pr(\mathbf{x}_{t+1} = (\Delta + 1, j)|\mathbf{x}_t = (\Delta, i), a_t = 1) = g_i p_j. \end{cases} \tag{6}$$

- Cost: The instantaneous cost $C(\mathbf{x}_t, a_t)$ at state $\mathbf{x}_t$ given action a_t in slot t is

$$C(\mathbf{x}_t, a_t) = \Delta_t + \omega a_t C_e. \tag{7}$$

For an MDP with infinite states and unbounded cost, it is not guaranteed to have a stationary and deterministic policy that attains the minimum average cost in general. Fortunately, we can prove the existence of stationary and deterministic policy in next sub-section.

3.2. The Existence of Stationary and Deterministic Policy

For rigorous mathematical analysis, this section is purposed to prove the existence of a stationary and deterministic optimal policy. According to [16], we first analyze the associated discounted cost problem of the original MDP. The expectation of discount cost with respect to discounted factor γ and initial state $\hat{\mathbf{x}}$ under a policy π is given by

$$V_{\pi,\gamma}(\hat{\mathbf{x}}) = \mathbb{E}_\pi\left[\sum_{t=0}^{\infty} \gamma^t C(\mathbf{x}_t, a_t) | \mathbf{x}_0 = \hat{\mathbf{x}}\right], \tag{8}$$

where a_t is the decision made in state $\hat{\mathbf{x}}$ under policy π, and $\gamma \in (0,1)$ is the discounted factor. We first verify that $V_{\pi,\gamma}(\hat{\mathbf{x}})$ is finite for any policy and all $\hat{\mathbf{x}} \in \mathcal{S}$.

Lemma 1. *Given $\gamma \in (0,1)$, for any policy π and all $\hat{\mathbf{x}} = (\hat{x}, \hat{h}) \in \mathcal{S}$, we have*

$$V_{\pi,\gamma}(\hat{\mathbf{x}}) = \mathbb{E}_\pi\left[\sum_{t=0}^{\infty} \gamma^t C(\mathbf{x}_t, a_t) | \mathbf{x}_0 = \hat{\mathbf{x}}\right] < \infty. \tag{9}$$

Proof. By definition, the instantaneous cost in state $\mathbf{x}_t = (\Delta_t, h_t)$ given action a_t is

$$C(\mathbf{x}_t, a_t) = \begin{cases} \Delta_t, & \text{if } a_t = 0, \\ \Delta_t + \omega C_e, & \text{if } a_t = 1. \end{cases} \tag{10}$$

Therefore, $C(\mathbf{x}_t, a_t) \le \Delta_t + \omega C_e$ holds. Combined with the fact that the AoI increases, at most, linearly at each slot for any policy, we have

$$\sum_{t=0}^{\infty} \gamma^t C(\mathbf{x}_t, a_t | \mathbf{x}_0 = (\hat{\Delta}, \hat{h}))$$
$$\le \sum_{t=0}^{\infty} \gamma^t (\hat{\Delta} + t + \omega C_e)$$
$$= \frac{1}{1-\gamma}\left(\hat{\Delta} + \frac{\gamma}{1-\gamma} + \omega\right) < \infty, \tag{11}$$

which completes the proof. $\square$

Let $V_\gamma(\hat{\mathbf{x}}) = \min_\pi V_{\pi,\gamma}(\hat{\mathbf{x}})$ denote the minimum expected discounted cost. By Lemma 1, $V_\gamma(\hat{\mathbf{x}}) = \min_\pi V_{\pi,\gamma}(\hat{\mathbf{x}}) < \infty$ holds for every $\hat{\mathbf{x}}$ and $\gamma \in (0,1)$.
According to [16] (Proposition 1), we have

$$V_\gamma(\hat{\mathbf{x}}) = \min_{a \in \mathcal{A}}\left\{C(\hat{\mathbf{x}}, a) + \gamma \sum_{\mathbf{x}' \in \mathcal{S}} \Pr(\mathbf{x}'|\hat{\mathbf{x}}, a) V_\gamma(\mathbf{x}')\right\}, \tag{12}$$

which implies that $V_\gamma(\hat{\mathbf{x}})$ satisfies the Bellman equation. $V_\gamma(\hat{\mathbf{x}})$ can be solved via a value iteration algorithm. In particular, we define $V_{\gamma,0}(\hat{\mathbf{x}}) = 0$, and for all $n \ge 1$, we have

$$V_{\gamma,n}(\hat{\mathbf{x}}) = \min_{a \in \mathcal{A}} Q_{\gamma,n}(\hat{\mathbf{x}}, a), \tag{13}$$

where

$$Q_{\gamma,n}(\hat{\mathbf{x}}, a) = \min_{a \in \mathcal{A}} \left\{ C(\hat{\mathbf{x}}, a) + \gamma \sum_{\mathbf{x}' \in \mathcal{S}} \Pr(\mathbf{x}'|\hat{\mathbf{x}}, a) V_{\gamma,n-1}(\mathbf{x}') \right\} \tag{14}$$

is related to the right-hand-side (RHS) of the discounted cost optimality equation. Then, $\lim_{n \to \infty} V_{\gamma,n}(\hat{\mathbf{x}}) = V_{\gamma}(\hat{\mathbf{x}})$ for every $\hat{\mathbf{x}}$ and γ.

Now, we can use the value iteration algorithm to establish the monotonic properties of $V_{\gamma}(\hat{\mathbf{x}})$

Lemma 2. *For all Δ and i, we have*

$$V_{\gamma}(\Delta, N - 1) \leq V_{\gamma}(\Delta, i), \tag{15}$$

and for all $\Delta_1 \leq \Delta_2$ and i, we have

$$V_{\gamma}(\Delta_1, i) \leq V_{\gamma}(\Delta_2, i). \tag{16}$$

Proof. See Appendix A. □

Based on Lemmas 1 and 2, we are ready to show that the MDP has a stationary and deterministic optimal policy in the following theorem.

Theorem 1. *For the MDP in (5), there exists a stationary and deterministic optimal policy π^* that minimizes the long-term average cost. Moreover, there exists a finite constant $\lambda = \lim_{\gamma \to 1}(1 - \gamma)V_{\gamma}(\mathbf{x})$ for all states $\mathbf{x}$, where λ is independent of the initial state, and a value function $V(\mathbf{x})$, such that*

$$\lambda + V(\mathbf{x}) = \min_{a \in \mathcal{A}} \left\{ C(\mathbf{x}, a) + \sum_{\mathbf{x}' \in \mathcal{S}} \Pr(\mathbf{x}'|\mathbf{x}, a) V(\mathbf{x}') \right\} \tag{17}$$

holds for all $\mathbf{x}$.

Proof. See Appendix B. □

3.3. Structural Analysis

According to Theorem 1, the optimal policy for the average cost problem satisfies the following equation

$$\pi^*(\mathbf{x}) = \arg \min_{a \in \mathcal{A}} Q(\mathbf{x}, a), \tag{18}$$

where

$$Q(\mathbf{x}, a) = C(\mathbf{x}, a) + \sum_{\mathbf{x}' \in \mathcal{S}} \Pr(\mathbf{x}'|\mathbf{x}, a) V(\mathbf{x}'). \tag{19}$$

Similar to Lemma 2, the monotonic property of the value function $V(\mathbf{x})$ is given in the following lemma.

Lemma 3. *Given the channel state i, for any $\Delta_2 \geq \Delta_1$, we have*

$$V(\Delta_2, i) \geq V(\Delta_1, i). \tag{20}$$

Proof. This proof follows the same procedure of Lemma 2, with one exception being that the value iteration algorithm is based on Equation (17). □

Moreover, based on Lemma 3, the property of the increment of the value function is established in following lemma.

Lemma 4. *Given the channel state i, for any $\Delta_2 \geq \Delta_1$, we have*

$$V(\Delta_2, i) - V(\Delta_1, i) \geq \Delta_2 - \Delta_1. \tag{21}$$

Proof. We first examine the relation between the state-action value functions, i.e., $Q(\Delta_2, i, a)$ and $Q(\Delta_1, i, a)$. Specifically, based on Lemma 3, we have

$$Q(\Delta_2, i, 0) - (\Delta_2 - \Delta_1) = \Delta_1 + \sum_{j=0}^{N-1} p_j V(\Delta_2 + 1, j)$$

$$\geq \Delta_1 + \sum_{j=0}^{N-1} p_j V(\Delta_1 + 1, j) = Q(\Delta_1, i, 0), \tag{22}$$

and

$$Q(\Delta_2, i, 1) - (\Delta_2 - \Delta_1)$$

$$= \Delta_1 + \omega C_e + q_i \sum_{j=0}^{N-1} p_j V(1, j) + g_i \sum_{j=0}^{N-1} p_j V(\Delta_2 + 1, j)$$

$$\geq \Delta_1 + \omega C_e + q_i \sum_{j=0}^{N-1} p_j V(1, j) + g_i \sum_{j=0}^{N-1} p_j V(\Delta_1 + 1, j)$$

$$= Q(\Delta_1, i, 1). \tag{23}$$

Since $V(\mathbf{x}) = \min_{a \in \mathcal{A}} Q(\mathbf{x}, a)$, we complete the proof. $\square$

Our main result is presented in the following theorem.

Theorem 2. *For any given channel state i, there exists a threshold β_i, such that when $\Delta \geq \beta_i$, the optimal action is to generate and transmit a new update, i.e., $\pi^*(\Delta, i) = 1$, and when $\Delta < \beta_i$, the optimal action is to remain idle, i.e., $\pi^*(\Delta, i) = 0$. Moreover, the optimal threshold β_i is a non-increasing function of channel state i, i.e., $\beta_i \geq \beta_j$ holds for all $i, j \in \mathbb{H}$ and $i \leq j$.*

Proof. See Appendix C. $\square$

According to Theorem 2, the sensor will not update the status until the AoI exceeds the threshold. Moreover, if the channel condition is not good, i.e., channel state i is small, the sensor will wait for a longer time before it samples and transmits the status update packet so as to reduce the energy consumption because of a higher probability of transmission failure.

Based on the threshold structure, we can reduce the computational complexity of the policy iteration algorithm to find the optimal policy. The details of the algorithm are presented in Algorithm 1.

3.4. Computing the Thresholds for a Special Case

In the above section, we have proven that the optimal policy has a threshold structure. Given the thresholds $(\beta_0, \beta_1, ..., \beta_{N-1})$, a Markov chain can be induced by the threshold policy. A special Markov chain is depicted in Figure 3, where the channel has two states. By leveraging the Markov chain, we first derive the average cost of the special case, which is summarized in the following theorem.

Algorithm 1 Policy iteration algorithm (PIA) based on the threshold structure.

1: Initialization: Set $k = 0$, iteration threshold ϵ, and initialize the value function $V_0(\mathbf{x}) = 0$ and policy $\pi(\mathbf{x}) = 0$ for all state $\mathbf{x} \in \mathcal{S}$
2: **repeat**
3: $\quad k \leftarrow k + 1$.
4: $\quad$ Based on last iterative value function $V_{k-1}(\mathbf{x})$, compute the current value function $V_k(\mathbf{x})$ by calculating the following equations.
5: $\quad V_k(\mathbf{x}) = \min\limits_{a \in \mathcal{A}} \left\{ C(\mathbf{x}, a) + \sum\limits_{\mathbf{x}' \in \mathcal{S}} \Pr(\mathbf{x}'|\mathbf{x}, a) V_{k-1}(\mathbf{x}') \right\}$
6: **until** $|V_k(\mathbf{x}) - V_{k-1}(\mathbf{x})| \leq \epsilon$ for all $\mathbf{x} \in \mathcal{S}$
7: **for** $\mathbf{x} = (\Delta, i) \in \mathcal{S}$ **do**
8: $\quad$ **if** $\mathbf{x}' = (\Delta - 1, i) \in \mathcal{S}$ and $\pi(\mathbf{x}') = 1$ **then**
9: $\quad\quad \pi(\mathbf{x}) \leftarrow 1$.
10: $\quad$ **else**
11: $\quad\quad \pi(\mathbf{x}) \leftarrow \arg\min\limits_{a \in \mathcal{A}} \left\{ C(\mathbf{x}, a) + \Pr\limits_{\mathbf{x}' \in \mathcal{S}} (\mathbf{x}'|\mathbf{x}, a) V(\mathbf{x}') \right\}$
12: $\quad$ **end if**
13: **end for**
14: $\pi^* \leftarrow \pi$
15: **return** the optimal policy π^*

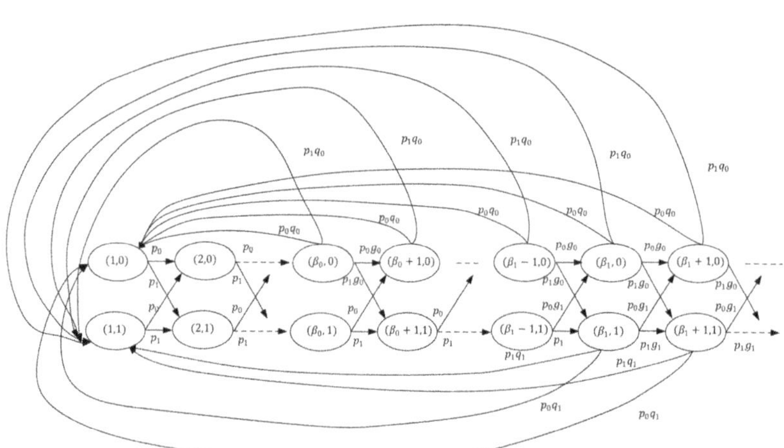

Figure 3. An illustration of established Markov chain with two channel states.

Theorem 3. *Let $\varphi(\mathbf{x})$ be the steady state probability of state $\mathbf{x}$ of the corresponding Markov chain with two states and β_0, β_1 be the threshold with respect to the channel state, respectively. The steady state probability is given by*

$$\varphi(i, j) = \begin{cases} p_j \varphi_1, & \text{if } 1 \leq i \leq \beta_1, \\ p_j s_0^{i - \beta_1} \varphi_1, & \text{if } \beta_1 < i \leq \beta_0, \\ p_j s_0^{\beta_0 - \beta_1} s_1^{i - \beta_1} \varphi_1, & \text{if } i > \beta_0, \end{cases} \tag{24}$$

where $\varphi_1 = \varphi(1, 0) + \varphi(1, 1)$, $s_0 = 1 - p_1 q_1$, $s_1 = 1 - p_0 q_0 - p_1 q_1$, and φ_1 satisfies following equation:

$$\varphi_1 = \frac{1}{\beta_1 + \frac{s_0 - s_0^{\beta_0 - \beta_1}}{1 - s_0} + s_0^{\beta_0 - \beta_1}\frac{s_1}{1 - s_1}}. \tag{25}$$

The average cost then is given by

$$C_{mc}(\beta_0, ..., \beta_{N-1})$$
$$= \varphi_1\left(\frac{\beta_1(\beta_1 + 1)}{2} + A + B + \omega C_e E\right), \tag{26}$$

where

$$A = \frac{s_0((\beta_1 + 1) - \beta_0 s_0^{\beta_0 - \beta_1})}{1 - s_0} + \frac{s_0^3 - s_0^{\beta_0 - \beta_1 + 1}}{(1 - s_0)^2}, \tag{27}$$

$$B = \frac{(\beta_0 + 1)s_1}{1 - s_1} + \frac{s_1^3}{1 - s_1}, \tag{28}$$

and

$$E = \left(s_0^{\beta_0 - \beta_1}\frac{1}{1 - s_1} + p_1\frac{1 - s_0^{\beta_0 - \beta_1}}{1 - s_0}\right). \tag{29}$$

Proof. See Appendix D. $\square$

Therefore, the closed form of the average cost is a function of thresholds. By linear search or gradient descent algorithm, the numerical solution of optimal thresholds can be obtained. However, computing its gradient directly requires a large amount of computation till convergence. Here, a nonlinear fractional programming (NLP) [17] based algorithm which can efficiently obtain the numerical solution is proposed.

Let $\mathbf{x} = (\beta_0, \beta_1)$. We can rewrite the cost function as a fractional form, where the numerator is denoted as $N(\mathbf{x}) = -C_{mc}(\mathbf{x})/\varphi_1$, and the denominator term is $N(\mathbf{x}) = 1/\varphi_1$. The solution to an NLP problem with the form in the following

$$\max\left\{\frac{N(\mathbf{x})}{D(\mathbf{x})}|\mathbf{x} \in A\right\} \tag{30}$$

is related to the optimization problem (31)

$$\max\{N(\mathbf{x}) - qD(\mathbf{x})|(\mathbf{x} \in A\}, \text{ for } q \in \mathbb{R}, \tag{31}$$

where the following assumption should also be satisfied:

$$D(\mathbf{x}) > 0, \text{ for all } \mathbf{x} \in A. \tag{32}$$

Define the function $F(q)$ with variable q as

$$F(q) = \max\{N(\mathbf{x}) - qD(\mathbf{x})|\mathbf{x} \in A\}, \text{ for } q \in \mathbb{R}. \tag{33}$$

According to [17], $F(q)$ is a strictly monotonic decreasing function and is convex over $\mathbb{R}$. Furthermore, we have $q_0 = N(\mathbf{x}_0)/D(\mathbf{x}_0) = \max\{N(\mathbf{x}) - qD(\mathbf{x})|\mathbf{x} \in A\}$ if, and only if,

$$F(q_0) = \max\{N(\mathbf{x}) - q_0 D(\mathbf{x})|\mathbf{x} \in A\} = 0. \tag{34}$$

Then, the algorithm can be described by two steps. The first step is to solve a convex optimization problem with a one dimensional parameter by a bisection method. The second step is to solve a high dimensional optimization problem by a gradient descent method.

According to [17], a bisection method can be used to solve the optimal q_0, under the assumption that the value of function $F(q)$ can be obtained exactly for given q. We will actually use the gradient descent algorithm to obtain the numerical solution of $F(q)$ since the global search method may not perform in polynomial time. As a trick, we alternate the optimization variables of thresholds (β_0, β_1) by the variables of the decrement of thresholds, i.e., $\mathbf{x} = (\beta_0 - \beta_1, \beta_1)$. To summarize, the numerical-based method for computing the optimal thresholds is given by Algorithm 2.

Algorithm 2 Numerical computation of the optimal thresholds.

Input: Iteration time k, error threshold δ
Output: Numerical result $\mathbf{x}^*$
1: Let $N(\mathbf{x}) = -C_{mc}(\mathbf{x})/\varphi_1$, and $D(\mathbf{x}) = 1/\varphi_1$. Define $F(q) = \max\{N(\mathbf{x}) - qD(\mathbf{x})|\mathbf{x} \geq 0\}$
2: Let the iteration starts with $i = 1$, search range $[a, b]$ of q.
3: **while** $i \leq k$ **do**
4:    $m = \frac{a+b}{2}$;
5:    **if** $F(m) * F(a) < 0$ **then**
6:       $b = m$;
7:    **else**
8:       $a = m$;
9:    **end if**
10:    **if** $\frac{b-a}{2} < \delta$ **then**
11:       $\mathbf{x}^* = \arg\min_{\mathbf{x}} F(m)$
12:       break;
13:    **end if**
14:    $i = i + 1$;
15: **end while**

---

4. Simulation Results and Discussions

In this section, the simulation results are presented to investigate the impacts of the system parameters. We also compare the optimal policy with the zero-wait policy and periodic policy, where the zero-wait policy immediately generates an update at each time slot and the periodic policy keeps a constant interval between two updates.

Figure 4 depicts the optimal policy for different AoI and channel states, where the number of channel states is 5. It can be seen that, for each channel state, the optimal policy has a threshold structure with respect to the AoI. In particular, when the AoI is small, it is not beneficial for the sensor to generate and transmit a new update because the energy consumption dominates the total cost. We can also see that the threshold is non-increasing with the channel state. In other words, if the channel condition is better, the threshold is smaller. This is because the success probability of packet transmission increases with the channel state.

Figure 5 illustrates the thresholds for the MDP with two channel states with respect to the weighting factor ω, in which the two dashed lines are obtained by PIA and the other two solid lines are obtained by the proposed numerical algorithm. Both of the thresholds grow with the increasing of ω. Since the energy consumption has more weight, it is not efficient to update when the AoI is small. On the contrary, when ω decreases, the AoI dominates and the thresholds decline. In particular, both of the thresholds equal 1 when $\omega = 0$. In this case, the optimal policy reduces to the zero-wait policy. We can also see that the value of the threshold for channel state 1 of the numerical algorithm is close to the optimal solution. In contrast, the value of the threshold for channel state 0 gradually deviates from the optimal value.

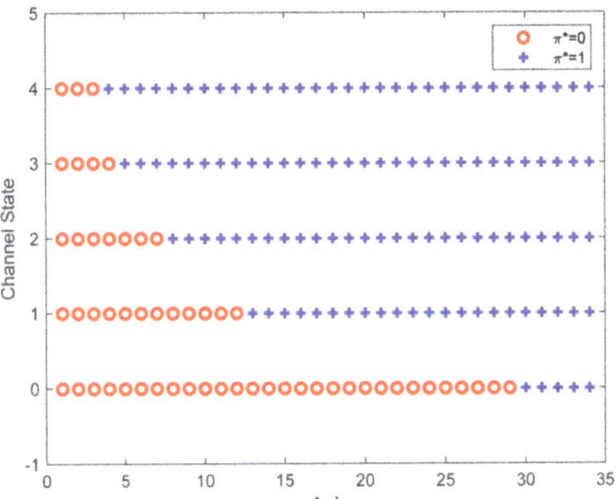

Figure 4. Optimal policy for different AoI and channel states ($q_0 = 0.1, q_1 = 0.2, q_2 = 0.3,$ $q_3 = 0.4, q_4 = 0.5, p_0 = 0.1, p_1 = 0.1, p_2 = 0.3, p_3 = 0.3, p_4 = 0.2, \omega = 10, C_e = 1$).

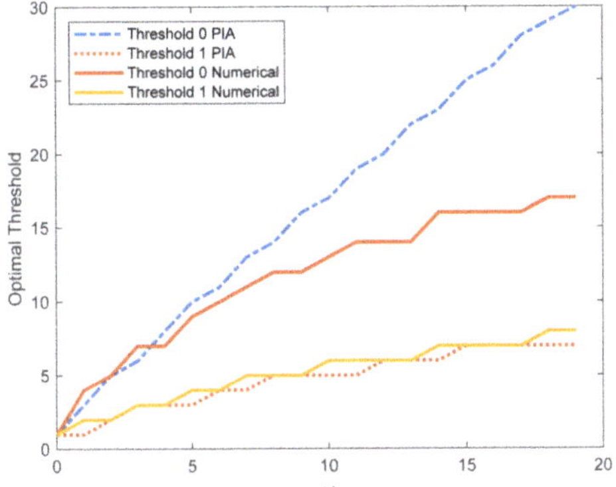

Figure 5. Optimal thresholds for two different channel states versus ω ($p_0 = 0.2, p_1 = 0.8,$ $q_0 = 0.2, q_1 = 0.5, C_e = 1$).

Figure 6 illustrates the performance comparison of four policies, i.e., the zero-wait policy, the periodic policy, the numerical-based policy, and the optimal policy, with respect to the weighting factor ω. It is easy to see that the optimal policy has the lowest average cost. As we see in Figure 6, the zero-wait policy has the same performance with the optimal policy when $\omega = 0$. As ω increases, the average cost of all three policies increases. However, the increment of the zero-wait policy is larger than the periodic policy and the optimal policy due to the frequent transmission in the zero-wait policy. Although the thresholds obtained by the PIA and the numerical algorithm are not exactly the same as shown in Figure 5, the performance of the numerical-based algorithm also coincides with the optimal policy. This is because the threshold for channel state 1 exists in the quadratic term of the cost function, while the threshold for channel state 0 exists in the negative exponential term of the cost function. As a result, the threshold for channel state 1 has a much more significant effect on the system performance.

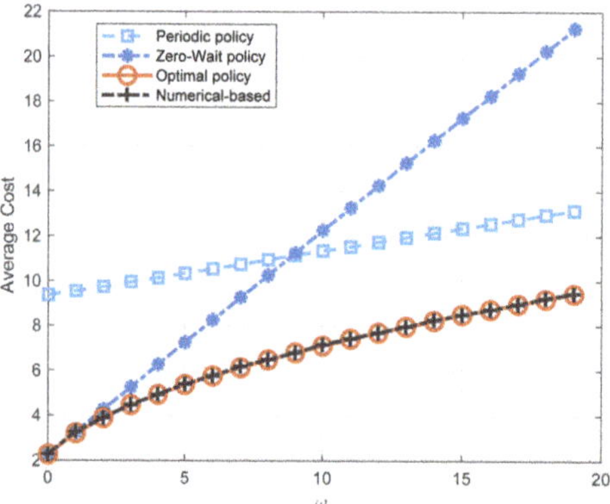

Figure 6. Comparison of the zero-wait policy, the periodic policy with period being 5, the numerical-based policy, and the optimal policy with respect to the weighting factor ω ($p_0 = 0.2, p_1 = 0.8$, $q_0 = 0.2, q_1 = 0.5, C_e = 1$).

Figure 7 compares the three policies with respect to the probability p_1 of the channel being in state 1. Since there is a higher probability that the channel has a good quality as p_1 increases, the average cost of all three policies decreases. We can see that, in the regime of p_1, the optimal policy has the lowest average cost, because it can achieve a good balance between the AoI and the energy consumption. We can also see that the cost of the periodic policy is greater than the zero-wait policy first, and smaller later. To further demonstrate these curves, we separate the energy consumption term and AoI term into different figures, i.e., Figures 8 and 9. We see that the update cost of the zero-wait policy is smaller than that of the periodic policy, but the AoI of the zero-wait policy has a smaller decrease with respect to p_1 than the periodic policy.

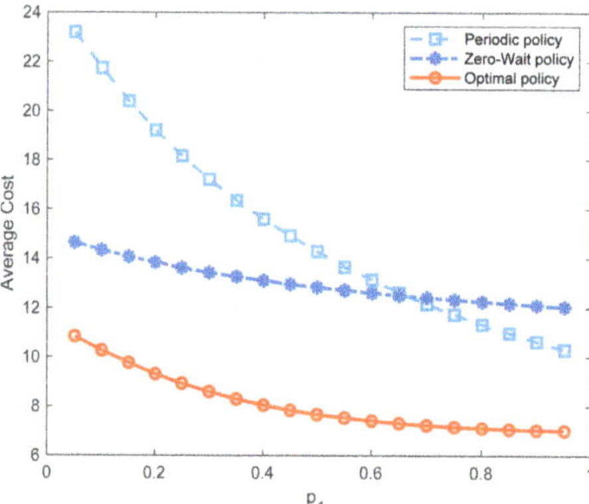

Figure 7. Comparison of the zero-wait policy, the periodic policy with period being 5, and the optimal policy with respect to p_1 ($q_0 = 0.2, q_1 = 0.5, \omega = 10, C_e = 1$).

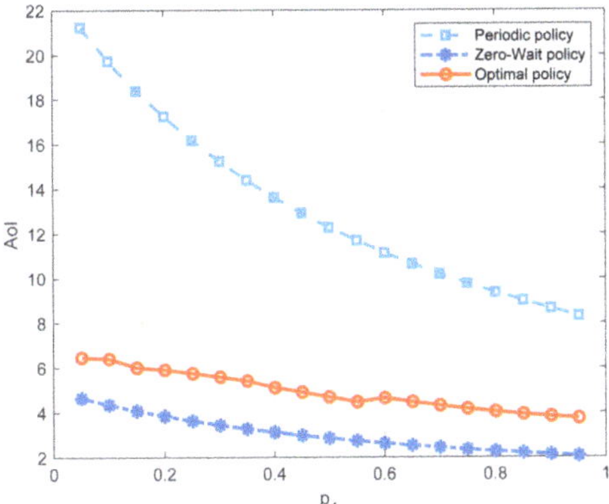

Figure 8. AoI comparison of the zero-wait policy, the periodic policy with period being 5, and the optimal policy with respect to p_1 ($q_0 = 0.2, q_1 = 0.5, \omega = 10, C_e = 1$).

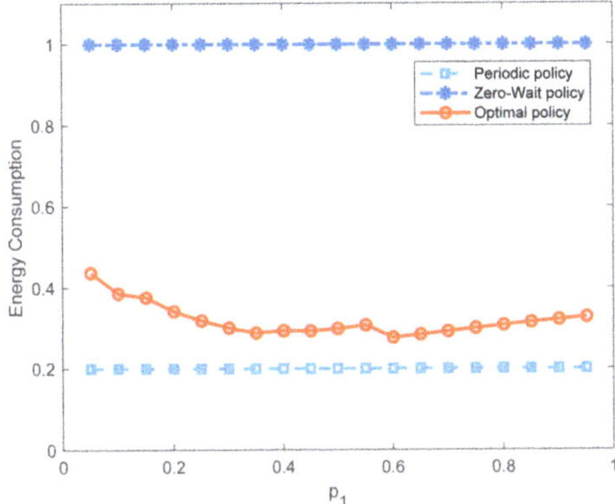

Figure 9. Energy consumption comparison of the zero-wait policy, the periodic policy with period being 5, and the optimal policy with respect to p_1 ($q_0 = 0.2, q_1 = 0.5, \omega = 10, C_e = 1$).

5. Conclusions

In this paper, we have studied the optimal updating policy in an IoT system, where the channel gain is quantized into multiple states and the channel state is fed back to the sensor before the decision making. The status update problem has been formulated as an MDP to minimize the long-term average of the weighted sum of the AoI and the energy consumption. By investigating the properties of the value function, it is proven that the optimal policy has a threshold structure with respect to AoI for any given channel state. We have also proven that the threshold is a non-increasing function of the channel state. Simulation results show the impacts of system parameters on the optimal thresholds and the average cost. Through comparisons, we have also shown that our proposed policy outperforms the zero-wait policy and the periodic policy. In our future research, the time-varying channel model will be further involved for guiding the future design of realistic IoT systems.

Author Contributions: Conceptualization, F.P., X.C., and X.W.; methodology, F.P. and X.W.; software, F.P.; validation, F.P., X.C., and X.W.; formal analysis, F.P. and X.W.; investigation, X.W.; writing—original draft preparation, F.P.; writing—review and editing, X.W. and X.C.; visualization, F.P. and X.W.; supervision, X.C. and X.W.; project administration, X.C.; funding acquisition, X.C. All authors have read and agreed to the published version of the manuscript.

Funding: This work was supported in part by the State's Key Project of Research and Development Plan under Grants (2019YFE0196400), in part by Guangdong R&D Project in Key Areas under Grant (2019B010158001), in part by Guangdong Basic and Applied Basic Research Foundation under Grant (2021A1515012631), in part by Key Laboratory of Modern Measurement & Control Technology, Ministry of Education, Beijing Information Science & Technology University (KF20201123202).

Institutional Review Board Statement: Not applicable.

Informed Consent Statement: Not applicable.

Data Availability Statement: Not applicable.

Conflicts of Interest: The authors declare no conflict of interest.

Appendix A. Proof of Lemma 2

Based on the value iteration algorithm, the induction method can be employed in following proof. Firstly, we initial that $V_{\gamma,0}(\mathbf{x}) = 0$, where both Equations (15) and (16) hold for all $\mathbf{x} \in \mathcal{S}$.

Appendix A.1. Proof of Equation (16)

When $n = 1$,

$$Q_{\gamma,1}(\Delta_1, i, 0) - Q_{\gamma,1}(\Delta_2, i, 0) = \Delta_1 - \Delta_2 \leq 0, \tag{A1}$$

and

$$Q_{\gamma,1}(\Delta_1, i, 1) - Q_{\gamma,1}(\Delta_2, i, 1) = \Delta_1 + \omega C_e - (\Delta_2 + \omega C_e) \leq 0, \tag{A2}$$

hold due to $\Delta_1 \leq \Delta_2$, and we have $V_{\gamma,1}(\Delta_1, i) \leq V_{\gamma,1}(\Delta_2, i)$.

Suppose that $V_{\gamma,K}(\Delta_1, i) \leq V_{\gamma,K}(\Delta_2, i)$ holds for $k \leq K$. Considering the case of $k = K + 1$,

$$Q_{\gamma,K+1}(\Delta_1, i, 0) - Q_{\gamma,K+1}(\Delta_2, i, 0) = \Delta_1 - \Delta_2 \leq 0, \tag{A3}$$

and

$$\begin{aligned}
&Q_{\gamma,K+1}(\Delta_1, i, 1) - Q_{\gamma,K+1}(\Delta_2, i, 1) \\
&= (\Delta_1 - \Delta_2) + \gamma \sum_{j \in \mathbb{H}} p_j g_i \big(V_{\gamma,K}(\Delta_1 + 1, j) - V_{\gamma,K}(\Delta_2 + 1, j) \big) \\
&\leq 0,
\end{aligned} \tag{A4}$$

hold for all i according to $\Delta_1 \leq \Delta_2$. Therefore, we have $V_{\gamma,K+1}(\Delta_1, i) \leq V_{\gamma,K+1}(\Delta_2, i)$. Since $\lim_{n \to \infty} V_{\gamma,n} = V_\gamma$, we have $V_\gamma(\Delta_1, i) \leq V_\gamma(\Delta_2, i)$.

Appendix A.2. Proof of Equation (15)

By the definition of function $Q_\gamma(\mathbf{x}, a)$, we have

$$Q_\gamma(\Delta, i, 0) = \Delta + \gamma \sum_{j \in \mathbb{H}} p_j V_\gamma(\Delta + 1, j), \tag{A5}$$

and

$$Q_\gamma(\Delta, i, 1) = \Delta + \omega C_e + \gamma \left(\sum_{j \in \mathbb{H}} p_j (g_i V_\gamma(\Delta + 1, j) + q_i V_\gamma(1, j)) \right). \tag{A6}$$

Therefore,

$$Q_\gamma(\Delta, N-1, 0) - Q_\gamma(\Delta, i, 0) \le 0, \tag{A7}$$

and

$$Q_\gamma(\Delta, N-1, 1) - Q_\gamma(\Delta, i, 1)$$
$$\overset{(a)}{=} \gamma \sum_{j \in \mathbb{H}} p_j(q_{N-1} - q_i)(V_\gamma(1, j) - V_\gamma(\Delta+1, j)) \le 0, \tag{A8}$$

hold for all i, where step (a) is due to Equation (16). Hence, we have $V_\gamma(\Delta, N-1) \le V_\gamma(\Delta, i)$. This completes the whole proof.

Appendix B. Proof of Theorem 1

Theorem 1 can be proven by verifying the conditions given in [16]. The conditions are listed as follows:

- (1): For every state $\mathbf{x}$ and discount factor γ, the discount value function $V_\gamma(\mathbf{x})$ is finite.
- (2): There exists a non-negative value L such that $-L \le h_\gamma(\mathbf{x})$ for all $\mathbf{x}$ and γ, where $h_\gamma(\mathbf{x}) = V_\gamma(\mathbf{x}) - V_\gamma(\hat{\mathbf{x}})$, and $\hat{\mathbf{x}}$ is a reference state.
- (3): There exists a non-negative value $M_{\mathbf{x}}$, such that $h_\gamma(\mathbf{x}) \le M_{\mathbf{x}}$ for every $\mathbf{x}$ and γ. For every $\mathbf{x}$, there exists an action $a_{\mathbf{x}}$ such that $\sum_{\mathbf{x}'} \Pr(\mathbf{x}'|\mathbf{x}, a_{\mathbf{x}})M_{\mathbf{x}'} < \infty$.
- (4): The inequality $\sum_{\mathbf{x}'} \Pr(\mathbf{x}'|\mathbf{x}, a)M_{\mathbf{x}'} < \infty$ holds for all $\mathbf{x}$ and a.

By Lemma 1, $V_\gamma(\hat{\mathbf{x}}) = \min_\pi V_{\pi,\gamma}(\hat{\mathbf{x}}) < \infty$ holds for every $\hat{\mathbf{x}}$ and γ. Hence, condition (1) holds. According to Lemma 2, by letting $\hat{\mathbf{x}} = (1, N-1)$ and $L = 0$, we have $h_\gamma(\mathbf{x}) \ge 0$, which verifies condition (2).

Before verifying condition (3), a lemma is given as follows:

Lemma A1. *Let us denote $\hat{\mathbf{x}} = (1, N-1)$ as the reference state and define the first time that an initial state $\mathbf{x}$ transits to $\hat{\mathbf{x}}$ as $K = \min\{k : k \le 1, \mathbf{x}_k = \hat{\mathbf{x}}\}$. Then, the expectation cost under the always-transmitting policy π_a, i.e., the sensor generates and transmits a new update in each slot, is*

$$C_{\mathbf{x},\hat{\mathbf{x}}}(\pi_a) = \mathbb{E}_{\pi_a}\left[\sum_{t=0}^{K-1} \gamma^t C(\mathbf{x}_t, a_t)|\mathbf{x}\right], \tag{A9}$$

where $C_{\mathbf{x},\hat{\mathbf{x}}}(\pi_a) < \infty$ holds for all $\mathbf{x}$.

Proof. Since $a_t = 1$ for all t, the probability that the state returns to $\hat{\mathbf{x}}$ from $\mathbf{x}$ after exactly K slot is given by

$$\Pr(K = k|\mathbf{x} = (\Delta, j)) = \begin{cases} p_{N-1}q_j, & \text{if } k = 1, \\ p_{N-1}g_j(1 - \sum_{i \in \mathbb{H}} p_i q_i)^{k-2}(\sum_{i \in \mathbb{H}} p_i q_i), & \text{otherwise.} \end{cases} \tag{A10}$$

Then, the expectation return cost from $\mathbf{x}$ to $\hat{\mathbf{x}}$ is expressed as

$$
\begin{aligned}
C_{\mathbf{x},\hat{\mathbf{x}}}(\pi_a) \\
&= \mathbb{E}\left[\sum_{t=0}^{K-1}\gamma^t C(\mathbf{x}_t, a_t)|\mathbf{x}\right] \\
&\overset{(a)}{\le} \sum_{k=1}^{\infty} \Pr(K=k|\mathbf{x}=(\Delta,j))\left[\sum_{m=0}^{k-1}(\Delta+m+\omega C_e)\right] \\
&< \infty,
\end{aligned}
\tag{A11}
$$

where step (a) is due to the fact that $C(\mathbf{x}_t, a_t) \le \Delta_t + \omega C_e$. $\square$

Considering a mixture policy π, in which it performs the always-transmitting policy π_a from initial state $\mathbf{x}$ until it enters the reference state $\hat{\mathbf{x}}$, it later performs the optimal policy π_γ that minimizes the discounted cost. Therefore, we have

$$
\begin{aligned}
V_\gamma(\mathbf{x}) \\
&\le \mathbb{E}_{\pi_a}\left[\sum_{t=0}^{K-1}\gamma^t C(\mathbf{x}_t, a_t)|\mathbf{x}\right] + \mathbb{E}_{\pi_b}\left[\sum_{t=K}^{\infty}\gamma^t C(\mathbf{x}_t, a_t)|\hat{\mathbf{x}}\right] \\
&\le C_{\mathbf{x},\hat{\mathbf{x}}}(\pi_a) + \mathbb{E}_\pi\left[\gamma^K V_\gamma(\hat{\mathbf{x}})\right] \\
&\le C_{\mathbf{x},\hat{\mathbf{x}}}(\pi_a) + V_\gamma(\hat{\mathbf{x}}),
\end{aligned}
\tag{A12}
$$

which implies that $h_\gamma(\mathbf{x}) \le C_{\mathbf{x},\hat{\mathbf{x}}}(\pi_a)$. Hence, let $\hat{\mathbf{x}} = (1, N-1)$ and $M_{\mathbf{x}} = C_{\mathbf{x},\hat{\mathbf{x}}}(\pi_a)$; condition (3) is verified.

On the other hand, $M_{\mathbf{x}} < \infty$ holds for all $\mathbf{x}$. The states that transit from $\mathbf{x}$ are finite. Thus, the weighted sum of finite $M_{\mathbf{x}}$ is also finite, i.e., $\sum_{\mathbf{x}'} P(\mathbf{x}'|\mathbf{x}, a)M_{\mathbf{x}'} < \infty$ holds for all $\mathbf{x}$ and a, which verifies condition (4). This completes the whole verification.

Appendix C. Proof of Theorem 2

Based on the definition of $Q(\Delta, i, a)$, we can obtain the difference between the state-action value function as follows:

$$
\begin{aligned}
Q(\Delta, i, 0) - Q(\Delta, i, 1) \\
&= \sum_{j=0}^{N-1} p_j V(\Delta+1, j) - q_i \sum_{j=0}^{N-1} p_j V(1, j) - g_i \sum_{j=0}^{N-1} p_j V(\Delta+1, j) - \omega C_e \\
&= q_i \sum_{j=0}^{N-1} p_j(V(\Delta+1, j) - V(1, j)) - \omega C_e \\
&\overset{(a)}{\ge} q_i \Delta - \omega C_e.
\end{aligned}
\tag{A13}
$$

where (a) is due to the property of the value function given in Lemma 4. We then discuss the difference between the state-action value function in two cases.

Case 1: $\omega = 0$.

In this case, $Q(\Delta, i, 0) - Q(\Delta, i, 1) \ge 0$ holds for any Δ and i. Therefore, the optimal policy is to update at each slot in spite of the channel state. In other words, the optimal thresholds are all equal to 1.

Case 2: $\omega > 0$.

We note that, given i, $q_i \Delta - \omega C_e$ increases linearly with Δ. Hence, there exists a positive integer $\hat{\beta}_i$, such that $\hat{\beta}_i$ is the minimum value that satisfies $q_i \hat{\beta}_i - \omega C_e \ge 0$. Therefore, if $\Delta \ge \hat{\beta}_i$, $Q(\Delta, i, 0) - Q(\Delta, i, 1) \ge q_i \hat{\beta}_i - \omega C_e \ge 0$ holds. This implies that there must exist a threshold β_i satisfying $1 \le \beta_i \le \hat{\beta}_i$. If $\Delta \ge \beta_i$, we have $Q(\Delta, i, 0) - Q(\Delta, i, 1) \ge 0$.

Altogether, the optimal policy has a threshold structure for $\omega \geq 0$. Then, we examine the non-increasing property of the thresholds. Firstly, we show that the difference between the state-action value function is monotonic with respect to the channel state by fixing the AoI. Assuming that $i, j \in \mathbb{H}$ and $i \leq j$, it is easy to obtain that

$$Q(\Delta, j, 0) - Q(\Delta, j, 1) - (Q(\Delta, i, 0) - Q(\Delta, i, 1))$$
$$= (q_j - q_i) \sum_{l=0}^{N-1} p_l(V(\Delta+1, l) - V(1, l)) \geq 0. \tag{A14}$$

Since $Q(\Delta, i, 0) - Q(\Delta, i, 1) \geq 0$ when $\Delta \geq \beta_i$, we have $Q(\Delta, j, 0) - Q(\Delta, j, 1) \geq 0$ according to (A14). This implies that the optimal threshold β_j corresponding to channel state j is no greater than β_i, i.e., $\beta_j \leq \beta_i$. This completes the whole proof.

Appendix D. Proof of Theorem 3

Assume that $\varphi(\mathbf{x})$ is the steady probability of state $\mathbf{x}$ in a Markov chain. The steady state probability $\varphi(\mathbf{x})$ satisfies the following global balance equation [18], i.e.,

$$\varphi(\mathbf{x}) = \sum_{\mathbf{x}' \in \mathcal{S}} \varphi(\mathbf{x}') \Pr(\mathbf{x}|\mathbf{x}'). \tag{A15}$$

Let $\varphi_1 = \varphi(1, 0) + \varphi(1, 1)$. We prove Equation (A23) by discussing three cases via mathematical induction.

Case 1: $1 < i \leq \beta_1$

Based on Equation (A15), we have

$$\varphi(2, j) = \varphi(1, 0)p_j + \varphi(1, 1)p_j$$
$$= p_j \varphi_1. \tag{A16}$$

Assuming that $\varphi(i, j) = p_j \varphi_1$ holds for all $i \leq k < \beta_1$, we examine $\varphi(k+1, j)$. We have

$$\varphi(k+1, j) = \varphi(k, 0)p_j + \varphi(k, 1)p_j$$
$$= p_j p_0 \varphi_1 + p_j p_1 \varphi_1$$
$$= p_j \varphi_1, \tag{A17}$$

which completes this segment of the proof.

Case 2: $\beta_1 < i \leq \beta_0$

Similarly, we have

$$\varphi(\beta_1 + 1, j) = p_j(1 - q_1)\varphi(\beta_1, 1) + p_j \varphi(\beta_1, 0)$$
$$= p_j \varphi_1((1 - q_1)p_1 + p_0)$$
$$= p_j \varphi_1 s_0, \tag{A18}$$

where $s_0 = 1 - p_1 q_1$. Assuming that $\varphi(i, j) = p_j \varphi_1 s_0^{i-\beta_1}$ holds for all $\beta_1 < i \leq k < \beta_0$, we examine $\varphi(k+1, j)$. We have

$$\varphi(k+1, j) = p_j(1 - q_1)\varphi(k, 1) + p_j \varphi(k, 0)$$
$$= p_j \varphi_1((1 - q_1)p_1 + p_0)s_0^{k-\beta_1}$$
$$= p_j \varphi_1 s_0^{k+1-\beta_1}, \tag{A19}$$

which completes this segment of the proof.

Case 3: $i > \beta_0$

Follow above discussion, we have

$$
\begin{aligned}
\varphi(\beta_0 + 1, j) &= p_j(1 - q_1)\varphi(\beta_0, 1) + p_j(1 - q_0)\varphi(\beta_0, 0) \\
&= p_j\varphi_1 s_0^{\beta_0 - \beta_1}((1 - q_1)p_1 + (1 - q_0)p_0) \\
&= p_j\varphi_1 s_0^{\beta_0 - \beta_1} s_1,
\end{aligned}
\tag{A20}
$$

where $s_1 = 1 - p_0 q_0 - p_1 q_1$. Assuming that $\varphi(i, j) = p_j \varphi_1 s_0^{\beta_0 - \beta_1} s_1^{i - \beta_0}$ holds for all $\beta_0 < i \le k$, we examine $\varphi(k + 1, j)$. We have

$$
\begin{aligned}
&\varphi(k + 1, j) \\
&= p_j(1 - q_1)\varphi(k, 1) + p_j(1 - q_0)\varphi(k, 0) \\
&= p_j\varphi_1 s_0^{\beta_0 - \beta_1}((1 - q_1)p_1 + (1 - q_0)p_0)s_1^{k - \beta_0} \\
&= p_j\varphi_1 s_0^{\beta_0 - \beta_1} s_1^{k + 1 - \beta_0}.
\end{aligned}
\tag{A21}
$$

Altogether, we obtain the steady state probability with respect to an unknown parameter φ_1. According to the fact that $\sum_{i=1}^{\infty} \sum_{j=0}^{N-1} \varphi(i, j) = 1$, we formulate an equation:

$$
\begin{aligned}
&\sum_{i=1}^{\infty} \sum_{j=0}^{N-1} \varphi(i, j) \\
&= \varphi_1 \left\{ \beta_1 + \sum_{i=\beta_1+1}^{\beta_0} s_0^{i-\beta_1} + \sum_{i=\beta_0+1}^{\infty} s_0^{\beta_0 - \beta_1} s_1^{i - \beta_0} \right\} \\
&= \varphi_1 \left\{ \beta_1 + \frac{s_0 - s_0^{\beta_1 - \beta_0}}{1 - s_0} + s_0^{\beta_1 - \beta_0} \frac{s_1}{1 - s_1} \right\} \\
&= 1,
\end{aligned}
\tag{A22}
$$

where the expression of φ_1 is obtained.

The average cost of a Markov chain is given by

$$
C_{mc} = \sum_{\mathbf{x} \in \mathcal{S}} \varphi(\mathbf{x}) C(\mathbf{x}, \pi^*(\mathbf{x})).
\tag{A23}
$$

Substituting (24) into (A23), we have

$$
\begin{aligned}
&C_{mc}(\beta_0, \beta_1) \\
&= \sum_{i=1}^{\infty} i \sum_{j=0}^{1} \varphi(i, j) + \sum_{i=\beta_0}^{\infty} \omega C_e \varphi(i, 0) + \sum_{i=\beta_1}^{\infty} \omega C_e \varphi(i, 1).
\end{aligned}
$$

Furthermore, the first term is given by

$$
\begin{aligned}
&\sum_{i=1}^{\infty} i \sum_{j=0}^{1} \varphi(i, j) \\
&= \sum_{i=1}^{\beta_1} i\varphi_i + \sum_{i=\beta_1+1}^{\beta_0} i s_0^{i - \beta_1} + s_0^{\beta_0 - \beta_1} \sum_{i=\beta_0+1}^{\infty} i s_1^{i - \beta_0} \\
&= \frac{\varphi_1 \beta_1(\beta_1 + 1)}{2} + \varphi_1 A + \varphi_1 B,
\end{aligned}
\tag{A24}
$$

where

$$
A = \frac{s_0((\beta_1 + 1) - \beta_0 s_0^{\beta_0 - \beta_1})}{1 - s_0} + \frac{s_0^3 - s_0^{\beta_0 - \beta_1 + 1}}{(1 - s_0)^2},
\tag{A25}
$$

and

$$B = \frac{(\beta_0 + 1)s_1}{1 - s_1} + \frac{s_1^3}{1 - s_1}.$$

(A26)

Furthermore, the sum of last two terms is given by

$$\sum_{i=\beta_0}^{\infty} \omega C_e \varphi(i, 0) + \sum_{i=\beta_1}^{\infty} \omega C_e \varphi(i, 1)$$

$$= \varphi_1 \omega C_e \left(s_0^{\beta_0 - \beta_1} \sum_{i=\beta_0}^{\infty} s_1^{i - \beta_0} + p_1 \sum_{i=\beta_1}^{\beta_0 - 1} s_0^{i - \beta_1} \right)$$

$$= \varphi_1 \omega C_e \left(s_0^{\beta_0 - \beta_1} \frac{1}{1 - s_1} + p_1 \frac{1 - s_0^{\beta_0 - \beta_1}}{1 - s_0} \right).$$

(A27)

This completes the proof.

References

1. Kaul, S.; Yates, R.; Gruteser, M. Real-time status: How often should one update? In Proceedings of the IEEE INFOCOM, Orlando, FL, USA, 25–30 March 2012; pp. 2731–2735.
2. Abd-Elmagid, M.A.; Dhillon, H.S. Average peak age-of-information minimization in UAV-assisted IoT networks. *IEEE Trans. Veh. Technol.* **2018**, *68*, 2003–2008. [CrossRef]
3. Farazi, S.; Klein, A.G.; Brown, D.R. Average age of information for status update systems with an energy harvesting server. In Proceedings of the IEEE INFOCOM 2018-IEEE Conference on Computer Communications Workshops (INFOCOM WKSHPS), Honolulu, HI, USA, 15–19 April 2018; pp. 112–117.
4. Costa, M.; Codreanu, M.; Ephremides, A. On the age of information in status update systems with packet management. *IEEE Trans. Inf. Theory* **2016**, *62*, 1897–1910. [CrossRef]
5. Sun, Y.; Uysal-Biyikoglu, E.; Yates, R.D.; Koksal, C.E.; Shroff, N.B. Update or wait: How to keep your data fresh. *IEEE Trans. Inf. Theory* **2017**, *63*, 7492–7508. [CrossRef]
6. Yun, J.; Joo, C.; Eryilmaz, A. Optimal real-time monitoring of an information source under communication costs. In Proceedings of the 2018 IEEE Conference on Decision and Control (CDC), Miami, FL, USA, 17–19 December 2018; pp. 4767–4772.
7. Imer, O.C.; Basar, T. Optimal estimation with limited measurements. In Proceedings of the 44th IEEE Conference on Decision and Control, Seville, Spain, 15 December 2005; pp. 1029–1034.
8. Rabi, M.; Moustakides, G.V.; Baras, J.S. Adaptive sampling for linear state estimation. *Siam J. Control Optim.* **2009**, *50*, 672–702. [CrossRef]
9. Gao, X.; Akyol, E.; Başar, T. On remote estimation with multiple communication channels. In Proceedings of the 2016 American Control Conference (ACC), Boston, MA, USA, 6–8 July 2016; pp. 5425–5430.
10. Chakravorty, J.; Mahajan, A. Remote-state estimation with packet drop. *IFAC-PapersOnLine* **2016**, *49*, 7–12. [CrossRef]
11. Lipsa, G.M.; Martins, N.C. Optimal state estimation in the presence of communication costs and packet drops. In Proceedings of the 2009 47th Annual Allerton Conference on Communication, Control, and Computing (Allerton), Monticello, IL, USA, 30 September–2 October 2009; pp. 160–169.
12. Huang, H.; Qiao, D.; Gursoy, M.C. Age-energy tradeoff in fading channels with packet-based transmissions. In Proceedings of the IEEE INFOCOM 2020-IEEE Conference on Computer Communications Workshops (INFOCOM WKSHPS), Toronto, ON, Canada, 6–9 July 2020; pp. 323–328.
13. Chakravorty, J.; Mahajan, A. Remote estimation over a packet-drop channel with Markovian state. *IEEE Trans. Autom. Control* **2019**, *65*, 2016–2031. [CrossRef]
14. Chen, X.; Yi, H.; Luo, H.; Yu, H.; Wang, H. A novel CQI calculation scheme in LTE/LTE-A systems. In Proceedings of the 2011 International Conference on Wireless Communications and Signal Processing (WCSP), Nanjing, China, 9–11 November 2011; pp. 1–5.
15. Toyserkani, A.T.; Strom, E.G.; Svensson, A. An analytical approximation to the block error rate in Nakagami-m non-selective block fading channels. *IEEE Trans. Wirel. Commun.* **2010**, *9*, 1543–1546. [CrossRef]
16. Sennott, L.I. Average cost optimal stationary policies in infinite state Markov decision processes with unbounded costs. *Oper. Res.* **1989**, *37*, 626–633. [CrossRef]
17. Dinkelbach, W. On nonlinear fractional programming. *Manag. Sci.* **1967**, *13*, 492–498. [CrossRef]
18. Chandry, K.M. *The Analysis and Solutions for General Queueing Networks*; Citeseer: Princeton, NJ, USA, 1974.

Article

Scheduling Strategy Design Framework for Cyber–Physical System with Non-Negligible Propagation Delay

Zuoyu An, Shaohua Wu *, Tiange Liu, Jian Jiao and Qinyu Zhang

Communication Engineering Research Centre, Harbin Institute of Technology (Shenzhen), Shenzhen 518055, China; 19s052043@stu.hit.edu.cn (Z.A.); 19s152063@stu.hit.edu.cn (T.L.); jiaojian@hit.edu.cn (J.J.); zqy@hit.edu.cn (Q.Z.)
* Correspondence: hitwush@hit.edu.cn

Abstract: Cyber–physical systems (CPS) have been widely employed as wireless control networks. There is a special type of CPS which is developed from the wireless networked control systems (WNCS). They usually include two communication links: Uplink transmission and downlink transmission. Those two links form a closed-loop. When such CPS are deployed for time-sensitive applications such as remote control, the uplink and downlink propagation delay are non-negligible. However, existing studies on CPS/WNCS usually ignore the propagation delay of the uplink and downlink channels. In order to achieve the best balance between uplink and downlink transmissions under such circumstances, we propose a heuristic framework to obtain the optimal scheduling strategy that can minimize the long-term average control cost. We model the optimization problem as a Markov decision process (MDP), and then give the sufficient conditions for the existence of the optimal scheduling strategy. We propose the semi-predictive framework to eliminate the impact of the coupling characteristic between the uplink and downlink data packets. Then we obtain the lookup table-based optimal offline strategy and the neural network-based suboptimal online strategy. Numerical simulation shows that the scheduling strategies obtained by this framework can bring significant performance improvements over the existing strategies.

Keywords: cyber–physical system; wireless networked control system; remote control; communication control co-design; age of information

Citation: An, Z.; Wu, S.; Liu, T.; Jiao, J.; Zhang, Q. Scheduling Strategy Design Framework for Cyber–Physical System with Non-Negligible Propagation Delay. *Entropy* **2021**, *23*, 714. https://doi.org/10.3390/e23060714

Academic Editors: Anthony Ephremides and Yin Sun

Received: 6 May 2021
Accepted: 2 June 2021
Published: 4 June 2021

Publisher's Note: MDPI stays neutral with regard to jurisdictional claims in published maps and institutional affiliations.

1. Introduction

In the recent past, applications of the wireless control networks have become more and more extensive, such as drone formations, autonomous vehicles, automatic factories, etc. Some of those scenarios implicate new requirements for remote control technology, which is a sub-topic of communication control co-design. Remote control technology originates from wireless control systems with long propagation delay such as far-sea monitoring and high-efficiency satellite IoT. The main cause of long propagation delay is the large-scale geographic distance. This feature makes it extremely challenging to design CPS under this scenario. In order to meet the need of remote control with propagation delay, that is, to maintain stable closed-loop control and reduce control costs, we propose a new framework to design uplink and downlink scheduling strategies.

As show in Figure 1, a typical CPS deployed under the single closed-loop control scenario contains a control system and a communication system. In the rest of this article, we use single-loop CPS to refer to this specific type of CPS. The communication process of a typical single-loop CPS can be divided into two parts: Uplink sensor transmission and downlink controller transmission. The uplink transmission is initiated by the sensor and sends the state update packet from the plant to the controller. The controller first uses this data to obtain a more accurate estimate of the factory status. Then the downlink transmission is initiated to send command information from the controller to the actuator located at the factory. The actuator acts on the factory to maintain the factory's stability.

Taking into account the characteristics of a control system, the command can only be generated with an accurate estimation, which means the downlink transmission must occur after a successful uplink transmission . Because of this fixed timing relationship, CPS has to work in half-duplex in most cases: namely, only one of the uplink sensor transmission and the downlink controller transmission can be activated to send a data packet in the same time slot. That means there is a problem of how to design a scheduling strategy between those two transmissions. Note that the uplink and downlink channels here are not just a single wireless channel, but a simplified modeling of a fixed routing link with multiple relays. This scenario is for some special remote control systems that use satellites as relays. Therefore, the propagation delay in our paper is essentially a collection of various delays contained in the entire relay link, including processing delay, transmission delay, propagation delay, etc. This unified modeling is used because the link characteristics of a fixed routing multi-relay link can be described by an equivalent link with a specific code error rate and propagation delay.

There are many related works about WNCS and CPS [1–4]. Focusing on the conflict of the accuracy requirements of control systems and the limited quantization level [5], proposed the application of dynamic quantization technology in the communication control co-design. Some works designed CPS with the limitation of wireless coding process, such as code length allocation [6,7], code length design [8,9] and adaptive code length adjustment [10]. Considering the fading characteristics of transmission channels, studies of adaptive transmit power adjustment technology by predicting the fast or slow fading of transmission channels are proposed in [11,12]. Some of the above studies include the idea of designing CPS for time-sensitive applications. Nowadays, the most widely used measure of timeliness is Age of Information (AoI) [13], which is defined as the time elapsed since a certain data packet was generated:

$$\Delta(t) = t - t'\tag{1}$$

where t represents the current time, t' represents the time when the packet was generated. It used to be very difficult to express the control performance measurement, that is, the system state mean square error (MSE) [14] when the control system and the communication system are combined. The proposal of AoI changed this situation. For example, the system state MSE of a linear time invariant system (LTI) can be simply expressed as a function of AoI. This improvement greatly reduces the difficulty of describing the overall system performance in the communication control co-design scenario [15,16].

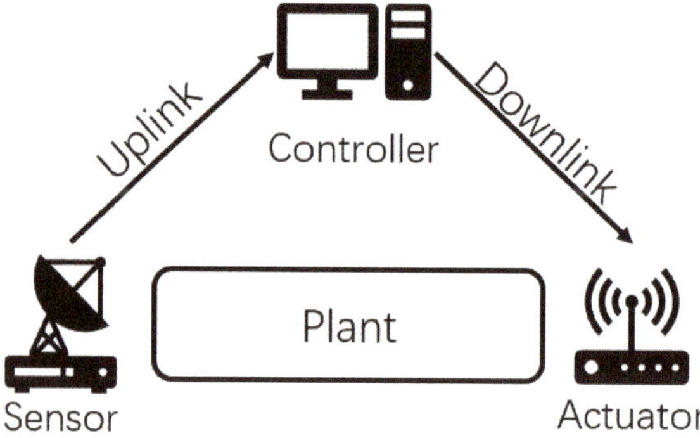

Figure 1. Cyber–physical system deployed under the single closed-loop control scenario.

Based on AoI, many related studies have been derived, such as the application of the HARQ mechanism for single-loop CPS to improve the overall timeliness [17,18], and the scheduling strategy aiming to minimize the long-term average MSE for single-loop

CPS without transmission delay [19]. Some studies about the multi-loop scheduling strategy design aiming at optimizing timeliness have also been proposed. Reference [20] focuses on the design of the data inter-arrival rate and code length allocation strategy. References [21,22] proposed the uplink scheduling strategy of multi-loop WNCS under the ideal assumption of downlink transmission. Furthermore, the authors of [23,24] discuss the application of data packet transmission result prediction technology in WNCS design.

The scenarios studied above concern mainly short-distance Industrial Internet of Things (IIoT), so the impact of uplink and downlink propagation delay on the closed-loop control performance of a CPS is generally ignored. Besides, the above studies only consider one of the two code error rates of the uplink and the downlink transmission. Under the remote control scenario, the code error rates and propagation delay of both links are not only non-negligible, but also have a huge impact on the overall performance of the single-loop CPS. Some works have studied the design of WNCS optimal control strategy under time-delay scenarios [25–27]. However, they do not consider the impact of the code error rate and the scheduling strategy which are issues that cannot be ignored in the design of communication systems in the field of communication engineering. To this end, we propose a new framework to obtain the optimal scheduling strategy while considering both the code error rates and propagation delay. This strategy can minimize the long-term average control cost.

Firstly, we model the single-loop CPS as an MDP problem and give the sufficient conditions for the stability of CPS. Secondly, we propose a heuristic semi-predictive framework to eliminate the impact of the coupling characteristic between the uplink and downlink data packets. Finally, we obtain the lookup table-based optimal offline strategy and the neural network-based suboptimal online strategy for the single-loop CPS. The whole process can be expanded according to actual deployment requirements with any fixed propagation delay as long as the sufficient condition is satisfied.

The rest of this paper is organized as follows: In Section 2, we provide the system model and formulate the optimization problem. In Section 3, we introduce the semi-predictive framework and transform the optimization problem into an MDP problem. In Section 4, we obtain the optimal offline strategy and the suboptimal online strategy. In Section 5, we show the numerical simulation results. We conclude this work in Section 6.

2. System Model

2.1. The Plant of the Single-Loop CPS

First, we model the plant in the single-loop CPS as a discrete-time LTI system:

$$X_{k+1} = AX_k + BU_k + Z_k, \forall k \tag{2}$$

where k represents the k-th time slot, $X_k \in \mathbb{R}$ represents the state of the plant at time slot k, $U_k \in \mathbb{R}$ represents the executed control command, $Z_k \in \mathbb{R}$ represents the normally distributed plant noise whose mean and variance are $\bar{z}$ and $\mathcal{K}$, respectively. $A \in \mathbb{R}$ represents the state transition coefficient, $B \in \mathbb{R}$ represents the command control coefficient. We assume that the plant state remains unchanged within a single time slot. The goal of CPS is to maintain X around 0.

2.2. The Communication Process of the Single-Loop CPS

In the previous subsection, we explained that the entire single-loop CPS works in the half-duplex mode. Now we will explain the communication process of the single-loop CPS. The entire system adopts a centralized scheduling scheme because this scheme is more suitable for single-loop CPS. Under this scheme, the scheduling decision of uplink and downlink transmission is completely determined by the remote controller. We use a_k to represent the scheduling decision made by the controller in the time slot k. If the controller schedules uplink transmission in the slot k, $a_k = 1$. If the controller schedules downlink transmission in the slot k, $a_k = 2$. We assume that the code error rate of the uplink and downlink transmission channels are $p_s, p_c \in (0, 1)$, respectively. Both code error rates are constant which means the uplink and downlink transmission fails with probability (p_s, p_c) in any time slot, respectively. Then we use δ_k to represent the transmission result of the packet sent in the time slot k. No matter which transmission is scheduled, if it succeeds, then $\delta_k = 1$. Otherwise, $\delta_k = 0$. Since the processing procedures of most actual CPS are digital, the packets that have experienced a certain delay will start to be processed in the next processing cycle after it is received; we model the propagation delay of the uplink and downlink channel integer time slots $d_{up}, d_{down} \in \mathbb{R}$, respectively. To simplify the analysis, we assume that the transmission of scheduling instructions and feedback information is ideal.

In addition to the variables described above, we define the following two parameters to describe the status of each part in a single-loop CPS:

(1) State Estimation Age τ_k: This is defined as the age of the latest valid uplink state update packet successfully received by the controller at the end of the time slot k. τ_k reflects the accuracy of the estimation maintained by the remote controller. Because of the uplink propagation delay, the minimum value of state estimation age is d_{up}. When the specific time slot is not considered, it is abbreviated as τ. Its update rule is as follows:

$$\tau_{k+1} = \begin{cases} d_{up} & \text{if } (a_j = 1)\,\&\,(\delta_j = 1) \\ \tau_k + 1 & \text{otherwise} \end{cases} \tag{3}$$

where $j = k - d_{up} + 1$.

(2) State Control Age φ_k: This is defined as the age of the uplink packet used to generate the latest successfully received downlink packet by the actuator at the end of the time slot k. This parameter represents the total time it takes for the entire CPS to complete a closed-loop control process. It reflects the degree of divergence of the plant's state. Because of the uplink and downlink propagation delay, the minimum value of the state control age is $d_{up} + d_{down}$. When the specific time slot is not considered, it is abbreviated as φ. Its update rule is as follows:

$$\varphi_{k+1} = \begin{cases} \tau_q + d_{down} & \text{if } (a_q = 2)\,\&\,(\delta_q = 1) \\ \varphi_k + 1 & \text{otherwise} \end{cases} \tag{4}$$

where $q = k - d_{down} + 1$. The abbreviations j and q will be used in the rest of this paper. Note that we set the initial values of τ_0 and φ_0 to be 2. These values can be arbitrarily selected within a reasonable range. This is because the long-term average cost we focus on is not affected by those initial values.

2.3. The Control Process of the Single-Loop CPS

In this subsection, we will explain the control process of the single-loop CPS in detail, which is mainly completed by the remote controller and the actuator. The task of the remote controller can be divided into three parts: Maintaining state estimation, generating control commands, and scheduling uplink and downlink transmissions, while the actuator has only one task: Executing the received control commands.

(1) Maintaining State Estimation: We assume that the sensor can sample the state of the plant without distortion. The uplink transmission cannot be scheduled in every time slot. What is more, the scheduled transmission can fail because of the code error occurring during its propagation process. So the remote controller cannot receive a new state update packet in every time slot. Under these circumstances, the remote controller has to update the estimation $\tilde{X}_k$ of the plant state X_k through the following process:

$$\tilde{X}_{k+1} = \begin{cases} g^{d_{up}}(X_j, k) & \text{if } (a_j = 1)\,\&\,(\delta_j = 1) \\ A\tilde{X}_k + BU_k & \text{otherwise} \end{cases} \tag{5}$$

where $g(X, k) = AX + BU_k$, $g^n(X, k) = g(g^{n-1}(X, k-1), k)$ $\forall n > 1$, and $g^1(X, k) = g(X, k)$. In this scenario, this estimation method has been proven to be optimal [28]. When a certain uplink transmission is successful, the remote controller can use the plant state $X_{k-d_{\text{up}}+1}$, which is the exact value for $d_{\text{up}} - 1$ time slots before, to obtain the state estimation $\tilde{X}_{k+1}$ of the next time slot. When the current time slot has no successful uplink transmission, the controller can only update $\tilde{X}_{k+1}$ with $\tilde{X}_k$. According to this process, we can derive the state estimation MSE of the remote controller as $\tilde{Q}_k$:

$$\tilde{Q}_k = \mathbb{E}[(\tilde{X}_k - X_k)^2] \tag{6}$$

Note that the state estimation error of the remote controller is entirely caused by the noise Z_k. By using the state estimation age τ_k, we can rewrite the state estimation MSE as a recursive function of the noise variance R:

$$\tilde{Q}_{k+1} = \begin{cases} f(d_{\text{up}}) & \text{if } (a_j = 1)\,\&\,(\delta_j = 1) \\ f(\tau_k + 1) & \text{otherwise} \end{cases} \tag{7}$$

where $f(x) = \sum_{i=1}^{x}(A^2)^{i-1} R$. Equation (6) uses the definition of AoI to derive the MSE of the estimation. This representation greatly reduces the difficulty of calculation. In the following part, we will use the same idea to derive the single-loop CPS control performance metrics.

(2) Control Command Generation and Execution: In each time slot, while the remote controller maintains the state estimation, it also uses the estimation to generate a control command $\tilde{U}_k$:

$$\tilde{U}_k = K\tilde{X}_k \tag{8}$$

where K is the command generation coefficient. The goal of this control process is to maintain the state around 0. Since the downlink transmission has a propagation delay of d_{down} time slots, we must ensure $BK = -A^{d_{\text{down}}}$. To simplify the analysis, we set $B = -A^{d_{\text{down}}}$, $K = 1$. Due to the code error rate and scheduling decisions, not every control command $\tilde{U}$ can be received by the actuator. Only those scheduled and successfully transmitted can be used by the actuator. Therefore, the control command executed by the actuator is U_{k+1}:

$$U_{k+1} = \begin{cases} \tilde{U}_q & \text{if } (a_q = 2)\,\&\,(\delta_q = 1) \\ 0 & \text{otherwise} \end{cases} \tag{9}$$

where $q = k - d_{\text{down}} + 1$. This control method shown by (8) and (9) is called single-step control, which is a common form in the field of classic cybernetics. Using this method, when a control command is successfully delivered to the actuator, the actual state value will return to a value as close to 0 as possible at one time. Such a process can maximize the effect of a single instruction.

(3) Single-Loop CPS Control Performance Metrics: Consistent with the estimation performance metrics, the control performance metrics is defined as the state MSE of the plant Q_k:

$$Q_k = \mathbb{E}[X_k^2] \tag{10}$$

Similar to $\tilde{Q}_k$, we can rewrite Q_k as a function of noise variance R and state control age φ:

$$Q_{k+1} = \begin{cases} f(\tau_q + d_{\text{down}}) & \text{if } (a_q = 2)\,\&\,(\delta_q = 1) \\ f(\varphi_k + 1) & \text{otherwise} \end{cases} \tag{11}$$

According to the control cost given by Equation (11), we can obtain the long-term average control cost, that is, the long-term average plant state MSE:

$$J = \lim_{K \to \infty} \frac{1}{K} \sum_{k=0}^{K} Q_k \tag{12}$$

Equation (12) reflects the state deviation in the field of classic cybernetics which is the core cost metrics we care about. Please note that this parameter used to be very difficult to quantify without the introduction of AoI. Under certain conditions, the limit contained in Equation (12) may not exist, and the problem is unsolvable. In order to prevent such situations, the sufficient condition for the stability of WNCS with propagation delay will be given later, namely equation (19). In this paper, the scheduling strategy will be designed on the premise that equation (19) is satisfied.

(3) Uplink and Downlink Scheduling Process: In the previous subsection, we introduced the control performance measurement of a single-loop CPS. Now we will describe the scheduling process in detail. It has been explained that a single-loop CPS has two communication scenarios—the uplink transmission and the downlink transmission—and we can only choose one of them in each time slot under half-duplex mode. According to the previous definition, the scheduling decision of time slot k is recorded as a_k. The set of scheduling decisions of all time slots is called a scheduling strategy:

$$\pi \triangleq (a_1, a_2, \ldots, a_k, \ldots) \in \Pi \tag{13}$$

where Π represents the set of all scheduling strategies. Different scheduling strategies can significantly affect the control performance of a single-loop CPS. Every scheduling strategy π has its corresponding long-term average control cost J_π. Among all scheduling strategies, there is an optimal strategy $\pi^* \in \Pi$, which satisfies:

$$J_{\pi^*} \leqslant J_\pi, \forall \pi \in \Pi \tag{14}$$

Therefore, we can construct the following optimization problem. The goal of this problem is to minimize the long-term average plant state MSE to obtain the optimal scheduling strategy while taking transmission propagation delay and code error rates of two wireless channels into account, namely

$$\min_{\pi} \lim_{K \to \infty} \frac{1}{K} \sum_{k=0}^{K} Q_k \tag{15}$$

3. Semi-Predictive Framework and MDP Modeling

In this section, we will introduce the coupling characteristic between the uplink and downlink data packets which is caused by their propagation delay. In the following paper, we will use the coupling characteristic to refer to the coupling characteristic between the uplink and downlink data packets to save space. We propose a semi-predictive framework to eliminate the effect of the coupling characteristics on the solution of optimization problem (15). Based on this framework, we remodel this optimization problem to an MDP problem. Note that the semi-predictive framework we proposed is suitable for any value of the uplink and downlink propagation delay. For the generality, we use $d_{\text{up}} = d_{\text{down}} = 1$ as an example to illustrate the scheduling strategy design process. In the actual applications with different propagation delay, we only need to modify the value of $d_{\text{up}}, d_{\text{down}}$ and adjust some parameters in the following modeling step to meet specific design requirements.

3.1. The Packet Outdate Problem

Section 2 introduced the control mechanism of a single-loop CPS. Through the above analysis, it is easy to see that state update packets and control command packets have strong coupling characteristic for single-step control methods. Actually, such a coupling characteristic exists in any closed-loop control scenario as long as there exists propagation delay. This characteristic will cause some successfully delivered packets to become outdated. As shown in Figure 2, the green and red arrows represent state update packets up_1 (left green arrow), up_2 (left red arrow) and the control command packets down_1 (right green arrow), down_2 (right red arrow), respectively. The command down_1 is generated by the controller using up_1, while down_2 is generated by the controller using up_2. During the period from the slot up_2 sent to the slot $down_2$ executed, if $down_1$ is executed successfully, both up_2 and $down_2$ become invalid. In time slot 4, $down_1$ is executed; the result is that the real state of the plant was returned to a value around 0. This process causes an interruption in the state estimation process which means the estimation updated by up_2 is no longer accurate, so up_2 is outdated. Since up_2 is outdated, the control command $down_2$ which was generated from it is also outdated. This is the main effect of the coupling characteristic and we named it the packet outdate problem.

As we can see, this problem is mainly caused by the discontinuity in the dynamic process of the plant. The discontinuity only occurs when a downlink control command is executed, which means the uplink state update packet will not cause this problem. When this happens, the outdated uplink and downlink data packets require different processing methods. For an outdated downlink packet, it only needs to be discarded. However, for an outdated uplink packet, we have to backtrack the state estimation before this outdated packet is used. We show the evolution of the state estimation age and state control age in Figure 2. It can be seen that the state estimation age has been backtracked by changing from $\tau(3) = 2$ to $\tau(4) = 4$. The state control age will not be updated like this.

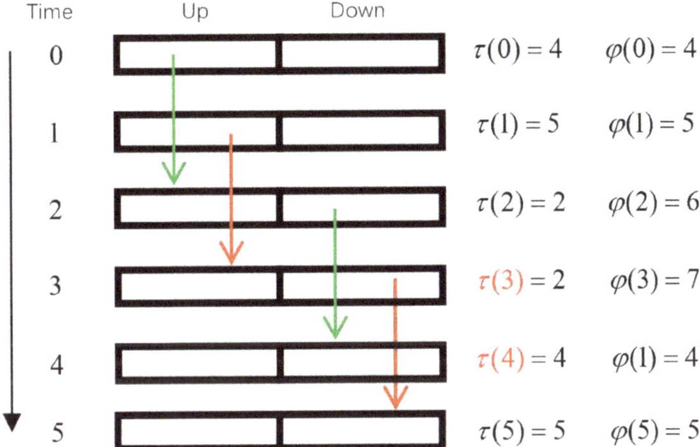

Figure 2. Analysis of Packet Outdated Phenomenon.

3.2. Main Idea of the Semi-Predictive Framework

In the previous subsection, we explained that the packet outdate problem has an impact on the update of the state estimation age, but this problem does not affect the update of the state control age. Therefore, when we try to construct a theoretical analysis framework, as long as the state control age is correct, the final analysis result can be guaranteed to be correct. In other words, the state estimation age of some time slots is allowed to deviate from the actual physical process. As long as it can be ensured that the state estimation age is accurate when the downlink data packet arrives at the actuator, the correct theoretical analysis can be guaranteed. It can be seen that it is possible to skip the state estimation age backtracking process in the theoretical analysis by using this feature. This is the main idea of the semi-predictive framework.

In the normal communication process, the decoding result of a data packet can only be determined after it arrives at the destination. For an uplink data packet, only after it arrives at the controller can it be known whether the data packet can be successfully decoded, while for a downlink packet, only after it arrives at the actuator can it be known whether the data packet can be successfully decoded. However, under the semi-predictive framework, we assume that the transmission result of a downlink packet is known as soon as the downlink packet is sent. Note that we do not predict the result of an uplink packet.

This is because the execution of the downlink command is the root cause of the packet outdated problem.

Take the case of Figure 2 as an example again; if we can foresee that the downlink control command packet $down_1$ can be successfully decoded and is not outdated, then during the period from its sending to its arrival, any packets sent or arrived can be directly discarded since they will be outdated by $down_1$. Through this process, the impact of the packet outdated problem is eliminated and state estimation age backtracking is avoided.

While the update process of the state estimation age under the semi-predictive framework is different from the actual physical process, the scheduling strategy obtained based on this framework can still be directly applied to an actual physical process. In the actual physical process, if a downlink data packet arrives at the actuator successfully and is not outdated, then the uplink and downlink transmissions scheduled during its transmission must be outdated. In other words, no matter what scheduling decision the controller made, those packets sent during this period will be outdated. In other words, those scheduling decisions can be arbitrary since they do not affect the final result. Assuming that the downlink control command packet $down_1$ in Figure 2 can be successfully decoded and not outdated, we will explain both age update processes under the semi-predictive framework and the actual physical process in detail.

(1) Semi-Predictive Framework: If $down_1$ can be successfully decoded and not outdated, then the controller knows that it does not matter whether it chooses uplink or downlink during the transmission of $down_1$ because those scheduled packets will be outdated anyway. Under these circumstance, a reasonable scheduling strategy is to regularly schedule one of the uplink and downlink transmissions during this period to consume time.

(2) Non-Predictive Framework (Actual Physical Process): In the actual physical process, during the transmission of $down_1$, the controller continues to schedule uplink or downlink transmissions according to a certain strategy. However, when $down_1$ is received and decoded successfully, the previous scheduled transmissions of the controller are all outdated. So in the end, the scheduled transmissions during this period only consume time and have no practical effect.

It can be observed that, under the semi-predictive framework and the actual non-predictive scheduling, the single-loop CPS transmission results are uniform; that is, it is accurate to use the semi-predictive framework in the theoretical design and directly apply the results to the real applications. This subsection qualitatively analyzes the unity of the semi-predictive framework and the actual physical process. In the next subsection, we will quantitatively illustrate how this framework corresponds to actual physical processes through MDP modeling.

3.3. MDP Modeling of the Semi-Predictive Framework

Based on the semi-predictive framework, we model the single-loop CPS with uplink and downlink propagation delay as an MDP process with the following four elements:

(1) State Space: The state space of this MDP is

$$\mathbb{S} \triangleq \{a'(-d_{\max}+1), \ldots, a'(-1), a'(0), D(0), \tau(0), \varphi(0)\} \tag{16}$$

where $d_{\max} = \max\{d_{\mathrm{up}}, d_{\mathrm{down}}\}$, $D(n) \in \{0, 1, \cdots, d_{\mathrm{down}} + 1\}$. $a(n)$ represents the scheduling decision made in the time slot n. $D(n)$ represents the time interval between the time slot when the latest valid downlink command packet (successfully transmitted and not outdated) in the time slot n was generated and the current time slot n. $\tau(n)$ and $\varphi(n)$ represent the state estimation age and the state control age at the time slot n, respectively. The time slot n is based on the current time slot: The time slot for which scheduling decisions are being made. Taking $a'(-1)$ as an example: It represents the transmission action taken in the previous time slot of the current time slot. We set both the uplink and downlink propagation delay to be 1 for illustration in the rest of this paper, so the corresponding state space is: $\mathbb{S} \triangleq \{a'(0), D(0), \tau(0), \varphi(0)\}$. In the subsequent sections of this paper, the state space is abbreviated as $\mathbb{S} \triangleq \{a', D, \tau, \varphi\}$ to save space.

(2) Action Space: The action space is $\mathbb{A} \triangleq \{0, 1\}$. This action space corresponds to the scheduling action a_k. If the controller schedules uplink transmission in the slot k, $a_k = 1$. If the controller schedules downlink transmission in the slot k, $a_k = 2$.

(3) State Transition Probability Matrix: The transition matrix is $P(s'|s, a)$. The state transition probability is the probability that the next state is s' by taking action a in the current state s. The transition probability is determined by the channel code error rate. According to the different parameter pairs: (a', D) in the state $\mathbb{S}$, the state transition matrix can be divided into five parts: $(a', D) = [(1, 1), (1, 2), (2, 0), (2, 1), (2, 2)]$. The complete construction rules are given in Appendix A.

(4) Cost Function: It can be seen from (4) and (11) that the cost function in a specific state is independent of the action. The cost function can be expressed as a function of the state control age φ_k:

$$C(s, a) = Q_k(s) = f(\varphi_k) \tag{17}$$

In the MDP modeling of the semi-predictive framework, the core parameter is $D(n)$. We limit its maximum value to $d_{\text{down}} + 1$ because we only need to track the downlink transmissions in the past d_{down} time slots to ensure that we do not miss any possible packet outdated problems. Besides, such process can help to reduce the scale of the state space. The update rule of $D(n)$ is as follows:

$$D_{k+1} = \begin{cases} 0 & \text{if } (a_k = 2) \& (\delta_k = 1) \\ \max(d_{\text{down}} + 1, D_k + 1) & \text{otherwise} \end{cases} \tag{18}$$

This updated process reflects the main idea of the semi-predictive framework and guarantees that it will not cause any differences between the state control ages of the theoretical analysis and the actual physical processes. In the next section, we will use the semi-predictive framework to design the optimal scheduling strategy.

4. Online and Offline Scheduling Strategies

In this section, we first give the sufficient condition for the existence of the optimal scheduling strategy. Then we use the relative value iteration algorithm to obtain the lookup table-based optimal offline strategy. Aiming at reducing the space complexity of the algorithm and saving space for storing the optimal offline strategy, we further propose a neural network-based suboptimal online strategy. For different uplink and downlink propagation delay, the acquisition process of both strategies is universal, which means that the semi-predictive framework has high practical application value.

4.1. Sufficient Conditions for the Strategies' Existence

Theorem 1. (*Sufficient conditions for the stability of multi-loop half-duplex CPS with fixed uplink and downlink propagation delay.*) *Assuming there are K single-loop CPS, all of them share the same controller and form up a multi-loop CPS. If the controller can only schedule L uplink transmissions or L downlink transmissions in each time slot, then for each single-loop CPS i, if the code error probability of its corresponding uplink and downlink channels satisfies*

$$\max\{p_{i,up}, p_{i,down}\} < \left(\frac{1}{(A_i)^2}\right)^{\lceil K/L \rceil}, i \in \{1, 2, \ldots, K\} \tag{19}$$

then there must exist a stationary deterministic scheduling strategy that can stabilize the multi-loop CPS. This stability remains as long as the uplink and downlink propagation delay are fixed, but the long term control performance metrics converge to a larger value with the increase of the propagation delay. When $K = 1$, $L = 1$ the above multi-loop CPS is just a single-loop CPS. The proof is given in Appendix B.

The essence of this sufficient condition is to link the instability of the control system with the reliability of the communication system. When the reliability of the communication system is higher than the instability of the control system, an optimal scheduling strategy can be found for the communication system to meet the needs of the control system. This condition can effectively guide the design of single- and multi-loop CPS.

4.2. Lookup Table-Based Optimal Offline Strategy

Since there is no theoretical upper limitation for the state estimation age and the state control age, the scale of the MDP state space is infinite, so it must be truncated before solving. We select $N = \max\{\tau, \varphi\}$ as the truncation condition, and use the relative value iteration algorithm to solve the MDP problem. When the value of N is appropriate, this truncation will have no effect on the control performance. Such a suitable N can be obtained by conducting Monte Carlo experiments. In this section, we take $N = 10$ as an example to show the resulting scheduling strategy in Figure 3.

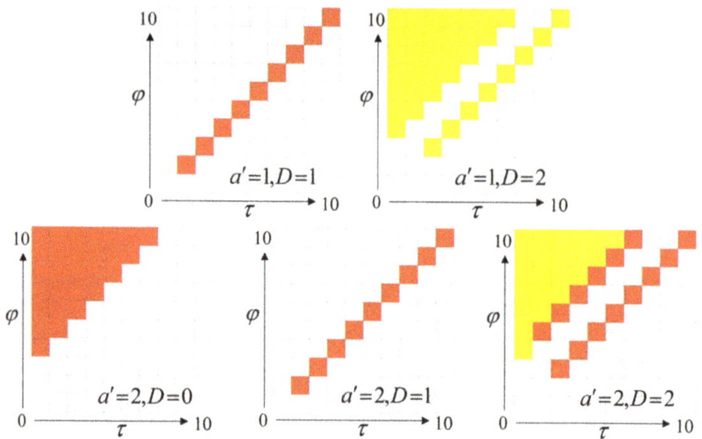

Figure 3. Optimal Off-line Policy with $N = 10$. Red squares represent action $a = 1$; yellow squares represent action $a = 2$.

In Figure 3, those red squares represent that the controller schedules uplink transmission in the corresponding state, and the yellow squares represent that the controller schedules downlink transmission in the corresponding state. As shown in Figure 3a,c,d, if $D = \{0,1\}$, no matter which transmission is scheduled, the related packet will be outdated. So under this circumstance, the scheduling strategy can choose any action arbitrarily. Since we chose the relative value iterative algorithm to solve the MDP problem, the strategy we obtained chooses to use uplink transmission to fill these unnecessary transmissions. Note that this part corresponds to the description of Section 3 part C. We take down_1 as an example again: In the actual physical process, it is not known that the next two transmissions are unnecessary transmissions after down_1 is sent. The controller does not know that $D = \{0,1\}$. Instead, it thinks that D is still equal to 2 at those time slots. Therefore, the controller continues to schedule according to the scheduling strategy. However, down_1 makes those two packets outdated when it is executed, while for those states whose $D = 2$, the controller can make a scheduling decision with the right state information. The entire process makes sure that the actual process is consistent with the theoretical process.

After obtaining this scheduling strategy, it is stored as a lookup table by the controller and does not require any extra calculation ability from the controller, so we call it an offline strategy. However, since the iterative algorithm is a model-based algorithm, as N gradually increases, the scale of the state space $N_S = 2 \cdot 3 \cdot N \cdot N = 6N^2$ in the MDP modeling increases exponentially. This leads to a sharp increase in the space complexity of the solving process and the lookup table could be too large to be stored. In order to solve these problems, we propose an improved scheme based on neural network in the next subsection.

4.3. Neural Network-Based Suboptimal Online Strategy

In Section 3, we remodeled the optimization problem to an MDP problem, and solved it to obtain the optimal offline strategy in the previous subsection. The optimal offline scheduling strategy based on the lookup table has two obvious shortcomings: The size of the lookup table increases linearly as the total number of states in the state space increases and the space complexity required in the calculation process increases exponentially as the total number of states increases. When the optimal offline strategy is actually deployed, there is no guarantee that the central controller has enough storage space to store the entire lookup table. It may even be impossible to perform calculations because the state space is too large. Therefore, here we design a new suboptimal online scheduling strategy based on neural network. The idea of this strategy is to replace the lookup table in the previous strategy with a neural network to save storage space. Neural network is a very ideal approximation function of lookup table, theoretically it can be approximated without error. That means in the theory of reinforcement learning, this strategy can achieve the performance of the optimal strategy. We will show that the performance of this suboptimal online strategy is very close to the performance of the optimal offline strategy in the next section.

In order to obtain this neural network, we use a the model-free algorithm called Deep Q Network (DQN). The algorithm continuously learns the hidden laws of the MDP problem by interacting with the environment and continuously trains the neural network to obtain better performance. We show the detailed process of the algorithm in Algorithm 1.

Algorithm 1: Deep Q Network Algorithm.

Data: State: $\mathbb{S} \triangleq \{a', D, \tau, \varphi\}$
Result: Action values: $A(s, a)$
initialization;
Initialize data set M;
Initialize evaluation network Q with random weights: θ;
Initialize target network $\hat{Q}$ with random weights: θ';
for $E = 1 : 2000$ **do**
    Initialize the environment;
    Set the origin state $s_1 \in \mathbb{S}$ randomly;
    for $t = 1 : 1000$ **do**
        Choose a random a_t with probability $1 - \varepsilon$;
        Otherwise choose $a_t = \arg\max_a Q(s, a_{t-1}; \theta)$;
        Execute a_t in the environment;
        Observe reward r_t and get new state s_{t+1};
        Store transition data (s_t, a_t, r_t, s_{t+1}) in M;
        Sample random mini batch of transitions (s_j, a_j, r_j, s_{j+1}) from M;
        Set $y_j = \begin{cases} r_j & \text{if episode ends at step } j + 1 \\ r_j + \gamma\max_{a'} \hat{Q}(s_{j+1}, a'; \theta') & \text{otherwise} \end{cases}$;
        Perform RMSprop on $(y_j - Q(s_j, a_j; \theta))^2$ with θ;
        Every 100 steps, set $\hat{Q} = Q$
    end
end

The structure of the neural network we obtained is shown in Figure 4: Four neurons in the input layer, fifty neurons in the hidden layer, and two neurons in the output layer. This neural network-based scheduling strategy is an online strategy which means that, in order to use this strategy, the current state s must be input to the neural network first. Then the controller needs to run real-time calculations to obtain the action values $A(s, a)$ for taking different actions in the current state. The action value represents how much reward can be obtained by taking the action, so the scheduling strategy is to select the action with the largest $A(s,a)$ among all actions.

DQN is a relatively mature reinforcement learning algorithm, so we only give the parameter settings of this algorithm and briefly introduce its training process. We run $E = 2000$ episodes, and each episode contains 1000 steps. In each step, this algorithm executes the greedy strategy with a probability of $\varepsilon = 0.7$, and the random strategy with a probability of $1 - \varepsilon = 0.3$. After each step, one state transition datum is stored in the data set. The scale of this data set is $M = 2048$, and it is updated in a loop covering manner. A new episode is automatically initialized every 1000 steps. In the meantime, the training process is performed every $T = 256$ steps, the algorithm selects $B = 512$ data from the data set for training. The optimizer we used is the Root Mean Square prop optimizer (RMSprop).

With the help of the DQN algorithm, we can obtain the neural network-based suboptimal online strategy. The controller only needs to store the node value of this network, and then calculates the action value in real time according to the current state in each time slot. In other words, this strategy saves a lot of storage space by consuming a small amount of computing ability of the controller. Such an advantage makes this strategy very meaningful in practical applications.

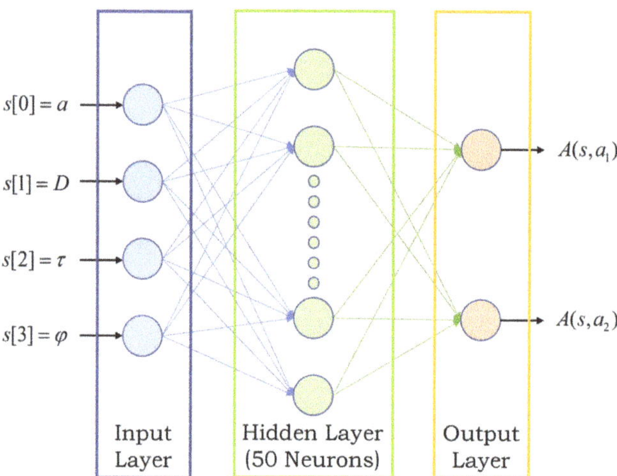

Figure 4. Neural Network Structure.

5. Numerical Simulation

In this section, we run the numerical simulation on those strategies we proposed and some existing strategies. We illustrate the advantages of the proposed strategies through comparison. First we introduce two benchmark strategies. The first is the switch scheduling strategy, that is, alternate uplink and downlink transmissions between each time slot; the second is the insist scheduling strategy, that is, continuous scheduling of uplink or downlink transmissions until success, then the transmission is exchanged.

The parameter settings in the numerical simulation are as follows: The state transition coefficient is $A = [1.1, 1.3]$, the code error rates of the uplink and downlink channels are $p_s = p_c = [0.1, 0.2]$, the specific values are marked on the curve obtained from the simulation. The initial state of the plant is $X_0 = 1$. The noise distribution is $\mathcal{N}(\bar{z} = 0, R = 1)$. The command control coefficient is $B = -A$. The initial state control variable is $s_0 = (a_0, D_0, \tau_0, \varphi_0) = (1, 1, 2, 2)$. The corresponding initial scheduling action is $a_0 = 1$. The initial state of the controller estimation is $\tilde{X}_o = 1$. The range of truncated state space is $N = \max\{\tau, \varphi\} = 20$. The plant noise follows normal distribution $\mathcal{N}\{\bar{z} = 0, R = 1\}$. Each strategy runs 500 episodes with 10,000 time slots each episode. The final long-term average plant state MSE is the average of the results of 500 episodes.

Figure 5 show the long-term average MSE of four strategies with $A = 1.3$ and $p_s = p_c = [0.1, 0.2]$. It can be seen that the MDP strategy, that is, the optimal offline strategy, has the best performance among all strategies, which also is the best performance that all possible scheduling strategies can achieve. While the performance of the neural network-based online strategy has slightly decreased, it is still significantly ahead of the existing strategies, and the performance gap between the optimal offline strategy and the suboptimal online strategy is very small. This gap can be eliminated in theory, but due to the limitations of deep reinforcement learning technology, it is currently difficult to fully achieve the optimal performance. It is relatively simple to obtain a suboptimal strategy with very close performance.

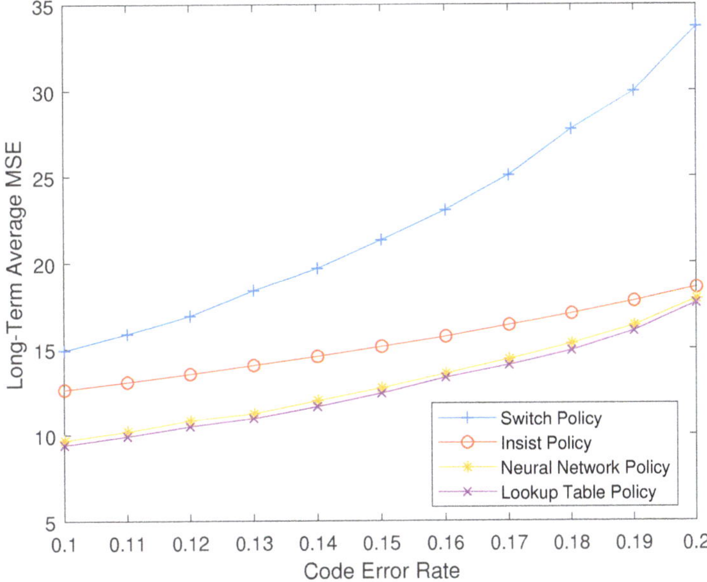

Figure 5. Long-term average plant state MSE of four policies with $A = 1.3$ and $p_s = p_c = [0.1, 0.2]$.

Figure 6 show the performance comparison between the optimal offline strategy and the two existing strategies under different state transition coefficient A. The suboptimal online strategy is not shown because it has been explained that the suboptimal strategy can theoretically approach the optimal. The state transition coefficient and the channel code error rates both reflect the instability of the control system and the reliability of the communication system in Equation (19). Combined with Figure 5, it can be seen that their influence on CPS is the same. A larger state transition coefficient or a higher channel code error rate lead to an increase in the long-term average plant state MSE, and when they exceed a certain limit and no longer satisfy Equation (19), the long-term average MSE of the CPS no longer converges, which means the single-loop CPS is unstable.

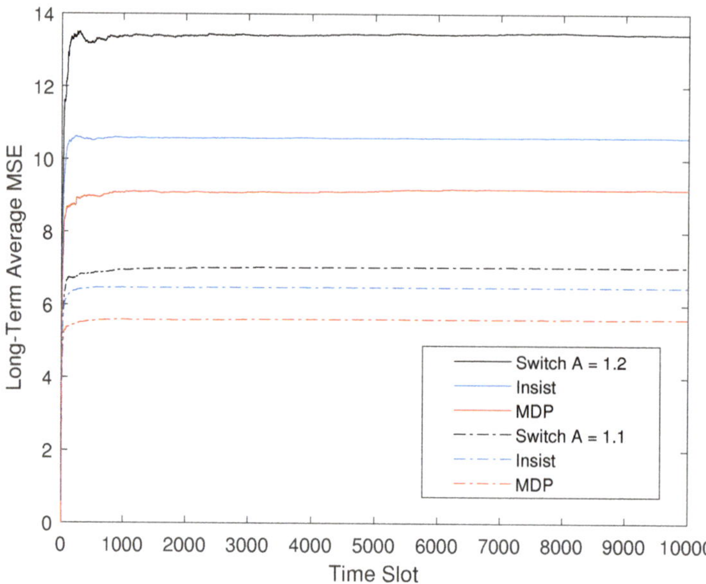

Figure 6. Long-term average plant state MSE of three policies with $p_s = p_c = 0.2$ and $A = \{1.1, 1.2\}$.

6. Conclusions

We proposed the semi-predictive framework to design scheduling strategies for single-loop CPS with uplink and downlink propagation delay. This framework can obtain the optimal offline strategy which is the upper bound on the performance among all strategies and a suboptimal online strategy with more practical application value. By adjusting the parameters, the semi-predictive framework can meet the need of any practical applications. We introduced the complete process of designing scheduling strategies under this framework by taking a specific situation as an example. The numerical simulation proved that the obtained strategies can effectively improve the performance of the existing strategies.

Author Contributions: Conceptualization, Z.A. and S.W.; methodology, Z.A. and T.L.; software, Z.A. and J.J.; validation, Z.A., S.W. and Q.Z.; formal analysis, Z.A. and T.L.; investigation, Q.Z.; resources, Z.A.; data curation, Z.A. and S.W.; writing—original draft preparation, Z.A.; writing—review and editing, Z.A.; visualization, Z.A.; supervision, Z.A.; project administration, Z.A.; funding acquisition, S.W. All authors have read and agreed to the published version of the manuscript.

Funding: This research was funded in part by the National Key Research and Development Program of China under Grant no. 2020YFB1806403, and in part by the National Natural Science Foundation of China under Grant nos. 61871147, 61831008, 62071141, 61371102, and in part by the Guangdong Science and Technology Planning Project under Grant no. 2018B030322004.

Institutional Review Board Statement: Not applicable.

Informed Consent Statement: Not applicable.

Data Availability Statement: Not applicable.

Acknowledgments: Z.A., S.W., T.L., J.J. and Q,Z. would like to thank Zehua Wang, Weihao Guo, Xiao Liang, Jiabao Kang and Dongrui Li for their fruitfulinsights and discussions.

Conflicts of Interest: The authors declare no conflict of interest.

Appendix A. Construction Rules of the State Transition Probability Matrix

Here we give the complete construction rules of the state transition matrix. Firstly, we give all the possible new states after a state transition as follows:

$$s'_1 = (1, 2, \tau + 1, \varphi + 1) \tag{A1}$$

$$s'_2 = (2, 2, \tau + 1, \varphi + 1) \tag{A2}$$

$$s'_3 = (2, 0, \tau + 1, \varphi + 1) \tag{A3}$$

$$s'_4 = (1, 2, 1, \varphi + 1) \tag{A4}$$

$$s'_5 = (2, 2, 1, \varphi + 1) \tag{A5}$$

$$s'_6 = (2, 0, 1, \varphi + 1) \tag{A6}$$

$$s'_7 = (1, 1, \tau + 1, \tau + 1) \tag{A7}$$

$$s'_8 = (2, 1, \tau + 1, \tau + 1) \tag{A8}$$

Secondly, we use R and R' to mark the transmission results. R represents the result of the downlink transmission scheduled in the next time slot. R' represents the result of the uplink transmission arrived in the next time slot. Note that R is known by prediction while R' is known by normal communication process. These abbreviations can help to simplify the expression of the rules.

We will give the construction rules in the form of $P[s'|s, c] = p$ which means that when the condition c is satisfied, the previous state s transfers to the new state s' with a probability of p.

When $s = (1, 1, \tau, \varphi)$:

$$\begin{aligned} P[s'_1|s, a = 1] &= 1 \\ P[s'_2|s, a = 2] &= 1 \end{aligned} \tag{A9}$$

When $s = (2, 0, \tau, \varphi)$:

$$\begin{aligned} P[s'_7|s, a = 1] &= 1 \\ P[s'_8|s, a = 2] &= 1 \end{aligned} \tag{A10}$$

When $s = (2, 1, \tau, \varphi)$:

$$\begin{aligned} P[s'_1|s, a = 1] &= 1 \\ P[s'_2|s, a = 2] &= 1 \end{aligned} \tag{A11}$$

When $s = (2, 2, \tau, \varphi)$:

$$\begin{aligned} P[s'_1|s, a = 1] &= 1 \\ P[s'_2|s, a = 2, R = 0] &= p_c \\ P[s'_2|s, a = 2, R = 1, \tau = \varphi] &= p_s \cdot (1 - p_c) \\ P[s'_3|s, a = 2, R = 1, \tau \neq \varphi] &= p_s \cdot (1 - p_c) \end{aligned} \tag{A12}$$

When $s = (1, 2, \tau, \varphi)$:

$$\begin{aligned} P[s'_1|s, a = 1] &= p_s \\ P[s'_2|s, a = 2, R' = 0, R = 0] &= p_s \cdot p_c \\ P[s'_2|s, a = 2, R' = 0, R = 1, \tau = \varphi] &= p_s \cdot (1 - p_c) \\ P[s'_3|s, a = 2, R' = 0, R = 1, \tau \neq \varphi] &= p_s \cdot (1 - p_c) \\ P[s'_4|s, a = 1, R' = 1] &= 1 - p_s \\ P[s'_5|s, a = 2, R' = 1, R = 0] &= (1 - p_s) \cdot p_c \\ P[s'_6|s, a = 2, R' = 1, R = 1] &= (1 - p_s) \cdot (1 - p_c) \end{aligned} \tag{A13}$$

Appendix B. Proof of Theorem 1

Appendix B.1. Scheduling 1 Subsystem per Time Slot without Delay

To prove sufficient conditions, we only need to prove that there exists a stationary deterministic strategy that can make multi-loop CPS stable. Here we prove that the round-robin insist scheduling strategy can keep the system stable. We first prove the case of $L = 1$. Round-robin means that in every K time slots, the controller schedules each subsystem once in turn, and the scheduling sequence is fixed from $i = 1$ to $i = K$. Insist refers to when scheduling each subsystem, continuously scheduling uplink or downlink transmission until it succeeds, then switch to another transmission. Therefore, the actions of a single subsystem under the round-robin insist scheduling strategy can be given in the form of the following time axis:

The time axis between two consecutive successful downlink scheduling is recorded as a control loop. It can be seen that the AoI evolution process of each control loop of one subsystem is:

(1) The initial estimation age is equal to the control age: $n'K$

(2) The current subsystem waits for the completion of the scheduling of other subsystems, that is, silence $(k - 1)$ time slots, and then schedules the uplink transmission when it is scheduled again. If the uplink transmission fails, the subsystem waits another (k-1) time slots and tries again until the uplink transmission is successful. This step takes mK time slots. At the end of this step, the estimated age is 0, and the control age is $(n' + m)K$;

(3) After the current subsystem silences for $(K - 1)$ time slots, it switches to schedule downlink transmission continuously until it succeeds. This step takes nK time slots. At the end of this step, the estimated age is equal to the control age: nK. Then it finishes a close control loop.

Note that the time slots included in a complete control loop are the time slots marked in red on the coordinate axis in Figure A1, that is, the control age ranges from $n'K$ to $(n' + m + n)K$. Each control loop has repeatability, so we only need to prove that the long-term average cost within the range of one control loop converges.

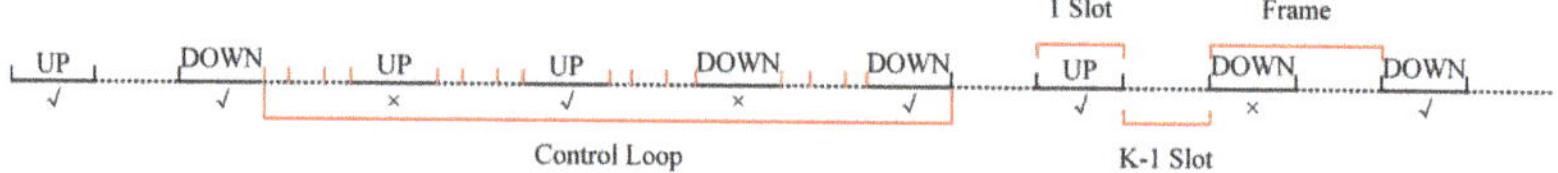

Figure A1. The round-robin insist scheduling strategy.

According to the channel error probability, the M uplink transmissions and N downlink transmissions in each control loop can be modeled as a geometric distribution with the probability of success being $(1 - p_s)$ and $(1 - p_c)$ respectively. M and N are different in each control loop, N' Represents the number of downlink transmissions in the previous loop of the current control loop. (n', m, n) are their specific observations. C_i and T_i represent the total cost and total time of the i-th control loop of the current subsystem respectively:

$$C_i = \sum_{q=0}^{(m+n)K-1} f(n'K + q) = \sum_{q=1}^{(m+n)K} f(n'K + q - 1) \tag{A14}$$

$$T_i = (m + n)K \tag{A15}$$

where $f(\varphi) = \sum_{i=1}^{\varphi} (A^2)^{i-1}$. Next, we can express the long-term average cost as:

$$J = \lim_{t \to \infty} \frac{C_1 + C_2 + \cdots + C_t}{T_1 + T_2 + \cdots + T_t} = \frac{\mathbb{E}[C]}{\mathbb{E}[T]} \tag{A16}$$

$$\mathbb{E}[C] = \sum_{n'}^{\infty} \sum_{m}^{\infty} \sum_{n}^{\infty} \begin{pmatrix} \mathbb{E}[C|N' = n', M = m, N = n] \\ \cdot \mathbb{P}[N' = n', M = m, N = n] \end{pmatrix} \tag{A17}$$

$$E[L] = \sum_{n'}^{\infty} \sum_{m}^{\infty} \sum_{n}^{\infty} \left((m + n) \cdot K \cdot \mathbb{P}[N' = n', M = m, N = n] \right) \tag{A18}$$

It can be seen that if $\mathbb{E}[C]$ is bounded, then J is bounded. According to the definition of C_i and the three geometrical distributions (N', M, N), which are independent of each other, we have:

$$\mathbb{E}[C|N' = n', M = m, N = n] = \sum_{q=0}^{(m+n)K-1} f(n'K + q) = \sum_{q=1}^{(m+n)K} f(n'K + q - 1) \quad (A19)$$

$$\begin{aligned}
&\mathbb{P}[N' = n', M = m, N = n] \\
&= \mathbb{P}[N' = n'] \cdot \mathbb{P}[M = m] \cdot \mathbb{P}[N = n] \\
&= (1 - p_c)p_c^{n'-1}(1 - p_s)p_s^{m-1}(1 - p_c)p_c^{n-1}
\end{aligned} \quad (A20)$$

Choose $p_{max} = \max\{p_s, p_c\}$, we can derive that:

$$\mathbb{E}[C] \leqslant \alpha_1 \cdot \sum_{n'}^{\infty}\sum_{m}^{\infty}\sum_{n}^{\infty} \left(\sum_{q=1}^{(m+n)K} f(n'K + q - 1) \cdot p_{max}^{n'+m+n} \right) \quad (A21)$$

where $\alpha_1 = (1 - p_c)p_c^{-1}(1 - p_s)p_s^{-1}(1 - p_c)p_c^{-1}$. Since $f(\cdot)$ is a strictly increasing function and (n', m, n) are all greater than 0, we can derive that:

$$\mathbb{E}[C] < \alpha_2 \cdot \sum_{n'}^{\infty}\sum_{m}^{\infty}\sum_{n}^{\infty} \left((n' + m + n) \cdot f(n'K + mK + nK) \cdot p_{max}^{n'+m+n} \right) \quad (A22)$$

where $\alpha_2 = K(1 - p_c)p_c^{-1}(1 - p_s)p_s^{-1}(1 - p_c)p_c^{-1}$. We abbreviate $n' + m + n$ as i, that is, $i = n' + m + n$. Considering $i \geqslant 3$, and when $i = n' + m + n$ is a fixed value, the possible combinations of $(n', m, n) \geqslant 1$ satisfy the mathematical relationship of $\sum_{n'}\sum_{m}\sum_{n}(1) < (n' + m + n)^3$, namely:

$$\sum_{n'}\sum_{m}\sum_{n}(n' + m + n) < (n' + m + n)^3 \cdot (n' + m + n) \quad (A23)$$

$$\sum_{n'}\sum_{m}\sum_{n}(i) < (i)^4 \quad (A24)$$

We can derive that:

$$\mathbb{E}[C] < \alpha_2 \cdot \sum_{i}^{\infty} \left(i^4 \cdot f(iK) \cdot p_{max}^{i} \right) \quad (A25)$$

Since there are always exist $p > p_{max}$ and $n < \infty$, satisfying $i^4 p_{max}^{i} < p^{i}, \forall i > n$. So we have:

$$\sum_{i}^{\infty} \left(i^4 \cdot f(iK) \cdot p_{max}^{i} \right) < \sum_{i}^{\infty} \left(f(iK) \cdot p^{i} \right) \quad (A26)$$

So if $\sum_{i}^{\infty} \left(f(iK) \cdot p^{i} \right) < \infty$, then $\sum_{i}^{\infty} \left(i^4 \cdot f(iK) \cdot p_{max}^{i} \right) < \infty$. Now seeking the conditions for the stability of the multi-loop CPS subsystem is transformed into seeking the conditions for the establishment of $\sum_{i}^{\infty} \left(f(iK) \cdot p^{i} \right) < \infty$. For $f(iK)$, we have:

$$f(iK) = \sum_{q=1}^{iK} (A^2)^{q-1} = 1 + A^2 + A^4 + \cdots + A^{2(iK-1)} = \frac{1 - (A^2)^{iK}}{1 - A^2} \quad (A27)$$

For $\sum_{i}^{\infty} \left(f(iK) \cdot p^{i} \right)$, we have:

$$\sum_{i}^{\infty}\left(f(iK)\cdot p^i\right) = \sum_{i}^{\infty}\left(\frac{1-\left(A^2\right)^{iK}}{1-A^2}\cdot p^i\right) = \frac{1}{1-A^2}\sum_{i}^{\infty}\left(\left(1-\left(A^2\right)^{iK}\right)\cdot p^i\right)$$

$$= \frac{1}{1-A^2}\left(\sum_{i}^{\infty}\left(p^i-\left(A^2\right)^{iK}p^i\right)\right) = \frac{1}{1-A^2}\left(\sum_{i}^{\infty}\left(p^i\right)-\sum_{i}^{\infty}\left(\left(A^2\right)^{iK}p^i\right)\right) \tag{A28}$$

So in order to ensure that $\frac{1}{1-A^2}\left(\sum_{i}^{\infty}\left(p^i\right)-\sum_{i}^{\infty}\left(\left(A^2\right)^{iK}p^i\right)\right) < \infty$ is satisfied, it is obvious that $p < 1$ and $A^{2K}p < 1$ must stand, that is, $p < \left(\frac{1}{A^2}\right)^K$. This completes the proof.

Appendix B.2. Scheduling L Subsystems per Time Slot without Delay

When each time slot can schedule L subsystems, the corresponding strategy can be set to multiple independent round-robin insist scheduling strategies. It can be ensured that the round-robin cycle of each subsystem does not exceed $\lceil K/L \rceil$, and the follow-up proof is consistent with Appendix B.1.

Appendix B.3. Scheduling L Subsystems per Time Slot with Delay

For a specific subsystem, we assume that the fixed delay for each transmission is D_i frames, which is equivalent to the uplink and downlink scheduling in the control lcfoop must be delayed by $D_i K$ time slots for AA reception, so the formula (A19) is modified as follows :

$$\mathbb{E}[C|N' = n', M = m, N = n]$$
$$= \sum_{q=0}^{(m+n+2D)K-1} f(n'K+DK+q)$$
$$= \sum_{q=1}^{(m+n+2D)K} f(n'K+DK+q-1) \tag{A29}$$

Since $\mathbb{E}[D] = \mathbb{E}[D_i] = D$ is a constant which has no effect on the subsequent proof, the proof process is consistent with Appendix B.1.

References

1. Zhang, X.M.; Han, Q.L.; Ge, X.; Ding, D.; Ding, L.; Yue, D.; Peng, C. Networked control systems: A survey of trends and techniques. *IEEE/CAA J. Autom. Sin.* **2020**, *7*, 1–17. [CrossRef]
2. Lin, J.; Yu, W.; Zhang, N.; Yang, X.; Zhang, H.; Zhao, W. A Survey on Internet of Things: Architecture, Enabling Technologies, Security and Privacy, and Applications. *IEEE Internet Things J.* **2017**, *4*, 1125–1142. [CrossRef]
3. Xu, H.; Yu, W.; Griffith, D.; Golmie, N. A Survey on Industrial Internet of Things: A Cyber–Physical Systems Perspective. *IEEE Access* **2018**, *6*, 78238–78259. [CrossRef]
4. Lu, C.; Saifullah, A.; Li, B.; Sha, M.; Gonzalez, H.; Gunatilaka, D.; Wu, C.; Nie, L.; Chen, Y. Real-Time Wireless Sensor-Actuator Networks for Industrial Cyber–Physical Systems. *Proc. IEEE* **2016**, *104*, 1013–1024. [CrossRef]
5. Liu, W.; Nair, G.; Li, Y.; Nesic, D.; Vucetic, B.; Poor, H.V. On the Latency, Rate, and Reliability Tradeoff in Wireless Networked Control Systems for IIoT. *IEEE Internet Things J.* **2021**, *8*, 723–733. [CrossRef]
6. Han, B.; Zhu, Y.; Jiang, Z.; Hu, Y.; Schotten, H.D. Optimal Blocklength Allocation Towards Reduced Age of Information in Wireless Sensor Networks. In Proceedings of the 2019 IEEE Globecom Workshops (GC Wkshps), Waikoloa, HI, USA, 9–13 December 2019; pp. 1–6.
7. Han, B.; Jiang, Z.; Zhu, Y.; Schotten, H.D. Recursive Optimization of Finite Blocklength Allocation to Mitigate Age-of-Information Outage. In Proceedings of the 2020 IEEE International Conference on Communications Workshops (ICC Workshops), Dublin, Ireland, 7–11 June 2020; pp. 1–6.
8. Li, D.; Wu, S.; Wang, Y.; Jiao, J.; Zhang, Q. Age-Optimal HARQ Design for Freshness-Critical Satellite-IoT Systems. *IEEE Internet Things J.* **2020**, *7*, 2066–2076. [CrossRef]
9. Parag, P.; Taghavi, A.; Chamberland, J. On Real-Time Status Updates over Symbol Erasure Channels. In Proceedings of the 2017 IEEE Wireless Communications and Networking Conference (WCNC), San Francisco, CA, USA, 19–22 March 2017; pp. 1–6.

10. Huang, K.; Liu, W.; Li, Y.; Savkin, A.; Vucetic, B. Wireless Feedback Control with Variable Packet Length for Industrial IoT. *IEEE Wirel. Commun. Lett.* **2020**, *9*, 1586–1590. [CrossRef]

11. Liu, C.-F.; Bennis, M. Data-Driven Predictive Scheduling in Ultra-Reliable Low-Latency Industrial IoT: A Generative Adversarial Network Approach. In Proceedings of the 2020 IEEE 21st International Workshop on Signal Processing Advances in Wireless Communications (SPAWC), Atlanta, GA, USA, 26–29 May 2020; pp. 1–5.

12. Eisen, M.; Gatsis, K.; Pappas, G.J.; Ribeiro, A. Learning in Wireless Control Systems Over Nonstationary Channels. *IEEE Trans. Signal Process.* **2019**, *67*, 1123–1137. [CrossRef]

13. Sun, Y.; Kadota, I.; Talak, R.; Modiano, E. *Age of Information: A New Metric for Information Freshness*; Morgan & Claypool: San Rafael, CA, USA, 2019.

14. Sinopoli, B.; Schenato, L.; Franceschetti, M.; Poolla, K.; Jordan, M.I.; Sastry, S.S. Kalman filtering with intermittent observations. *IEEE Trans. Autom. Control* **2004**, *49*, 1453–1464. [CrossRef]

15. Champati, J.P.; Mamduhi, M.H.; Johansson, K.H.; Gross, J. Performance Characterization Using AoI in a Single-loop Networked Control System. In Proceedings of the IEEE INFOCOM 2019—IEEE Conference on Computer Communications Workshops (INFOCOM WKSHPS), Paris, France, 29 April–2 May 2019; pp. 197–203.

16. Park, P.; Araújo, J.; Johansson, K.H. Wireless networked control system co-design. In Proceedings of the 2011 International Conference on Networking, Sensing and Control, Delft, The Netherlands, 11–13 April 2011; pp. 486–491.

17. Huang, K.; Liu, W.; Li, Y.; Vucetic, B. To Retransmit or Not: Real-Time Remote Estimation in Wireless Networked Control. In Proceedings of the ICC 2019—2019 IEEE International Conference on Communications (ICC), Shanghai, China, 21–23 May 2019; pp. 1–7.

18. An, Z.; Wu, S.; Wang, Y.; Jiao, J.; Zhang, Q. HARQ Based Joint Uplink-Downlink Optimal Scheduling Strategy for Single-Loop WNCS. In Proceedings of the 2020 International Conference on Wireless Communications and Signal Processing (WCSP), Nanjing, China, 21–23 October 2020; pp. 7–12.

19. Huang, K.; Liu, W.; Li, Y.; Vucetic, B.; Savkin, A. Optimal Downlink—Uplink Scheduling of Wireless Networked Control for Industrial IoT. *IEEE Internet Things J.* **2020**, *7*, 1756–1772. [CrossRef]

20. Wang, X.; Chen, C.; He, J.; Zhu, S.; Guan, X. AoI-Aware Control and Communication Co-design for Industrial IoT Systems. *IEEE Internet Things J.* **2020**, *8*, 8464–8473. [CrossRef]

21. Jiang, Z.; Krishnamachari, B.; Zhou, S.; Niu, Z. Can Decentralized Status Update Achieve Universally Near-Optimal Age-of-Information in Wireless Multiaccess Channels? In Proceedings of the 2018 30th International Teletraffic Congress (ITC 30), Vienna, Austria, 3–7 September 2018; pp. 144–152.

22. Ayan, O.; Vilgelm, M.; Kellerer, W. Optimal Scheduling for Discounted Age Penalty Minimization in Multi-Loop Networked Control. In Proceedings of the 2020 IEEE 17th Annual Consumer Communications & Networking Conference (CCNC), Las Vegas, NV, USA, 10–13 January 2020; pp. 1–7.

23. Girgis, A.M.; Park, J.; Liu, C.-F.; Bennis, M. Predictive Control and Communication Co-Design: A Gaussian Process Regression Approach. In Proceedings of the 2020 IEEE 21st International Workshop on Signal Processing Advances in Wireless Communications (SPAWC), Atlanta, GA, USA, 26–29 May 2020; pp. 1–5.

24. Chang, B.; Zhang, L.; Li, L.; Zhao, G.; Chen, Z. Optimizing Resource Allocation in URLLC for Real-Time Wireless Control Systems. *IEEE Trans. Veh. Technol.* **2019**, *68*, 8916–8927. [CrossRef]

25. Maity, D.; Mamduhi, M.H.; Hirche, S.; Johansson, K.H.; Baras, J.S. Optimal LQG Control Under Delay-Dependent Costly Information. *IEEE Control Syst. Lett.* **2019**, *3*, 102–107. [CrossRef]

26. Mamduhi, M.H.; Maity, D.; Baras, J.S.; Johansson, K.H. A Cross-Layer Optimal Co-Design of Control and Networking in Time-Sensitive Cyber–Physical Systems. *IEEE Control Syst. Lett.* **2021**, *5*, 917–922. [CrossRef]

27. Maity, D.; Baras, J.S. Minimal Feedback Optimal Control of Linear-Quadratic-Gaussian Systems: No Communication is also a Communication. *IFAC-PapersOnLine* **2020**, *53*, 2201–2207. [CrossRef]

28. Schenato, L.; Sinopoli, B.; Franceschetti, M.; Poolla, K.; Sastry, S.S. Foundations of Control and Estimation Over Lossy Networks. *Proc. IEEE* **2007**, *95*, 163–187. [CrossRef]

MDPI
St. Alban-Anlage 66
4052 Basel
Switzerland
Tel. +41 61 683 77 34
Fax +41 61 302 89 18
www.mdpi.com

Entropy Editorial Office
E-mail: entropy@mdpi.com
www.mdpi.com/journal/entropy